GW01066316

Artificial Intelligence Solutions for Cyber-Physical Systems

Edited by Pushan Kumar Dutta,
Pethuru Raj,
B. Sundaravadivazhagan, and
Chithirai Pon Selvan

AN AUERBACH BOOK

First edition published 2025
by CRC Press
2385 NW Executive Center Drive, Suite 320, Boca Raton FL 33431

and by CRC Press
4 Park Square, Milton Park, Abingdon, Oxon, OX14 4RN

CRC Press is an imprint of Taylor & Francis Group, LLC

ISBN: 9781032686721 (hbk)
ISBN: 9781032694368 (pbk)
ISBN: 9781032694375 (ebk)

DOI: 10.1201/9781032694375

Typeset in Times
by codeMantra

Artificial Intelligence Solutions for Cyber-Physical Systems

Smart manufacturing environments are revolutionizing the industrial sector by integrating advanced technologies, such as the Internet of Things (IoT), artificial intelligence (AI), and robotics, to achieve higher levels of efficiency, productivity, and safety. However, the increasing complexity and interconnectedness of these systems also introduce new security challenges that must be addressed to ensure the safety of human workers and the integrity of manufacturing processes. Key topics include risk assessment methodologies, secure communication protocols, and the development of standard specifications to guide the design and implementation of HCPS. Recent research highlights the importance of adopting a multi-layered approach to security, encompassing physical, network, and application layers. Furthermore, the integration of AI and machine learning techniques enables real-time monitoring and analysis of system vulnerabilities, as well as the development of adaptive security measures.

Artificial Intelligence Solutions for Cyber-Physical Systems discusses such best practices and frameworks as NIST Cybersecurity Framework, ISO/IEC 27001, and IEC 62443 of advanced technologies. It presents strategies and methods to mitigate risks and enhance security, including cybersecurity frameworks, secure communication protocols, and access control measures. The book also focuses on the design, implementation, and management of secure HCPS in smart manufacturing environments. It covers a wide range of topics, including risk assessment, security architecture, data privacy, and standard specifications, for HCPS. The book highlights the importance of securing communication protocols, the role of artificial intelligence and machine learning in threat detection and mitigation, and the need for robust cybersecurity frameworks in the context of smart manufacturing.

Pushan Kumar Dutta is Assistant Professor Grade III in the Electronics and Communication Engineering Department of ASETK in Amity University, Kolkata, India.

Pethuru Raj is Chief Architect at Reliance Jio Platforms Ltd. (JPL) Bangalore, India.

B. Sundaravadivazhagan is an experienced researcher and educator in the field of information and communication engineering.

Chithirai Pon Selvan has extensive experience in teaching engineering students and has worked in academia for over 23 years.

Contents

Contributors ix

Chapter 1 Enhancing the Power of Cyber-Physical Systems Enabled with AI: An Introduction—Facts and Myths along with Modular Approach 1

Shivani Trivedi, Vanshika Aggarwal, and Rohit Rastogi

Chapter 2 AI for Secure and Resilient Cyber-Physical Systems 40

Pawan Whig, Anant Aggarwal, Veeramani Ganeshan, Venugopal Reddy Modhugu, and Ashima Bhatnagar Bhatia

Chapter 3 Power of Emotions in AI: Strengthening the Bond of Human-Machine with Heart 64

Asha Sharma and Aditya Mishra

Chapter 4 Advancing Manufacturing Excellence: ML and AI-Driven Threat Detection Strategies 73

Almas Begum, Alex David, Sivagami S., and Carmel Mary Belinda M. J.

Chapter 5 Enhancing Resilience in the Integration of Cybersecurity for Smart Manufacturing... 92

R. Vijayakumari, Phaneendra Varma Chintalapati, K. Baskar, and Karamath Ateeq

Chapter 6 Leveraging Artificial Intelligence and Machine Learning for Advanced Threat Detection in Smart Manufacturing 101

Gnanasankaran Natarajan, Sundaravadivazhagan Balasubramanian, Elakkiya Elango, and Rakesh Gnanasekaran

Chapter 7 Integrating Cybersecurity Threats into Smart Manufacturing: Best Practices and Frameworks 120

G. Ramya and K. G. Srinivasagan

Chapter 8 Proactive Risk Management in Smart Manufacturing: A Comprehensive Approach to Risk Assessment and Mitigation 139

Kuwata Muhammed Goni, Aliyu Mohammed, Shanmugam Sundararajan, and Sulaiman Ibrahim Kassim

Chapter 9 Entrepreneurial Strategies for Mitigating Risks in Smart Manufacturing Environments 165

M. Ashok Kumar, Aliyu Mohammed, Pethuru Raj, and B. Sundaravadivazhagan

Chapter 10 Fortifying Smart Manufacturing against Cyber Threats: A Comprehensive Guide to Cybersecurity Integration, Best Practices, and Frameworks 180

Aliyu Mohammed, M. Ashok Kumar, Pethuru Raj, and M. Sangeetha

Chapter 11 Tagging Blockchain Technology Fostering Panacea for Data Privacy, Cloud Computing, and Integrity: Legal Framework in India and at a Globe 206

Bhupinder Singh and Christian Kaunert

Chapter 12 Enhancing Human-Centered Security in Industry 4.0: Navigating Challenges and Seizing Opportunities 214

Aliyu Mohammed, Shanmugam Sundararajan, and Senthil Kumar

Chapter 13 Using AI for Student Success: Early Warning, Performance Analytics, and Automated Grading in Education Cyber-Physical Systems 236

Debosree Ghosh

Chapter 14 Securing Industrial Internet of Things (IIoT): A Review of Technologies, Strategies, Challenges, and Future Trends 244

Adam A. Alli, Kassim Kalinaki, Mugigayi Fahadi, and Lwembawo Ibrahim

Chapter 15 Strategic Management of Intelligent Robotics and Drones in Contemporary Industrial Operations: An Assessment of Roles and Integration Strategies....264

Ashok Kumar Manoharan, Aliyu Mohammed, Pethuru Raj, and Sundaravadivazhagan Balasubramanian

Chapter 16 Digital Twins as New Paradigm for Bridging the Gap from Personalized Medicine to Specific Rural Public Health: Exploring Foundations, Legal Angels, and Technologies for Health 5.0—Future Research Directions.... 286

Bhupinder Singh

Chapter 17 Privacy-Preserving Strategies for Enhanced Big Data Analytics in Evolving Healthcare Environments: A 5G and Beyond Perspective....310

Hemanth Kumar and Dinesh Nilkant

Chapter 18 Securing the Digital Realm: Unleashing Hybrid Optimization for Deep Neural Network Intrusion Detection....339

Thupakula Bhaskar, BJ Dange, SN Gunjal, and HE Khodke

Chapter 19 AI for Industrial IoT: A Review of Emerging Trends and Advanced Research....354

Meet Kumari

Chapter 20 The Role of Human-Centric Solutions in Tackling Challenges and Unlocking Opportunities in Industry 4.0.... 365

Krishnaveni, Swathi, Eleanor Schwartz, and Sangeetha

Chapter 21 Strategies for Managing Risk and Mitigation in the Era of Smart Manufacturing....376

S. Krishnaveni, Faizan Ahmad, Mohamed Akbar, and S. Sangeetha

Chapter 22 Cultivating a Security-Conscious Smart Manufacturing Workforce: A Comprehensive Approach to Workforce Training and Awareness....385

Aliyu Mohammed, Ashok Kumar Manoharan, Pethuru Raj Chelliah, and Sulaiman Ibrahim Kassim

Chapter 23 Data Analytics for Pandemic: A Covid-19 Case Study in Kolkata 404

Supratim Bhattacharya, Saberi Goswami, Poulami Chowdhury, Prashnatita Pal, and Jayanta Poray

Chapter 24 Deep Learning Techniques for DDoS Assault 419

M. Srisankar and Dr. K. P. Lochanambal

Chapter 25 Study on Blockchain-Based Framework for Central Bank Digital Currency Design: Opportunities, Risk, and Challenges 432

Shivangi Verma

Chapter 26 NLP-Based Chatbot for Handling Mental Health-Related Issues 439

Abhinav Juneja, Yonis Gulzar, Sapna Juneja, Junaid Mohsin, Himanshi Mishra, Harshita Shukla, and Vishal Jain

Index 451

Contributors

Anant Aggarwal is Research Scientist at Threws, Delhi, India.

Vanshika Aggarwal (B.Tech CSE Third Year) is at ABES Engineering College, Ghaziabad, Delhi-NCR, India.

Faizan Ahmad is in the Department of Computer Science and Engineering (AIML), Noida Institute of Engineering and Technology, Greater Noida, Uttar Pradesh, India.

Mohamed Akbar is Lecturer in the Department of Information Technology, University of Technology and Applied Sciences, Muscat, Sultanate of Oman.

Adam A. Alli is Senior Lecturer in the Department of Computer Science, Islamic University in Uganda (IUIU), Mbale, Uganda.

Karamath Ateeq is Senior Faculty, in the Department of School of Computing, Skyline University College, University City, Sharjah, United Arab Emirates.

Sundaravadivazhagan Balasubramanian is Professor in the Department of Information Technology, University of Technology and Applied Sciences, Al Mussanah, Oman.

K. Baskar is in the Department of AI&DS, Kongunadu College of Engineering and Technology, Trichy, Tamil Nadu, India.

Almas Begum is Professor in the Department of Computer Science and Engineering, Saveetha School of Engineering, Saveetha Institute of Medical and Technical Sciences, Chennai, Tamil Nadu, India.

Thupakula Bhaskar is Associate Professor at the Sanjivani College of Engineering (Autonomous), Kopargaon, Maharashtra, India.

Ashima Bhatnagar Bhatia is Assistant Professor at the Vivekananda Institute of Professional Studies – Technical Campus, Delhi, India.

Supratim Bhattacharya is Research Scholar, Techno India University in the Department of Computer Science & Engineering, Techno India University, Kolkata, West Bengal, India.

Phaneendra Varma Chintalapati is Assistant Professor in the Department of Computer Science and Engineering, Shri Vishnu Engineering College for Women (A), Bhimavaram, West Godavari District, Andhra Pradesh, India.

Poulami Chowdhury is Professor at Charnock Health Institute, India.

B. J. Dange is Associate Professor at the Sanjivani College of Engineering (Autonomous), Kopargaon, Maharashtra, India.

Alex David is Professor in the Department of Computer Science and Engineering, Vel Tech Rangarajan Dr Sagunthala R&D Institute of Science and Technology, Chennai, Tamil Nadu, India.

Elakkiya Elango is Guest Lecturer at Government Arts College for Women, Sivaganga, Tamil Nadu, India.

Mugigayi Fahadi is Lecturer in the Department of Computer Science, Islamic University in Uganda (IUIU), Mbale, Uganda.

Veeramani Ganeshan Sr Lead Mobile and OTT Engineer at CBS News and Stations, Paramount USA

Debosree Ghosh is Assistant Professor in the Department of Computer Science and Technology, Shree Ramkrishna Institute of Science and Technology, Kolkata, West Bengal, India.

Rakesh Gnanasekaran is Assistant Professor in the Department of Computer Science, Thiagarajar College, Madurai, Tamil Nadu, India.

Kuwata Muhammed Goni is PhD Research Scholar in the Faculty of Entrepreneurship and Business, University of Malaysia Kelantan, Malaysia.

Saberi Goswami is Research Scholar at the Techno India University, Kolkata, West Bengal, India.

S. N. Gunjal is Assistant Professor at the Sanjivani College of Engineering (Autonomous), Kopargaon, Maharashtra, India.

Lwembawo Ibrahim is Senior Lecturer in the Department of Computer Science, Islamic University in Uganda (IUIU), Mbale, Uganda.

Carmel Mary Belinda M. J. is Professor in the Department of Computer Science and Engineering, Saveetha School of Engineering, Saveetha Institute of Medical and Technical Sciences, Chennai, Tamil Nadu, India.

Vishal Jain is Associate Professor in the School of Engineering and Technology, Sharda University, Greater Noida, Uttar Pradesh, India.

Abhinav Juneja is HOD & Professor at KIET Group of Institutions, Ghaziabad, Delhi-NCR, India.

Sapna Juneja is Professor at KIET Group of Institutions, Ghaziabad, Delhi-NCR, India.

Kassim Kalinaki is Lecturer in the Department of Computer Science, Islamic University in Uganda (IUIU), Mbale, Uganda; Borderline Research Laboratory, Kampala, Uganda.

Sulaiman Ibrahim Kassim is Lecturer in the Department of Business Administration, Federal University Dutse, Jigawa State, Nigeria.

Christian Kaunert is Professor of International Security, Dublin City University, Ireland; Professor of Policing and Security, Director, International Centre for Policing and Security, University of South Wales, UK.

H. E. Khodke is Assistant Professor at the Sanjivani College of Engineering (Autonomous), Kopargaon, Maharashtra, India.

S. Krishnaveni is Associate Professor in the Department of Computational Intelligence, SRM Institute of Science and Technology, Kattankulathur, Tamil Nadu, India.

M. Ashok Kumar is Lecturer in the Department of Computer Science and Software Engineering, Skyline University Nigeria, Kano, Nigeria.

Senthil Kumar is Head in the Department of Management, Skyline University Nigeria, Kano, Nigeria.

Meet Kumari is Assistant Professor in the Department of ECE, UIE, Chandigarh University, Gharuan, Mohali, Punjab, India.

K. P. Lochanambal is Assistant Professor, Department of Computer Science, Government Arts College, Udumalpet, Bharathiar University, Tamil Nadu, India.

Aditya Mishra is Student in the Department of Electronics and Communications Engineering (ECE), College of Technology and Engineering, Maharana Pratap University of Agriculture and Technology, Udaipur, Rajasthan, India.

Himanshi Mishra is Student at the KIET Group of Institutions, Ghaziabad, Delhi-NCR, India.

Venugopal Reddy Modhugu is at Oracle Cloud Infrastructure (OCI), Oracle, Senior Member of Technical Staff at Oracle, Falls Church, Virginia, United States.

Aliyu Mohammed is PhD Research Scholar in the Department of Business Administration, Skyline University Nigeria, Kano, Nigeri

Junaid Mohsin is Student at the KIET Group of Institutions, Ghaziabad, Delhi-NCR, India.

Gnanasankaran Natarajan is Assistant Professor, Department of Computer Science, Thiagarajar College, Madurai, Tamil Nadu, India.

Dinesh Nilkant is Professor and Pro-Vice-Chancellor, Faculty of Management Studies, CMS Business School, JAIN (Deemed-to-be University), Bengaluru, Karnataka, India.

Swathi P. is Assistant Professor in the Department of Management, Kristu Jayanti College (Autonomous), Bengaluru, Karnataka, India.

Prashnatita Pal is Assistant Professor in the Department of Electronics & Telecommunication St. Thomas College of Engineering & Technology, Kolkata, West Bengal, India.

Jayanta Poray is Assistant Professor in the Department of Computer Science & Engineering, Techno India University, Kolkata, West Bengal, India.

Pethuru Raj is Chief Architect at Edge AI Division, Reliance Jio Platforms Ltd, Bangalore, Karnataka, India

G. Ramya is Assistant Professor, Information Technology Department, National Engineering College, Kovilpatti, Tamil Nadu, India

Rohit Rastogi is Associate Professor, Department of CSE, ABES Engineering College, Ghaziabad, Uttar Pradesh, India.

Hemanth Kumar S. is Professor in the Faculty of Management Studies, CMS Business School, JAIN (Deemed-to-be University), Bengaluru, Karnataka, India.

Sivagami S. is Assistant Professor (SG), Department of Computer Science and Engineering, Saveetha School of Engineering, Saveetha Institute of Medical and Technical Sciences, Chennai, Tamil Nadu, India.

M. Sangeetha is Assistant Professor, Department of Computer Science, SNS Institutions, Coimbatore, Tamil Nadu, India

S. Sangeetha is Assistant Professor in the Department of Information Technology, Kongunadu College of Engineering and Technology, Thottiam, Trichy, Tamil Nadu, India.

Eleanor Schwartz is Senior Specialist in the Department of Management, Marketing, and Humanities, New York Institute of Technology, Long Island and New York City, United States.

Asha Sharma is Assistant Professor, Department of Accountancy and Business Statistics, University College of Commerce & Management Studies, Mohanlal Sukhadia University, Udaipur, Rajasthan, India.

Harshita Shukla is Student at the KIET Group of Institutions, Ghaziabad, Delhi-NCR, India.

Bhupinder Singh is Professor in the Sharda School of Law, Sharda University, Greater Noida, Uttar Pradesh, India.

K. G. Srinivasagan is Professor and Head of the Information Technology Department, National Engineering College, Kovilpatti, Tamil Nadu, India.

M. Srisankar is Research Scholar in the Department of Computer Science, Government Arts College, Bharathiar University, Udumalpet, Tamil Nadu, India.

Shanmugam Sundararajan is the Head of the Department of Entrepreneurship, Skyline University Nigeria, Kano State, Nigeria.

Shivani Trivedi is Assistant Professor, ABES Engineering College, Ghaziabad, Delhi-NCR, India.

Shivangi Verma is Assistant Professor, Department of Commerce & Management, Ramanand Institute of Pharmacy and Management, Haridwar, Uttarakhand, India.

R. Vijayakumari is Assistant Professor in the Department of Computer Science, Krishna University, Machilipatnam, Krishna District, Andhra Pradesh, India.

Pawan Whig is Senior IEEE Member and Dean Research at the Vivekananda Institute of Professional Studies, Technical Campus, Delhi, India.

1 Enhancing the Power of Cyber-Physical Systems Enabled with AI

An Introduction—Facts and Myths along with Modular Approach

Shivani Trivedi, Vanshika Aggarwal, and Rohit Rastogi

Cyber-physical systems (CPS) are systems that tightly integrate physical components with computational and networking elements. They are becoming increasingly prevalent in a wide range of applications, such as transportation, healthcare, and manufacturing. Artificial intelligence (AI) has the potential to significantly improve the performance and capabilities of CPS. For example, AI may be used to automate jobs that are now done by humans, increase the efficiency and accuracy of decision-making, and more. Hence, the motivation is to explore that the synergy lies in understanding and harnessing the unprecedented potential it presents. As AI evolves to exhibit more human- like cognitive abilities, CPS seamlessly merges the digital and physical realms.

The scope of study of this chapter is to bring new capabilities for monitoring, controlling, and optimizing processes; AI has the potential to transform CPS. AI techniques like machine learning, natural language processing, and computer vision can be used to extract insights from data, produce predictions, and automate tasks. It also covers the fundamentals of AI, the difficulties of integrating AI into CPS, and some of the most significant uses of AI in CPS in the upcoming years as CPS become more complicated and networked, AI is anticipated to play a significant role in CPS.

This study, first of all, gives an idea of the CPS. The author team gave a brief introduction of what a CPS is following which they have explained its characteristics, applications, and the integration of physical and digital realms. The author team has then given the introduction of AI including its evolution. They have then explained the integration of AI into CPS, its benefits and the challenges that were being faced due to the integration of AI into CPS. While coming towards the end of the chapter, they've given some of the real-life scenarios of the AI transforming CPS. The author team conducted literature research and examined five publications on the relevant

DOI: 10.1201/9781032694375-1

subjects to support the study. This literature review offers in-depth knowledge about the integration of AI into CPS. It also discussed some of the difficulties that resulted from this synergy. The author team has presented a framework showing the integration of AI into CPS in which they have represented the synergy of AI and CPS. They have collected data by reading papers and then presented the framework representing the working, process and benefits of the integrated systems. Further, the chapter discusses the limitations and the future scope in the study which includes the improvements that can be done in the future in the chapter. One of the most crucial sections of the research investigations, the recommendation section, makes recommendations for particular applications to deal with the problems and limitations noted in the evaluation. The research's novel aspects are discussed in the novelty section. The final section, the conclusion, summarizes the major findings of the study and represents the ultimate judgement.

1 INTRODUCTION

In the age of technological acceleration, the fusion of CPS and AI emerges as a transformative force with the potential to reshape industries, societies, and the very essence of human interaction with the digital world. In this chapter, it embarks on a captivating journey to unravel the intricacies of this convergence.

1.1 Defining Cyber-Physical System

A CPS, often known as an intelligent system, is a device that controls or keeps an eye on a mechanism using computer-based algorithms. CPS are made up of intricately entwined physical and software components that can operate at various spatial and temporal scales, display a variety of distinctive behavioural modalities, and interact with one another in context-dependent ways. CPS integrate theories from multiple disciplines, including design, process science, mechatronics, and cybernetics. "Embedded system" is a common term used to describe process control. The emphasis in embedded systems is typically more on the computational components and less on a close relationship between the computational and physical components. Although CPS and the internet of things (IoT) have a similar basic architecture, CPS exhibits a stronger fusion and coordination of physical and computational aspects (as per Figure 1.1) (Hu et al., 2022; Cyber Physical System from Wikipedia, 2023).

1.1.1 The Integration of Physical and Digital Realms

The integration of physical and digital realms epitomizes the fusion of tangible reality with virtual intelligence. CPS seamlessly intertwine sensors, actuators, and physical devices with computational power and AI-driven analytics. This synergy allows for real-time data exchange, informed decision-making, and adaptive responses, revolutionizing sectors from manufacturing to healthcare. Applications range from autonomous vehicles navigating city streets to smart cities optimizing energy usage. However, challenges including security, interoperability, and ethical considerations underscore the need for thoughtful implementation. This integration blurs boundaries, enabling unprecedented innovations that leverage the combined strengths of the physical and digital worlds, propelling society into a new era of transformative possibilities.

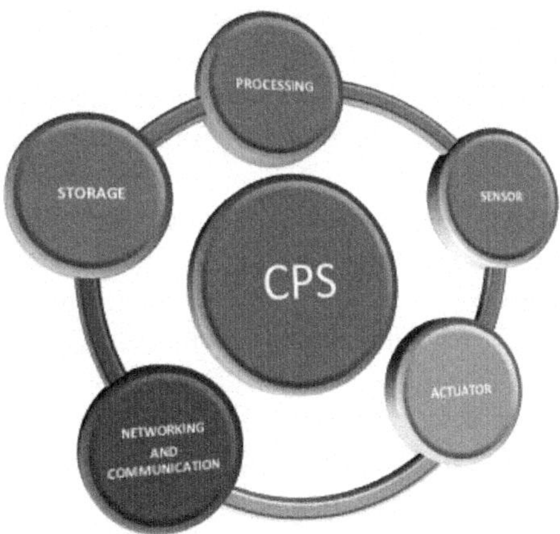

FIGURE 1.1 Cyber-Physical Systems (CPS).

1.1.2 Characteristics and Components of CPS

A. Characteristics of CPS

Integration: CPS seamlessly combine physical components with digital intelligence, bridging the gap between the physical and virtual worlds.

Real-Time Interaction: CPS enable real-time data exchange and decision-making, allowing for immediate responses and adaptive behaviours.

Interconnectedness: Components within CPS communicate and collaborate, creating a networked ecosystem that shares information for enhanced functionality.

Autonomy: CPS exhibit autonomous capabilities, where physical devices can make decisions and take actions based on AI algorithms without direct human intervention.

Sensing and Actuation: CPS involve sensors that capture data from the physical environment and actuators that execute actions in response to digital inputs.

Feedback Loop: CPS create closed-loop systems, where actions based on digital analysis influence the physical environment, creating a continuous cycle of interaction.

Adaptability: CPS can adapt to changing conditions in real-time, optimizing processes and responses for efficiency and effectiveness.

B. Components of CPS

Physical Entities: These include sensors, actuators, machinery, robots, vehicles, and any tangible devices that interact with the physical environment.

Sensors: Sensors gather data from the physical world, such as temperature, pressure, motion, or environmental conditions.

Actuators: Actuators execute physical actions based on digital inputs, such as turning on a motor, adjusting a valve, or changing a physical state.

Computational Intelligence: This includes AI algorithms, machine learning models, and data analytics tools that process sensor data to derive insights and make decisions.

Communication Networks: These networks facilitate the transmission of data between physical components and computational intelligence.

Software Infrastructure: Software platforms and frameworks manage data flow, communication, and orchestration between physical and digital components.

Control Systems: These systems implement algorithms that control and regulate the behaviour of physical entities based on data analysis.

User Interfaces: Interfaces allow humans to interact with CPS, providing insights, controls, and feedback through graphical interfaces or other means.

Data Storage: CPS generate and process massive amounts of data, necessitating storage solutions to store and retrieve relevant information.

Security Measures: As CPS involve the exchange of sensitive data, security protocols, encryption, and authentication mechanisms are vital to safeguard against cyber-physical threats.

Feedback Mechanisms: These refer to mechanisms that allow CPS to adjust their behaviour based on the outcomes of their actions, contributing to self-improvement and optimization.

CPS amalgamate these components and characteristics, creating a dynamic and transformative ecosystem that revolutionizes various domains by leveraging the strengths of both the physical and digital worlds (Habib & Chimsom, 2022).

1.1.3 Applications across Industries of CPS

Certainly, here are some applications of CPS across various industries:

a. **Manufacturing and Industry 4.0**
   - **Smart Factories:** CPS enhance manufacturing processes with real-time monitoring, predictive maintenance, and optimized production lines.
   - **Supply Chain Management:** CPS-driven sensors track inventory, monitor shipment conditions, and ensure efficient logistics.

b. **Healthcare**
   - **Telemedicine and Remote Monitoring:** CPS facilitate remote patient monitoring (RPM) and telehealth, enabling timely interventions and reducing hospitalization.
   - **Smart Medical Devices:** CPS-powered medical devices can monitor patient vitals, administer medication, and even perform diagnostics.

c. **Transportation**
   - **Autonomous Vehicles:** CPS enable self-driving cars, trucks, and drones, enhancing safety and efficiency in transportation.
   - **Traffic Management:** Smart traffic lights and intelligent road systems based on CPS optimize traffic flow and reduce congestion.

d. **Energy and Utilities**
   - **Smart Grids:** CPS monitor energy consumption in real-time, optimizing distribution and reducing energy wastage.
   - **Renewable Energy Management:** CPS help manage and optimize the generation, storage, and distribution of renewable energy sources.

e. **Agriculture**
   - **Precision Farming:** To optimize irrigation, fertilizing, and planting, CPS-driven sensors keep an eye on the health of the crops, weather patterns, and soil conditions.
   - **Livestock Management:** CPS track animal health and behaviour, enabling timely care and enhancing overall livestock management.

f. **Aerospace and Defence**
   - **Unmanned Aerial Vehicles (UAVs):** CPS power drones for surveillance, reconnaissance, and delivery in defence and civilian applications.
   - **Aircraft Maintenance:** CPS monitor aircraft systems for real-time diagnostics and predictive maintenance, ensuring safety and operational efficiency.

g. **Smart Cities**
   - **Urban Mobility:** CPS support smart traffic management, public transportation optimization, and parking solutions.
   - **Environmental Monitoring:** CPS track air quality, noise levels, and waste management, contributing to cleaner and more sustainable cities.

h. **Home Automation**
   - **Smart Homes:** CPS-driven devices control lighting, temperature, security, and appliances for energy efficiency and user convenience.
   - **Health Monitoring:** CPS-enabled wearables and home devices monitor health metrics and provide alerts for medical attention.

i. **Environmental Monitoring**
   - **Natural Disaster Management:** CPS facilitate early detection of earthquakes, floods, and other disasters, enabling rapid response and evacuation.
   - **Wildlife Conservation:** CPS support tracking and monitoring of endangered species, helping protect biodiversity.

j. **Retail and Customer Service**
   - **Smart Retail:** CPS enhance customer experiences with personalized shopping, inventory management, and efficient checkout processes.
   - **Chatbots and Customer Support:** CPS-powered AI chatbots provide instant customer assistance and support.

The applications of CPS span diverse sectors, transforming industries by combining real-world data, computational intelligence, and autonomous decision-making to optimize operations, enhance safety, and create innovative solutions to complex challenges (as per Figure 1.2) (Habib & Chimsom, 2022).

1.2 The Evolution of Artificial Intelligence

Since its conception, AI has undergone substantial development. Early AI focused on rule-based systems and symbolic reasoning in the mid-20th century. The field later embraced machine learning, with neural networks becoming prominent in the 1980s. However, progress slowed during the "AI winter." A breakthrough came with deep learning in the 2010s, empowering AI to excel in image recognition, natural language processing, and more. This led to the development of AI-driven applications

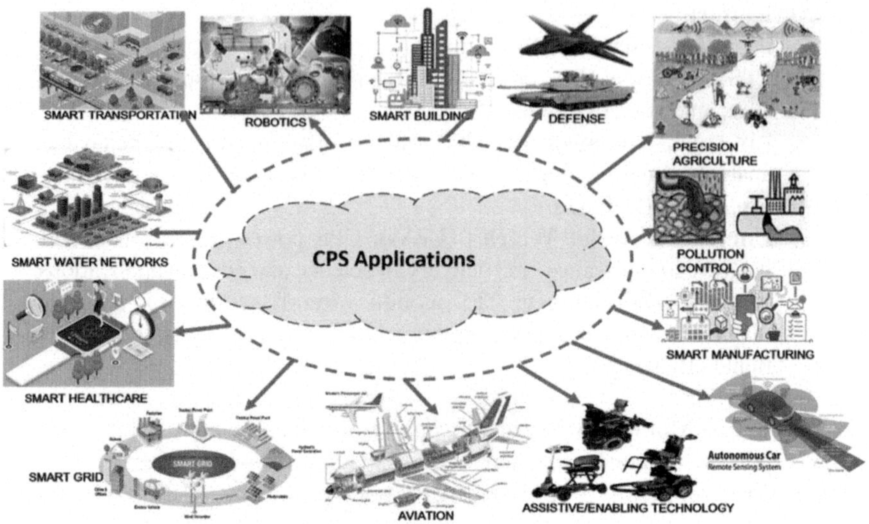

FIGURE 1.2 Applications of CPS.

like virtual assistants and recommendation systems. The integration of AI with big data propelled its capabilities. Ethical concerns and debates about AI's impact on jobs and society emerged as AI systems grew more complex. Ongoing research involves explainable AI, reinforcement learning, and the quest to achieve artificial general intelligence (AGI) (Spector et al., 2006).

1.2.1 From Rule Based System to Machine Learning

The evolution of AI from rule-based systems to machine learning represents a significant shift in AI paradigms:

Rule-Based Systems (Early AI): In the early stages of AI development, rule-based systems were prevalent. These systems relied on explicit programming of human-defined rules to perform tasks. They were limited in handling complex and uncertain situations due to their rigid nature.

Machine Learning Emergence (1980s–1990s): Machine learning gained attention as a new approach to AI. It involved developing algorithms that allowed computers to learn patterns and relationships from data. Neural networks, decision trees, and Bayesian networks were some of the techniques used during this period.

Neural Networks and the AI Winter (Late 1990s): Neural networks, especially deep learning, showed promise but faced challenges due to limitations in computing power and available data. This led to the "AI winter," a period of reduced funding and interest in AI research.

Resurgence of Machine Learning (2010s): Advances in computing power and the availability of massive datasets led to a resurgence in machine learning. Deep learning, supported by neural networks with many layers, demonstrated exceptional performance in image and speech recognition, language translation, and other tasks.

Deep Learning Dominance (2010s–Present): Deep learning became the dominant approach in AI. Convolutional neural networks (CNNs) revolutionized image analysis, while recurrent neural networks (RNNs) improved sequential data processing. Transfer learning and pre-trained models accelerated progress by allowing models to learn from one task and apply knowledge to another.

Shift to Data-Driven AI (Present): Machine learning models evolved to become data-driven, capable of extracting insights from vast amounts of data. Reinforcement learning, which enables machines to learn through trial and error, gained traction in training AI agents to make decisions.

Integration of AI into Everyday Life (Present): AI-powered technologies, such as virtual assistants, recommendation systems, and autonomous vehicles, have become integral to daily life, demonstrating the practical applications of machine learning in diverse domains.

Challenges and Future Directions (Present-Future): Even though machine learning has produced impressive results, there are still issues to be resolved, such as ethical issues, bias reduction, and the pursuit of AGI, or robots that can reason like humans. For more interpretable and flexible AI systems, researchers are investigating explainable AI and hybrid techniques that integrate rule-based systems with machine learning.

The journey from rule-based systems to machine learning represents a shift from explicit programming of rules to data-driven learning, enabling AI systems to handle complexity, uncertainty, and real-world variability more effectively (as per Figure 1.3) (Cohen et al., 2021).

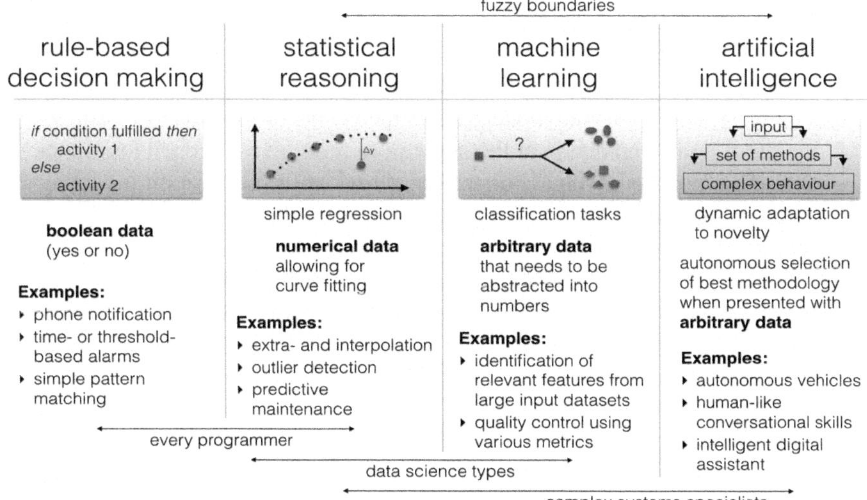

FIGURE 1.3 Rule-based AI vs Machine Learning.

1.2.2 Deep Learning and Neural Network

Deep learning and neural networks are subfields of AI and machine learning (ML) that have gained significant attention and success in recent years. They involve the use of computational models inspired by the structure and function of the human brain to process and analyse complex data.

Neural Networks: A neural network is a type of computing model made up of layers of interconnected nodes called neurons. Every neuron receives information, uses an internal function to process it, and then creates an output. Typically, the layers fall into the following categories:

Input Layer: Receives the raw input data and passes it to the subsequent layers.

Hidden Layers: These layers process the input data through weighted connections and activation functions. Deep neural networks have multiple hidden layers, hence the term "deep" learning.

Output Layer: Produces the final prediction or output of the network.

Deep Learning: Deep learning is a subfield of machine learning that focuses on employing neural networks with numerous hidden layers to model and solve complex patterns in data. It has had great success in a number of fields, including speech and picture identification, natural language processing, and autonomous driving.

The primary factors contributing to its success include the availability of large datasets, powerful hardware (such as GPUs), and advancements in training algorithms.

Key Concepts in Deep Learning:

a. **Activation Function:** A function that is applied to each neuron's output to add nonlinearity to the network. The rectified linear unit (ReLU), sigmoid, and hyperbolic tangent (tanh) are frequently used activation functions.
b. **Weights and Biases:** Each neuron-to-neuron connection has corresponding weights and biases that control the strength of the connection and the contribution of each neuron to the total output.
c. **Training:** Deep learning models are trained using large datasets by adjusting the weights and biases of the network to minimize a defined loss function. This process involves forward and backward propagation, updating parameters through optimization algorithms like gradient descent.
d. **Backpropagation:** The process of calculating the loss function's gradients in relation to the model's parameters. These gradients show how each parameter needs to be changed in order to reduce the loss.
e. **Overfitting:** When a deep learning model does well with training data but poorly with fresh, untried data. To reduce overfitting, regularization strategies and early halting are employed.
f. **Convolutional Neural Networks (CNNs):** Specialized neural networks designed for image processing tasks, utilizing convolutional layers that capture local patterns in data.
g. **Recurrent Neural Networks (RNNs):** Neural networks designed to work with sequential data, incorporating loops to maintain a sense of memory about previous inputs.

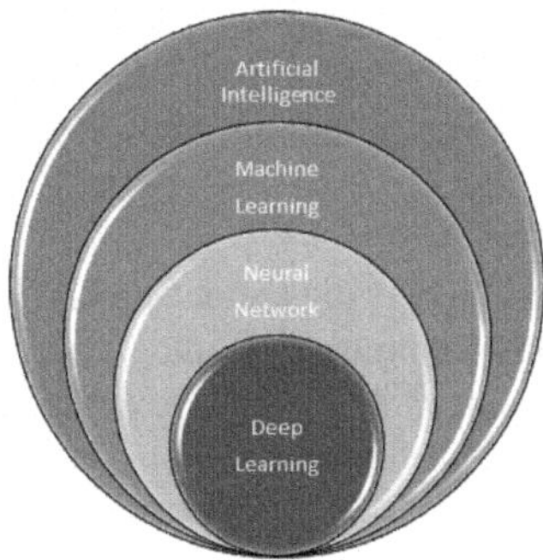

FIGURE 1.4 Hierarchy of Artificial Intelligence (AI).

h. **Long Short-Term Memory (LSTM) Networks:** A kind of RNN that can recognize long-distance dependencies in sequential data and solves the vanishing gradient problem.
i. **Generative Adversarial Networks (GANs):** Neural network pairs, one of which produces data and the other of which evaluates it. GANs are employed in the creation of images and the transfer of style.

Deep learning and neural networks have revolutionized various industries, enabling machines to perform tasks that were previously challenging for traditional algorithms. However, they require substantial computational resources, careful tuning, and expertise to achieve optimal results.

Figure 1.4 shows the hierarchy of AI. AI comprises all three of the previously described subsets: deep learning, neural networks, and machine learning.

1.3 Key Concepts of AI in CPS

The integration of AI into CPS brings forth transformative concepts that revolutionize system behaviour and capabilities. One fundamental concept is real-time adaptation, where AI algorithms enable CPS to dynamically adjust control strategies based on changing data, ensuring optimal performance in dynamic environments. Predictive maintenance is another pivotal idea, with AI analysing sensor data to predict equipment failures in advance, minimizing downtime and maintenance costs. Anomaly detection is crucial for identifying unusual patterns in sensor data, enhancing CPS' ability to detect malfunctions or cyber threats. AI empowers CPS with autonomous decision-making, enabling systems like self-driving vehicles to make real-time choices based on collected data. Optimization, both in terms of energy distribution in smart grids and resource management in various contexts, is facilitated by AI algorithms, resulting in efficient and effective system operation. Human-machine interaction reaches new heights as AI-driven interfaces offer intuitive and user-friendly interactions, enhancing control and usability. Simulation and testing benefit from AI-based simulations that model CPS behaviour, aiding design and reducing development time. Adaptive learning enables CPS to continuously learn from data and adapt to changing environments, contributing to system resilience. Fault tolerance is improved with AI, ensuring that CPS can continue functioning even in the presence of failures. These key concepts collectively reshape CPS, making them more adaptive, efficient, and capable of addressing real-world challenges (Radanliev et al., 2021).

1.3.1 The Synergy between AI and CPS

The synergy between AI and CPS brings about powerful capabilities and transformative impacts across various domains. This integration enhances the performance, efficiency, adaptability, and safety of CPS by leveraging AI techniques. Here are some ways in which AI and CPS synergize:

a. **Real-Time Decision Making:** AI algorithms process data from sensors and other sources in real time, allowing CPS to make informed decisions rapidly. This is crucial in applications like autonomous vehicles, where split-second decisions can impact safety.
b. **Predictive Analytics:** AI analyses historical and real-time data to predict future events and conditions. In CPS, this translates to predictive maintenance, where AI predicts equipment failures before they occur, minimizing downtime and costs.
c. **Adaptive Control:** CPS integrated with AI can dynamically adjust control strategies based on changing conditions. For example, in smart grids, AI optimizes energy distribution based on real-time demand and availability.
d. **Optimization:** AI optimizes complex CPS operations by finding the best configurations for achieving objectives such as energy efficiency, resource allocation, and traffic management.
e. **Anomaly Detection and Fault Tolerance:** AI algorithms detect anomalies in sensor data that might indicate malfunctions or cyberattacks. In case of component failures, AI can adapt the system's behaviour to ensure continued operation.
f. **Human-Machine Interaction:** AI-powered interfaces enable more intuitive interactions between humans and CPS. Natural language processing and gesture recognition make control and communication more user-friendly.
g. **Autonomous Systems:** AI enables CPS to function autonomously, making decisions based on real-time data and predefined rules. This is seen in self-driving cars, drones, and smart manufacturing systems.
h. **Energy Management:** In CPS like smart buildings, AI optimizes energy consumption by analysing occupancy patterns and adjusting heating, cooling, and lighting systems accordingly.
i. **Healthcare and Well-being:** AI-enhanced CPS can monitor patients' health status and provide personalized care. Wearable devices integrated with AI can detect anomalies and alert medical professionals.
j. **Industrial Automation:** AI optimizes production processes by analysing sensor data, predicting maintenance needs, and ensuring efficient resource usage in manufacturing environments.
k. **Smart Cities:** The integration of AI with CPS enables urban systems to manage traffic, energy, waste, and infrastructure more efficiently, improving the quality of life for residents.

l. **Simulation and Testing:** AI-based simulations allow designers to model and test CPS behaviour before physical implementation. This reduces development time and costs.
m. **Environmental Impact:** AI-driven CPS can optimize energy consumption, reduce emissions, and manage resources effectively, contributing to sustainability goals.
n. **Remote Monitoring and Control:** AI-enabled CPS can be remotely monitored and controlled, allowing for safer operation in hazardous environments or remote locations.
o. **Data-Driven Insights:** The massive amounts of data generated by CPS can be analysed by AI to extract valuable insights, leading to data-driven decisions and improvements.
p. **Ethical Considerations:** AI can help CPS adhere to ethical standards by ensuring transparency, fairness, and accountability in decision-making processes.

The convergence of AI and CPS has the potential to transform entire businesses, improve public services, and enhance our quality of life as a whole. However, it also presents difficulties like data security and privacy, as well as the requirement for qualified employees who can develop, implement, and maintain these intricate systems (Radanliev et al., 2021).

1.3.2 Enhancing CPS Capabilities with AI

Enhancing CPS capabilities through the integration of AI yields a paradigm shift in how these systems operate, adapt, and respond to complex environments. AI empowers CPS with real-time adaptability, allowing them to dynamically adjust control strategies based on evolving data inputs. This leads to optimized performance and efficiency, especially in scenarios such as autonomous vehicles navigating unpredictable traffic patterns. By leveraging predictive maintenance capabilities, AI analyses sensor data to foresee equipment failures, enabling timely maintenance actions that minimize downtime and reduce costs.

Anomaly detection, a critical feature of AI-enhanced CPS, provides early identification of irregular patterns in sensor readings, enhancing system robustness against malfunctions or security breaches. CPS achieve new levels of autonomy with AI-driven decision-making, as they autonomously interpret data and respond to changing conditions. This autonomy is harnessed in various applications, including smart grids efficiently managing energy distribution and smart factories orchestrating intricate manufacturing processes.

Moreover, AI's optimization prowess refines resource allocation within CPS, ensuring optimal use of limited resources like energy, bandwidth, and computational power. Human-machine interaction becomes more intuitive, with AI-enabled interfaces understanding natural language and gestures, facilitating seamless communication between users and CPS. Simulations enriched by AI-driven models enable precise testing and validation, accelerating CPS development cycles.

Adaptive learning mechanisms empower CPS to acquire knowledge from data over time, leading to systems that evolve and improve their performance. AI's capacity for fault tolerance ensures that CPS continue to operate effectively despite component failures or disruptions, enhancing system reliability and stability.

In essence, AI amplifies CPS' capabilities by infusing them with real-time intelligence, predictive insights, autonomous decision-making, optimal resource management, adaptive learning, and fault resilience. This synergy propels CPS into a new era, where responsiveness, efficiency, and adaptability redefine their potential across domains as diverse as transportation, manufacturing, energy management, and beyond (Radanliev et al., 2021).

1.3.3 Real-Time Interactions and Decision-Makings

Real-time interactions and decision-making in AI-CPS enable immediate responses to changing conditions. Through seamless data exchange between physical components and AI algorithms, systems like autonomous vehicles, smart grids, and healthcare monitors can swiftly adapt. Real-time decisions, powered by AI, optimize processes such as predictive maintenance, emergency response, and supply chain management. Challenges include managing latency, ensuring data accuracy, and maintaining algorithm efficiency. Balancing human oversight with autonomous decision-making is crucial for ethical and safe operation. Overall, this integration enhances efficiency, adaptability, and responsiveness in various domains.

1.3.4 Benefits and Challenges of Integration

There are many advantages to the combination of AI with CPS, but there are also certain issues that must be resolved. Let's examine the advantages and difficulties of this integration (as per Figure 1.5).

Benefits of Integration

a. **Improved Decision-Making:** AI can analyse large volumes of real-time data from CPS components, such as sensors and actuators, to make informed decisions. This enhances the overall system's ability to respond to changing conditions quickly and accurately.
b. **Optimized Performance:** AI can optimize the operation of CPS by adjusting parameters and configurations in real time, leading to improved efficiency and reduced energy consumption. For example, in a smart grid, AI can balance energy generation and consumption more effectively.
c. **Predictive Maintenance:** By examining data from CPS devices, AI systems can anticipate equipment breakdowns and maintenance requirements. Unplanned downtime can be avoided as a result, and maintenance costs can be decreased while equipment longevity is enhanced.
d. **Adaptive and Self-Learning Systems:** The integration of AI enables CPS to adapt to dynamic environments and learn from new data, allowing them to improve over time. This self-learning capability can lead to more resilient and efficient systems.
e. **Enhanced Automation:** AI can enable higher levels of automation by enabling CPS to make autonomous decisions based on real-time data. This is particularly valuable in manufacturing, transportation, and logistics.

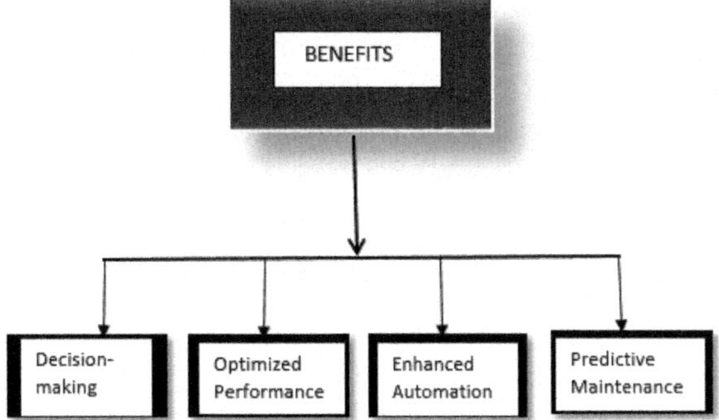

FIGURE 1.5 Benefits of Integration of AI and CPS.

f. **Real-Time Monitoring and Control:** AI-CPS integration enables real-time monitoring of processes and environments. This can be used for quality control, safety monitoring, and overall system optimization.

Challenges of Integration

a. **Complexity:** Integrating AI with CPS requires dealing with the complexities of both domains. It involves understanding and managing interactions between software, hardware, sensors, actuators, and algorithms, which can be challenging.
b. **Safety and Security:** The integration of AI and CPS raises concerns about safety and security. Malicious actors could exploit vulnerabilities in AI-CPS systems, potentially leading to physical harm or data breaches.
c. **Data Quality and Reliability:** AI's performance heavily relies on data quality and reliability. In CPS environments, sensor data might be noisy or incomplete, which can impact the accuracy of AI algorithms.
d. **Interoperability:** Different components of CPS might come from various manufacturers and follow different standards. Ensuring seamless communication and interoperability among these components can be challenging.
e. **Ethical Considerations:** AI-CPS systems can make autonomous decisions that have ethical implications. Deciding who is responsible when these systems make mistakes or cause harm is a complex issue.
f. **Regulations and Standards:** The integration of AI and CPS might outpace the development of regulations and standards to ensure their safe and ethical use. Clear guidelines are needed to prevent misuse and ensure compliance.
g. **Human-Machine Interaction:** As AI-CPS systems become more autonomous, defining how humans interact with these systems and ensuring effective collaboration is essential.
h. **Lack of Expertise:** Developing and implementing AI-CPS systems requires expertise in both AI and CPS, which is a relatively niche skill set. Finding individuals with the necessary knowledge can be a challenge.

In conclusion, the integration of AI and CPS holds great promise for enhancing system performance, efficiency, and adaptability. However, addressing the challenges related to complexity, safety, data quality, interoperability, ethics, and expertise is crucial to fully realize the potential benefits of this integration (Gupta et al., 2020).

1.4 Challenges and Considerations in AI-CPS Integration

Integrating AI with CPS presents numerous challenges and considerations due to the complex and interconnected nature of these systems. AI-CPS integration has the potential to enhance the efficiency, reliability, and autonomy of CPS, but it brings several key challenges that need to be addressed. Here are some of the main challenges and considerations.

1.4.1 Real-Time Data Processing

Real-time data processing is a critical aspect of AI-CPS integration and presents its own set of challenges and considerations. When dealing with real-time data in CPS, there are several factors to keep in mind:

a. **Low Latency Requirements:** Many CPS applications, such as autonomous vehicles, industrial automation, and medical devices, have strict latency requirements. Real-time AI algorithms must process data quickly and provide timely responses to ensure the system operates effectively and safely.
b. **Data Volume:** Real-time CPS generate large volumes of data from sensors and actuators. Efficient data handling and processing become crucial to avoid bottlenecks and delays in the system.
c. **Resource Constraints:** CPS often operate with limited computational resources, including processing power, memory, and energy. Real-time AI algorithms must be optimized to run efficiently within these constraints.
d. **Sensor Noise and Variability:** Sensor data in CPS can be noisy and subject to variability due to environmental conditions. AI algorithms must be robust enough to handle noisy input and make accurate decisions.
e. **Distributed Systems:** Many CPS are distributed systems with multiple components and sensors spread across different locations. Real-time data must be efficiently collected, processed, and communicated between these components.
f. **Predictive Modelling:** Real-time AI may require predictive modelling to anticipate future states or events based on historical data. Developing accurate predictive models in real-time is challenging but essential for proactive decision-making.
g. **Concurrency and Parallelism:** Real-time AI algorithms often need to handle multiple tasks concurrently. Designing algorithms that can take advantage of parallel processing capabilities is essential for meeting real-time requirements.
h. **Feedback Loops:** CPS may rely on feedback loops to control physical processes. Real-time AI should seamlessly integrate with these feedback mechanisms to make dynamic adjustments as needed.

i. **Fault Tolerance:** Real-time systems should be resilient to failures. This includes the ability to detect and recover from faults in data sources or processing components without causing system-wide disruptions.
j. **Security:** Real-time data processing must consider security aspects such as data integrity, authentication, and protection against real-time cyber threats.
k. **Adaptive Algorithms:** Some CPS applications require AI algorithms that can adapt to changing conditions in real-time. Algorithms must continuously learn and evolve without compromising stability or performance.
l. **Testing and Validation:** Validating real-time AI algorithms can be challenging, as traditional testing methods may not adequately simulate real-time conditions. Specialized testing environments and simulation tools may be necessary.
m. **Regulatory Compliance:** Depending on the industry and application, there may be regulatory requirements related to real-time data processing and decision-making. Ensuring compliance with these regulations is essential.
n. **Human Interaction:** Real-time AI-CPS may involve human interaction. The user interface should provide timely feedback and be intuitive to use.
o. **Energy Efficiency:** Real-time data processing can consume a significant amount of energy in CPS devices. Optimizing algorithms for energy efficiency is crucial, especially in battery-powered systems.

Addressing these challenges in real-time data processing for AI-CPS integration requires a combination of expertise in real-time systems, AI, and domain-specific knowledge. It often involves a trade-off between speed, accuracy, resource utilization, and adaptability, depending on the specific CPS application. Careful design, testing, and validation processes are necessary to ensure that real-time AI-CPS systems meet performance and safety requirements (Lee et al., 2020).

1.4.2 Data Reliability and Quality

Data reliability and quality are paramount considerations in the integration of AI with CPS. Poor data quality or unreliable data sources can significantly impact the performance, safety, and effectiveness of AI-CPS systems. Here are some challenges and considerations related to data reliability and quality in AI-CPS integration:

a. **Sensor Accuracy and Calibration:** CPS rely on sensors to collect data from the physical environment. Ensuring the accuracy and proper calibration of these sensors is essential for reliable data. Sensor drift, noise, and calibration issues can introduce errors into the data.
b. **Data Integrity:** Data may be corrupted or tampered with during transmission or storage. Implementing data integrity checks and encryption mechanisms can help protect against data corruption and unauthorized access.
c. **Data Preprocessing:** Raw sensor data often requires preprocessing to remove outliers, filter noise, and handle missing values. The quality of preprocessing algorithms directly affects the reliability of the data used by AI systems.
d. **Data Synchronization:** In distributed CPS, data from multiple sensors and components must be synchronized accurately in time. Inaccurate synchronization can lead to erroneous conclusions and actions by AI algorithms.

e. **Redundancy and Backup:** To enhance data reliability, redundant sensors and data sources can be employed. This redundancy can help in detecting and mitigating sensor failures or data inconsistencies.
f. **Data Anomalies and Outliers:** AI algorithms need to be robust to handle data anomalies and outliers gracefully. Detecting and handling such anomalies is crucial to maintaining system reliability.
g. **Data Labelling and Ground Truth:** Supervised AI models often require labelled training data. Ensuring the accuracy and consistency of data labelling, as well as establishing a reliable ground truth, is essential for training AI models.
h. **Data Validation and Verification:** Implementing validation mechanisms to verify data accuracy and consistency in real-time is crucial. This includes methods to detect and flag erroneous data before it influences AI decision-making.
i. **Environmental Variability:** The physical environment in which CPS operates can be dynamic and subject to various external factors. AI models must be adaptable to handle environmental variability and maintain reliability.
j. **Data Volume and Throughput:** High-throughput CPS generates vast amounts of data that must be processed in real time. Designing data pipelines and storage systems capable of handling these volumes while ensuring data quality is a challenge.
k. **Bias and Fairness:** Biased data can lead to biased AI models. Ensuring fairness and mitigating bias in the data used for training and inference is essential, especially in applications with societal impacts.
l. **Data Privacy and Consent:** Compliance with data privacy regulations and obtaining appropriate consent for data collection are important considerations, especially when dealing with sensitive data in AI-CPS.
m. **Continuous Monitoring:** Real-time monitoring of data quality and reliability is necessary to detect issues promptly and trigger corrective actions.
n. **Feedback Loops:** CPS often use AI for control and decision-making. Feedback loops that account for data reliability and quality should be implemented to prevent erroneous actions.
o. **Human Oversight:** Human operators should have the ability to override AI decisions in cases where data reliability or quality is in question.

Addressing these challenges requires a holistic approach that combines domain expertise, data engineering, and AI modelling. It's essential to establish data governance practices, employ data validation and monitoring tools, and conduct thorough testing and validation to ensure that AI-CPS systems operate with reliable and high-quality data. Additionally, ongoing maintenance and updates are necessary to adapt to changing conditions and evolving data quality requirements (Yaacoub et al., 2020).

1.4.3 Safety and Security Concerns

Safety and security concerns are paramount in the integration of AI with CPS. The convergence of AI and CPS can introduce new vulnerabilities and risks that must be

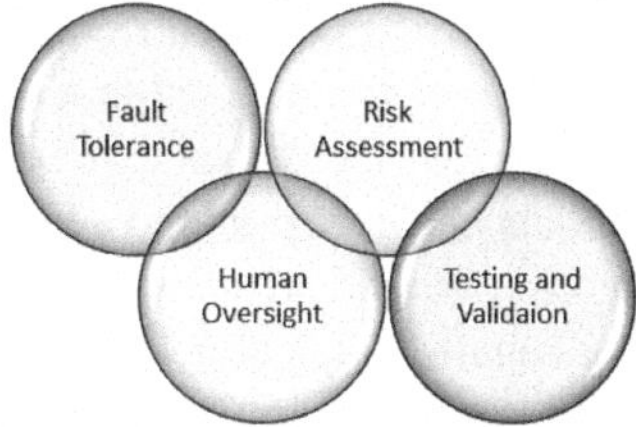

FIGURE 1.6 Safety Concerns of Integration.

carefully managed to ensure the safety of these systems and protect against potential threats. Here are some key safety and security considerations.

Safety Concerns

a. **Safety-Critical Applications:** Many CPS are used in safety-critical domains such as autonomous vehicles, medical devices, and industrial automation. Failures in AI algorithms or system components can have severe consequences, including harm to human life. Ensuring the safety of these systems is of utmost importance.
b. **Fault Tolerance:** CPS should be designed to withstand faults and errors gracefully. This includes the ability to detect and recover from failures in AI components or sensors to prevent accidents.
c. **Risk Assessment:** A thorough risk assessment is essential to identify potential safety hazards and their associated risks in AI-CPS integration. This assessment should consider both known and novel risks introduced by AI.
d. **Safety Standards and Regulations:** Adherence to safety standards and regulations specific to the industry is crucial. For example, the automotive industry follows ISO 26262, while medical devices must comply with regulatory standards like ISO 13485.
e. **Explainability and Transparency:** Understanding why and how AI systems make decisions is vital, especially in safety-critical applications. Transparent AI models can be more easily validated and trusted.
f. **Testing and Validation:** Rigorous testing and validation procedures are required to demonstrate that AI-CPS systems meet safety requirements. This includes simulation-based testing, real-world testing, and validation against safety benchmarks.
g. **Human Oversight:** Human operators should have the ability to intervene and override AI decisions in critical situations. Ensuring human-machine collaboration is crucial for safety.

Figure 1.6 shows the safety concerns that are being faced due to integration of AI and CPS.

Security Concerns

a. **Cybersecurity Threats:** The integration of AI with CPS exposes these systems to potential cyberattacks, including data breaches, unauthorized access, and malicious manipulation of data or AI algorithms.

b. **Data Security:** Protecting sensitive data collected by CPS, such as patient health records or vehicle telemetry data, is essential. Encryption, access control, and secure data storage mechanisms are necessary.
c. **Model Security:** AI models are susceptible to adversarial attacks, in which nefarious individuals modify input data to trick AI systems. It is essential to provide model robustness against such attacks.
d. **Secure Communication:** Secure communication protocols must be implemented to protect data transmitted between CPS components and AI systems from interception and tampering.
e. **Authentication and Authorization:** To guarantee that only authorized personnel may access and control CPS systems, robust authentication and authorization processes are required.
f. **Patch Management:** Regularly updating and patching both hardware and software components is essential to address security vulnerabilities. This includes AI model updates and security patches.
g. **Intrusion Detection and Response:** Implementing intrusion detection systems to monitor for unauthorized access and timely response mechanisms to mitigate security incidents.
h. **Security Audits and Compliance:** Regular security audits and compliance assessments help identify vulnerabilities and ensure that the system aligns with industry-specific security standards.
i. **Supply Chain Security:** Ensuring the security of the supply chain, including third-party components and software, is critical to prevent the introduction of vulnerabilities at the source.
j. **Incident Response Plans:** Developing and rehearsing incident response plans to react effectively to security breaches or safety incidents is essential.

Balancing safety and security in AI-CPS integration requires a comprehensive and interdisciplinary approach. Collaboration between domain experts, cybersecurity specialists, AI engineers, and compliance officers is essential to address both safety and security concerns effectively. Additionally, ongoing monitoring, threat intelligence, and adaptation to emerging threats are crucial to maintain the integrity of AI-CPS systems (Lyu et al., 2019).

1.4.4 Ethical Implications

The integration of AI with CPS carries significant ethical implications that must be carefully considered and addressed. These ethical concerns are particularly important due to the potential impact of AI-CPS systems on individuals, society, and the environment. Here are some key ethical implications:

a. **Bias and Fairness:** AI-CPS systems can inherit biases present in training data, which can result in unfair or discriminatory outcomes. Addressing bias and ensuring fairness in decision-making is essential to avoid perpetuating existing inequalities.
b. **Privacy:** The collection and analysis of data in CPS, especially in contexts like healthcare and smart cities, raise concerns about individual privacy.

Proper data anonymization, consent mechanisms, and data protection measures are necessary to safeguard privacy rights.

c. **Transparency and Accountability:** AI-CPS systems often involve complex algorithms that make decisions autonomously. Ensuring transparency in how these decisions are made and establishing mechanisms for accountability when things go wrong is crucial.
d. **Autonomy and Human Oversight:** The increasing autonomy of AI-CPS systems raises questions about the level of human oversight and control. Striking the right balance between automation and human intervention is an ethical challenge.
e. **Safety and Risk Management:** Ethical considerations encompass ensuring the safety of AI-CPS systems. This includes risk assessment, mitigating potential harm, and defining acceptable levels of risk in safety-critical applications.
f. **Informed Consent:** In medical and healthcare applications, patients and users should provide informed consent for AI-based diagnostics and treatments. The extent of consent and disclosure of AI involvement should be clear.
g. **Accountability and Liability:** Determining liability in case of AI-CPS failures can be complex. Clear rules for assigning responsibility and liability are necessary to protect individuals and organizations.
h. **Algorithmic Decision-Making:** AI-CPS systems make decisions with far-reaching consequences. Ensuring that these decisions align with ethical principles and societal values is a significant challenge.
i. **Data Ownership and Control:** Clarifying who owns and controls the data generated and processed by AI-CPS systems is essential to prevent data exploitation and maintain user autonomy.
j. **Societal Impact:** Assessing the broader societal impact of AI-CPS systems, including their effects on employment, inequality, and access to technology, is crucial for ethical decision-making.
k. **Environmental Impact:** The energy consumption and environmental footprint of AI-CPS systems should be considered, especially as sustainability becomes a more significant concern.
l. **Dual-Use Concerns:** AI-CPS technology can have dual-use applications, such as in military and surveillance contexts. Ethical considerations involve weighing the potential benefits against the risks of misuse.
m. **Education and Awareness:** Ensuring that stakeholders, including developers, policymakers, and users, are educated about the ethical implications of AI-CPS is essential to foster responsible development and deployment.
n. **Global Ethical Standards:** The development and deployment of AI-CPS systems often cross international borders. Establishing global ethical standards and norms can help guide responsible innovation.
o. **Ethical Review Boards:** In some cases, independent ethical review boards or committees may be necessary to evaluate and approve the deployment of AI-CPS systems, particularly in sensitive domains.

Addressing these ethical implications requires a proactive approach that includes ethical impact assessments, clear guidelines and regulations, ongoing monitoring,

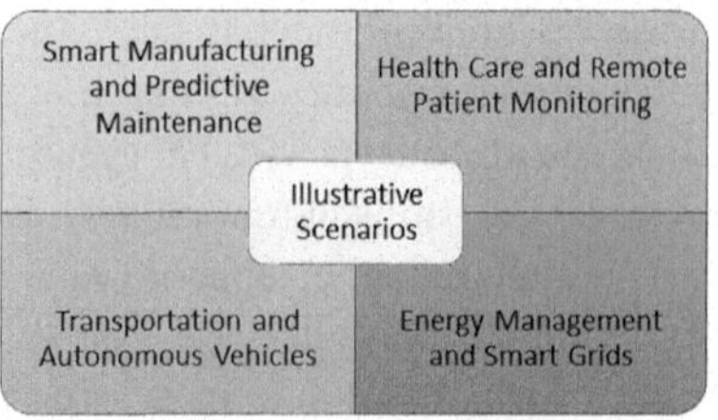

FIGURE 1.7 Illustrative Scenarios: AI Transforming CPS.

and stakeholder engagement. It also involves open and transparent discussions among technologists, ethicists, policymakers, and the public to ensure that AI-CPS systems align with societal values and ethical principles (Ramasamy et al., 2022).

1.5 Illustrative Scenarios: AI transforming CPS

AI has the potential to transform CPS across various domains. Here are some illustrative scenarios that highlight how AI can bring about significant changes in CPS (as per Figure 1.7):

1.5.1 Smart Manufacturing and Predictive Maintenance

Smart manufacturing and predictive maintenance are two interconnected concepts that leverage AI and data analytics to optimize industrial processes and reduce downtime in manufacturing facilities. Here's an overview of both concepts and their implications:

1.5.1.1 Smart Manufacturing

Smart manufacturing, often referred to as Industry 4.0 or the industrial internet of things (IIoT), involves the integration of advanced technologies, including AI and CPS, into manufacturing processes to make them more efficient, flexible, and data-driven. Key components and implications of smart manufacturing include:

- **Sensors and IoT Devices:** Sensors and IoT devices are deployed throughout the manufacturing facility to collect real-time data from equipment, machines, and production lines. These sensors generate vast amounts of data.
- **Data Integration:** Data from various sources are integrated into a centralized platform or system, often called a Manufacturing Execution System (MES) or Manufacturing Operations Management (MOM) system.
- **AI and Analytics:** AI algorithms analyse the collected data to identify patterns, anomalies, and opportunities for optimization. Machine learning models can predict equipment failures and production issues.
- **Predictive Maintenance:** AI-driven predictive maintenance models use historical data and real-time sensor data to predict when equipment is likely to fail. This enables proactive maintenance, reducing unplanned downtime.

- **Quality Control:** AI can be used for real-time quality control by analysing sensor data to detect defects or deviations in the manufacturing process.
- **Automation and Robotics:** Smart manufacturing often incorporates advanced automation and robotics to increase efficiency and reduce labour costs.
- **Customization:** Smart manufacturing allows for greater product customization and flexibility in production, often referred to as "mass customization."

1.5.1.2 Predictive Maintenance

Predictive maintenance is a critical component of smart manufacturing. It involves using AI and data analytics to predict when equipment or machinery is likely to fail so that maintenance can be performed just in time to prevent breakdowns. Key aspects and implications of predictive maintenance include:

- **Condition Monitoring:** Sensors continuously monitor the condition of machinery and equipment, collecting data on temperature, vibration, pressure, and other relevant parameters.
- **Data Analysis:** AI algorithms analyse this data to identify trends, patterns, and anomalies that may indicate early signs of equipment degradation or impending failure.
- **Failure Prediction:** Based on historical data and machine learning models, predictive maintenance systems can predict when a piece of equipment is likely to fail, giving maintenance teams advance notice.
- **Reduced Downtime:** By addressing maintenance needs proactively, rather than reactively, organizations can reduce unplanned downtime, improve equipment reliability, and extend the lifespan of machinery.
- **Cost Savings:** Predictive maintenance can result in significant cost savings by reducing the need for emergency repairs, minimizing spare parts inventory, and optimizing maintenance schedules.
- **Efficiency and Productivity:** It ensures that equipment is available and operational when needed, leading to improved production efficiency and higher overall productivity.
- **Safety:** Predictive maintenance can enhance workplace safety by reducing the risk of accidents associated with equipment failures.

In summary, the integration of AI and predictive maintenance into smart manufacturing environments enables manufacturers to operate more efficiently, reduce costs, and achieve higher levels of equipment reliability. It represents a transformative approach to industrial processes that leverages data-driven insights to optimize operations and deliver tangible business benefits (Singh et al., 2023).

1.5.2 Health Care and Remote Patient Monitoring

Healthcare and RPM are areas where the integration of AI with CPS is making significant advancements. These technologies are transforming healthcare delivery, improving patient outcomes, and enhancing the management of chronic diseases. Here's an overview of healthcare and RPM in the context of AI-CPS integration.

1.5.2.1 Healthcare

- **Electronic Health Records (EHRs):** AI-CPS integration enhances the management of electronic health records. AI can extract valuable insights from patient data, improving diagnosis and treatment planning.
- **Medical Imaging:** AI-powered CPS are used for medical imaging interpretation. Machine learning algorithms can assist radiologists in detecting and diagnosing conditions like cancer from X-rays, MRIs, and CT scans.
- **Drug Discovery:** AI and CPS help accelerate drug discovery by analysing vast datasets, identifying potential drug candidates, and predicting their efficacy.
- **Telehealth and Telemedicine:** AI-CPS systems support telehealth consultations by enabling remote video conferencing, monitoring vital signs, and providing diagnostic support.
- **Predictive Analytics:** AI-CPS solutions predict patient admission rates, disease outbreaks, and resource needs, improving resource allocation in healthcare facilities.
- **Hospital Management:** AI-CPS optimizes hospital operations, from patient scheduling and resource allocation to bed management and inventory control.
- **Personalized Medicine:** AI analyses genetic and clinical data to personalize treatment plans, ensuring that therapies are tailored to individual patients' unique profiles.
- **Remote Surgery:** AI-assisted robotic surgery systems enhance surgical precision and provide surgeons with real-time feedback.

1.5.2.2 Remote Patient Monitoring

RPM involves the continuous monitoring of patients outside traditional healthcare settings, often in the comfort of their homes. AI-CPS integration into RPM offers several advantages:

- **Wearable Devices:** Patients wear AI-enhanced wearables that collect data on vital signs, activity levels, and other health metrics.
- **Data Transmission:** Data collected by wearables are transmitted to a central platform for analysis in real time.
- **Alerts and Notifications:** AI algorithms analyse patient data and send alerts to healthcare providers when anomalies or concerning trends are detected.
- **Chronic Disease Management:** RPM is particularly effective in managing chronic diseases like diabetes, heart disease, and hypertension. AI-CPS can adjust treatment plans based on real-time data, improving disease management.
- **Reduced Hospitalization:** Timely intervention through RPM can reduce hospital admissions, healthcare costs, and the burden on healthcare facilities.
- **Medication Adherence:** AI-CPS solutions can remind patients to take their medications and provide feedback on adherence.
- **Post-Surgical Monitoring:** After surgery, RPM allows physicians to remotely monitor patients' recovery and intervene if complications arise.

- **Aging in Place:** RPM enables elderly individuals to age in place by providing continuous monitoring and support, reducing the need for institutional care.
- **Early Detection:** RPM helps in early detection of deteriorating health conditions, enabling timely interventions and potentially saving lives.
- **Improved Patient Engagement:** Patients become active participants in their care as they have access to real-time health data and insights.

The integration of AI and CPS into healthcare and RPM enhances patient care, reduces healthcare costs, and enables more efficient resource allocation. However, it also raises ethical and privacy concerns related to the collection and sharing of patient data. Ensuring data security, patient consent, and compliance with healthcare regulations is crucial in these applications (Bays et al., 2023).

1.5.3 Transportation and Autonomous Vehicles

Transportation and autonomous vehicles represent a field where the integration of AI with CPS has the potential to revolutionize mobility, improve safety, and reduce environmental impacts. Here's an overview of transportation and autonomous vehicles in the context of AI-CPS integration:

1.5.3.1 Autonomous Vehicles

- **Sensors and Perception:** Autonomous vehicles are equipped with various sensors, including LiDAR, radar, cameras, and ultrasonic sensors. These sensors continuously collect data about the vehicle's surroundings.
- **Sensor Fusion:** AI-CPS systems integrate data from multiple sensors and use sensor fusion techniques to create a comprehensive and accurate perception of the vehicle's environment.
- **Machine Learning and Computer Vision:** AI algorithms, such as deep learning and computer vision, process sensor data to identify objects, pedestrians, road signs, and lane markings.
- **Localization and Mapping:** Autonomous vehicles use simultaneous localization and mapping (SLAM) algorithms to create and update maps of their surroundings and determine their precise location within those maps.
- **Path Planning:** AI-CPS systems generate safe and efficient driving paths by considering real-time data, traffic conditions, and vehicle dynamics.
- **Control Systems:** Autonomous vehicles rely on advanced control systems, often incorporating AI, to execute precise manoeuvres, such as steering, braking, and accelerating.
- **Safety and Redundancy:** AI-CPS systems prioritize safety, often with redundant systems and fail-safe mechanisms to mitigate risks and prevent accidents.
- **Communication:** Vehicles can communicate with each other (V2V) and with infrastructure (V2I), enhancing traffic management and safety.
- **Connected Services:** AI-CPS integration enables connected services like over-the-air updates, remote diagnostics, and vehicle-to-cloud data sharing.

1.5.3.2 Transportation

- **Traffic Management:** AI-CPS systems optimize traffic flow through smart traffic lights, congestion management, and predictive analytics to reduce traffic jams and improve commute times.
- **Public Transit:** AI-enhanced public transit systems provide real-time updates, optimize bus and train schedules, and offer improved passenger experiences.
- **Ride-Sharing and Mobility as a Service (MaaS):** AI-CPS integration powers ride-sharing and MaaS platforms, enabling efficient and personalized transportation services.
- **Logistics and Delivery:** AI-CPS systems optimize delivery routes, automate last-mile delivery with drones and autonomous vehicles, and enhance supply chain management.
- **Environmental Impact:** AI-CPS integration contributes to eco-friendly transportation by optimizing routes to reduce fuel consumption and enabling electric and autonomous vehicles.
- **Predictive Maintenance:** Predictive maintenance using AI-CPS helps reduce downtime and maintenance costs for public transportation fleets.
- **Traffic Safety:** AI-CPS systems assist in accident prevention through features like automated emergency braking, adaptive cruise control, and lane-keeping assistance.
- **Infrastructure Management:** Smart infrastructure, such as smart highways and adaptive traffic control systems, improve transportation efficiency.
- **Parking Solutions:** AI-CPS systems assist drivers in finding available parking spaces and help parking garages optimize space utilization.
- **Regulation and Policy:** AI-CPS integration necessitates the development of regulations and policies to ensure safety, liability, and privacy considerations are addressed.

The integration of AI and CPS into transportation and autonomous vehicles holds the promise of safer, more efficient, and environmentally friendly mobility solutions. However, it also raises concerns about data privacy, cybersecurity, and the need for industry standards and regulations to ensure the responsible deployment of these technologies (Albasir et al., 2023).

1.5.4 Energy Management and Smart Grids

Energy management and smart grids represent areas where the integration of AI with CPS can bring about substantial improvements in energy efficiency, sustainability, and grid reliability. Here's an overview of energy management and smart grids in the context of AI-CPS integration:

1.5.4.1 Smart Grids

- **Sensors and Data Collection:** Smart grids incorporate sensors and IoT devices at various points in the electrical grid to monitor voltage, current, power quality, and equipment status.
- **Data Communication:** These sensors transmit data in real-time to a centralized control system, which forms the core of the smart grid infrastructure.

- **AI-Powered Analytics:** AI algorithms analyse the vast amount of data collected from sensors to detect anomalies, identify areas with high energy consumption, and optimize grid operations.
- **Demand Response:** AI-CPS systems enable demand response programmes where electricity consumers can adjust their usage based on real-time pricing or grid conditions, reducing peak loads and improving grid stability.
- **Energy Storage Integration:** Smart grids integrate energy storage solutions like batteries and capacitors, and AI manages the storage systems to store excess energy during low demand periods and release it during peak demand.
- **Renewable Energy Integration:** AI-CPS integration helps manage the variability of renewable energy sources like solar and wind by forecasting energy generation and optimizing grid operations accordingly.
- **Grid Resilience:** Smart grids can quickly detect and respond to grid disruptions, including equipment failures and natural disasters, by rerouting power flows and minimizing downtime.
- **Electric Vehicle Charging:** AI-CPS systems optimize electric vehicle charging, taking into account user preferences, grid load, and renewable energy availability.
- **Grid Balancing:** AI-CPS algorithms balance supply and demand in real-time, reducing energy wastage and optimizing energy distribution.
- **Fault Detection:** AI-CPS systems detect faults and issues in the grid, enabling rapid fault isolation and minimizing outages.

1.5.4.2 Energy Management

- **Building Energy Management:** AI-CPS integration into building systems monitors and controls heating, cooling, lighting, and other energy-consuming devices to reduce energy consumption.
- **Predictive Maintenance:** AI analyses data from CPS sensors to predict equipment failures and optimize maintenance schedules, reducing downtime and maintenance costs.
- **Energy Auditing:** AI-CPS systems conduct energy audits to identify opportunities for energy efficiency improvements in industrial and commercial facilities.
- **Energy Conservation:** AI-CPS integration into homes and commercial buildings allows for more efficient use of energy by adjusting HVAC systems, lighting, and appliances based on occupancy and user preferences.
- **Grid Interaction:** Some AI-CPS systems allow for grid interaction by feeding surplus energy back into the grid, earning credits or reducing energy bills through net metering.
- **Energy Optimization:** AI-CPS systems use optimization algorithms to minimize energy consumption in various applications, from transportation to manufacturing.
- **Load Forecasting:** AI predicts future energy demand patterns, allowing utilities to optimize power generation and distribution accordingly.
- **Energy Trading:** In peer-to-peer energy trading systems, AI-CPS systems enable users to buy and sell excess renewable energy directly with one another, reducing dependence on centralized utilities.

The integration of AI with CPS into energy management and smart grids enhances energy efficiency, reduces carbon emissions, and improves the resilience and reliability of energy systems. It also facilitates the integration of renewable energy sources and promotes sustainable energy practices. However, challenges such as data privacy, cybersecurity, and regulatory considerations must be addressed to ensure the responsible deployment of these technologies (Inderwildi et al., 2020).

2 LITERATURE REVIEW

The constantly changing technological landscape, new consumer and business demands, and market trends have been introduced. The background summary has been presented below.

Plakhotnikov et al. (2020) proposed that CPS are integrated physical and digital systems that work together to run a process effectively and safely. The effectiveness of the information processing in the systems heavily influences the calibre of CPS' work. The use of AI for CPS is therefore essential. Deep learning algorithms and other AI techniques are being employed more frequently to enhance the performance of CPS' constituent parts. These techniques can be used to carry out tasks including process optimization, fault detection, and real-time prediction. CPS will advance with AI as both fields do. Numerous industries, including manufacturing, transportation, and healthcare, stand to benefit from the integration of AI with CPS (Plakhotnikov et al., 2020).

Sakhnini et al. (2020) proposed that CPS are a growing trend since they are employed in so many crucial fields like healthcare and industrial control systems. Security issues are raised by the integration of CPS with internet networks, though. This paper highlights some of the security issues surrounding CPS as well as clever security measures. The paper also explores AI-based techniques for improved CPS performance and security, and it offers case studies and proof of concept experiments in virtual settings (Sakhnini et al., 2020).

Radanliev et al. (2021) reviewed the challenges of using AI in CPS, both now and in the future. A survey of the literature and a taxonomic analysis of IoT-connected and linked CPS were done by the authors. The authors examined articles from academia and business that were released between 2010 and 2020. They discovered that the AI decision-making process in CPS is developing hierarchically and cascadingly. This is a result of the CPS' greater use of IoT devices. In order to create summary maps that were utilized to establish the hierarchical cascade conceptual framework, the authors modified taxonomy approach. According to the authors, the development of AI decision-making is inevitable and independent. They think that this progression will produce CPS that are more resilient and supported by both technical and human automation (Radanliev et al., 2021).

Veith et al. (2019) proposed that the traditional methods for CPS analysis rely on analytical techniques that change depending on whether liveness or safety considerations are taken into account. Different methods, including contracts and stochastic modelling, are used to abstract complexity. These strategies, meanwhile, are frequently insufficient to address the ambiguity and complexity of CPS. The ambiguity of CPS can be addressed using AI-based methods. The requirement for vast amounts of data and the difficulties in assuring safety are just two issues that these approaches can bring about. The examination of CPS using conventional methods is contrasted with the investigation of complex systems by AI researchers. According to the authors, a promising new method of CPS analysis that can take into account the complexity and uncertainty of these systems is reinforcement learning (Veith et al., 2019).

Battina (2016) proposed the application of AI-enhanced automation for DevOps in the modelling of CPS is covered in the article. The authors contend that a more effective engineering process is required given the growing complexity of CPS development and operation. They suggest a model-based system that automates the ongoing production of CPS using AI. The difficulties of using AI to CPS modelling are then covered by the authors. They contend that CPS are intricate systems with a variety of demands. The authors put out a methodology based on models to overcome the problems of applying AI to CPS modelling. The framework develops models of CPS using MDE ideas and methods. The ongoing development of CPS can then be automated with the use of AI models that have been trained using these models (Battina 2016).

Salau et al. (2022) and his team proposes that the advancement of CPS and IoT is being driven by wireless and AI technology. Low latency, throughput, and scheduling difficulties are faced by wireless networks for CPS and IoT. Effective wireless CPS/IoT methods have been developed using AI techniques, particularly ML algorithms. In this paper, the use of ML paradigms, including transfer learning, distributed learning, and federated learning, is examined in relation to wireless networking for CPS and IoT. The paper also discusses issues with CPS/IoT wireless networks that exist now and in the future (Salau et al., 2022) (Table 1.1).

3 FRAMEWORK AND PROPOSED MODEL

The proposed framework (as per Figure 1.8) is presented with multiple layers for an AI-CPPS (artificial intelligence-enabled cyber-physical production system). There are three layers to it: The AI-CPPS framework consists of three layers: (A) the process layer on top, which involves simulating, executing, and analysing processes that are monitored, controlled, or otherwise modified by advanced AI methods; (B) the semantic modelling and integration layer, which connects the processes to the underlying systems by creating a semantic description to integrate the components; and (C) the CPS and human actors layer, which focuses on the systems and ways humans interact with them. The suggested framework thus offers a general strategy that may be used with any physical component or actor, as well as any communication protocol. Additionally, we pinpoint broad issues from the literature that each layer in the following should address. The diagram shows a three-layer structure for an AI-CPPS. Let's dissect each of these layers to determine what it does:

A. Process Layer: This top layer is responsible for managing and controlling the core processes within the AI-CPPS. It involves the simulation, execution, and analysis of processes that are monitored, controlled, or adapted using advanced AI methods. This layer is focused on optimizing and enhancing the performance of processes within the production system using AI

TABLE 1.1
Tabular Summary for Literature Review–Based Papers

S.No.	Paper, Author Name	Summary	Methodology, Dataset, Algo	Concluding Remarks/Findings	Gap
1	Plakhotnikov, D. P., and Kotova, E. E.	Cyber-physical systems (CPS) play a significant role in the information era, and artificial intelligence can enhance their performance on all fronts. The article demonstrates the real-world application in a CPS.	Oracle is the name of the database server used to process a CPS. Using the built-in ETL features of the Qlik analytic platform, processing data is uploaded to the analytics server. Python data tools for Qlik are used to implement automated machine learning, automatic machine learning, and deep learning on the server.	CPS play a significant role in the information era, and artificial intelligence can enhance their performance on all fronts. The paper demonstrates the usefulness of a genuine CPS.	This paper does not provide the adequate method for the security i.e. does not provide proper security to the system.
2	Sakhnini, J., and Karimipour, H.	Security issues are raised by the integration of CPS with internet networks, though. This paper highlights some of the security issues surrounding CPS as well as clever security measures.	The suggested methodology converts functional and side channel-based parametric behaviour into the corresponding formal model, performs functional verification to guarantee accurate and reliable formal modelling, and then examines the safety and security properties to find any potential security vulnerabilities.	The author of the study provided a quick summary of the many security flaws and attacks at various CPS tiers as well as the potential attack models with regard to the attacker's intent, level of access, and skills. The author also provides a quick overview of the most recent static and adaptive security methods for CPS.	The system model provided does not solves the security issues and is also prone to the system attacks.

S.No.	Paper, Author Name	Summary	Methodology, Dataset, Algo	Concluding Remarks/Findings	Gap
3	Radanliev, P., De Roure, D., Van Kleek, M., Santos, O., and Ani, U.	This paper reviews the challenges of using artificial intelligence in CPS. A survey of the literature and a taxonomic analysis of IoT-connected and linked CPS were done by the authors.	The authors examined articles from academia and business that were released between 2010 and 2020.The authors modified taxonomy approach in order to establish the hierarchical cascade conceptual framework.	The authors developed a hierarchical cascade framework to examine how AI decision-making has evolved in CPS. A hierarchical cascade framework was created by the authors to study how AI decision-making has changed in CPS. The findings of the new framework are important because they demonstrate that this evolution is autonomous as a result of the increasing integration of IoT devices into CPS. This evolution is also inevitable because only AI can analyse the volume of data generated in low-latency, near-real-time, and as a result, only AI can derive value from new and emerging forms of big data.	Accurate reviews on cyber risk analytics and financial assessment of cyber risk from CPS are not provided by the author.
4	Veith, E. M., Fischer, L., Tröschel, M., and Nieße, A.	The examination of CPS using conventional methods is contrasted with the investigation of complex systems by AI researchers.	The author and his team provide a framework that integrates formal synthesis and machine learning in this area, The framework put forward treats synthesis as a language acquisition issue. The authors discuss how a programme synthesis differs from the standard ML methodology. They performed a parameter synthesis as well.	Beginning with the fundamentals of temporal logic and the requirements that serve to formally characterize the CPS' components at hand, the author of this paper offered a survey of the literature on approaches for CPS creation and analysis. They have suggested techniques for synthesis of programmes. They have expanded the viewpoint, including MAS as communicative, distributed problem-solvers that now not only can, but do manage enormous CPS, and they have covered simulation frameworks for CPS.	The STL formulæ that are being used by the author in the given framework lack concrete signal or time values. They provide an approximate value or a value that is near to the limit of actual parameter values.

(*Continued*)

TABLE 1.1
(Continued)

S.No.	Paper, Author Name	Summary	Methodology, Dataset, Algo	Concluding Remarks/Findings	Gap
5	Battina, D. S.	This paper proposes the application of AI-enhanced automation for DevOps in the modelling of CPS is covered in the article. The difficulties of using AI to CPS modelling are then covered by the authors	It identifies the DevOps pipeline's core jobs that can be enhanced by AI. Create AI models to carry out these functions. Integrate the DevOps process with the AI models. Analyse the DevOps pipeline's performance with the addition of AI. The study suggests a model-based framework for DevOps with AI enhancements, which primarily consists of a controller, a model of the DevOps pipeline, and a set of AI models.	Automation with AI enhancements has the potential to dramatically increase DevOps' efficacy and efficiency. Adaptable, scalable, extendable, and simple to use. Careful preparation and execution can help you overcome the difficulties of deploying AI-augmented DevOps. The paper also covers how AI models can be used to automate a number of DevOps pipeline tasks, including code review, testing, deployment, and monitoring.AI models can also be used to improve the quality of software releases by identifying and fixing bugs early in the development process and to lower operational costs by automating tasks that are currently done manually.	DevOps with AI enhancements can be used to enhance the creation and management of CPS. In the DevOps pipeline, AI can be utilized to automate processes, boost quality, and cut costs. In the DevOps pipeline, AI can be utilized to automate processes, boost quality, and cut costs.
6	Salau, B. A., Rawal, A., and Rawat, D. B.	This paper proposed that the advancement of CPS and IoT is being driven by wireless and AI technology. The paper also discusses issues with CPS/IoT wireless networks that exist now and in the future	There is no such model being described in the paper as this is a review paper which proposes the survey of the integration of AI and ML techniques within the realm of wireless networking for CPS	This article provides a comprehensive survey of the integration of AI and ML techniques within the realm of wireless networking for CPS and internet of things (IoT) systems, encompassing three major facets of ML applications in wireless technologies.	There is a need for research on real-time AI processing for low-latency applications and improving AI explainability and trustworthiness in critical CPS contexts.

technologies. It can include tasks such as predictive maintenance, process optimization, quality control, and real-time monitoring.

B. Semantic Modelling and Integration Layer: This middle layer plays a crucial role in connecting the processes to the underlying systems. It involves designing semantic descriptions or models that help integrate various components and systems within the AI-CPPS. These semantic models provide a standardized way to represent and communicate information between different elements of the system. It ensures that different parts of the AI-CPPS can understand and interact with each other effectively.

C. Cyber-Physical Systems and Human Actors Layer: This bottom layer focuses on the physical components of the system and how humans interact with them. This layer deals with the actual hardware, sensors, actuators, machines, and the physical environment where production processes take place. It also addresses the role of human actors within the system, including their interactions with machines, decision-making.

The author team used a thorough assessment of the literature to develop the conceptual framework (as per Figure 1.8), which is based on more than 90 of the most authoritative works on the subject. The most prominent concepts were those that featured in several articles, and each article's relationships with those concepts were recorded. This makes it possible to develop a conceptual framework in a new way,

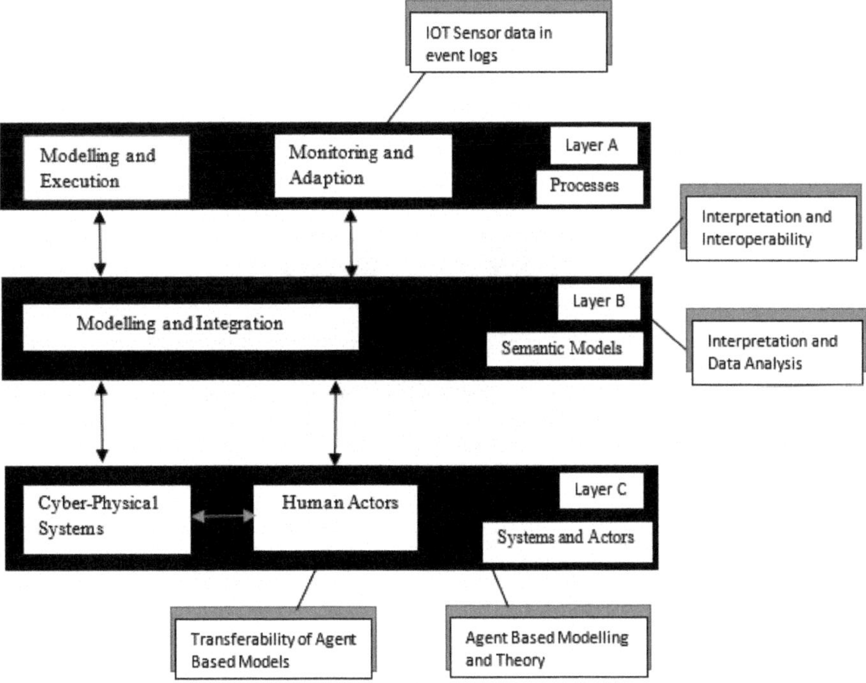

FIGURE 1.8 Reconceptualized Model for Self-Adaptive Cyber-Physical Process Systems.

based on complex socio-economic, organizational, and policy challenges that have been mentioned in over 90 of the best works on the subject that have been published in the last ten years.

The researcher team provided a modified 4C architecture—four layers of CPS architecture—that hierarchically integrates these new and evolving notions. With the rising integration of connected devices (IoT), the conceptual diagram provides new insight into why cognitive growth in CPS is inevitable and autonomous. The grounded theory approach is used to build the hierarchical cascading in for linking developing notions. The emerging concepts identified in the literature review are first shown in summary maps, after which the categories are related using a taxonomy approach and organized in a hierarchy from most closely related to least closely related. The hierarchy in a framework is then cascaded using conceptual design. The framework investigates the acceleration (beginning to automate) of autonomous AI in CPS through automated and semi-automated methods. The updated structure depicts the modifications. Large volumes of data are generated by linked devices and are captured and stored in a variety of heterogeneous formats (e.g. real-time, analytical, spatiotemporal, and high-dimensional data). The new framework shown (as per Figure 1.9) describes the steps used to acquire, store, process, analyse, and utilize the new data in a low-latency, near real-time manner.

4 NOVELTIES AND RECOMMENDATIONS

Enhancing the power of CPS with AI introduces several novelties and innovative concepts:

- **Autonomous Decision-Making:** AI-driven CPS can make autonomous decisions based on real-time data and learning algorithms, reducing the need for human intervention in critical processes.

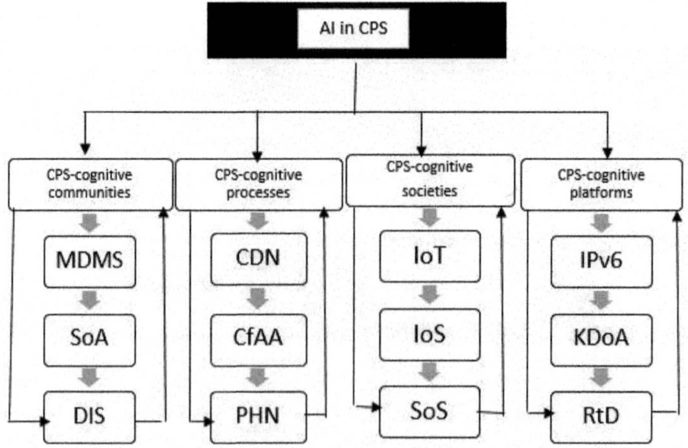

FIGURE 1.9 A Framework Describing the Development of AI in CPS.

- **Predictive Maintenance:** AI enables CPS to predict equipment failures and schedule maintenance proactively, preventing costly downtime and improving operational efficiency.
- **Human-AI Collaboration:** Novel interfaces and interaction models facilitate seamless collaboration between humans and AI, allowing for more intuitive control and oversight of CPS.
- **Edge AI:** The integration of AI algorithms at the edge of CPS devices allows for faster processing and real-time decision-making without relying heavily on cloud resources, enhancing system responsiveness.
- **Explainable AI (XAI):** Novel techniques in XAI enable AI-driven CPS to provide understandable explanations for their decisions, increasing transparency and trust.
- **Energy Efficiency:** AI optimization algorithms help CPS reduce energy consumption and environmental impact, making them more sustainable and cost-effective.
- **Swarm Intelligence:** In applications like robotics and smart grids, AI-driven CPS can exhibit swarm behaviour, where multiple autonomous entities collaborate to achieve complex objectives.
- **Digital Twins:** Creating digital twins of physical systems using AI allows for simulations and testing in a virtual environment, enabling better understanding and optimization of CPS performance.
- **Biologically Inspired AI:** Concepts from biology, such as neural networks inspired by the human brain, are being applied to AI-enhanced CPS to create more adaptable and robust systems.
- **Blockchain Integration:** Combining AI with blockchain technology enhances the security and transparency of data and transactions within CPS, critical for sectors like supply chain management and finance.
- **Ethical AI Frameworks:** The development of novel ethical frameworks for AI in CPS addresses complex moral dilemmas, ensuring responsible and fair system behaviour.
- **Quantum AI:** Research in quantum computing and AI promises to unlock new capabilities in data processing, optimization, and cryptography, which can benefit CPS in various domains.

These novelties demonstrate the dynamic and ever-evolving nature of AI-enabled CPS, pushing the boundaries of what's possible in automation, efficiency, and intelligence across a wide range of industries and applications.

We highly recommend this chapter for its critical insights into the convergence of AI and CPS. This chapter serves as an essential gateway to understanding the cutting-edge developments in this field. Here's why it deserves your attention:

Firstly, the chapter provides a comprehensive introduction to the field, making it accessible to both newcomers and seasoned professionals. Whether you're a student looking to grasp the fundamentals or a practitioner seeking to stay updated, this chapter lays the groundwork for a deeper understanding. Secondly, it tackles prevalent myths and misconceptions surrounding AI in CPS. Debunking these myths is crucial as it clears the path for informed decision-making and adoption. By dispelling common misconceptions, readers can approach the subject with greater clarity and confidence. Thirdly, the chapter introduces a modular approach to AI-CPS, a framework that promises scalability, adaptability, and efficiency. This approach is not only forward-thinking but also immensely practical in a rapidly evolving technological landscape.

In sum, this chapter sets the stage for a journey into the intersection of AI and CPS. Its introductory, myth-busting, and modular components collectively make it a must-read for anyone interested in harnessing the potential of these transformative technologies.

5 FUTURE RESEARCH DIRECTIONS AND LIMITATIONS

While artificial intelligence offers tremendous potential for improving various industries and aspects of our lives, like any technology, there are limitations and challenges that need to be considered. Here are some key future directions and limitations when it comes to enhancing CPS with AI.

5.1 Limitations

- **Data Privacy and Security:** CPS generate and rely on vast amounts of data. Ensuring the privacy and security of this data is a significant challenge. Unauthorized access, data breaches, and cyberattacks can compromise the integrity of the system and the privacy of individuals involved.
- **Reliability and Safety:** AI systems, while powerful, can sometimes be unpredictable. Ensuring the reliability and safety of CPS when AI is involved is essential, especially in critical applications like autonomous vehicles or medical devices. Failures or misinterpretations by AI could lead to accidents or harm.
- **Ethical Concerns:** The use of AI in CPS raises ethical questions. Decisions made by AI systems can have profound impacts on individuals and society. It's crucial to address issues related to fairness, bias, accountability, and transparency in AI-driven CPS.
- **Interoperability:** Many CPS are composed of various components and systems from different vendors. Ensuring that AI-enhanced components can seamlessly integrate with existing infrastructure and other components can be challenging.
- **Scalability:** Implementing AI in CPS may require significant computational resources. Ensuring scalability to handle the growing volume of data and complexity of systems can be a limitation, both in terms of hardware requirements and cost.
- **Regulatory and Legal Challenges:** Developing and deploying AI-enhanced CPS can be hampered by a lack of clear regulations and standards. Navigating legal and regulatory frameworks, especially when these systems cross international boundaries, can be complex.
- **Human-Machine Interaction:** While AI can automate many tasks, humans are still an integral part of CPS. Ensuring effective and safe interaction

between humans and AI-driven systems is a challenge. This includes designing user interfaces that are intuitive and provide meaningful feedback.

- **Training and Maintenance:** AI models need continuous training and maintenance to stay effective and up-to-date. This can be resource-intensive and require ongoing investment.
- **Limited Understanding of AI Decision-Making:** AI models like deep neural networks can be considered black boxes, making it challenging to understand their decision-making processes. This lack of transparency can be a limitation, especially in applications where decisions need to be explained or justified.
- **Energy Efficiency:** Energy consumption can be a significant concern, particularly in embedded systems or CPS deployed in remote or resource-constrained environments. Optimizing AI algorithms for energy efficiency is an ongoing challenge.
- **Cost:** Developing and implementing AI in CPS can be expensive, particularly for small and medium-sized enterprises. The cost of hardware, software, data collection, and training can be a barrier to adoption.
- **Human Resistance to Change:** People may be resistant to the integration of AI into CPS, fearing job displacement or distrusting AI-driven decisions. Addressing these concerns and ensuring a smooth transition can be a limitation.

5.2 Future Directions

- **Edge Computing and AI:** The integration of AI with edge computing is likely to become more prevalent. This allows AI models to process data closer to where it's generated, reducing latency and enabling real-time decision-making in CPS. This is especially crucial for applications like autonomous vehicles and industrial automation.
- **Explainable AI (XAI):** Developing AI models that can provide understandable explanations for their decisions will become increasingly important. This will help build trust in AI-driven CPS, especially in critical domains where transparency is essential.
- **Federated Learning:** To address data privacy concerns, federated learning techniques will gain prominence. This approach allows AI models to be trained across decentralized data sources without sharing the raw data, making it suitable for applications in healthcare, finance, and more.
- **Quantum Computing:** As quantum computing technology matures, it could significantly accelerate AI training and optimization, potentially revolutionizing AI-enhanced CPS by solving complex problems more efficiently.
- **Ethics and Regulation:** Expect increased focus on ethical guidelines and regulations governing AI in CPS. Governments and international organizations will likely play a more active role in setting standards and ensuring responsible AI deployment.
- **Human-AI Collaboration:** Future CPS will place a greater emphasis on human-AI collaboration. This involves designing systems where humans

and AI work together seamlessly, with AI augmenting human capabilities rather than replacing them.

- **AI for Resilience:** AI can help enhance the resilience of CPS, allowing them to adapt to unforeseen events or disruptions. This is crucial in applications like smart cities, where CPS must withstand natural disasters or cyberattacks.
- **AI for Sustainability:** The integration of AI into CPS can contribute to sustainability efforts. For instance, in energy management, AI can optimize resource utilization, reduce waste, and lower environmental impact.
- **Interdisciplinary Research:** Collaboration between experts in AI, engineering, cybersecurity, and other fields will be essential. Interdisciplinary research will drive innovations in AI-enhanced CPS, addressing complex challenges.
- **Robustness and Security:** Ensuring the robustness and security of AI-enhanced CPS will remain a top priority. Developments in adversarial AI and advanced security measures will be essential to protect against cyber threats.
- **Customization and Personalization:** CPS will become more personalized, adapting to individual preferences and needs. AI will play a significant role in tailoring services, such as healthcare treatments or smart home automation, to users' requirements.
- **AI-Driven Simulation:** Advanced simulations powered by AI will enable CPS to model and predict real-world scenarios more accurately. This is crucial for applications like traffic management, climate modelling, and disaster response.
- **Global Collaboration:** Expect to see increased international collaboration in AI research and development, particularly in domains with global challenges like climate change, healthcare, and disaster response.
- **AI-Driven Innovation Ecosystems:** Ecosystems of startups, research institutions, and established companies will continue to drive AI innovation in CPS. These ecosystems will foster rapid development and adoption of new technologies.

6 CONCLUSION

6.1 Significance of Understanding AI-CPS Synergy

Understanding the synergy between AI and CPS is of paramount significance. It paves the way for transformative advancements across industries, from autonomous transportation and smart cities to healthcare and manufacturing. Harnessing this synergy empowers us to create more efficient, adaptable, and resilient systems, improving the quality of life and enhancing safety. Additionally, it provides a foundation for addressing complex global challenges, making it a pivotal area of research and innovation with far-reaching implications for our increasingly interconnected world.

6.2 Implications for Researcher and Practitioners

For researchers, delving into the synergy of AI and CPS offers a rich landscape for exploration, demanding a deep understanding of both domains. Investigating new AI algorithms, CPS architectures, and ethical considerations will be vital in shaping the future of technology.

Practitioners stand to benefit from this understanding by applying AI-CPS integration into optimize processes, increase automation, and enhance decision-making across various industries. However, they must also prioritize security, reliability, and ethical practices to ensure the successful deployment and sustainability of AI-enhanced CPS solutions. Collaboration between researchers and practitioners will be essential in translating theoretical insights into real-world innovations.

ACKNOWLEDGEMENTS

The author team pays their respects to God. They then thank each of their parents for everything they have done for them. They then express their sincere appreciation to ABES management, director ABESEC, HOD CSE, and all other senior coordinators for their extraordinary support to their team. In the end, they pay their due respect and thanks to each and everyone involved, directly or indirectly in this work.

REFERENCES

Albasir, A., Naik, K., & Manzano, R. (2023). Toward improving the security of IoT and CPS devices: An AI approach, *Digital Threats: Research and Practice*, *4*(2), 1–30. https://dl.acm.org/doi/full/10.1145/3497862

Battina, D. S. (2016). AI-augmented automation for DevOps, a model-based framework for continuous development in cyber-physical systems, *International Journal of Creative Research Thoughts (IJCRT), ISSN*, 2320–2882. https://papers.ssrn.com/sol3/papers.cfm?abstract_id=4004315

Bays, H. E., Fitch, A., Cuda, S., Gonsahn-Bollie, S., Rickey, E., Hablutzel, J., …, & Censani, M. (2023). Artificial intelligence and obesity management: An Obesity Medicine Association (OMA) Clinical Practice Statement (CPS) 2023, *Obesity Pillars*, *6*, 100065. https://www.sciencedirect.com/science/article/pii/S2667368123000116

Cohen, S. (2021). The evolution of machine learning: Past, present, and future, in *Artificial intelligence and deep learning in pathology*, Elsevier (pp. 1–12). https://www.sciencedirect.com/science/article/abs/pii/B9780323675383000014

Cyber Physical System, Wikipedia (Access date: 3 Sept. 2023). https://en.wikipedia.org/wiki/Cyber-physical_system#cite_ref-aut_2-0

Gupta, R., Tanwar, S., Al-Turjman, F., Italiya, P., Nauman, A., & Kim, S. W. (2020). Smart contract privacy protection using AI in cyber-physical systems: Tools, techniques and challenges, *IEEE Access*, *8*, 24746–24772. https://ieeexplore.ieee.org/stamp/stamp.jsp?tp=&arnumber=8976143

Habib, M. K., & Chimsom, C. (2022). CPS: Role, characteristics, architectures and future potentials, *Procedia Computer Science*, *200*, 1347–1358. https://www.sciencedirect.com/science/article/pii/S1877050922003453

Hu, J., Lennox, B., & Arvin, F. (2022). Robust formation control for networked robotic systems using negative imaginary dynamics, *Automatica*. https://www.sciencedirect.com/science/article/pii/S0005109822000802

Inderwildi, O., Zhang, C., Wang, X., & Kraft, M. (2020). The impact of intelligent cyber-physical systems on the decarbonization of energy, *Energy & Environmental Science*, *13*(3), 744–771. https://pubs.rsc.org/en/content/articlelanding/2020/ee/c9ee01919g/unauth

Lee, J., Azamfar, M., Singh, J., & Siahpour, S. (2020). Integration of digital twin and deep learning in cyber-physical systems: Towards smart manufacturing, *IET Collaborative Intelligent Manufacturing*, *2*(1), 34–36. https://ietresearch.onlinelibrary.wiley.com/doi/full/10.1049/iet-cim.2020.0009

Lyu, X., Ding, Y., & Yang, S. H. (2019). Safety and security risk assessment in cyber-physical systems, *IET Cyber-Physical Systems: Theory & Applications*, *4*(3), 221–232. https://ietresearch.onlinelibrary.wiley.com/doi/full/10.1049/iet-cps.2018.5068

Plakhotnikov, D. P., & Kotova, E. E. (2020, May). The use of artificial intelligence in cyber-physical systems, in *2020 XXIII International Conference on Soft Computing and Measurements (SCM)* (pp. 238–241). IEEE. https://ieeexplore.ieee.org/abstract/document/9198749

Radanliev, P., De Roure, D., Van Kleek, M., Santos, O., & Ani, U. (2021). Artificial intelligence in cyber physical systems, *AI & Society*, *36*, 783–796. https://link.springer.com/article/10.1007/s00146-020-01049-0

Ramasamy, L. K., Khan, F., Shah, M., Prasad, B. V. V. S., Iwendi, C., & Biamba, C. (2022). Secure smart wearable computing through artificial intelligence-enabled internet of things and cyber-physical systems for health monitoring, *Sensors*, *22*(3), 1076. https://www.mdpi.com/1424-8220/22/3/1076

Sakhnini, J., & Karimipour, H. (2020). AI and security of cyber physical systems: Opportunities and challenges, *Security of Cyber-Physical Systems: Vulnerability and Impact*, 1–4. https://link.springer.com/chapter/10.1007/978-3-030-45541-5_1

Salau, B. A., Rawal, A., & Rawat, D. B. (2022). Recent advances in artificial intelligence for wireless internet of things and cyber–physical systems: A comprehensive survey, *IEEE Internet of Things Journal*, *9*(15), 12916–12930 (Access date: 1 Aug. 2022). doi: 10.1109/JIOT.2022.3170449, https://ieeexplore.ieee.org/abstract/document/9763022?casa_token=EoFSL0b40p8AAAAA:kz7hNUqLzSBLu21F1ZwjUTA0cI67Zu8Bt3o1S5-2IEiMbt5fxYquU4KmSfajWTbie9zqpQHkv5o

Singh, A., Madaan, G., Hr, S., & Kumar, A. (2023). Smart manufacturing systems: A futuristics roadmap towards application of industry 4.0 technologies, *International Journal of Computer Integrated Manufacturing*, *36*(3), 411–428. https://www.tandfonline.com/doi/abs/10.1080/0951192X.2022.2090607

Spector, L. (2006). Evolution of artificial intelligence, *Artificial Intelligence*, *170*(18), 1251–1253. https://www.sciencedirect.com/science/article/pii/S0004370206000907

Veith, E. M., Fischer, L., Tröschel, M., & Nieße, A. (2019, December). Analyzing cyber-physical systems from the perspective of artificial intelligence, in *Proceedings of the 2019 International Conference on Artificial Intelligence, Robotics and Control* (pp. 85–95). https://dl.acm.org/doi/abs/10.1145/3388218.3388222

Yaacoub, J. P. A., Salman, O., Noura, H. N., Kaaniche, N., Chehab, A., & Malli, M. (2020). Cyber-physical systems security: Limitations, issues and future trends, *Microprocessors and Microsystems*, *77*, 103201. https://www.sciencedirect.com/science/article/pii/S0141933120303689

ADDITIONAL READINGS

- The Role of Artificial Intelligence in Cyber Physical Systems. https://ts2.space/en/the-role-of-artificial-intelligence-in-cyber-physical-systems-2/
- Artificial Intelligence and Cyber-Physical Systems: A Review and Perspectives for the Future in the Chemical Industry. https://www.mdpi.com/2673-2688/2/3/27

- Artificial Intelligence and Cyber-Physical Systems. https://encyclopedia.pub/entry/14862
- Cyber-Physical Systems: The New and Emerging Systems of Intelligence. https://technative.io/cyber-physical-systems-the-new-and-emerging-systems-of-intelligence/
- Let's Get Cyber-Physical: The Expanding Role of CPS. https://www.electronicdesign.com/technologies/embedded/article/21250006/luos-lets-get-cyberphysical-the-expanding-role-of-cps
- The Cyber-Physical Infrastructure. https://assets.publishing.service.gov.uk/government/uploads/system/uploads/attachment_data/file/1053955/cyber-physical-infrastructure-vision.pdf

2 AI for Secure and Resilient Cyber-Physical Systems

Pawan Whig, Anant Aggarwal, Veeramani Ganeshan, Venugopal Reddy Modhugu, and Ashima Bhatnagar Bhatia

1 INTRODUCTION

In an age defined by the relentless intertwining of technology and society, the emergence of artificial intelligence (AI) has catalyzed a profound transformation, birthing an innovative paradigm known as cyber-physical systems (CPS). This convergence of ethereal digital intelligence with tangible material reality encapsulates a groundbreaking force that reverberates across a diverse spectrum of industries – from the expansive realms of energy and transportation to the intricate domains of healthcare and manufacturing [1, 2]. The intricate dance between the ethereal and the corporeal engenders a juxtaposition of potential and peril, wherein CPS emergesas both a catalyst for unprecedented efficiency and a canvas upon which the intricate tapestry of cyber threats and operational vulnerabilities is woven. The symphony of AI's infiltration into the very fabric of society has given rise to a novel epoch – an epoch wherein the boundaries between the digital and the physical become fluid, and the abstract becomes the tangible. The metamorphosis engendered by this symbiotic union ushers in a duality of promises and challenges, reflecting the dichotomy of its existence. CPS emerge as the embodiment of this duality, a realm where algorithms orchestrate machinery, where data flows seamlessly through the veins of industrial processes, and where the boundaries between virtual realms and concrete landscapes blur.

1.1 Background and Motivation

Within the pages of this discourse, we journey through the intricate labyrinth of AI for secure and resilient CPS. The landscape is illuminated by the radiance of technological marvels, each tethered to the sublime potential of AI to foster security, efficiency, and resolute resilience [3]. Yet, this journey is not one of mere celebration; it is also one of discernment. It is the odyssey of exploring the intricacies woven into the fabric of our modern existence by the convergence of AI and CPS, and the unwavering motivation to unravel the multidimensional tapestry of challenges that arises as a consequence [4]. The symphonic march of AI's integration into the very

DOI: 10.1201/9781032694375-2

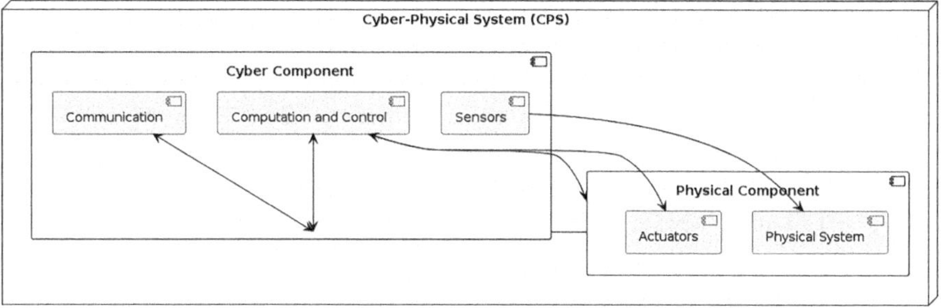

FIGURE 2.1 Cyber-Physical Systems.

essence of CPS is driven by an indomitable impetus – the impetus to enhance the efficacy and versatility of systems that straddle the realms of the virtual and the physical. This union has not only redefined the contours of innovation but has also beckoned forth a set of challenges that demand both intellectual prowess and ethical introspection. As we embark on this journey, we do so with the dual purpose of illuminating the path that AI for CPS security charts and recognizing the clarion call that the challenges it presents enunciate [5].

1.2 Definition of Cyber-Physical Systems

In the realm of definition, CPS stand as a testament to the culmination of human ingenuity in the quest for interconnectedness [6]. This nomenclature, resonant with technological vitality, encapsulates the fusion of computational acumen, networked interconnectivity, and the physical corporeality of systems. At the heart of CPS resides an intricate orchestration where digital cognition harmoniously dances with tangible embodiment as shown in Figure 2.1. This liaison bequeaths unprecedented functionality and responsiveness, permeating diverse sectors ranging from the dynamic realm of smart energy grids to the intricacies of healthcare and the orchestration of modern transportation [7].

1.3 Importance of Security and Resilience in CPS

The profoundness of the fusion inherent in CPS is matched only by the labyrinthine nature of the challenges it beckons forth. As the digital fabric entwines itself with the physical substratum, it unveils the potential for untold opportunities but also unfurls avenues for vulnerabilities hitherto unseen [8]. The intrinsic interplay begets vulnerabilities that transcend conventional paradigms, amplifying the potency of cyber threats while augmenting the impact of operational failures. The tenet of security and resilience thus stands as an immutable cornerstone within the architecture of CPS. The symphony of CPS' existence is tempered by the symphony of security that safeguards its intricate dance. The symbiosis is not merely a functional requirement; it is an ethical obligation and a pragmatic necessity that underscores the ethical and practical duty to fortify the bedrock upon which the edifice of modern society stands. As the narrative unfurls, the exploration shall delve into the nuanced interplay of AI technologies and CPS safeguards, entwining the realms of AI-driven threat detection, predictive foresight, and adaptive defense mechanisms. From the contours of AI's embrace within CPS security to the resonant symphony of real-world case studies, this discourse traverses the labyrinthine path carved by the convergence of AI and CPS. The journey concludes not merely as an elucidation of the present but as a clarion call for collaborative endeavors in the pursuit of fortified security in the cyber-physical landscape [9].

2 AI TECHNOLOGIES FOR CPS SECURITY

CPS security involves the protection of systems that tightly integrate physical processes with networked computing and control components [10]. With the increasing complexity and connectivity of CPS, ensuring their security has become a critical challenge. Various AI technologies can be employed to enhance CPS security. Here are some AI technologies that can be applied to enhance CPS security:

1. **Anomaly Detection**: AI-powered anomaly detection systems can monitor the behavior of CPS components and networks to identify unusual activities or deviations from normal behavior. Machine learning algorithms can learn patterns of normal operation and raise alerts when anomalies are detected, potentially indicating a security breach.
2. **Intrusion Detection and Prevention Systems (IDPS)**: These systems use AI to detect and prevent unauthorized access or attacks on CPS components. They can analyze network traffic, system logs, and sensor data to identify known attack patterns as well as previously unseen threats.
3. **Predictive Maintenance**: AI can analyze sensor data from CPS components to predict when maintenance is required. By identifying potential failures before they happen, it can prevent attacks that exploit vulnerabilities in poorly maintained systems.
4. **Security Analytics**: AI technologies like machine learning and data analytics can be used to analyze large volumes of data generated by CPS components. This analysis can help identify trends, patterns, and potential security threats that might not be obvious through manual analysis.
5. **Federated Learning**: In scenarios where data privacy is crucial, federated learning allows AI models to be trained across distributed CPS components without the need to centralize data. This can enhance security by keeping sensitive data local while still benefiting from AI-powered insights.
6. **Behavioral Profiling**: AI can build profiles of normal behavior for different CPS components. When deviations from these profiles occur, it can signal potential security breaches. This can be particularly effective in detecting insider threats.
7. **Secure Communication and Encryption**: AI can be used to enhance encryption techniques for securing communication between CPS components and networks. This ensures that data remains confidential and unaltered during transit.

8. **Vulnerability Assessment and Penetration Testing**: AI can aid in identifying potential vulnerabilities within CPS components and networks. It can simulate attacks to identify weak points, helping organizations proactively address security flaws.
9. **Automated Response**: AI can be used to develop automated response systems that can take action in real-time when security threats are detected. This can include isolating compromised components, shutting down specific processes, or reconfiguring the network to mitigate the threat.
10. **Machine Learning for Access Control**: ML algorithms can help in dynamic access control by analyzing contextual information and user behavior to determine whether access requests should be granted or denied.
11. **Blockchain and Distributed Ledgers**: While not solely AI, combining blockchain and AI can enhance CPS security by providing tamper-resistant records and enabling secure transactions and interactions between CPS components.

2.1 Machine Learning and Deep Learning

In the expansive realm of fortifying CPS security, the symbiotic amalgamation of AI technologies unfolds as an intricate symphony of innovation and resilience. At the forefront of this symphony stand machine learning and its more intricate sibling, deep learning [11, 12]. These technologies, imbued with the remarkable ability to process vast datasets and unveil intricate patterns, form the keystone of a paradigm shift in fortifying the security of CPS. Machine learning algorithms, ranging from the classical ensemble methods to the profound neural networks, reconfigure the landscape of CPS security. The autonomy these algorithms exhibit in distinguishing between normal system behavior and aberrant anomalies ushers forth a new era of threat detection. The tapestry of deep learning's complex neural architectures further amplifies this discernment by enabling hierarchical feature extraction. This hierarchical approach equips AI to unveil latent threats that often evade the purview of conventional security mechanisms [13, 14].

2.2 Natural Language Processing for Threat Intelligence

In the intricate ecosystem of CPS security, the resonance of natural language processing (NLP) reverberates as a sentinel of paramount significance. As the torrent of digital information surges ceaselessly, the role of NLP emerges as pivotal in sifting through this deluge to distill actionable insights [15]. NLP's ability to transmute unstructured textual and contextual data into structured knowledge finds a profound application in CPS threat intelligence. The fusion of NLP with CPS security enables the deciphering of threat intelligence reports, cyber forums, and vulnerability databases. This synthesis extracts the quintessence of linguistic nuances, orchestrating real-time insights into the evolving threat landscape. By equipping security practitioners with an arsenal of knowledge derived from myriad sources, NLP forges a proactive stance against emergent threats, empowering decision-makers to navigate

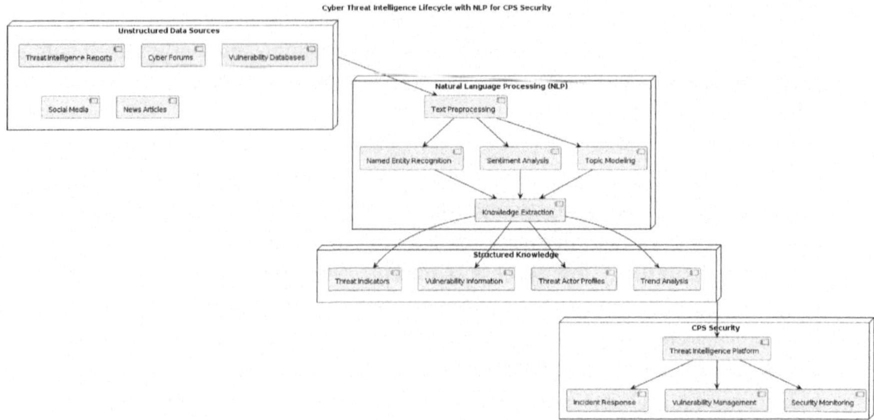

FIGURE 2.2 Cyber Threat Intelligence Lifecycle.

the ever-shifting sands of cyber vulnerabilities [16]. The Cyber Threat intelligence Lifecycle is shown in Figure 2.2.

2.3 Computer Vision for Anomaly Detection

The landscape of AI in CPS security expands beyond the boundaries of the purely computational. At the crossroads of AI and CPS emerges computer vision, an enigmatic fusion that empowers security mechanisms with the potency of visual perception. Within this fusion, anomaly detection becomes a tour de force – an intersection where AI discerns deviations from expected scenarios in visual data streams, a feat that often eludes conventional monitoring methodologies. Computer vision's transformative impact manifests in realms such as manufacturing, where minute deviations in production processes can foreshadow catastrophic outcomes. By parsing streams of visual data and discerning anomalies, computer vision establishes itself as an unparalleled guardian, vigilant in its observation of the CPS landscape.

2.4 Reinforcement Learning for Adaptive Defense

At the pinnacle of the AI-driven symphony for CPS security stands reinforcement learning, a pillar that champions the cause of adaptive defense strategies. Reinforcement learning's intrinsic modus operandi of learning optimal actions through iterative interaction with environments aligns seamlessly with the dynamic contours of cyber threats. It ushers in a future where CPS systems evolve their defense mechanisms autonomously, adapting in real time to the changing tactics of adversaries [17–19]. As adaptive defense strategies evolve, the fabric of CPS security acquires a sentient quality. Reinforcement learning empowers systems to autonomously modify their defense postures based on evolving threat landscapes. This adaptability encompasses not only known threats but also unforeseen adversities, transforming CPS into an ecosystem that thrives amidst perpetual change. The narrative unfolds, traversing the realm of threat landscapes, adaptive intrusion detection, AI-fortified risk assessment, predictive analytics, and the embodiment of real-world case studies, resonating with the clarion call for collaborative efforts in advancing the resilient architecture of CPS security.

3 THREAT LANDSCAPE IN CPS

The threat landscape in CPS is complex and evolving, encompassing a wide range of potential risks and vulnerabilities due to the integration of physical processes with digital technologies. Here are some key threats that characterize the CPS threat landscape:

1. **Cyber Attacks**: Malicious actors can exploit vulnerabilities in CPS components, networks, and software to gain unauthorized access, disrupt operations, steal sensitive data, or cause physical harm. These attacks can target both the cyber and physical aspects of CPS.
2. **Insider Threats**: Threats originating from within an organization, such as employees, contractors, or partners, who abuse their access to CPS systems for malicious purposes. Insider threats can be particularly difficult to detect due to the legitimate access these individuals have.
3. **Denial of Service (DoS) and Distributed DoS (DDoS) Attacks**: Attackers can overwhelm CPS networks, components, or services with excessive traffic, causing disruptions in operations and affecting the reliability of critical processes.
4. **Physical Attacks**: Physical attacks on CPS components, such as tampering with sensors, actuators, or control systems, can have direct consequences on the physical processes they control. Attackers might physically manipulate devices or systems to cause damage or safety hazards.
5. **Malware and Ransomware**: Malicious software can infect CPS components, compromising their functionality or integrity. Ransomware attacks can encrypt critical data or systems, demanding a ransom for their release.
6. **Supply Chain Attacks**: Attackers can compromise CPS components during the manufacturing or distribution process, introducing vulnerabilities or backdoors that can be exploited later.
7. **Communication Interception**: Eavesdropping on communication between CPS components can expose sensitive information and potentially enable attackers to manipulate data and commands exchanged between these components.
8. **Zero-Day Exploits**: Attackers can exploit unknown vulnerabilities (zero-day vulnerabilities) in CPS systems before they are patched or mitigated, giving them an advantage in infiltrating and disrupting systems.
9. **Insecure Remote Access**: Insecurely configured remote access points can provide entry points for attackers to infiltrate CPS networks and components.
10. **Lack of Security Updates**: Failure to apply security updates and patches promptly can leave CPS components vulnerable to known exploits.

11. **Human Error**: Misconfigurations, improper use of systems, and mistakes by personnel can lead to security vulnerabilities and operational disruptions.
12. **Data Integrity Attacks**: Manipulating data within CPS systems can lead to incorrect decisions being made based on compromised or falsified data, potentially causing safety hazards.
13. **Emerging Threats**: As CPS technologies evolve, new types of threats can emerge, such as attacks on machine learning models used within CPS systems or exploits targeting emerging communication protocols.
14. **Convergence of IT and OT**: The convergence of information technology (IT) and operational technology (OT) networks can introduce new attack vectors as traditionally isolated systems become interconnected.

Addressing the CPS threat landscape requires a multidisciplinary approach involving cybersecurity experts, engineers, policy makers, and industry stakeholders. It's crucial to implement defense-in-depth strategies that combine technical solutions, employee training, incident response plans, and ongoing monitoring to mitigate risks and ensure the security and resilience of CPS environments. In the intricate realm of CPS, the canvas of security is brushed with a tapestry of threats that transcend conventional boundaries. This chapter delves deep into the multifaceted tapestry of the threat landscape that CPS grapples with. As the digital and the physical intertwine, a complex mosaic emerges, where cyber vulnerabilities can cascade into tangible repercussions, reshaping not only digital landscapes but also the fabric of the physical world.

3.1 Types of Threats and Attacks

The threat landscape of CPS is an ever-evolving realm, reflecting the relentless creativity and adaptability of adversaries. From classical malware and denial-of-service assaults to more sophisticated supply chain attacks and zero-day vulnerabilities, the typologies of threats in CPS encompass a spectrum of tactics. These threats do not confine themselves to the realms of the digital; rather, they infiltrate the physical plane with unprecedented consequences. Advanced persistent threats (APTs) exemplify the tenacity of modern cyber adversaries. These orchestrated campaigns, often state-sponsored, transcend the boundaries of conventional attacks, infiltrating CPS systems to sow seeds of disruption and espionage. Their nuanced orchestration underscores the escalating complexities of cyber threats in the CPS landscape.

3.2 Potential Impact of Cyber Threats on Physical Systems

The symphony of cyber threats resonates beyond the digital realm, permeating into the tangible world that CPS straddles. A breach in an industrial control system can unleash chaos, halting production lines, jeopardizing worker safety, and imperiling supply chains. In the healthcare sector, compromised medical devices could transmute from mere data breaches to life-threatening scenarios for patients. The coalescence of the virtual and the physical illuminates the profound implications of cyber threats in CPS – implications that stretch far beyond conventional data breaches.

3.3 Case Studies of Notable CPS Breaches

The annals of CPS security are marked by watershed moments, where cyber threats transcended digital confines to reverberate in the physical world. The Stuxnet worm, a harbinger of state-sponsored cyber warfare, penetrated Iran's nuclear facilities, demonstrating the potential of cyber attacks to disrupt physical infrastructure. Similarly, the compromise of Ukraine's power grid showcased the vulnerability of critical infrastructure to digital intrusions, cascading into power outages that affected tens of thousands. These case studies underscore the domino effect – the transformative power of cyber threats in CPS. As the digital and the physical converge, vulnerabilities and threats become amplified, necessitating adaptive defense strategies and the potency of AI technologies to preserve the integrity of CPS ecosystems. The journey continues, weaving through the realms of AI-driven intrusion detection, risk assessment fortified by machine intelligence, predictive analytics for resilience enhancement, and adaptive defense strategies that harness the prowess of AI in protecting the intricate dance of CPS.

4 AI-ENHANCED INTRUSION DETECTION

AI-enhanced intrusion detection is a cybersecurity approach that leverages AI technologies to improve the accuracy, speed, and effectiveness of detecting and responding to security breaches and unauthorized activities within computer systems, networks, and other digital environments. Traditional intrusion detection systems (IDS) often rely on predefined rules and signatures to identify known attack patterns. AI-enhanced intrusion detection goes beyond this by incorporating machine learning and other AI techniques to detect both known and previously unknown threats.

Here's how AI can enhance intrusion detection:

1. **Anomaly Detection**: AI algorithms can learn the baseline behavior of a system, network, or user and identify deviations from this baseline. Anomalies might indicate unauthorized access, data exfiltration, or other malicious activities.
2. **Behavioral Profiling**: AI can build profiles of normal behavior for users, devices, and processes. Deviations from these profiles can trigger alerts, helping identify insider threats or compromised accounts.
3. **Advanced Pattern Recognition**: AI algorithms, particularly deep learning models, can learn complex patterns and correlations that might be missed by traditional rule-based systems. This includes identifying subtle attack behaviors that don't fit known signatures.
4. **Adaptive Learning**: AI-enhanced IDS can continuously adapt to changing attack techniques. As attackers evolve their tactics, the system can learn from new data and adjust its detection methods accordingly.
5. **Reduced False Positives**: AI can improve the accuracy of intrusion detection by reducing false positives. Machine learning models can learn to differentiate between normal variations and actual threats.

6. **Unknown Threat Detection**: AI can identify previously unknown threats based on behavioral anomalies or other indicators that might not match known attack patterns.
7. **Faster Response Times**: AI-powered systems can quickly analyze vast amounts of data in real-time, enabling rapid detection and response to security incidents.
8. **Automated Threat Hunting**: AI-enhanced systems can automatically search for signs of potential threats, helping security teams proactively identify and address vulnerabilities.
9. **Enhanced Network Visibility**: AI can analyze network traffic in depth, providing insights into traffic patterns, potential threats, and unusual activities.
10. **User and Entity Behavior Analytics (UEBA)**: UEBA solutions use AI to analyze user behavior and identify deviations from normal patterns. This can help detect compromised accounts and insider threats.
11. **Dynamic Rule Generation**: AI can generate and adapt rules dynamically based on evolving threat landscapes, reducing the need for manual rule maintenance.
12. **Scalability**: AI-powered systems can scale to analyze and process large volumes of data from diverse sources, which is crucial in modern complex IT environments.

However, implementing AI-enhanced intrusion detection also poses challenges, such as the need for substantial training data, the potential for adversarial attacks against AI models, and the requirement for ongoing model maintenance. Additionally, human expertise is still crucial for interpreting alerts, responding to incidents, and refining detection strategies. AI-enhanced IDS offer significant advantages in identifying and mitigating security threats, but they should be part of a comprehensive cybersecurity strategy that combines technical solutions, human expertise, and best practices.

4.1 Traditional Intrusion Detection Systems

The symphony of AI's integration into the realm of CPS crescendos with the pivotal act of intrusion detection, where AI's capabilities unfold as a transformative force. Traditionally, IDS have relied on rule-based and signature-based mechanisms to identify known threats. However, as the orchestration of cyber threats becomes increasingly sophisticated and elusive, the limitations of these deterministic approaches become palpable. AI's advent reshapes this landscape, bestowing intrusion detection with an unprecedented agility. No longer confined to predefined rules, AI-infused intrusion detection algorithms evolve to discern the subtleties of anomalous behavior. The symphony of machine learning algorithms orchestrates a profound shift from deterministic to probabilistic detection, ushering in a new epoch where emergent threats are identified through nuanced patterns and evolving behaviors.

4.2 Role of Machine Learning in Intrusion Detection

The heart of AI's transformative influence on intrusion detection rests within the realm of machine learning. This domain, endowed with the capacity to discern intricate relationships within data, bequeaths to CPS security an indispensable tool. Machine learning algorithms, informed by historical data, learn to differentiate between benign and malicious activities with a discernment that transcends human intuition. These algorithms continuously refine their discernment, adapting to emergent threat behaviors with unwavering precision. The symphony of machine learning envelops the CPS landscape, orchestrating the preemptive identification of vulnerabilities and threats that imperil the integrity of systems.

4.3 Anomaly Detection and Behavioral Profiling

The essence of AI's intrusion detection prowess lies in its mastery of anomaly detection – a modality that transcends the constraints of predefined rules. The digital footprints of CPS systems are labyrinthine, and anomalies often elude simplistic categorization. Behavioral profiling, a testament to AI's cognitive prowess, transforms intrusion detection into an art of profound discernment. By comprehending the intricate behavioral norms of CPS components, AI unveils deviations that bear the hallmarks of potential threats. This profound sensitivity to behavioral nuances underscores the AI-driven intrusion detection's capacity to traverse the spectrum of known and unknown threats, positioning itself as a vanguard against contemporary and nascent adversaries.

4.4 Real-Time Threat Detection in CPS Environments

In the realm of CPS, the fabric of security is interlaced with the demand for real-time responsiveness. The urgency of threat detection and mitigation is not constrained by the cadence of human observation; it operates at the pace of digital interconnectivity. Herein, AI unfurls as a sentinel, capable of real-time threat detection that operates with the urgency requisite for CPS landscapes. By processing data streams in real-time and identifying anomalies in the nascent stages, AI-infused intrusion detection augments the resilience of CPS ecosystems. The symphony of AI's responsiveness ensures that potential threats are met with swift and calibrated responses, forestalling the cascade of impacts that could emanate from a compromised CPS environment. The narrative unfurls, traversing the domains of AI-driven risk assessment, predictive analytics, and adaptive defense strategies, culminating in a resonant crescendo that underscores the significance of AI in fortifying secure and resilient CPS.

5 AI-SUPPORTED RISK ASSESSMENT

AI-supported risk assessment involves the use of AI technologies to enhance the accuracy, efficiency, and depth of risk assessment processes across various domains, such as cybersecurity, finance, healthcare, and more. It leverages AI's capabilities in data analysis, pattern recognition, predictive modeling, and decision-making to provide insights into potential risks, their impacts, and possible mitigation strategies. Here's how AI can enhance risk assessment:

1. **Data Analysis and Aggregation**: AI can process and analyze large volumes of structured and unstructured data from diverse sources to identify patterns and correlations. This allows organizations to gain a comprehensive view of potential risks.
2. **Early Warning Systems**: AI can identify early indicators of risks based on historical data and current trends. This helps organizations take proactive measures to mitigate risks before they escalate.
3. **Predictive Modeling**: By analyzing historical data, AI can build predictive models that forecast potential future risks and their likelihood. This enables organizations to allocate resources effectively and plan risk mitigation strategies.
4. **Scenario Analysis**: AI can simulate various scenarios to assess potential risks and their impacts. This helps organizations understand how different factors interact to influence risk outcomes.
5. **Automated Risk Identification**: AI can automatically identify risks by analyzing data and comparing it to predefined risk profiles. This reduces the need for manual risk identification and speeds up the assessment process.
6. **Quantitative Risk Analysis**: AI can assist in assigning quantitative values to risks, such as financial losses or impact on operations, helping organizations prioritize risks based on their potential severity.
7. **Continuous Monitoring**: AI-enabled systems can provide real-time monitoring of data streams, allowing organizations to detect and respond to emerging risks in a timely manner.
8. **Natural Language Processing**: NLP techniques enable AI systems to analyze and extract insights from textual information, such as reports, news articles, and regulatory documents, enriching risk assessment with contextual information.
9. **Pattern Recognition**: AI can identify subtle patterns in data that human analysts might overlook, aiding in the identification of hidden or emerging risks.
10. **Vulnerability Assessment**: In cybersecurity, AI can scan networks and systems to identify vulnerabilities that could be exploited by malicious actors, enhancing risk assessment and prioritization.
11. **Portfolio Risk Management**: In finance, AI can assess risks across investment portfolios, helping investors make informed decisions and balance risk and return.
12. **Healthcare Risk Assessment**: AI can analyze patient data to identify patterns that could indicate health risks, contributing to personalized medicine and early disease detection.
13. **Supply Chain Risk Management**: AI can analyze supply chain data to identify potential disruptions, helping organizations optimize their supply chain strategies and enhance resilience.

While AI offers numerous benefits for risk assessment, it's important to note that AI models are not immune to biases, and their accuracy depends on the quality and relevance of the data they are trained on. A successful AI-supported risk assessment strategy requires collaboration between domain experts, data scientists, and risk management professionals to ensure that the AI systems are effective, reliable, and aligned with the organization's goals and values.

5.1 Quantifying Cybersecurity Risks in CPS

Within the realm of securing CPS, the symphony of AI's influence crescendos in the arena of risk assessment. The intricate fusion of AI with the art of risk evaluation ushers in an era where the complexity of CPS vulnerabilities finds equilibrium in the prowess of machine intelligence. Traditional risk assessment approaches, tethered to intuitive evaluations, falter in the face of CPS's dynamic intricacies. Herein, AI emerges as a sentinel, a guardian that quantifies and navigates the labyrinthine contours of cybersecurity risks. The symphony of AI-supported risk assessment hinges upon the synthesis of data-driven insights and probabilistic algorithms. This amalgamation imparts precision to the assessment process, enabling decision-makers to transcend intuitive approximations and embrace a quantitative understanding of vulnerabilities. The result is an intricate risk profile that embodies the manifold dimensions of CPS security, spanning from the purely digital to the tangible.

5.2 AI-Driven Risk Assessment Models

The AI-driven transformation of risk assessment finds resonance in predictive modeling. Machine learning algorithms, trained on historical data, embark on a journey of extrapolation, forecasting potential risk scenarios. The symphony of AI's predictive capabilities augments the risk assessment palette, affording decision-makers a panoramic perspective on vulnerabilities. This dynamic profiling, anchored in the predictive potential of AI, enables organizations to anticipate and allocate resources with surgical precision. The narrative of risk assessment transcends the boundaries of static evaluations. Instead, AI introduces a cadence of adaptability, where evolving threats are met with evolving defenses. The symphony of AI's predictive models becomes a strategic maneuver in the choreography of CPS security, orchestrating a forward-looking perspective that combats vulnerabilities proactively.

5.3 Incorporating Threat Intelligence into Risk Analysis

The resounding synergy of AI and risk assessment harmonizes with the ethereal realm of threat intelligence. AI becomes the conduit through which a deluge of disparate data sources is distilled into actionable insights. The symphony of AI's information synthesis encompasses security bulletins, forums, and social media, parsing linguistic nuances and uncovering patterns that might elude the human gaze. In the dynamic arena of CPS security, threat landscapes evolve with breathtaking swiftness. Here, AI's integration amplifies the agility of risk assessment, empowering it with the potency of real-time insights. Decision-makers navigate these shifting sands fortified by the symphony of AI-facilitated threat intelligence, assuring that the orchestration of defenses remains harmoniously aligned with emergent adversities.

5.4 Case Study: Risk Assessment in a Smart Grid System

The embodiment of AI-supported risk assessment finds tangible manifestation in the realm of a smart grid system. As a quintessential example of CPS, the smart grid's vulnerabilities traverse both the cyber and physical domains. By quantifying the potential impact of cyber threats on the electricity distribution infrastructure, AI-driven risk assessment paints an intricate portrait of vulnerabilities. This case study is emblematic of the symphony that emerges from the fusion of AI and risk assessment. It resonates with the potential to preclude cascading failures, proactively allocate resources, and forge a resilient architecture that thrives amidst the complexities of modern CPS landscapes. As the narrative unfolds, the symphony evolves to the realm of predictive analytics, enveloping the domains of resilience enhancement, adaptive defense strategies, and real-world case studies. The harmonious interplay of AI and CPS security paints a profound tableau of safeguarded landscapes and fortified horizons.

6 PREDICTIVE ANALYTICS FOR RESILIENCE

Predictive analytics for resilience involves using data analysis, statistical algorithms, and machine learning techniques to forecast potential challenges, disruptions, and risks in various systems, industries, or contexts. The goal is to enhance the ability of organizations and systems to proactively prepare for and respond to adverse events, ensuring continuity of operations and minimizing negative impacts. Here's how predictive analytics can be applied to enhance resilience:

1. **Early Warning Systems**: Predictive analytics can identify early indicators of potential disruptions or crises. By analyzing historical data and monitoring real-time information, organizations can receive advance warnings, allowing them to take pre-emptive measures.
2. **Supply Chain Resilience**: Predictive analytics can assess supply chain data to anticipate potential disruptions due to factors like supplier issues, geopolitical events, or transportation delays. This helps organizations adjust their supply chain strategies to minimize disruptions.
3. **Demand Forecasting**: Accurate demand forecasting using predictive analytics allows businesses to allocate resources efficiently, preventing overstocking or understocking of products.
4. **Natural Disasters and Weather Events**: Predictive analytics can analyze weather data to forecast severe weather events and their potential impact on operations. This is crucial for industries like agriculture, energy, and logistics.
5. **Healthcare Capacity Planning**: Predictive analytics can model patient admission rates, helping hospitals and healthcare systems plan for surges in demand and allocate resources effectively.

6. **Financial Risk Management**: In finance, predictive analytics can assess market trends and economic indicators to predict financial risks and market fluctuations, helping investors make informed decisions.
7. **Cybersecurity Threats**: Predictive analytics can analyze network traffic and system logs to identify potential cyber threats before they lead to data breaches or system disruptions.
8. **Infrastructure Maintenance and Asset Management**: By analyzing sensor data and historical maintenance records, predictive analytics can forecast when equipment might fail, allowing for proactive maintenance and preventing costly downtimes.
9. **Energy Grid Management**: Predictive analytics can help energy providers anticipate demand fluctuations and optimize energy distribution, ensuring a stable power supply.
10. **Agricultural Resilience**: In agriculture, predictive analytics can analyze climate and soil data to forecast crop yields, enabling farmers to plan for potential challenges such as droughts or pests.
11. **Transportation and Logistics**: Predictive analytics can optimize transportation routes and schedules, minimizing delays and disruptions in the supply chain.
12. **Epidemic and Pandemic Planning**: Predictive analytics can model disease spread based on factors like population density and mobility patterns, aiding in pandemic response and resource allocation.

To implement effective predictive analytics for resilience:

- **Quality Data**: Accurate and relevant data is crucial. Ensure data is cleaned, normalized, and integrated from various sources.
- **Advanced Analytics**: Employ statistical methods, machine learning, and AI techniques to extract meaningful insights from data.
- **Domain Expertise**: Collaborate with subject matter experts to ensure analytics align with domain-specific nuances.
- **Continuous Learning**: Models should adapt as new data becomes available and the environment evolves.
- **Interdisciplinary Collaboration**: Resilience requires input from multiple disciplines, including data science, domain experts, and decision-makers.

By leveraging predictive analytics, organizations can make informed decisions, allocate resources effectively, and enhance their ability to respond to disruptions, ultimately building greater resilience.

6.1 Predictive Maintenance and Fault Detection

In the intricate tapestry of fortifying CPS, the integration of AI begets a symphony of predictive analytics that resonates with the essence of resilience. At the heart of this symphony lies the realm of predictive maintenance and fault detection – an orchestration where AI's predictive prowess becomes a sentinel that safeguards CPS systems from impending disruptions. Predictive maintenance, an ode to AI's anticipatory capabilities, transcends traditional maintenance schedules. By analyzing real-time data from CPS components, AI discerns patterns that herald the specter of impending failures. The symphony of AI's insights empowers organizations to intervene proactively, forestalling disruptions and preserving the continuous cadence of system operations.

6.2 Utilizing AI for Early Anomaly Prediction

As the intricate dance between AI and CPS resilience unfolds, the art of early anomaly prediction emerges as a resounding chorus. The symphony of predictive analytics unfurls a canvas where AI parses data streams to unveil subtle deviations that often bear the omens of impending disruptions. Traditional monitoring mechanisms, constrained by human observation, pale in comparison to AI's capacity to discern these incipient anomalies. The essence of AI's anomaly prediction lies in its capacity to transcend the constraints of human cognition. Through machine learning algorithms, AI gains the ability to decipher intricate patterns and forecast disruptions with unwavering precision. This predictive potency not only enhances operational efficiency but also fosters an ecosystem of fortified resilience.

6.3 Enhancing System Robustness through Predictive Analytics

Within the realms of CPS, robustness becomes a cornerstone of resilience. The symphony of predictive analytics redefines this robustness, infusing it with the cadence of adaptability. AI, through its mastery of data synthesis, comprehends the dynamic interplay between system components, orchestrating insights that fortify the CPS architecture against vulnerabilities.

Predictive analytics becomes a sentinel that evaluates potential scenarios and their impacts. By forecasting potential system weaknesses and vulnerabilities, AI empowers organizations to augment system robustness proactively. The symphony of resilience embraces not only reactive responses to disruptions but also anticipatory maneuvers that thwart vulnerabilities before they cascade into crises.

6.4 Application of Predictive Analytics in Industrial Automation

The resonant echo of AI's predictive analytics finds embodiment in the realm of industrial automation. Here, AI's anticipatory capabilities converge with the intricacies of manufacturing processes, yielding a symphony that redefines the contours of operational efficiency and resilience. By forecasting potential equipment failures, AI enables manufacturers to orchestrate interventions that stave off production halts and disruptions. The symphony of predictive analytics transforms industrial automation from a realm of reaction to a domain of orchestrated anticipation. Through AI's insights, manufacturers navigate the complexities of CPS landscapes with a nuanced understanding of potential vulnerabilities. This proactive orchestration empowers industries to harmonize operational efficiency with the resonant symphony of resilience. The journey continues, traversing the landscapes of adaptive defense strategies, real-world case studies, and the intricacies of AI's interplay with human expertise. The harmonious intermingling of AI and CPS resilience paints a tapestry of secure landscapes, where disruption is met with anticipation, and vulnerability yields to orchestrated fortification.

7 ADAPTIVE DEFENSE STRATEGIES

Adaptive defense strategies, often referred to as adaptive security or adaptive cyber defense, involve a dynamic and proactive approach to cybersecurity that continuously adjusts and evolves in response to emerging threats and changing attack techniques. Instead of relying solely on static security measures, adaptive defense strategies aim to anticipate, detect, and respond to threats by leveraging real-time data, advanced analytics, automation, and threat intelligence. These strategies are designed to enhance an organization's resilience against increasingly sophisticated cyber threats. Here are key components of adaptive defense strategies:

1. **Continuous Monitoring and Analysis**: Adaptive defense involves real-time monitoring of network and system activity. By collecting and analyzing data from various sources, organizations can identify unusual patterns, behaviors, or anomalies that might indicate a security breach.
2. **Behavioral Analytics**: Organizations use machine learning and AI to build models of normal behavior for users, devices, and applications. Deviations from these established patterns can trigger alerts and responses, helping to identify insider threats or compromised accounts.
3. **Threat Intelligence Integration**: Adaptive defense strategies incorporate threat intelligence feeds that provide up-to-date information about emerging threats, attack techniques, and malicious actors. This information helps organizations tailor their defenses to specific risks.
4. **Automated Incident Response**: Automation plays a crucial role in adaptive defense. When anomalies or threats are detected, automated responses can be triggered to isolate affected systems, block suspicious activities, and initiate incident response procedures.
5. **User and Entity Behavior Analytics (UEBA)**: UEBA solutions analyze user behavior to detect unusual or risky activities. By identifying anomalous actions, organizations can prevent or mitigate potential breaches.
6. **Deception Technologies**: Adaptive defense strategies might include the use of deception technologies, where decoy systems and data are deployed to confuse and divert attackers while real threats are detected.
7. **Dynamic Network Segmentation**: Networks are segmented into smaller, isolated parts to contain threats and limit lateral movement by attackers. Segmentation can be dynamically adjusted based on threat intelligence.
8. **Threat Hunting**: Security teams actively search for signs of potential threats that might not trigger automated alerts. Threat hunting involves using analytics and investigative techniques to find indicators of compromise.
9. **Response Playbooks**: Adaptive defense includes predefined response playbooks that outline steps to take when specific threats or scenarios are

identified. These playbooks guide incident response teams in a structured manner.

10. **Scalable Architecture**: Adaptive defense strategies are designed to scale with growing data volumes and evolving threats. This might involve cloud-based security solutions that can dynamically adjust resources based on demand.
11. **Continuous Training and Learning**: Security teams need ongoing training to keep up with evolving threat landscapes and new attack techniques. Adaptive defense encourages continuous learning and skill development.
12. **Collaboration and Communication**: Adaptive defense requires collaboration between security teams, IT teams, and management. Effective communication ensures timely responses and informed decision-making.
13. **Feedback Loops**: Organizations continuously evaluate the effectiveness of their adaptive defense strategies through feedback loops. Insights from incidents and responses are used to refine and optimize the approach.

Adaptive defense recognizes that cyber threats are constantly evolving, and a rigid, one-size-fits-all security approach may not be sufficient. By embracing agility, automation, and intelligence, organizations can better prepare for and respond to the changing nature of cyber risks.

7.1 Dynamic Cybersecurity Measures for CPS

In the unfolding symphony of fortifying CPS, the harmonious integration of AI with defense strategies emerges as a crescendo of resilience. At the core of this symphony lie adaptive defense strategies – an avant-garde approach that transcends conventional security paradigms. The dynamic nature of CPS landscapes, where threats and vulnerabilities mutate with relentless agility, necessitates an orchestration of defenses that mirrors this dynamism. Adaptive defense strategies harmonize with AI's cognitive acumen, coalescing into a dynamic equilibrium where security postures evolve in real-time. The symphony of adaptive defenses envisions an ecosystem where AI-driven insights recalibrate the defense architecture based on emergent threat landscapes. This orchestration imbues CPS systems with the potency to counteract adversaries with calibrated precision, embodying an ethos of resilience that mirrors the symphony of change.

7.2 AI-Enabled Adaptive Access Control

At the crossroads of AI and adaptive defense strategies stands the sentinel of adaptive access control. Traditional access control mechanisms often falter when confronted with the multidimensional dynamics of CPS environments. Here, AI-infused defenses orchestrate a dynamic interplay where access permissions evolve with the context of users, devices, and emergent threats. The symphony of AI's integration augments access control with an awareness that transcends conventional binary permissions. By assimilating diverse data streams – user behavior, system status, threat intelligence – AI tailors access privileges in real-time. This orchestration unfolds as an adaptive dance, where the symphony of secure access aligns harmoniously with the dynamic contours of CPS landscapes.

7.3 Self-Healing Systems and Autonomous Response

Within the evolving realm of adaptive defense strategies, the embodiment of self-healing systems emerges as a paragon of resilience. The symphony of AI-driven orchestration envisions CPS systems endowed with the capacity to discern and rectify anomalies autonomously. This self-healing orchestration transcends mere detection; it manifests as an autonomous response mechanism that restores the system's integrity.

The essence of self-healing systems resides in the symbiotic interplay between AI's cognition and the nuances of CPS landscapes. By perceiving deviations and anomalies, AI embarks on an autonomous journey of remediation. This orchestration is a testament to the symphony of resilience, where AI infuses systems with an intrinsic ability to rebound from disruptions and adversities.

7.4 Case Study: Adaptive Defense in Autonomous Vehicles

The integration of AI and adaptive defense strategies finds tangible embodiment in the realm of autonomous vehicles. As the symphony of AI-driven orchestration unfolds, the dynamic nature of vehicular landscapes becomes apparent. Here, adaptive defenses transcend traditional boundaries, orchestrating an interplay of threat intelligence, behavioral profiling, and dynamic access control. In this case study, adaptive defense strategies metamorphose autonomous vehicles into bastions of resilience. AI's insights empower vehicles to autonomously discern anomalous behaviors, recalibrate access controls, and even orchestrate responses to mitigate potential threats. The symphony of autonomous vehicle defense resonates as a testament to the transformative potential of AI-infused defenses in CPS landscapes. As the narrative advances, the symphony of AI's integration into CPS security evolves to the realm of real-world case studies, enfolding the domains of AI-driven healthcare IoT security, resilience enhancement in smart cities, and the fortification of industrial IoT landscapes through AI-enabled incident response. The symphony of adaptive defense strategies echoes the resilience of fortified landscapes, where security and adaptability harmonize into a symphony of protection.

8 REAL-WORLD CASE STUDIES

A few real-world case studies that highlight the application of advanced technologies and strategies in cybersecurity and resilience:

1. **Equifax Data Breach (2017)**: In 2017, Equifax, a major credit reporting agency, suffered a massive data breach that exposed the personal and financial information of over 143 million individuals. This breach was a result of a known vulnerability in the Apache Struts framework, which was not patched in a timely manner. The case underscores the importance

of vulnerability management and the potential consequences of failing to address known security issues promptly.

2. **WannaCry Ransomware Attack (2017)**: The WannaCry ransomware attack was a global cyber incident that infected hundreds of thousands of computers across 150 countries. It exploited a vulnerability in outdated Windows systems. Organizations that had not applied the available security patches were particularly vulnerable. This case highlighted the significance of regular patching and updates to prevent widespread attacks.
3. **NotPetya Cyber Attack (2017)**: NotPetya was a destructive malware that targeted organizations primarily in Ukraine but spread globally. It disguised itself as ransomware but was primarily designed to cause damage rather than generate ransom payments. NotPetya exploited weaknesses in software supply chains and led to significant operational disruptions. The incident emphasized the importance of supply chain security and the need to verify the integrity of software updates.
4. **Maersk Cyber Attack (2017)**: Maersk, a major shipping company, was severely impacted by the NotPetya attack. Its global operations were disrupted, and many of its IT systems were paralyzed. The case highlighted the potential cascading effects of cyber attacks on interconnected systems and the need for robust disaster recovery and business continuity plans.
5. **SolarWinds Supply Chain Attack (2020)**: The SolarWinds attack involved the compromise of the SolarWinds software update mechanism, leading to the distribution of malicious updates to thousands of organizations. This highly sophisticated attack targeted both public and private sectors and involved a long reconnaissance phase. It underscored the importance of monitoring and verifying third-party software providers and the challenges of detecting advanced threats.
6. **COVID-19 Pandemic and Remote Work Security Challenges**: The COVID-19 pandemic forced many organizations to rapidly transition to remote work, exposing them to new security challenges. Cybercriminals exploited the situation by launching phishing attacks, targeting remote access vulnerabilities, and taking advantage of the increased use of personal devices for work purposes. This case highlighted the importance of adapting cybersecurity strategies to changing circumstances.

These case studies illustrate the evolving nature of cyber threats and the need for adaptive defense strategies, proactive risk management, continuous training, and collaboration among stakeholders. They also emphasize the importance of integrating cybersecurity and resilience considerations into organizational culture and decision-making processes.

8.1 AI-Driven Security in Healthcare IoT Devices

The symphony of AI in the context of CPS crescendos to a harmonious resonance in the realm of healthcare IoT devices. This case study unravels a tapestry where AI emerges as the guardian of patient safety and data integrity. As healthcare embraces the digital realm, the confluence of AI and CPS security is vividly exemplified in the interconnected web of medical devices that permeate modern healthcare ecosystems. The symphony of AI-driven security unfurls as these medical devices, from pacemakers to insulin pumps, pulse with real-time data streams. AI's vigilant gaze discerns normal behavior from anomalies that bear the signs of potential compromise. The orchestration of AI's anomaly detection amplifies patient safety by identifying deviations in patient vitals and system communications.

8.2 Resilience Enhancement in Smart Cities Infrastructure

In the architectural canvas of modern urban landscapes, smart cities emerge as a resonant testament to the fusion of digital intelligence with physical reality. Here, AI-infused resilience resonates with an orchestration that safeguards urban life in its multifaceted dimensions. The symphony of AI and CPS security reverberates as smart grids manage energy distribution, traffic systems optimize flow, and waste management systems harmonize urban rhythms. The symphony of resilience enhancement unfolds as AI anticipates traffic congestion, enables dynamic allocation of resources, and optimizes energy distribution. In the fabric of smart cities, AI-infused resilience is the symphony that empowers urban landscapes to thrive amidst complexity, resource constraints, and the dynamic interplay of urban life.

8.3 Industrial IoT Security and AI-Enabled Incident Response

As industries pivot toward automation, the realm of industrial IoT security beckons as a poignant embodiment of AI's symphony within CPS. Here, the convergence of AI and security unfolds as an eloquent narrative where AI is both a guardian and a responder. The symphony of AI's integration orchestrates a harmonious interplay where the critical systems that underpin industries are fortified against threats and respond autonomously to breaches. The symphony of AI-enabled incident response envisions an ecosystem where the fusion of AI's behavioral analysis and real-time monitoring provides insights into potential breaches. The response, orchestrated autonomously, mitigates disruptions and orchestrates a course of action that minimizes the impact of cyber threats. This orchestration resonates with the ethos of resilience, empowering industries to traverse the dynamic landscapes of modern manufacturing with unwavering confidence. As the narrative of real-world case studies advances, it traverses the intricacies of AI's interplay with CPS security, illuminating the potential that resides within their fusion. This symphony of tangible manifestations transcends theoretical constructs, demonstrating the transformative capacity of AI in fortifying CPS against a spectrum of contemporary and emerging threats.

8.3.1 Challenges and Future Directions

Certainly, the field of cybersecurity and resilience faces a multitude of challenges and is continually evolving to address new threats and technologies. Here are some challenges and future directions that the field is likely to encounter:

8.3.1.1 Challenges

1. **Sophisticated Threat Landscape:** Cyber threats are becoming more sophisticated, including APTs, nation-state attacks, and ransomware-as-a-service.

These threats can bypass traditional security measures, requiring more advanced defenses.

2. **Skills Shortage:** There is a shortage of skilled cybersecurity professionals. Organizations struggle to find and retain talent capable of effectively countering advanced threats.
3. **IoT and OT Security:** The increasing integration of internet of things (IoT) devices and OT introduces new attack surfaces and challenges for securing industrial systems.
4. **Supply Chain Vulnerabilities:** Cyber attackers are increasingly targeting supply chains, exploiting vulnerabilities in third-party software and services.
5. **Data Privacy and Regulation:** Stricter data privacy regulations (such as GDPR and CCPA) impose challenges for organizations in handling and protecting user data while complying with legal requirements.
6. **Insider Threats:** Malicious or negligent actions by insiders remain a significant concern, necessitating improved user behavior analytics and access controls.
7. **Adversarial AI:** Attackers are using AI to create more convincing phishing attacks, evade detection, and exploit vulnerabilities, leading to the rise of adversarial AI techniques.
8. **Lack of Standardization:** The lack of standardized cybersecurity practices across industries and regions makes it difficult to establish consistent defense measures.

8.3.1.2 Future Directions

1. **AI-Driven Security:** AI and machine learning will play a larger role in threat detection, incident response, and decision-making, both on the defender's side and for attackers (AI-enhanced attacks).
2. **Zero Trust Architecture:** The concept of zero trust, where no one is trusted by default and every user/device must be verified, will gain prominence to counter insider and lateral movement threats.
3. **Quantum Computing and Cryptography:** As quantum computing advances, it could potentially break current encryption methods, leading to the need for quantum-resistant cryptographic techniques.
4. **Automated Threat Response:** Automation will continue to play a crucial role in rapid threat detection and response, particularly for mitigating common attacks.
5. **Secure IoT and OT:** Security measures will focus more on securing IoT and OT devices, including better authentication, encryption, and segmentation of networks.
6. **Blockchain for Security:** Blockchain technology may be applied to enhance data integrity, supply chain security, and identity verification.
7. **Cybersecurity Regulations:** More stringent regulations will likely be introduced to improve cybersecurity practices and data protection.
8. **Integrated Risk Management:** Organizations will shift toward holistic approaches that integrate risk management, cybersecurity, and business continuity planning.

9. **Behavioral Biometrics:** Behavioral patterns of users and devices can be used as an additional layer of authentication and threat detection.
10. **Cybersecurity Awareness:** Ongoing cybersecurity training and awareness programs will be essential to empower individuals to identify and mitigate risks.
11. **Cloud Security:** As cloud adoption continues, improving cloud security practices will be crucial to prevent misconfigurations and unauthorized access.

The future of cybersecurity and resilience will require continuous adaptation to new technologies and threats. Collaboration, interdisciplinary approaches, and a proactive mindset will be key in ensuring robust defenses against evolving cyber risks.

8.4 Ethical Considerations of AI in CPS Security

As the symphony of AI unfurls within the realm of securing CPS, the composition resonates with a harmonious interplay of technological innovation and ethical contemplation. The orchestration of AI within CPS security is not devoid of ethical considerations; it is accompanied by a symphony of complexities that intertwine technological advancement with societal implications. The symphony of ethical considerations embraces the fine balance between security and privacy. AI's data-hungry appetite can infringe upon individual privacy, raising questions about the collection, utilization, and retention of personal data. The deployment of AI-driven security mechanisms necessitates a harmonization where robust protection against cyber threats coalesces with the principles of transparency, consent, and data ownership.

8.5 Interplay between AI and Human Expertise

As the symphony of AI's integration with CPS security reverberates, a harmonious interplay emerges between AI's cognitive prowess and human expertise. The symphony's score is characterized by collaboration rather than displacement, where AI serves as an instrument that augments human intelligence. The ethos of this interplay lies in the orchestration of a partnership that leverages AI's capabilities to empower human experts. The symphony unfolds as human experts navigate the intricacies of CPS security, supported by AI's insights. AI's predictive analytics forecast potential vulnerabilities, while human expertise contextualizes these insights within the broader operational and strategic dimensions. This partnership resonates as a harmonious duet, where AI's insights harmonize with human intuition to forge an orchestration that confronts cyber threats with acuity.

8.6 Anticipated Evolution of AI-Enhanced CPS Security

The symphony of AI-infused CPS security, while echoing the present, resounds with the vibrations of the future. The evolutionary cadence of this symphony is marked by an amalgamation of AI's unfolding potential with the escalating dynamics of cyber threats. The symphony's narrative envisions a landscape where AI becomes not only an instrument of defense but also a proactive force that reshapes the contours of CPS security. The orchestration of AI's evolution in CPS security encompasses an integration with quantum computing, amplifying the potency of threat detection and encryption mechanisms. Moreover, the symphony anticipates the fusion of AI with blockchain, orchestrating secure and transparent data exchanges across CPS landscapes. The exploration of AI in secure enclaves and federated learning exemplifies the symphony's quest for adaptive and decentralized defenses. As the symphony of challenges and future directions unfolds, it traverses the realms of ethical contemplation, the harmonious interplay of AI and human expertise, and the orchestration of AI's anticipated evolution within the CPS security domain. This symphony, a tapestry of innovation and reflection, resounds with a resonant call for a collaborative endeavor where technology aligns harmoniously with ethical considerations to forge a secure and resilient future.

9 CONCLUSION

The journey through the intricate terrain of "AI for Secure and Resilient Cyber-Physical Systems" has illuminated the profound symbiosis between AI technologies and CPS protection mechanisms. From AI-driven threat detection to predictive analytics and adaptive defense strategies, these elements have woven together to construct a formidable bulwark against multifaceted threats. The narrative resounds with the message that the integration of AI is not a mere technological appendage, but a transformative enabler of security and resilience. AI's presence fortifies CPS security by enabling proactive threat detection, foreseeing potential challenges, and orchestrating agile defense responses. This harmonious interplay inaugurates an era of adaptive protection that embraces the dynamism of modern threats. As the journey culminates, it resonates with a resounding call for collaborative endeavors. The quest for secure and resilient CPS landscapes rests upon the shoulders of multidisciplinary collaboration. Engineers, data scientists, ethicists, and policy makers must converge their expertise to forge an interconnected fabric that weaves AI's capabilities into the very essence of CPS. This collaborative mandate is not a mere aspiration; it is an exigent imperative for a safer and more resilient future.

REFERENCES

1. Zhu, Q., & Xu, Z. (2020). Cross-Layer Design for Secure and Resilient Cyber-Physical Systems: A Decision and Game Theoretic Approach. *In Advances in Information Security* Vol. 81. (pp. 1–209). Springer.
2. Xu, Z., & Zhu, Q. (2015, December). A cyber-physical game framework for secure and resilient multi-agent autonomous systems. In *2015 54th IEEE Conference on Decision and Control (CDC)* (pp. 5156–5161). IEEE.
3. Gupta, R., Tanwar, S., Al-Turjman, F., Italiya, P., Nauman, A., & Kim, S. W. (2020). Smart contract privacy protection using AI in cyber-physical systems: Tools, techniques and challenges. *IEEE Access*, 8, 24746–24772.
4. Latif, S. A., Wen, F. B. X., Iwendi, C., Li-Li, F. W., Mohsin, S. M., Han, Z., & Band, S. S. (2022). AI-empowered, blockchain and SDN integrated security architecture for IoT network of cyber physical systems. *Computer Communications*, 181, 274–283.

5. Barbeau, M., Carle, G., Garcia-Alfaro, J., & Torra, V. (2019). Next generation resilient cyber-physical systems. arXiv preprint arXiv:1907.08849.
6. Radanliev, P., De Roure, D., Van Kleek, M., Santos, O., & Ani, U. (2021). Artificial intelligence in cyber physical systems. *AI & Society*, 36, 783–796.
7. Olowononi, F. O., Rawat, D. B., & Liu, C. (2020). Resilient machine learning for networked cyber physical systems: A survey for machine learning security to securing machine learning for CPS. *IEEE Communications Surveys & Tutorials*, 23(1), 524–552.
8. Das, A. K., Bera, B., Saha, S., Kumar, N., You, I., & Chao, H. C. (2021). AI-envisioned blockchain-enabled signature-based key management scheme for industrial cyber–physical systems. *IEEE Internet of Things Journal*, 9(9), 6374–6388.
9. Jovanov, I., & Pajic, M. (2019). Relaxing integrity requirements for attack-resilient cyber-physical systems. *IEEE Transactions on Automatic Control*, 64(12), 4843–4858.
10. Mohanty, S. P. (2020). Advances in Transportation Cyber-Physical System (T-CPS). *IEEE Consumer Electronics Magazine*, 9(4), 4–6.
11. Jin, A. S., Hogewood, L., Fries, S., Lambert, J. H., Fiondella, L., Strelzoff, A., ... & Linkov, I. (2022). Resilience of cyber-physical systems: Role of AI, digital twins, and edge computing. *IEEE Engineering Management Review*, 50(2), 195–203.
12. Adil, M., Khan, M. K., Jadoon, M. M., Attique, M., Song, H., & Farouk, A. (2022). An AI-enabled hybrid lightweight Authentication scheme for intelligent IoMT based cyber-physical systems. *IEEE Transactions on Network Science and Engineering*, 10(5), 2719–2730. 1 Sept.–Oct. 2023, doi: 10.1109/TNSE.2022.3159526.
13. Whig, P., Kouser, S., Velu, A., & Nadikattu, R. R. (2022). Fog-IoT-assisted-based smart agriculture application. In *Demystifying Federated Learning for Blockchain and Industrial Internet of Things* (pp. 74–93). IGI Global. doi: 10.4018/978-1-6684-3733-9.ch005.
14. Whig, P., Velu, A., Nadikattu, R.R. and Alkali, Y.J. (2023). Computational Science Role in Medical and Healthcare-Related Approach. In *Handbook of Computational Sciences* A.A. Elgnar, M. Vigneshwar, K.K. Singh and Z. Polkowski (eds.). https://doi.org/10.1002/9781119763468.ch12.
15. Whig, P., Velu, A., & Bhatia, A. B. (2022). Protect nature and reduce the carbon footprint with an application of blockchain for IIoT. In *Demystifying Federated Learning for Blockchain and Industrial Internet of Things* (p. 20). IGI Global. doi: 10.4018/978-1-6684-3733-9.ch007.
16. Whig, P., Velu, A., & Naddikatu, R. R. (2022). The economic impact of AI-enabled blockchain in 6G-based industry. In: Dutta Borah, M., Singh, P., Deka, G.C. (eds.) *AI and Blockchain Technology in 6G Wireless Network* (pp. 205–224). Springer, Singapore. https://doi.org/10.1007/978-981-19-2868-0_10
17. Whig, P., Velu, A., & Nadikattu, R. R. (2022). Blockchain platform to resolve security issues in IoT and smart networks. In *AI-Enabled Agile Internet of Things for Sustainable FinTech Ecosystems* (pp. 46–65). IGI Global. doi: 10.4018/978-1-6684-4176-3.ch003.
18. Whig, P., Velu, A., & Ready, R. (2022). Demystifying federated learning in artificial intelligence with human-computer interaction. In *Demystifying Federated Learning for Blockchain and Industrial Internet of Things* (pp. 94–122). IGI Global. doi: 10.4018/978-1-6684-3733-9.ch006.
19. Whig, P., Velu, A., & Sharma, P. (2022). Demystifying federated learning for blockchain: A case study. In *Demystifying Federated Learning for Blockchain and Industrial Internet of Things* (p. 23). IGI Global. doi: 10.4018/978-1-6684.-3733-9.ch008.

3 Power of Emotions in AI

Strengthening the Bond of Human-Machine with Heart

Asha Sharma and Aditya Mishra

1 INTRODUCTION

"Fifth Industrial Revolution" (5IR) was not well-established as a separate idea or commonly acknowledged. Nonetheless, the notion of harmonious human–machine cooperation aligns with wider debates concerning the direction of technology and industry.

The continuous incorporation of digital technologies, automation, and artificial intelligence (AI) into many facets of business and society has come to be known as the "Fourth Industrial Revolution" (4IR). The merging of technology, which makes it harder to distinguish between the digital, biological, and physical domains, is what defines this revolution.

The notions of advanced technology integration, industrial evolution, and human–machine collaboration are likely to continue being important topics in discussions about the future of technology and society.

A key component of the changing dynamic between people and AI or machines is the emotional intelligence of human-machine bonds. Although AI systems and robots that can identify and react to human emotions are becoming more and more popular, machines do not actually have emotions. This is frequently called "affective computing." A continuing problem in AI research and development is finding the correct balance between harnessing emotion's power for beneficial human–machine interactions and steering clear of potential dangers.

If humans form bonds with volleyballs and bottle corks, it's futile to assume we can design digital objects that avoid activating human emotions. We're hard wired to emotionally connect to the things, digital or not, in our possession.

But if we apply the insights learned from object attachment research, it's not the emotional connection between human and machine that is the problem. It's what motivates that connection.

We have an ethical obligation to consider the nature of the emotional connection that digital objects provoke. Does the connection with the machine enable the user's greater well-being? Or does it isolate, manipulate, or negatively impact the user's sense of self?

In the 4IR, technology became ubiquitous, and putting trust in technology was critical, because it encouraged trial and usage. As technological boundaries continue

DOI: 10.1201/9781032694375-3

to be pushed in the 5IR, the ethical and humane use of technology will become paramount (Noble et al. 2022).

1.1 The Human–Machine Relationship

Emotions are powerful tools to strengthen human–machine bonds. The bonds between man and machine are influenced by the power of emotion in a number of ways.

Enhancement of User Experience: Using emotion recognition technologies, users can have a better computer experience – that both developers and users to collaborate and cooperate in the creation of user preferred UI design would be able to enhance user satisfaction (Hui and See 2015).

Humanising Technology: Robots creation with emotional expression capabilities or emotional intelligence is a unique technology. It can improve level of interactions of reliability and comfort. It is especially crucial in situations where humans and robots coexist or help each other out on a daily basis and increase attention to an apparent need for "humanising" hospital environments (Bates 2018).

Customisation: More individualised relationships can be facilitated by emotionally intelligent technology. AI systems could, for example, customise learning materials, content recommendations, or therapeutic interventions according to a person's emotional state. Customisation in mobile health apps is associated with increased intentions to engage in physical activity for people who have a larger need for autonomy, according to an interaction effect between customisation and need for autonomy (Bol et al. 2019).

Applications in Healthcare: Emotion-sensing technology is being investigated for use in mental health monitoring and other applications. AI systems are able to evaluate an individual's emotional state by examining their facial expressions, voice tones, and other physiological cues.

Robotics and Companion AI: By identifying and reacting to users' emotions, emotionally intelligent robots or companion AI systems seek to build relationships with their users. This is especially important in sectors like support for people with special needs or elder care. This involves making certain that robots are sensitive to cultural variations and do not control or take advantage of human emotions. While the effects of industrial robots have been felt for some time, it is still too early to determine how service robots will affect homes and workplaces (Torresen 2018).

Ethical Considerations: The creation of ethical AI requires an awareness of and attention to the emotional components of human–machine interaction. Evidence points to AI models' large-scale deployment and embedding of social and human biases. But the real culprit should be the underlying data rather than the algorithm (Naik et al. 2022).

2 LITERATURE AND IMPORTANCE OF THE RESEARCH

The law of Newton for the attraction of bodies is derived with the help of the concept of gravitons. The expression for the gravitational constant is obtained through the momentum of gravitons and the absorption coefficient. Calculations of the values of the coefficient of absorption and of the energy power of flows of gravitons in the space were made. It is shown that during the movement with constant speed the law of inertia is acting (Fedosin 2009).

The objective of this task is to see if there is a presence of emotions in those models, and analyse how authors that have created them consider their impact in consumer choices. In this paper, the most important models of consumer behaviour are analysed. This review is useful to consider unproblematic background knowledge in the literature. The order that has been established for this study is chronological.

That is why we ask questions such as: what are emotions? Are there different types of emotions? What components do they have? Which theories exist about them? In this study, we will review the main theories and components of emotion analysing the cognitive factor and the different emotional states that are generally recognisable with a focus in the classic debate as to whether they occur before the cognitive process or the affective process.

Emotions are present in all consumer decision making processes, meaning that purchase decisions have never been purely cognitive or as they traditionally have been defined, rational. Human beings, in all kinds of decisions, has "always" used neural systems related to emotions along with neural systems related to cognition, regardless of the type of purchase or the product or service in question. Therefore, all purchase decisions are, at the same time, cognitive and emotional. This chapter presents an analysis of the main contributions of researchers in this regard.

The majority of employees at local commercial banks claimed that accounting information is critical in management decision making, and the study also discovered that accounting information of high quality may lead to excellent decision making. The study recommends that accounting information be kept well for future managerial decision making, that managers who are interested in accounting information should have knowledge of accounting principles, that the interpretation of accounting information requires a higher level of accounting knowledge, that knowledge and skills are required, and that the accounting section be well staffed. In terms of practical implications, the study's goal was to contribute to academic disciplines and recognise the importance of accounting data in management decision-making (Chol and Duku 2021).

Three databases were explored and only papers published in English that concentrated on information disclosure, factors influencing information disclosure, strategic management accounting, strategic management accounting information disclosure (SMAID), corporate governance, competitor accounting, customer accounting, and were published in peer reviewed journals were included in the review. The review suggested multiple factors as determinants of the levels of SMAID (Asiedu and Opoku 2022).

The outcome of casual relationship analysis revealed that successful managerial accounting practices had the most direct effect on decision making effectiveness and followed by top management support had a direct effect towards accounting competency. Lastly, top management support had a direct and indirect effect on successful managerial accounting practices through accountant competency. The results implied that manager should focus on building accountant competency in order to create successful managerial accounting practice and to get the valuable information for right decision making (IGNAT and ȘARGU 2022).

This article aims to evaluate the factors affecting the application of strategic management accounting in manufacturing enterprises in Hanoi, thereby making recommendations for these enterprises to increase the use of strategic management accounting information to help business managers make business decisions (Ha, Linh, and Anh 2023).

The main objective of this study was to determine the effect of Accounting Information Systems on the decision-making process in the case of Addis Ababa City Electric Utility related to inventory management, internal control system, bill collection (sales), and financial statements. The finding of this study showed that accounting information systems have a positive and significant effect on inventory management, financial statement, bill collection, and internal control system in the decision-making process. As a result, the researcher came to the conclusion that the accounting information system significantly and favourably influences the decision-making process. For better decision-making, the study recommends businesses employ accounting information systems (Yigrem et al. 2023).

The purpose of this chapter is to assess the relation between accounting comparability and earning management method selection. Comparability is a qualitative characteristic of accounting information. It makes information users' able to identify and understand similarities and differences between two set of information and use them in their decision making. Accrual earning management is a potential factor which can seriously damage comparability. We defined comparability as a characteristic of accounting system outputs and the assessed companies' return and earning which were in an industry. We examined a sample of 72 companies of the Tehran Stock Exchange in 6 industries during 2005–2015. Findings show that accounting comparability has no relation with real and accrual earning management (Rahmani and Ghashghaei 2017).

3 DATA AND METHODOLOGY

The use of partial least squares structural equation model (PLS-SEM) as a standard method for examining intricate interactions between latent and observable variables has grown (Sarstedt et al. 2020). To identify the Power of Emotion and Human–Machine Bonds, the technique has been applied.

Table 3.1 indicates the variables applied in the study. Power of emotion in human–machine bonds is dependent variables and theses six variables, i.e. enhancement of user experience, humanising technology, customisations, applications in healthcare, robotics and companion AI, and ethical considerations, have been selected as independent variables on the base of review of literature.

3.1 Research Objective

The main objective of the study is to measure the power of emotion for human–machine bonds to achieve goals. It is focused to find the impact of power of emotion in strengthening the man and the machine bonds.

TABLE 3.1
Variable used for Relationship between Power of Emotion and Human–Machine Bonds

Variable	Definition
Dependent variable	
PEHMB	Power of emotion in human–machine bonds
Independent variables:	Human–machine bonds
EE	Enhancement of user experience
HT	Humanising technology
CZ	Customisations
AH	Applications in healthcare
RAI	Robotics and companion AI
EC	Ethical considerations

Source: Own compiled based on review of literature.

3.2 Hypothesis

Based on the review of the literature, research gap has been found and the following hypothesis has been framed:

H_{01} There is no significant impact of power of emotion in strengthening the human–machine bonds.

4 RESULT AND DISCUSSION

Consequently, the route coefficient remains significant even if the bootstrap confidence interval does not contain a zero value. That would be regarded as criterion 02. Figure 3.1 displays the path diagram.

Using a structural equation model (SEM), a quantitative, correlational, and explanatory empirical study is performed to determine causal links among variables (Sarstedt et al. 2020). Based on the Smart PLS output, hypotheses were tested and results summarised according to the respective hypotheses. Main hypothesis (H1) was formulated to test the relationship between Powers of emotion in human–machine bonds. Using this method, latent variables may be established and research models can be constructed. Latent variables are those that are deduced from other observed variables (indicators) but are not explicitly observed. PLS is a model validation technique that is primarily focused on causal-predictive analysis. It is typically employed in scenarios with a high degree of complexity and little theoretical knowledge

The R2 values of each endogenous component should be evaluated by the researchers in order to gauge the in-sample predictive potential of the model. R2 values of 0.25, 0.50, and 0.75, indicate that the corresponding endogenous variables are weak, moderate, and strong, respectively. Consequently, the evaluation of the coefficient of determination (R2) is one of the primary components of the structural model assessment. The primary component of benchmark decision-making and long-term

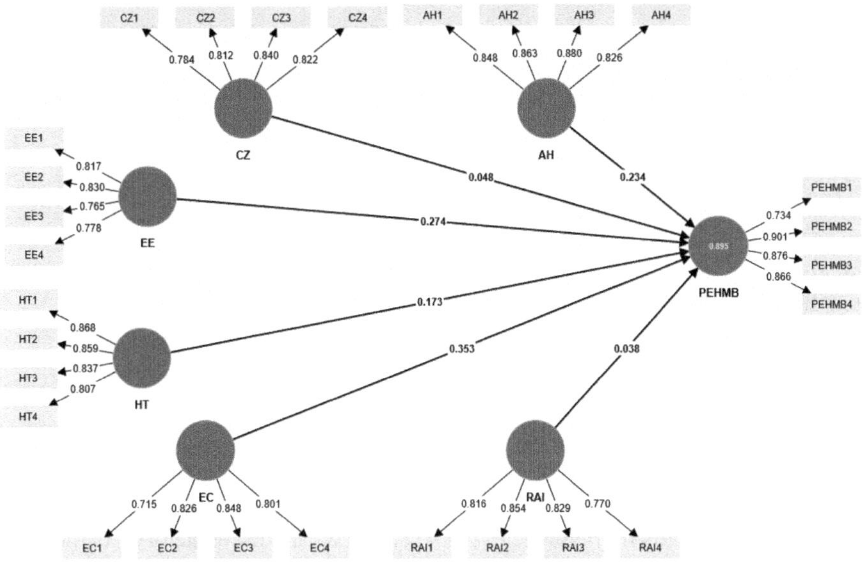

FIGURE 3.1 Power of Emotion and Human–Machine Bonds.

TABLE 3.2
Construct Reliability and Validity

	Cronbach's Alpha	Composite Reliability (rho_a)	Composite Reliability (rho_c)	Average Variance Extracted (AVE)
AH	0.877	0.877	0.915	0.73
CZ	0.831	0.83	0.887	0.663
EC	0.81	0.817	0.876	0.638
EE	0.81	0.814	0.875	0.637
HT	0.864	0.866	0.908	0.711
PEHMB	0.868	0.888	0.91	0.717
RAI	0.835	0.838	0.89	0.669

vision (dependent variable) in the current study is the power of emotion. Based on the estimated structural model presented in Figure 3.1, a strong level of overall R2 (0.895) is determined. In this instance, it implies that the six indicators—that is, the machine and human strength indicators. Together, these factors can account for 89.5% of the variance in the endogenous construct: improving the user experience, humanising technology, customisation, applications in healthcare, robots and companion AI, and ethical issues. The PLS diagram (see Figure 3.1) displays the R2 value, which is 0.895, inside the blue circle representing the power of emotion in human–machine bonds.

Statistical results have been shown in Table 3.2.

TABLE 3.3
Discriminant Validity

	AH	CZ	EC	EE	HT	PEHMB	RAI
AH							
CZ	0.901						
EC	0.657	0.683					
EE	0.671	0.785	0.81				
HT	0.682	0.833	0.758	0.895			
PEHMB	0.844	0.857	0.981	0.963	0.9		
RAI	0.849	0.775	0.902	0.706	0.683	0.872	

Source: Smart PLS output.

Scale dependability as determined by Cronbach's Alpha coefficients is displayed in Table 3.2. If it falls in the range of 0.6–0.8, it is deemed satisfactory. As the value is discovered to be approximately 0.8, it is determined to be very well with the full variable. Therefore, data discriminant validity has been tested once data reliability has been determined.

Table 3.3 represents the construct validity or discriminant validity. Discriminant validity assesses the extent to which measures of different construct are distinct and not highly correlated. Result indicates that no significant relationship have been found between the various variables.

5 CONCLUSION

Using the power of emotion, Model to inform decisions would be flawless. The basis for this work is the idea that the ball should always be in the other person's court.

Directors and executives can no longer act unethically, and there is no possibility of fraudulent activity. Making decisions with the best interests of the community, people, planet, environment, economy, natural resources, and each and every one of them in mind would optimise outcomes not only for business.

The chapter offered several viewpoints on the development emotions for enhancement of the bonding between human and machine for good output. It focused on robotics and AI in the future, including a discussion of the moral dilemmas raised by the advancement of these technologies, role of customisation, use of AI on health care, and enhancement of user experience. Robotics and AI system designers should be mindful of ethical issues, and autonomous systems themselves need to understand the moral ramifications of their decisions.

As a firm starts to use emotion to improve decision-making and long-term vision, there won't be any passivity to harm anyone. With a strong sense of reason, while drafting any policy and strategy in other people's courts. Every decision made with great emotion will always be the best one. Great decision-making considers both logic and emotion. When companies factor in customer profitability and satisfaction alongside profit maximization, they can better align with stakeholder expectations.

CONFLICT OF INTEREST STATEMENT

The authors declare no conflict of interest.

ACKNOWLEDGEMENT

No fund has been raised in regard to write this chapter.

REFERENCES

Asiedu, Michael Amoh, and Mustapha Osman Opoku. 2022. "A Review of Multi-Theoretic Determinants of Strategic Management Accounting Information Disclosure." December. https://doi.org/10.5281/ZENODO.7491601.

Bates, Victoria. 2018. "'Humanizing' Healthcare Environments: Architecture, Art and Design in Modern Hospitals." *Design for Health* 2 (1). https://doi.org/10.1080/24735132.2018.1436304.

Bol, Nadine, Nina Margareta Høie, Minh Hao Nguyen, and Eline Suzanne Smit. 2019. "Customization in Mobile Health Apps: Explaining Effects on Physical Activity Intentions by the Need for Autonomy." *Digital Health* 5. https://doi.org/10.1177/2055207619888074.

Chol, Majak Michael, and Anthony Duku. 2021. "Influence of Accounting Information on Managerial Decision Making Case of Local Commercial Banks in South Sudan." August. https://doi.org/10.5281/ZENODO.5172166.

Fedosin, Sergey G. 2009. "Model of Gravitational Interaction in the Concept of Gravitons." March. https://doi.org/10.5281/ZENODO.890886.

Ha, Tran Thi Thu, Nguyen Thi Linh, and Nguyen Hoai Anh. 2023. "Factors Influencing the Application of Strategic Management Accounting: A Study of Manufacturing Enterprises in Hanoi." June. https://doi.org/10.5281/ZENODO.8063535.

Hui, Sarah Low Tze, and Swee Lan See. 2015. "Enhancing User Experience Through Customisation of UI Design." *Procedia Manufacturing* 3. https://doi.org/10.1016/j.promfg.2015.07.237.

Ignat, Gabriela, and Nicu Şargu. 2022. "Accounting Management Decision Tool in Business." May. https://doi.org/10.5281/ZENODO.6518041.

Naik, Nithesh, B. M. Zeeshan Hameed, Dasharathraj K. Shetty, Dishant Swain, Milap Shah, Rahul Paul, Kaivalya Aggarwal, et al. 2022. "Legal and Ethical Consideration in Artificial Intelligence in Healthcare: Who Takes Responsibility?" *Frontiers in Surgery* 9. https://doi.org/10.3389/fsurg.2022.862322.

Noble, Stephanie M., Martin Mende, Dhruv Grewal, and A. Parasuraman. 2022. "The Fifth Industrial Revolution: How Harmonious Human–Machine Collaboration Is Triggering a Retail and Service [R]Evolution." *Journal of Retailing* 98 (2): 199–208. https://doi.org/10.1016/j.jretai.2022.04.003.

Rahmani, Ali, and Fatemeh Ghashghaei. 2017. "The Relation between Accounting Comparability and Earning Management." December. https://doi.org/10.22059/ACCTGREV.2018.231579.1007589.

Sarstedt, Marko, Christian M. Ringle, Jun Hwa Cheah, Hiram Ting, Ovidiu I. Moisescu, and Lacramioara Radomir. 2020. "Structural Model Robustness Checks in PLS-SEM." *Tourism Economics* 26 (4). https://doi.org/10.1177/1354816618823921.

Torresen, Jim. 2018. "A Review of Future and Ethical Perspectives of Robotics and AI." *Frontiers in Robotics and AI*. https://doi.org/10.3389/frobt.2017.00075.

Yigrem, Minwyelet Wendyifraw, Cherinet Yigrem, Hanna Yeshinegus, and Achamyeleh W. Yigrem. 2023. "The Effect of Accounting Information System on the Decision-Making Process of Addis Ababa City Electric Utility's." January. https://doi.org/10.5281/ZENODO.7546006.

4 Advancing Manufacturing Excellence
ML and AI-Driven Threat Detection Strategies

Almas Begum, Alex David, Sivagami S., and Carmel Mary Belinda M. J.

1 INTRODUCTION

The development of manufacturing techniques has always been a crucial link between several facets of human growth, including productivity, innovation, and technological improvements [1]. Throughout history, the field of manufacturing has experienced a significant evolution, starting from the rudimentary craftsmanship of handcrafted tools to the sophisticated automation processes employed in contemporary factories. The integration of sophisticated technology, data-centric analysis, and networked systems has led to the emergence of a contemporary epoch referred to as smart manufacturing in recent years. Smart manufacturing, commonly known as Industry 4.0, signifies a fundamental transformation that surpasses the scope of simple automation [2]. The concept is a comprehensive methodology in which interconnected machinery, intelligent systems, and real-time data converge to optimise production, increase decision-making processes, and establish a manufacturing ecosystem that is more adaptable and responsive. The transformation discussed herein is propelled by the amalgamation of the internet of things (IoT), artificial intelligence (AI), cyber-physical systems, and data analytics inside the industrial domain [3].

Figure 4.1 showcases the integration of data streams, sensors, AI algorithms, and human expertise, illuminating the interconnected ecosystem of smart manufacturing. It illustrates the constantly changing flow of data and insights that support operational effectiveness and threat identification in the manufacturing setting.

The shift from conventional production to intelligent manufacturing has yielded unparalleled advantages. The implementation of real-time monitoring, predictive maintenance, and adaptive manufacturing processes has resulted in a substantial enhancement of productivity alongside a reduction in operational expenses [4]. Nevertheless, the progress in these areas also presents novel obstacles and susceptibilities that require urgent consideration. The origins of smart manufacturing can be attributed to the digitalisation of industrial processes that commenced around the latter part of the 20th century [5]. The initial endeavours in computer-integrated

DOI: 10.1201/9781032694375-4

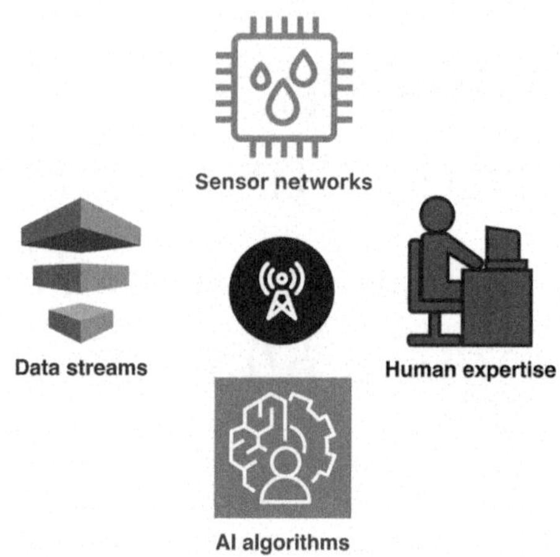

FIGURE 4.1 Ecosystem of Smart Manufacturing.

manufacturing laid the foundation for the emergence of Supervisory Control and Data Acquisition (SCADA) systems, enabling the remote supervision and regulation of industrial operations [6]. The aforementioned preliminary measures establish the foundation for the ultimate incorporation of digital technologies into the manufacturing sector.

The emergence of the industrial internet of things (IIoT) can be attributed to the advancements in processing power and the widespread availability of connectivity [7]. This event signified the commencement of a novel epoch when sensors, actuators, and gadgets could be coupled to collect data in real-time and offer valuable insights into manufacturing processes. The data produced by these interconnected devices has become crucial for smart manufacturing, as it facilitates the optimisation of processes, minimisation of downtime, and improvement of overall efficiency [8]. The advent of AI and machine learning (ML) has empowered manufacturing systems with the capacity to analyse extensive quantities of data, enabling the identification of patterns, anomalies, and correlations that were previously beyond the realm of human capabilities. This development marked the advent of heightened cognitive capabilities and self-governance, empowering machines to promptly formulate judgements and adaptions by leveraging data-derived discernments.

The Significance of Efficient Threat Detection in the Context of Smart Manufacturing: The advent of smart manufacturing presents a myriad of advantages, yet it also brings forth a host of obstacles that must not be overlooked. With the growing integration and reliance on digital technology, factories are exposed to a diverse range of vulnerabilities, encompassing both internal and external threats [9]. The presence of cybersecurity vulnerabilities in smart industrial environments is a substantial risk. The interconnectivity of devices and systems results in an expanded attack surface, rendering it appealing to malevolent entities aiming to capitalise on vulnerabilities. A cyber assault against a smart manufacturing facility has the potential to result in disruptions to production, compromises of sensitive data, and consequential harm to physical equipment [10].

In conjunction with the presence of cybersecurity vulnerabilities, the intricate nature of smart manufacturing systems gives rise to operational hazards [11]. The occurrence of a failure in a specific component within the interconnected network has the potential to initiate a chain reaction, thereby affecting numerous processes and potentially resulting in significant financial losses due to the unavailability of services. The timely identification and management of operational threats holds significant significance in ensuring the continuous and seamless functioning of production processes. The significance of efficient threat detection in the context of smart manufacturing cannot be emphasised enough. The ramifications of not swiftly detecting and addressing threats are diverse and extensive. The potential consequences encompass a range of significant outcomes, including financial losses, reputational damage, impaired safety, and diminished competitive advantage. With the increasing integration of digital technologies into production systems, the conventional methods of security and risk management prove inadequate, hence demanding a fundamental change in the identification, assessment, and mitigation of hazards [12].

This work aims to investigate the potential of ML and AI in the analysis of data, detection of abnormalities, and mitigation of hazards in real-time. By adopting and harnessing the capabilities of technology, we can effectively negotiate the intricate intricacies of smart manufacturing, so guaranteeing a future that is both secure and resilient for this transformative industrial domain [13].

2 A CONVERGENCE OF INNOVATION AND VULNERABILITIES IN SMART MANUFACTURING

Manufacturing has expanded beyond its traditional definition in the contemporary industrialisation scenario. Industry 4.0, sometimes referred to as smart manufacturing, has become a disruptive force that combines cutting-edge technology, data-driven insights, and networked systems [14]. Manufacturing companies can now increase productivity, streamline operations, and make well-informed decisions like never before because to this convergence, which has unleashed a whole new world of opportunities [15].

Smart manufacturing is fundamentally an advancement above conventional automation [16]. It is a comprehensive strategy that uses real-time data interchange, intelligent machines, and digitalisation to build a dynamic and responsive manufacturing ecosystem. The blending of the physical and digital worlds has made it possible for machines to communicate easily with one another and work together, adapt, and self-optimise [17].

2.1 IoT, Robotics, and Cyber-Physical Systems Are Components of Smart Manufacturing

The elements that mould smart manufacturing's essence are vital to its structure. These elements cover a wide range of technological marvels, each of which is essential to transforming conventional factories into intelligent, data-driven organisations.

IoT (Internet of Things): The IoT serves as the framework for smart manufacturing. In the production environment, sensors and gadgets are everywhere, gathering real-time data on anything from machine performance to environmental conditions [18]. This deluge of information enables manufacturers to keep track of workflows, anticipate maintenance requirements, and optimise production plans.

Automation and Robotics: Robots are no longer just tools; they now play a crucial role as participants in the manufacturing process [19]. They handle a variety of jobs, from tedious material handling to repetitious assembly. When robotic systems and AI algorithms work together, they can adapt to changing circumstances and learn from mistakes, which increases overall effectiveness and precision.

Cyber-Physical Systems: The blending of analogue and digital processes is embodied by CPS. It encompasses the incorporation of computational power, sensors, and actuators into physical objects to create systems that interact with both the physical and digital worlds. CPS lays the groundwork for in-the-moment data interchange and decision-making, connecting the virtual and physical facets of manufacturing [20].

2.2 Identifying Vulnerabilities: Operational Risks and Cybersecurity Risks

Although the beginning of smart manufacturing shows promise, it is not without flaws. These sophisticated systems' interconnectedness opens up more of a surface area for potential dangers. To protect the integrity and security of smart manufacturing environments, it is essential to recognise and comprehend these risks.

Cybersecurity Threats: Cybercriminals have a golden opportunity thanks to the growth of connected systems and devices [21]. Significant hazards include unauthorised access to crucial systems, data breaches, and ransomware attacks. Data loss, intellectual property theft, and impaired operational integrity can all be consequences of a breach in the manufacturing environment. To minimise and reduce these hazards as manufacturing processes become more digitalised, effective cybersecurity measures are essential.

Operations Disruption: Because smart manufacturing systems are interconnected, a problem in one area of the network might have a knock-on effect that affects other processes. Cascading failures can occur as a result of a single device or component failing, which can cause production to stop and cost money [22]. Operational vulnerabilities can result from human error, software bugs, and hardware malfunctions, underlining the importance of resilient and adaptable systems.

2.3 Early Threat Detection's Importance

The saying "prevention is better than cure" has special meaning in the world of smart manufacturing. The cornerstone of guaranteeing the seamless and unhindered operation of smart manufacturing systems is the capacity to recognise and respond to risks in their early phases.

Reducing Impact: Early danger detection enables manufacturers to take action before a possible threat develops into a serious emergency. The impact on production processes and overall operational efficiency is minimised by prompt discovery of anomalies, breaches, or disruptions [23].

Improving Decision-Making: Early threat detection generates data-driven insights that provide manufacturers the knowledge they need to make wise decisions. These insights inform resource allocation, maintenance planning, and process improvement, improving the efficiency of manufacturing operations as a whole.

Avoiding Catastrophic Situations: Early threat identification occasionally prevents disastrous outcomes. Manufacturers can avoid significant interruptions, guarantee employee safety, safeguard assets, and maintain the organisation's reputation by spotting and fixing vulnerabilities before they result in system-wide failures.

The symbiotic relationship between innovation and vulnerabilities in the dynamic environment of smart manufacturing emphasises the necessity for a pro-active approach to threat identification. Harnessing the potential of cutting-edge technologies while being alert to possible threats is crucial as we move through this period of upheaval [24]. Manufacturers may fully utilise smart manufacturing while assuring a secure and resilient future by adopting early threat detection measures.

3 UTILISING ARTIFICIAL INTELLIGENCE AND MACHINE LEARNING FOR THREAT DETECTION

The role of human intervention in danger detection is drastically changing in the era of smart manufacturing, when data flows like a digital river and machines interact on their own. AI and ML have emerged as the forerunners of this transition, changing the threat detection and mitigation environment. ML and AI provide the promise of not only detecting dangers but also anticipating and preventing them through the integration of sophisticated algorithms and intelligent systems, representing a substantial advancement in the field of industrial security [25] and illustrated in Figure 4.2.

3.1 Artificial Intelligence and Machine Learning: Definitions in the Context of Manufacturing

Systems can learn from data and make intelligent decisions thanks to two interrelated fields of computer science called ML and AI. ML and AI have the potential to revolutionise production by allowing computers to evaluate massive volumes of data, find hidden patterns, and make educated predictions. Manufacturing facilities are better positioned to be more responsive, flexible, and resilient in the face of possible threats.

The term "Machine Learning" refers to a variety of methods that let computers learn from data without having to be explicitly programmed. These methods include unsupervised learning, which reveals hidden structures within data, and supervised

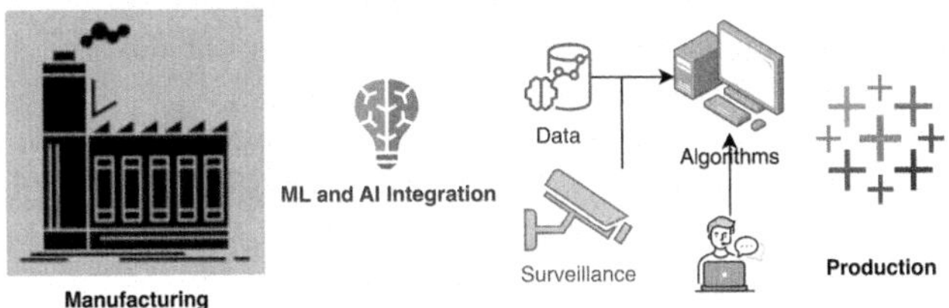

FIGURE 4.2 ML and AI Impact on Threat Detection.

learning, which trains models using labelled data to produce predictions or classifications. Machines can learn with the help of reinforcement learning by interacting with their surroundings and receiving feedback. On the other hand, the term "artificial intelligence" refers to the broader idea of computers displaying intelligence similar to that of humans. It includes a variety of technological advancements, such as robotics, ML, and NLP [26]. With the use of AI, machines may mimic human cognitive processes like learning, thinking, problem-solving, and decision-making.

3.2 The Effectiveness of Pattern Recognition and Predictive Analysis

The capacity of ML and AI to identify patterns and abnormalities inside massive datasets is one of their outstanding characteristics. Human analysts would find it difficult and time-consuming to perform this work. Systems are given the ability to predict future outcomes based on prior data thanks to predictive analysis, a subset of ML. Predictive analysis can foresee probable security breaches or operational interruptions in the context of threat detection by spotting trends and deviations [27].

Another fundamental component of ML is pattern recognition, which enables systems to find recurrent patterns in data that may be signs of anomalies or dangers. Patterns may appear in sensor data, production logs, or historical records in the context of manufacturing [28]. These patterns provide as crucial cues for spotting possible threats and vulnerabilities even though they are frequently imperceptible to human observers.

3.3 Automating Threat Detection to Transform It

The revolutionary change that ML and AI bring to threat detection in smart manufacturing is driven by automation. Manual monitoring is a common component of the conventional threat detection strategy, where human operators watch over operations and react to deviations. However, this strategy is inadequate due to the complexity and size of contemporary industrial systems.

Manufacturing facilities can attain a degree of awareness and reactivity that exceeds human capabilities by automating danger detection processes with ML and AI. Continuously monitoring data streams, real-time anomaly detection, and quick reaction generation are all capabilities of algorithms. Through automation, threats are identified more quickly, human error is reduced, and quick action may be taken before a threat worsens. Additionally, as ML and AI systems are self-learning, they can adapt and change over time. They improve their models when they come across new data and scenarios, increasing accuracy and efficiency. Through this system of ongoing development, threat detection skills are kept current and resistant to changing security risks [29].

A new era of threat detection has begun with the fusion of ML and AI with the complex environment of smart manufacturing. These innovations enable manufacturing facilities to use proactive, automated solutions rather than merely reactive ones. In an environment that is becoming more networked and data-rich, ML and AI give enterprises the capabilities they need to protect their operations, assets, and data through predictive analysis, pattern recognition, and revolutionary automation.

4 OBTAINING AND GETTING READY DATA FOR ANALYSIS

Data has become the lifeblood of intelligent decision-making, process optimisation, and—critically—effective danger detection in the dynamic world of smart manufacturing. Utilising the power of data requires a thorough and methodical procedure for gathering and getting it ready for analysis. This process, which entails a meticulous orchestration of data sources, preprocessing methods, and quality assurance mechanisms, serves as the foundation upon which accurate and insightful threat detection tactics are developed.

4.1 Smart Manufacturing Data Sources: Sensors, Devices, and Operational Logs

An array of networked sensors, gadgets, and operational logs create veritable data treasure troves in smart manufacturing environments. Together, these various data sources offer a rich tapestry of perceptions into every aspect of the manufacturing process. The data produced within a smart factory is extensive and diverse, ranging from temperature and pressure sensors to machine performance indicators.

By recording measurements of many factors in real-time, sensors play a crucial function. They keep an eye on things like pressure, vibration, humidity, and temperature. These measurements are frequently recorded at high frequency, enabling a detailed comprehension of manufacturing procedures.

Devices that are a part of the industrial ecosystem add new layers of data. Data generated by automation systems, robotics, and equipment reflects their operational condition, energy usage, and efficiency. This information provides a window into the condition and operation of vital production assets.

Operational logs record timestamps and events that take place during the production process, acting as a digital record of all actions [30]. These logs provide insight into activity sequences, mistakes, and abnormalities, allowing for a greater comprehension of process dynamics.

4.2 The Data Preprocessing Pipeline: Normalisation, Cleaning, and Transformation

While there is a large amount of data coming in from smart manufacturing environments, there is frequently noise, inconsistencies, and abnormalities that might obstruct useful insights [30]. The careful processes in the data preparation pipeline serve to clean up and get the data ready for analysis. The three key phases of this pipeline are cleansing, transformation, and normalisation.

Data cleaning: At this stage, incorrect or missing data points are located and fixed. It is discussed how outliers and anomalies might influence analytical results. Using statistical methods, missing data can be imputed, guaranteeing a complete dataset for study.

Data Transformation: Data transformation entails transforming raw data into an analytically-ready structure. Aggregating data over predetermined time periods, identifying pertinent traits, or encoding categorical variables are a few examples of how to do this. The efficiency of the succeeding analysis processes is improved by transformation.

Data Normalisation: Data is normalised to a common scale to allow for fair comparisons across various variables. It ensures that the data is suitable for ML algorithms and avoids features with greater magnitudes from dominating the analysis.

4.3 Data Quality Assurance for Precise Threat Detection

The calibre of the data utilised for analysis determines the precision and dependability of threat detection [31]. The axiom "garbage in, garbage out" emphasises the significance of data quality. A combination of exacting validation, verification, and quality assurance procedures are required to ensure data quality.

Data validation: Validation makes sure that the data gathered meets predetermined standards. Data consistency, integrity, and conformity to expected formats are all checked in this process. Validation stops errors and inconsistencies from spreading throughout the entire analytical process.

Data Verification: Cross-referencing data with outside sources or ground truth data is referred to as data verification. Verification procedures verify the data's agreement with actual observations and the precision of the measurements.

Quality Assurance: Continuous monitoring and upkeep of data quality fall under the purview of quality assurance. To find and fix data discrepancies, periodic audits, error monitoring, and feedback loops are used.

Manufacturers lay a strong foundation for precise threat detection by carefully curating and preparing data. The canvas on which ML and AI algorithms paint patterns and insights is clean, converted, and normalised data. Data collection, preprocessing, and quality assurance remain crucial for generating accurate and significant threat detection results as the industrial sector continues its transition into the era of smart manufacturing.

5 FEATURE ENGINEERING: CREATING DATA STRUCTURES FOR POWERFUL ANALYSIS

The function of feature engineering is comparable to that of a master sculptor moulding raw materials into a work of art in the world of data-driven insights and sophisticated analytics. The art of turning raw data into features that are useful, instructive, and actionable is known as feature engineering, and it is a key component of data preparation [32]. This complex procedure has the potential to improve model performance, increase the effectiveness of analysis, and reveal hidden patterns in the data. Mastering the discipline of feature engineering is crucial for gaining insightful knowledge and improving threat detection in the context of smart manufacturing, where the abundance of data is both a difficulty and an opportunity. To simplify the analysis process, feature engineering is fundamentally a strategic process that comprises choosing, converting, and producing features from raw data. The objective is to extract the key information from the data and transform it into useful attributes that capture important data and omit irrelevant material. By enhancing the dataset and giving analytical models the tools they need to produce precise predictions and classifications, feature engineering enhances the dataset.

Domain expertise and a thorough understanding of the data and its context are prerequisites for feature engineering. As practitioners must choose which traits to develop or extract according on their relevance to the topic at hand, it requires a delicate mix of creativity and logic.

5.1 Time-Series Data, Spectral Analysis, and Image Processing: Extracting Useful Insights

Data can take many different forms in the complex world of smart manufacturing, and each one necessitates a different strategy for feature engineering. Utilising time-series data, using spectral analysis, and using image processing are three noteworthy strategies that have a significant influence [33].

Time-Series Information: Time-dependent data are produced by many manufacturing processes, including sensor readings and production records. Extraction of statistical measures (mean, variance, etc.) over various time periods, detection of trends, and identification of cyclic patterns are all part of feature engineering techniques for time-series data. This makes it possible for models to recognise temporal connections and create forecasts based on past behaviour.

Spectral Analysis: When dealing with signals that change in frequency over time, spectral analysis is crucial. It entails converting data to the frequency domain in order to expose hidden patterns that might not be visible in the temporal domain. Extraction of prominent frequencies, detection of harmonics, and measurement of signal strength are some examples of features that can be engineered by spectrum analysis.

Image Processing: Image processing techniques are crucial to feature engineering in manufacturing processes involving visual data, such as defect identification or quality control inspections. Edge detection, texture analysis, and object recognition are a few examples of the tasks involved in extracting features from photos. These characteristics give models useful information they can use to make wise judgements.

5.2 Consider Dimension Reduction and Feature Importance While Choosing Features

The problem of dimensionality becomes more obvious as industrial data complexity rises. High-dimensional data creates computational and analytical challenges that impair model performance by causing overfitting [34]. This problem is solved by feature engineering using dimensionality reduction methods.

Dimensionality Reduction: Techniques, such as Principal Component Analysis (PCA) and t-SNE (t-distributed Stochastic Neighbour Embedding), reduce the amount of variation in high-dimensional data while condensing it into a lower-dimensional space. This highlights the most important aspects while also simplifying analysis.

Feature Importance: Some factors have a greater impact on results than others. Permutation importance and tree-based algorithms like Random Forest, which quantify the contribution of each feature to model performance, are examples of feature significance techniques. By making it easier to choose the most important attributes for analysis, model effectiveness is increased.

Feature engineering fills the role of a choreographer in the complex dance between data and analysis, coordinating the harmonic interaction of unstructured data and valuable attributes. It is the keystone that converts data into an analytically friendly format, allowing models to decipher complex patterns, spot abnormalities, and foresee possible risks inside the manufacturing ecosystem. Mastering the art of feature engineering is a crucial talent for maximising the value of data-driven insights and ensuring efficient threat identification in the age of smart manufacturing as the manufacturing landscape continues to change.

6 SUPERCHARGING THREAT DETECTION WITH SUPERVISED LEARNING

The search for efficient danger detection has brought together cutting-edge technologies and fresh ideas in the quickly changing world of smart manufacturing. Supervised learning stands out among them as a potent technology with the potential to enhance danger detection abilities. The concepts of supervised learning, which are based on pattern recognition and predictive modelling, give industrial facilities the capacity to not only recognise current dangers but also foresee and stop foreseeable disturbances [35]. This chapter goes deeply into the field of supervised learning, examining its workings, uses, and contribution to the advancement of danger detection in the context of smart manufacturing.

6.1 Including Classification and Regression in Supervised Learning

Fundamentally, supervised learning is a type of ML where algorithms learn from labelled training data to produce predictions or categorical judgements. In order for the algorithm to be able to generalise its predictions to cases that have not yet been seen, this learning process requires collecting patterns and relationships from the data. By enabling the automatic detection of anomalies, deviations, and possible dangers inside industrial processes, supervised learning plays a crucial role in the area of threat detection.

Classification: Categorisation Assigning examples to predetermined groups or classes is done using the supervised learning technique of classification. Classification algorithms can determine if a manufacturing process is running normally or if it is displaying symptoms of a potential threat in the context of threat detection. Anomalies, flaws, or strange patterns can be found and categorised, allowing for quick action.

Regression: Another aspect of supervised learning is regression, which includes making predictions about continuous numerical values based on input features. Regression models can predict aspects of production like product quality, energy usage, and equipment wear. These models forecast future values by using existing data, enabling producers to take proactive measures and avoid operational problems.

Testing Sets, Validation Sets, and Training Data: The availability of top-notch training data is the cornerstone of supervised learning. Input samples and labels that match to them, representing the desired results, make up training data. The algorithm learns to recognise patterns and relationships using this data as its foundation.

The training set, validation set, and testing set are the three sets that are created once the training data is established.

- Training Equipment: The portion of the data used to train the algorithm is known as the training set. It comprises input features and the labels that go with them, giving the algorithm the data it needs to understand and modify its internal settings.
- The model's hyperparameters are adjusted and the model's performance during training is evaluated using the validation set. It helps avoid overfitting, a condition in which the model becomes too tuned to the training data and has trouble generalising to new data.
- The testing set is a distinct, secret dataset used to assess the model's effectiveness after training. It evaluates how well the model generalises to fresh, untested situations and offers details on its practical effectiveness.

Ultimately, supervised learning is a powerful friend in the quest for accurate danger detection in smart manufacturing. Manufacturers can detect and avoid future disruptions thanks to supervised learning, which gives them the tools to do so through classification, regression, and rigorous training data management. Real-world success stories, from supply chain optimisation to quality control, highlight its revolutionary impact. The use of supervised learning has the potential to improve danger detection, increase operational effectiveness, and contribute to the creation of a manufacturing ecosystem that is smarter, safer, and more robust as the manufacturing landscape continues to change.

7 TECHNIQUES FOR UNSUPERVISED LEARNING AND ANOMALY DETECTION

The capacity to spot abnormalities is crucial in the field of smart manufacturing, where networked systems and data-driven insights combine [36]. Processes can be disrupted, quality can be compromised, and a manufacturing facility's overall operating effectiveness can be negatively impacted by anomalies, which are frequently signs of possible dangers or irregularities. These hidden outliers can now be found and early intervention made possible by unsupervised learning approaches, particularly anomaly detection. This chapter examines the fundamentals of anomaly detection, highlights practical applications in the context of smart manufacturing, and examines important unsupervised learning techniques.

7.1 The Essence of Anomaly Detection: Detecting the Unknown

The art of finding instances or patterns within a dataset that significantly depart from the predicted norm is known as anomaly detection [37]. Anomaly detection operates in an unsupervised environment, in contrast to supervised learning, which uses labelled data to train models; this makes it especially ideal for situations where the nature of anomalies is unknown or continually changing. Anomaly detection is essential for spotting abnormalities, flaws, or potential dangers that conventional methods could miss.

K-Means and DBSCAN Are Two Techniques for Clustering: A fundamental method of unsupervised learning called clustering combines comparable data points based on shared traits. Clustering can be used to find instances in the context of anomaly detection that do not fit into any particular cluster [38]. The clustering methods K-Means and DBSCAN are both often employed.

- *K-Means Clustering:* This technique divides data into groups based on similarities. Since anomalies are different from typical data, they frequently form smaller clusters. Anomalies can be highlighted by locating instances far from cluster centres or with low cluster membership.
- *DBSCAN (Density-Based Spatial Clustering of Applications with Noise):* Data density is used by DBSCAN to identify clusters. By labelling anomalies as noise points or placing them in clusters with a small number of members, anomalies—points isolated from dense clusters—can be found.

Isolation Forest, One-Class SVM, Outlier Detection: Techniques for finding outliers concentrate on locating specific data points that differ from the predicted distribution. These methods are excellent at finding irregularities in sparse or complicated data. The Isolation Forest and the One-Class SVM are two noteworthy strategies.

- The Isolation Forest When compared to typical data, the Isolation Forest isolates anomalies in a relatively short amount of time by creating a random forest of isolation trees. This method is effective for real-time threat identification since anomalies are isolated more quickly due to their distinctiveness.
- Support Vector Machine (SVM) of One Class: A binary classification algorithm called One-Class SVM distinguishes between regular data and anomalies. This technique may recognise cases that are distant from the decision border as anomalies by learning the boundaries of normal data.

7.2 Applications of Unsupervised Learning in Smart Manufacturing in the Real World

Unsupervised learning techniques have a ripple effect on smart manufacturing, improving threat identification, operational effectiveness, and quality control, and particularly anomaly detection [39].

- Predictive Maintenance: By identifying anomalies in sensor data, unsupervised learning can predict equipment breakdowns. Alerts can be set off by deviations from regular operating circumstances, enabling prompt repair and minimising downtime.

- Quality Control: Anomalies frequently indicate flaws or irregularities in production procedures. Anomaly patterns in sensory data or product qualities can be found using unsupervised learning approaches, guaranteeing that defective items are found early in the production process.
- Optimisation of the Supply Chain: Supply chain data anomaly detection can identify anomalies like delayed shipments or stock shortages. Manufacturers can improve inventory control and streamline logistics by spotting these irregularities.
- Cybersecurity Threats: Cybersecurity present a substantial danger in the era of Industry 4.0. Unsupervised learning improves the security posture of industrial systems by detecting unusual network behaviours or unauthorised access attempts.
- Energy Administration: Energy usage patterns that are unusual can indicate problems or inefficiency. Manufacturers can reduce their energy use by using unsupervised learning algorithms to spot anomalous energy usage patterns.

Anomaly detection and unsupervised learning approaches serve as watchful sentinels in the intricate web of smart manufacturing, uncovering hidden anomalies that could otherwise go undetected. Manufacturers may proactively identify and mitigate dangers, optimise processes, and guarantee the smooth operation of a resilient and responsive smart manufacturing ecosystem by utilising the power of clustering, outlier detection, and other unsupervised methods.

8 THREAT IDENTIFICATION IN REAL-TIME AND ADAPTIVE RESPONSE

The capacity to quickly recognise hazards and take immediate action in response to them has grown to be an essential requirement in the dynamic environment of smart manufacturing. Real-time danger identification and adaptive response mechanisms are now possible thanks to the convergence of data-driven insights, networked systems, and intelligent algorithms [40]. The importance of real-time threat detection is examined in depth in this chapter, along with the difficulties and advantages of streaming data analysis, the use of AI models in real-time settings, and case studies that show real-time threat detection in action. We discover the revolutionary effect of real-time threat identification on the robustness and effectiveness of smart manufacturing ecosystems as we explore this world.

8.1 The Importance of Immediate Threat Detection

Delayed threat detection might have serious repercussions in a time of increased digitisation and connectivity. Anomalies, disruptions, or cyberattacks that are not anticipated have the potential to cause extensive operational breakdowns that result in production halts, quality problems, and financial losses. When it comes to preserving the security, continuity, and integrity of smart manufacturing processes, real-time threat detection is of the utmost significance.

The capacity to quickly analyse data streams as they appear, spot anomalies, and set off prompt responses is essential for real-time threat detection. Manufacturers can reduce the impact of threats, stop an escalation, and adjust to changing conditions in a proactive way thanks to this adaptability.

8.2 Streaming Data Analysis: Opportunities and Challenges

Analysis of streaming data, where data flows continuously rather than in discrete batches, is key to real-time threat detection. While streaming data analysis gives unmatched prospects for quick threat detection, it also poses several difficulties:

- Volume and Velocity: Since streaming data is frequently large and moves quickly, effective data processing techniques are needed to keep up with the information flow.
- Latency: To identify dangers as soon as they materialise, real-time analysis requires low-latency processing. Processing delays may result in missed possibilities for intervention.
- Data Reliability: As anomalies may be hidden by jittery or inaccurate measurements, it is essential to guarantee the quality and accuracy of streaming data.
- Scalability: Scalability is required in the infrastructure supporting streaming data analysis to account for changing data volumes and expanding manufacturing processes.

Despite these difficulties, real-time threat detection via streaming data analysis now has never been more possible. It gives producers the ability to react quickly to potential threats, change course when necessary, and increase the overall agility of smart production systems.

8.3 AI Model Deployment in Real-Time Environments

The use of AI models in real-time settings is essential for turning insights into practical solutions. Real-time threat detection relies on AI models, whether they are based on ML or deep learning.

- Edge Computing: Edge computing involves installing AI models on nearby hardware or sensors to enable real-time network analysis. Because of the decreased latency and improved responsiveness, it is appropriate for situations where quick action is essential.
- Solutions Based on the Cloud: Scalable infrastructure for managing and deploying AI models is provided by cloud platforms. Streams of data are uploaded to the cloud for processing, where AI models spot dangers and launch reactions that are then transmitted back to the production environment.
- Hybrid Approaches: To capitalise on the advantages of both strategies, hybrid solutions integrate edge computing and cloud-based analysis. At the

edge, real-time threat detection is possible, with cloud processing handling complicated analysis and storing past data.

8.3.1 Use Case: Manufacturing Sensors: Anomaly Detection

8.3.1.1 Input

Sensor Data: It includes current information from multiple sensors in a production facility, such as pressure and temperature measurements.

8.3.1.2 Process

Real-Time Analysis: The system continuously gathers sensor data from production equipment, such as pressure and temperature readings.

Anomaly Detection: An anomaly detection model or algorithm is applied to the gathered sensor data.

Analysis of Irregularities: The programme examines the sensor data to look for anomalies, highlighting situations in which pressure or temperature measurements drastically differ from average.

Early Detection Alert: The system sounds an alert when anomalies that might point to an equipment malfunction or failure are found.

8.3.1.3 Output

Alert Generated: The timely generation of an alert signifies the early identification of anomalies in temperature and pressure measurements.

Intervention Initiated: Maintenance staff or automated systems act quickly to address the problem after getting the alarm.

Prevention of Equipment Breakdowns: Prompt action reduces downtime, ensures uninterrupted and seamless production, and prevents equipment breakdowns.

8.3.1.4 Result

Stability of Operations: Rapid response in the event of an anomaly reduces the likelihood of unplanned equipment breakdowns and ensures uninterrupted manufacturing operations.

Sample Code:

```
import matplotlib.pyplot as plt

def detect_anomaly(sensor_data):
    temperature = sensor_data.get("temperature")
    pressure = sensor_data.get("pressure")

    # Checking if temperature or pressure readings are
irregular
    if temperature > 100 or pressure < 20:
        return True  # Return True if anomaly detected
    else:
        return False  # Return False if no anomaly detected

# Generating multiple sets of sensor data
```

```
sensor_data_list = [
    {"temperature": 95, "pressure": 25},
    {"temperature": 105, "pressure": 18},
    {"temperature": 90, "pressure": 22},
    {"temperature": 98, "pressure": 30},
    {"temperature": 88, "pressure": 21}
]

anomalies_detected = []  # List to store anomaly detection
results

# Detecting anomalies in each set of sensor data
for idx, sensor_data in enumerate(sensor_data_list):
    result = detect_anomaly(sensor_data)
    anomalies_detected.append((idx + 1, result))  # Storing
the result and index of data set

# Plotting anomalies detected
x_values = [data[0] for data in anomalies_detected]  # Indices
of sensor data sets
y_values = [1 if data[1] else 0 for data in anomalies_
detected]  # Binary values indicating anomalies

plt.figure(figsize=(8, 6))
plt.scatter(x_values, y_values, color='red', marker='o')
plt.xlabel('Sensor Data Sets')
plt.ylabel('Anomaly Detected')
plt.title('Anomaly Detection in Sensor Data Sets')
plt.yticks([0, 1], ['No Anomaly', 'Anomaly'])
plt.ylim(-0.5, 1.5)
plt.grid(True)
plt.savefig('Figure 07.png',dpi=1000)
plt.show()

Sensor Data Set 1: Anomaly Detected - False
Sensor Data Set 2: Anomaly Detected - True
Sensor Data Set 3: Anomaly Detected - False
Sensor Data Set 4: Anomaly Detected - False
Sensor Data Set 5: Anomaly Detected - False
```

This use case demonstrates how early anomaly identification made possible by real-time sensor data analysis in a manufacturing facility allows for timely interventions that guarantee uninterrupted production and prevent equipment breakdowns, eventually optimising efficiency and lowering operating costs.

9 CONCLUSION

The combination of AI and ML is a game-changer in the dynamic field of smart manufacturing, as it has the potential to completely change danger detection techniques.

A chapter has explored the complex fabric of contemporary manufacturing, shedding light on the significant contribution that AI-driven analytics provide to strengthening threat detection in an ever-changing landscape. The epilogue emphasises how crucial ML and AI have been in transforming conventional surveillance paradigms. Real-time threat detection and preemptive mitigation are made possible by the revolutionary move from traditional approaches to AI-driven analytics. Manufacturing ecosystems may take a proactive approach to preventing operational disruptions by quickly identifying anomalies through the seamless integration of ML and AI technologies. Furthermore, the comparison between AI-driven systems and conventional methods demonstrates the latter's higher efficacy and efficiency in protecting production processes. This thorough investigation highlights not just the development of technology but also the overall symbiosis between technical innovation and human skill, resulting in a method for danger detection in smart manufacturing that is robust, adaptive, and forward-looking.

REFERENCES

1. Javaid, Mohd, Abid Haleem, Ravi Pratap Singh, Rajiv Suman, and Ernesto Santibañez Gonzalez. "Understanding the adoption of Industry 4.0 technologies in improving environmental sustainability." *Sustainable Operations and Computers* 3 (2022): 203–217.
2. Agolla, Joseph Evans. "Human capital in the smart manufacturing and Industry 4.0 revolution." *Digital Transformation in Smart Manufacturing* 2 (2018): 41–58.
3. Vermesan, Ovidiu, and Joël Bacquet, eds. *Cognitive Hyperconnected Digital Transformation: Internet of Things Intelligence Evolution*. River Publishers, 2017. (1st ed., 336 pages). eBook ISBN: 9781003337584
4. Lee, Jay, Jun Ni, Jaskaran Singh, Baoyang Jiang, Moslem Azamfar, and Jianshe Feng. "Intelligent maintenance systems and predictive manufacturing." *Journal of Manufacturing Science and Engineering* 142, no. 11 (2020): 110805.
5. Tao, Fei, Qinglin Qi, Ang Liu, and Andrew Kusiak. "Data-driven smart manufacturing." *Journal of Manufacturing Systems* 48 (2018): 157–169.
6. Rafeeq, Mohammed, Asif Afzal, and Sree Rajendra. "Remote supervision and control of air conditioning systems in different modes." *Journal of the Institution of Engineers (India): Series C* 100 (2019): 175–185.
7. Munirathinam, Sathyan. "Industry 4.0: Industrial internet of things (IIOT)." *Advances in Computers* 117, no. 1 (2020): 129–164.
8. Ren, Shan, Yingfeng Zhang, Yang Liu, Tomohiko Sakao, Donald Huisingh, and Cecilia MVB Almeida. "A comprehensive review of big data analytics throughout product lifecycle to support sustainable smart manufacturing: A framework, challenges and future research directions." *Journal of Cleaner Production* 210 (2019): 1343–1365.
9. Ani, Uchenna P. Daniel, Hongmei He, and Ashutosh Tiwari. "Review of cybersecurity issues in industrial critical infrastructure: Manufacturing in perspective." *Journal of Cyber Security Technology* 1, no. 1 (2017): 32–74.
10. Pandey, Shipra, Rajesh Kumar Singh, Angappa Gunasekaran, and Anjali Kaushik. "Cyber security risks in globalized supply chains: Conceptual framework." *Journal of Global Operations and Strategic Sourcing* 13, no. 1 (2020): 103–128.
11. Tuptuk, Nilufer, and Stephen Hailes. "Security of smart manufacturing systems." *Journal of Manufacturing Systems* 47 (2018): 93–106.
12. Zio, Enrico. "The future of risk assessment." *Reliability Engineering & System Safety* 177 (2018): 176–190.

13. Rane, Nitin Liladhar. "Multidisciplinary collaboration: Key players in successful implementation of ChatGPT and similar generative artificial intelligence in manufacturing, finance, retail, transportation, and construction Industry." OSF Preprints (2023). https://doi.org/10.31219/osf.io/npm3d
14. Ng, Tan Ching, Sie Yee Lau, Morteza Ghobakhloo, Masood Fathi, and Meng Suan Liang. "The application of Industry 4.0 technological constituents for sustainable manufacturing: A content-centric review." *Sustainability* 14, no. 7 (2022): 4327.
15. Industry Agenda. "Industrial internet of things: Unleashing the potential of connected products and services." *White Paper, in Collaboration with Accenture* 34 (2015): 40.
16. Tao, Fei, Qinglin Qi, Ang Liu, and Andrew Kusiak. "Data-driven smart manufacturing." *Journal of Manufacturing Systems* 48 (2018): 157–169.
17. Lu, Yuqian, Xun Xu, and Lihui Wang. "Smart manufacturing process and system automation–A critical review of the standards and envisioned scenarios." *Journal of Manufacturing Systems* 56 (2020): 312–325.
18. Malik, Praveen Kumar, Rohit Sharma, Rajesh Singh, Anita Gehlot, Suresh Chandra Satapathy, Waleed S. Alnumay, Danilo Pelusi, Uttam Ghosh, and Janmenjoy Nayak. "Industrial Internet of Things and its applications in Industry 4.0: State of the art." *Computer Communications* 166 (2021): 125–139.
19. Pfeiffer, Sabine. "Robots, Industry 4.0 and humans, or why assembly work is more than routine work." *Societies* 6, no. 2 (2016): 16.
20. Wei, Fuyin, Cyril Alias, and Bernd Noche. "Applications of digital technologies in sustainable logistics and supply chain management." In: Melkonyan, A., Krumme, K. (eds.) *Innovative Logistics Services and Sustainable Lifestyles.* Springer, Cham. (2019): 235–263. https://doi.org/10.1007/978-3-319-98467-4_11
21. Gangwar, Suraj, and Vinayak Narang. "A survey on emerging cyber crimes and their impact worldwide." In *Research Anthology on Combating Cyber-Aggression and Online Negativity*, IGI Global, (2022): 13. doi: 10.4018/978-1-6684-5594-4.ch080
22. Xing, Liudong. "Cascading failures in internet of things: Review and perspectives on reliability and resilience." *IEEE Internet of Things Journal* 8, no. 1 (2020): 44–64.
23. Ani, Uchenna P. Daniel, Hongmei He, and Ashutosh Tiwari. "Review of cybersecurity issues in industrial critical infrastructure: Manufacturing in perspective." *Journal of Cyber Security Technology* 1, no. 1 (2017): 32–74.
24. Anderegg, William R.L., Anna T. Trugman, Grayson Badgley, Christa M. Anderson, Ann Bartuska, Philippe Ciais, Danny Cullenward, et al. "Climate-driven risks to the climate mitigation potential of forests." *Science* 368, no. 6497 (2020): eaaz7005.
25. Bécue, Adrien, Isabel Praça, and João Gama. "Artificial intelligence, cyber-threats and Industry 4.0: Challenges and opportunities." *Artificial Intelligence Review* 54, no. 5 (2021): 3849–3886.
26. Lee, Changhun, and Chiehyeon Lim. "From technological development to social advance: A review of Industry 4.0 through machine learning." *Technological Forecasting and Social Change* 167 (2021): 120653.
27. Aljohani, Abeer. "Predictive analytics and machine learning for real-time supply chain risk mitigation and agility." *Sustainability* 15 (2023): 15088. https://doi.org/ 10.3390/su152015088
28. Yu, Wenjin, Tharam Dillon, Fahed Mostafa, Wenny Rahayu, and Yuehua Liu. "A global manufacturing big data ecosystem for fault detection in predictive maintenance." *IEEE Transactions on Industrial Informatics* 16, no. 1 (2019): 183–192.
29. Sridhar, Siddharth, Adam Hahn, and Manimaran Govindarasu. "Cyber–physical system security for the electric power grid." *Proceedings of the IEEE* 100, no. 1 (2011): 210–224.
30. Suriadi, Suriadi, Robert Andrews, Arthur HM ter Hofstede, and Moe Thandar Wynn. "Event log imperfection patterns for process mining: Towards a systematic approach to cleaning event logs." *Information Systems* 64 (2017): 132–150.

31. Alzubi, Omar A. “A deep learning-based frechet and dirichlet model for intrusion detection in IWSN.” *Journal of Intelligent & Fuzzy Systems* 42, no. 2 (2022): 873–883.
32. Gardner, Josh, and Christopher Brooks. “Student success prediction in MOOCs.” *User Modeling and User-Adapted Interaction* 28 (2018): 127–203.
33. Gómez, Cristina, Joanne C. White, and Michael A. Wulder. “Optical remotely sensed time series data for land cover classification: A review.” *ISPRS Journal of Photogrammetry and Remote Sensing* 116 (2016): 55–72.
34. Advani, Madhu S., Andrew M. Saxe, and Haim Sompolinsky. “High-dimensional dynamics of generalization error in neural networks.” *Neural Networks* 132 (2020): 428–446.
35. Akinosho, Taofeek D., Lukumon O. Oyedele, Muhammad Bilal, Anuoluwapo O. Ajayi, Manuel Davila Delgado, Olugbenga O. Akinade, and Ashraf A. Ahmed. “Deep learning in the construction industry: A review of present status and future innovations.” *Journal of Building Engineering* 32 (2020): 101827.
36. Ghahramani, Mohammadhossein, Yan Qiao, Meng Chu Zhou, Adrian O’Hagan, and James Sweeney. “AI-based modeling and data-driven evaluation for smart manufacturing processes.” *IEEE/CAA Journal of Automatica Sinica* 7, no. 4 (2020): 1026–1037.
37. Pang, Guansong, Chunhua Shen, and Anton van den Hengel. “Deep anomaly detection with deviation networks.” In *Proceedings of the 25th ACM SIGKDD International Conference on Knowledge Discovery & Data Mining*, (KDD ’19), August 4–8, 2019, Anchorage, AK, USA. ACM, New York, NY, USA. (2019): 10. https: //doi.org/10.1145/3292500.3330871
38. Ahmed, Mohiuddin, Abdun Naser Mahmood, and Md Rafiqul Islam. “A survey of anomaly detection techniques in financial domain.” *Future Generation Computer Systems* 55 (2016): 278–288.
39. Gamal, Mohamed, Ahmed Donkol, Ahmed Shaban, Francesco Costantino, G. Di, and Riccardo Patriarca. “Anomalies detection in smart manufacturing using machine learning and deep learning algorithms.” In *Proceedings of the International Conference on Industrial Engineering and Operations Management, Rome, Italy*, pp. 1611–1622. 2021.
40. Ibrahim, Muhammad Sohail, Wei Dong, and Qiang Yang. “Machine learning driven smart electric power systems: Current trends and new perspectives.” *Applied Energy* 272 (2020): 115237.

5 Enhancing Resilience in the Integration of Cybersecurity for Smart Manufacturing

R. Vijayakumari, Phaneendra Varma Chintalapati, K. Baskar and Karamath Ateeq

1 INTRODUCTION

The surge in investments in smart manufacturing is yielding substantial gains in both productivity and quality, attributed to the seamless integration of internet of things (IoT), cloud computing, data analytics, artificial intelligence (AI), and machine learning into manufacturing processes. The convergence of all these technologies has ushered in a new era of smart manufacturing, enhancing productivity and quality globally. Smart manufacturing, propelled by the integration of these technologies, is rapidly becoming a cornerstone of economic uplift for countries worldwide. While smart manufacturing offers economic advantages, the foremost concern revolves around the security of data, given the extensive amount of information involved and the increasing risk of data breaches. The primary challenge in smart manufacturing lies in mitigating the security risks associated with data breaches, as the integration of IoT, amplifies the volume of sensitive information at stake. The rapid adoption of new technologies in manufacturing has created a significant challenge, as companies grapple with understanding and addressing the emerging threats, particularly in the context of the heightened network-based risks.

Conventional manufacturing operated within well-defined networks, where equipment on the shop floor connected seamlessly, relying solely on physical security measures to safeguard the framework. However, the advent of virtual manufacturing has transformed these once localized networks into distributed systems spanning remote locations. In contrast to conventional manufacturing's focus on physical security through system isolation, the paradigm shifts towards smart manufacturing underscores the centrality of network security as a primary risk factor. Smart manufacturing's transformative impact on productivity and quality is accompanied by a critical concern – the vulnerability of extensive data networks, making data security the forefront issue in the industry. As countries worldwide embrace smart manufacturing for economic advancement, the critical Achilles' heel remains the security of data, posing a formidable challenge amid the rapid integration of advanced technologies.

DOI: 10.1201/9781032694375-5

The paradigm shifts from conventional to smart manufacturing brings forth a heightened risk landscape, where the primary focus shifts from physical access control to safeguarding against potential network vulnerabilities and cyber threats. The production community has started facing the unique challenge – cyber-attacks on their system of manufacturing. The challenge they confront is the uncertainty regarding the timing, target severity, and cascading effects of potential attacks, along with the absence of a clear strategy for managing such incidents within their production schedule. Addressing this issue involves bolstering the cybersecurity infrastructure in smart manufacturing.

This chapter aims to analyse specific cybersecurity threats and propose strategies to mitigate them seamlessly within the production schedule, ultimately enhancing the integration of cybersecurity systems in manufacturing and enhancing overall resilience. The initial action involves precisely defining a smart manufacturing system, outlining the criteria for cyber-attacks on these systems, and elucidating the mechanisms through which cybersecurity operates to fortify the system's resilience. Industry resilience is characterized by the system's ability to return to the required state following a security breach and recover from disruptions by isolating affected components and redistributing tasks among non-affected components.

By integrating cybersecurity measures into the system, the manufacturing unit's resilience can be significantly heightened. Subsequently, the discussion will focus on strategies to mitigate these threats, exploring the tangible impacts of their implementation within the system to augment overall productivity. In the envisioned system, data flow is secured through unidirectional gates, ensuring comprehensive security for operational networks. An integrated model is proposed which gives an assurance to manufacturing system through cybersecurity and resilience mechanisms. It proposes the challenges involved in resilient system and emphasis is also made on the reaction of the system to unexpected events.

The chapter is organized as follows: The introduction provides an overview of smart manufacturing, explores associated risks and challenges, and highlights the integration of resilient systems to bolster cybersecurity in this context. The subsequent section comprises a comprehensive literature review, analysing studies conducted by researchers to enhance resilient mechanisms in cyber manufacturing systems. Following this, the third section delves into the manufacturing system, elucidating its layout and operational functions. The fourth section outlines various types of cybersecurity attacks that pose threats to the system. Moving forward, the fifth section details strategies to fortify the resilient mechanism within cyber manufacturing systems. The chapter concludes by summarizing key findings and insights, providing a cohesive wrap-up of the discourse on smart manufacturing cybersecurity, and concludes with a references section. This structured approach ensures a logical flow and organization of information, guiding the reader through the essential components of the chapter.

2 LITERATURE REVIEW

Sheth et al. (1) introduced a comprehensive five-part framework encompassing the external and internal environment, manufacturing processes, technology, and demand-supply networks, emphasizing the importance of integrating these networks. Espinoza-Zelaya et al. (2) defined a resilience mechanism aimed at reducing the probability, detection time, and impact of cyber-attacks, as well as minimizing the time required for recovery. Ren et al. (3) explored state-of-the-art technologies addressing security issues in smart manufacturing and discussed existing strategies for enhancing security in manufacturing industries. Rahman et al. (4) presented the outcomes of their framework through a case study in an additive manufacturing company, validated using Yager's recursive rule.

Babiceanu et al. (5) proposed a Software-Defined Networking (SDN)-based cybersecurity-resilience protection mechanism tailored for virtual manufacturing. Tuptuk and Hailes (6) delved into industrial system security, addressing current vulnerabilities, anticipating future cyber-attacks, highlighting system weaknesses, outlining upcoming security challenges, and emphasizing the pivotal role of cybersecurity in smart manufacturing. Faleiro et al. (7) innovatively enhanced cyber resilience by introducing the concept of digital twins. Their approach involves a holistic analysis that integrates digital twin technology into cybersecurity frameworks. Kusiak (8) outlined the evolutionary stages of smart manufacturing, emphasizing the crucial roles of autonomy, intelligent decision-making, and collaborative work. They categorized smart manufacturing evolution into four distinct stages. Babiceanu and Seker (9) proposed an SDN-based modelling environment designed to ensure cybersecurity and resilience for manufacturing applications, providing a robust framework for addressing security concerns.

Mihalache et al. (10) conducted a comprehensive review focusing on threats and vulnerabilities affecting systems, along with an assessment of existing methods for enhancing resilience in smart manufacturing. Leng et al. (11) advocated for the use of block chain systems to mitigate cybersecurity threats in manufacturing. They proposed ten metrics for implementing block chain applications in manufacturing systems. Camarinha et al. (12) analysed sustainable and resilient manufacturing and collaborative networks, outlining the distribution of sustainability facets across various entities in manufacturing. They identified challenges in the context of smart manufacturing. Babiceanu and Seker (13) proposed trustworthy solutions for manufacturing systems, employing resilient systems framework for cybersecurity component modelling to ensure data preservation. Riel et al. (14) introduced a novel logic to address security challenges in the design and operation of manufacturing systems.

Lees et al. (15) offered a practical perspective on cybersecurity resilience in industrial environments, including recommendations for implementation. Mitcheltree et al. (16) emphasized the imperative to enhance resilience and security in manufacturing environments to mitigate the adverse effects of cyber threats. Mullet et al. (17) provided a step-by-step conceptual guide for manufacturing companies to navigate cybersecurity threats. Their approach encompassed classical cybersecurity, countermeasures as well as innovative solutions. Bac et al. (18) introduces key concepts for enabling smart manufacturing in Industry 5.0, exemplifying the explainable artificial intelligence (XAI) approach through an illustrative example focused on anomaly detection for cybersecurity threats. Zou et al. (19) explore autonomous mobility system scenarios, specifically addressing the identification of threats before potential attacks occur.

Zelaya and Omar (20) showcased the resilience of a Cyber-Physical System (CMS) capable of withstanding cyber-attacks. They advocated for a manufacturing-centric approach, utilizing production key performance indicators as targets to enhance resilience. Konstantinou et al. (21) implemented the principles of chaos engineering to mitigate the adverse effects of data breaches. Their approach aimed to predict environmental changes and minimize the severity of the resultant effects on the system. Colabianchi et al. (22) categorized existing models based on their impact on security, privacy, and safety. Additionally, they developed frameworks and metrics to evaluate and compare the resilience capabilities of systems.

3 MANUFACTURING SYSTEMS

The primary objective of a manufacturing system is to consistently deliver cost-effective, high-quality products. All sectors within manufacturing are focused on continuous process improvement to establish sustainable and secure manufacturing systems. Figure 5.1 depicts the hierarchical structure of the smart manufacturing system, comprising five layers with top management positioned at the highest tier and field operators at the lowest. At the apex, top-level management makes decisions regarding the manufacturing process and defines the workflow for product production. The subsequent layer, Plant Management level, focuses on decisions related to the workshop where products are manufactured. The Process control level oversees and supervises design, workflow and manufacturing processes, followed by IT-based employees (PLC control level) responsible for virtual aspects of manufacturing. The bottommost layer consists of machine operators, individuals engaged in executing the physical operations.

Observing Figure 5.1, it becomes evident that the industrial environment is inherently open and susceptible to various security threats, including eavesdropping, denial of service, and man-in-the-middle attacks, all of which pose risks to product quality. Unlike network-based attacks, those on the manufacturing system are particularly perilous as they can result in physical damage to the products. To safeguard the manufacturing process effectively, security measures must be integrated from the design stage onward. The system's openness and configurability contribute to its complexity, potentially leading to intricate system behaviours. The components within the manufacturing system exhibit heterogeneity, incorporating multiple devices. Managers and engineers operate under real-time constraints, demanding careful attention to ensure continuous processes.

The virtual manufacturing unit engages in intricate interactions with the physical system, where any system failure may cascade into product failures. It is imperative to comprehend the system's vulnerabilities before addressing its security. Traditionally, manufacturing systems are designed under the assumption of isolation, lacking inherent security measures. While software developers prioritize security requirements in their implementations, aspects such as testing, patch management, and coding for manufacturing systems often go overlooked in terms of security considerations.

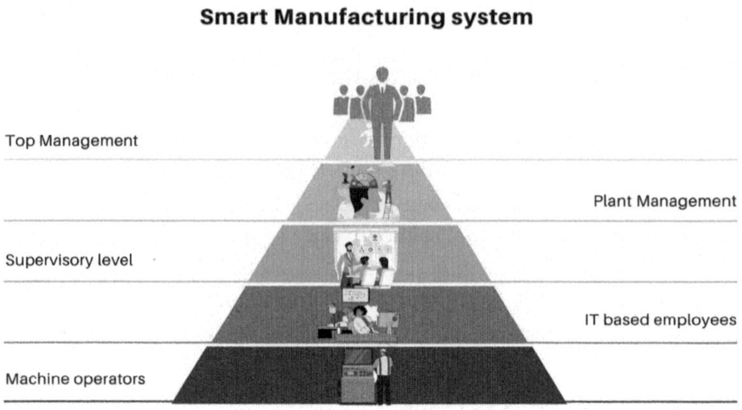

FIGURE 5.1 Smart Manufacturing System.

4 CYBERSECURITY ATTACKS

This section delves into the potential attack on the networked system, with implications extending to the vulnerability of manufacturing systems. The severity of the attack may render it inconspicuous without extensive knowledge of the manufacturing system, and in some cases, attackers might establish a virtual system capable of disrupting the entire manufacturing process. The impact of an attack can vary, ranging from minimal effects to system failure, and the connection between system failure and a possible attack may not always be apparent. Moreover, attacks are frequently underreported due to concerns about potential business damage.

Attacks on a system can take various forms, such as denying access to networks or computational resources, monitoring networks to extract sensitive information, positioning an attacker between communicating devices and control systems to intercept information, injecting false information through unwanted commands, introducing time delays into the control system leading to disturbances and potential system crashes, and tampering with or altering data, resulting in undesired changes to the system.

In replay attacks, adversaries reuse data to trick a system into accepting it as valid, potentially leading to serious issues such as unauthorized access, data manipulation, or unintended actions. In spoofing attacks, adversaries falsify their identity to gain unauthorized access to a system. Side-channel attacks involve analysing patterns like power consumption, execution time, electromagnetic radiation, or acoustic signals instead of directly attacking the network system, providing entry points into computational systems. Covert-channel attacks are linked to systems with varying access privileges. These attacks exploit unintended communication channels to transfer information between processes or entities discreetly.

Zero-day attacks pose a formidable challenge for organizations, targeting undisclosed vulnerabilities in software and catching them off guard with limited knowledge to defend against such exploits. Exploiting weaknesses unknown to software vendors, attackers make it challenging for organizations to predict which part of the system might be the focus of the attack. In some instances, attackers may gain physical access to equipment, allowing them to manipulate data, modify signals, or introduce errors

FIGURE 5.2 Cyber-Attacks in Industry 4.0.

into the received input. Machine learning techniques are widely employed in manufacturing systems for network monitoring. Tampering with the data within the machine learning processes can have severe consequences, potentially compromising the system's integrity. Figure 5.2 illustrates various types of cyber-attacks in smart manufacturing, providing a visual representation of the threat landscape.

5 ENHANCING RESILIENCE IN MANUFACTURING SYSTEM

Resilience in a cyber-manufacturing system entails the ability to achieve its objectives even in the face of deliberate disruptions, effectively reducing the likelihood of failure and ensuring rapid recovery from cyber-attacks. It should play a proactive role in preventing both internal and external threats, integrated seamlessly into the system's design. Routine execution within the system is essential to promptly identify cyber and physical device threats. The design must prioritize addressing component failures, isolating them to prevent systemic collapse. The resilience framework should demonstrate the system's capacity to withstand cyber-attacks, allowing for continuous operation to enhance productivity, or alternatively, facilitate swift recovery to avert a complete system failure. In brief, a resilient cyber manufacturing system should reduce the chance of failure, withstand the effects of failure, and recover from failure quickly before the entire system collapses.

In the design phase of a cyber-manufacturing system, it is imperative to integrate the resilience mechanism to create a system that is resistant to exploitation by attackers. This proactive approach allows for early assessment of potential threats, enabling the implementation of precautionary measures to forestall further damage. By designing a virtual manufacturing model that incorporates cybersecurity and resilience mechanisms, a comprehensive framework encompassing hardware, software, operations, and maintenance is established. The integrated resilient mechanism within this virtual model automatically derives system configurations and failure modes, enhancing the system's capacity to withstand and recover from cyber threats. Researchers emphasize that threats to a cyber-manufacturing system can emanate not only from external sources but also from insiders with system access. To bolster security, these researchers advocate for the utilization of block chain technology to store ledgers, access rules, and topology information, thereby enhancing network security.

Recognizing that attackers persistently find ways to breach security mechanisms, the design of a resilient cyber-manufacturing system should prioritize continuous monitoring. The swift detection of threats is crucial as they can directly impact the system, potentially leading to total failure. Integration with a reactive protection concept facilitates the identification of responses that do not disrupt industrial system operations. Given that a cyber-manufacturing system involves both physical and digital data, it is essential to note that conventional IT security practices often focus solely on digital threats. A resilient cyber-manufacturing system should be designed to provide alerts for both physical and digital threats, ensuring a comprehensive approach to threat detection and response.

A critical consideration is to minimize the impact of an attack as early as possible. The duration between the onset of a threat and its impact on the system is the window during which system losses occur. The primary function of a resilient system is to actively reduce this timeframe, swiftly mitigating the effects of an attack on the system. During this critical period, the cyber-manufacturing system must make real-time decisions to minimize the impact of the attack. To ensure continuous production, affected parts can be swiftly replaced with redundant components. The resilient mechanism should incorporate a robust recovery system, enabling a rapid transition of the system back to its normal state. Additionally, the recovery mechanism should analyse and absorb the methods used in the attack, preventing the recurrence of the same type of attack in the future.

In summary, manufacturers should conduct comprehensive assessments across the entire system to identify vulnerabilities, prioritize risks, and comprehend the potential impact on operations. This foundational understanding is crucial for the development of an effective resilient mechanism. Additionally, security measures, such as firewalls, endpoint protection, and intrusion detection systems, must be implemented in the manufacturing system to restrict access to authorized personnel and enhance overall cybersecurity.

Human error significantly influences risk management, necessitating employee training and awareness programmes. These initiatives should focus on instilling best practices, adhering to security protocols, identifying and mitigating threats, and overall enhancing the company's cybersecurity. Collaboration between cybersecurity experts and employees is essential, involving the sharing of information on potential threats, methods for identification, preventive measures to pre-empt attacks, and strategies for swift recovery to normalcy after a security threat.

6 CONCLUSION

This chapter has provided a comprehensive and systematic approach in the enhancement of resilience in cyber manufacturing system. It also underscores the proactive measures to safeguard the industrial system. The landscape of smart manufacturing is illuminated, highlighting the risks and challenges inherent in the rapidly advancing technological ecosystem. The recognition of these challenges sets the stage for the central theme of the chapter -the integration of resilient cybersecurity measures. A panoramic view of existing research is provided, revealing the dynamic efforts to fortify resilient mechanisms in cyber manufacturing systems. This synthesis of scholarly work not only informs current practices but also lays the groundwork for future developments in the field. Delving into the manufacturing system's layout and operational intricacies, the chapter establishes a foundational understanding essential for identifying vulnerabilities and potential points of cyber-attacks. By recognizing the specific nuances of smart manufacturing processes, the chapter emphasizes the need for tailored cybersecurity strategies to fortify resilience effectively. The exploration of various cybersecurity threats serves as a stark reminder of the diverse challenges faced by smart manufacturing systems. The intricate interplay of cyber threats demands a nuanced approach to defend, underlining the necessity of resilient mechanisms that can adapt to evolving attack vectors. The chapter also details the strategies for fortifying resilient mechanisms, offers actionable insights for practitioners and decision-makers. From continuous monitoring to swift response and recovery measures, the strategies presented are integral to creating a robust defence against cyber threats. By amalgamating cybersecurity measures with resilience strategies, organizations can not only mitigate risks but also respond effectively to evolving cyber threats. The integration of these practices is pivotal for ensuring the sustained efficiency, security, and adaptability of smart manufacturing systems in an ever-changing technological landscape.

REFERENCES

1. Sheth, Ananya, and Andrew Kusiak. "Resiliency of smart manufacturing enterprises via information integration." *Journal of Industrial Information Integration* 28 (2022): 100370.
2. Espinoza-Zelaya, Carlos, and Young Bai Moon. "Resilience enhancing mechanisms for cyber-manufacturing systems against cyber-attacks." *IFAC-PapersOnLine* 55, no. 10 (2022): 2252–2257.
3. Ren, Anqi, Dazhong Wu, Wenhui Zhang, Janis Terpenny, and Peng Liu. "Cyber security in smart manufacturing: Survey and challenges." In *IIE Annual Conference. Proceedings*, pp. 716–721. Institute of Industrial and Systems Engineers (IISE), 2017.
4. Rahman, Sazid, Niamat Ullah Ibne Hossain, Kannan Govindan, Farjana Nur, and Mahathir Bappy. "Assessing cyber resilience of additive manufacturing supply chain leveraging data fusion technique: A model to generate cyber resilience index of a supply chain." *CIRP Journal of Manufacturing Science and Technology* 35 (2021): 911–928.
5. Babiceanu, Radu F., and Remzi Seker. "Cyber resilience protection for industrial internet of things: A software-defined networking approach." *Computers in Industry* 104 (2019): 47–58.
6. Tuptuk, Nilufer, and Stephen Hailes. "Security of smart manufacturing systems." *Journal of Manufacturing Systems* 47 (2018): 93–106.
7. Faleiro, Rajiv, Lei Pan, Shiva Raj Pokhrel, and Robin Doss. "Digital twin for cybersecurity: Towards enhancing cyber resilience." In *Broadband Communications, Networks, and Systems: 12th EAI International Conference, BROADNETS 2021, Virtual Event*, October 28–29, 2021, Proceedings 12, pp. 57–76. Springer International Publishing, 2022.

8. Kusiak, Andrew. "Smart manufacturing." In: Nof, S.Y. (eds) *Springer Handbook of Automation Springer Handbooks*. pp. 973–985. Springer, Cham: Springer International Publishing, 2023. https://doi.org/10.1007/978-3-030-96729-1_45
9. Babiceanu, Radu F., and Remzi Seker. "Cybersecurity and resilience modelling for software-defined networks-based manufacturing applications." In: Borangiu, T., Trentesaux, D., Thomas, A., Leitão, P., Oliveira, J. (eds) *Service Orientation in Holonic and Multi-Agent Manufacturing: Proceedings of SOHOMA 2016*, Studies in Computational Intelligence, vol 694. Springer, Cham. Springer International Publishing, pp. 167–176. 2017. https://doi.org/10.1007/978-3-319-51100-9_15
10. Mihalache, S.F., Pricop, E., Fattahi, J. (2019). Resilience Enhancement of Cyber-Physical Systems: A Review. In: Mahdavi Tabatabaei, N., Najafi Ravadanegh, S., Bizon, N. (eds) *Power Systems Resilience. Power Systems*. Springer, Cham. https://doi.org/10.1007/978-3-319-94442-5_11
11. Leng, Jiewu, Shide Ye, Man Zhou, J. Leon Zhao, Qiang Liu, Wei Guo, Wei Cao, and Leijie Fu. "Blockchain-secured smart manufacturing in industry 4.0: A survey." *IEEE Transactions on Systems, Man, and Cybernetics: Systems* 51, no. 1 (2020): 237–252.
12. Camarinha-Matos, L.M., Rocha, A.D. & Graça, P. Collaborative approaches in sustainable and resilient manufacturing. *J Intell Manuf* 35, 499–519 (2024). https://doi.org/10.1007/s10845-022-02060-6
13. Babiceanu, Radu F., and Remzi Seker. "Trustworthiness requirements for manufacturing cyber-physical systems." *Procedia Manufacturing* 11 (2017): 973–981.
14. Riel, Andreas, Christian Kreiner, Georg Macher, and Richard Messnarz. "Integrated design for tackling safety and security challenges of smart products and digital manufacturing." *CIRP Annals* 66, no. 1 (2017): 177–180.
15. Lees, Michael J., Melissa Crawford, and Christoph Jansen. "Towards industrial cybersecurity resilience of multinational corporations." *IFAC-PapersOnLine* 51, no. 30 (2018): 756–761.
16. Mitcheltree, Christina Marie, Godfrey Mugurusi, and Halvor Holtskog. "Cyber security culture as a resilience-promoting factor for human-centered machine learning and zero-defect manufacturing environments." In *International Conference on Flexible Automation and Intelligent Manufacturing*, pp. 741–752. Cham: Springer Nature Switzerland, 2023.
17. Mullet, Valentin, Patrick Sondi, and Eric Ramat. "A review of cybersecurity guidelines for manufacturing factories in industry 4.0." *IEEE Access* 9 (2021): 23235–23263.
18. Bac, T.P., Ha, D.T., Tran, K.D., Tran, K.P. (2023). Explainable Articial Intelligence for Cybersecurity in Smart Manufacturing. In: Tran, K.P. (eds) *Artificial Intelligence for Smart Manufacturing. Springer Series in Reliability Engineering*. Springer, Cham. https://doi.org/10.1007/978-3-031-30510-8_10
19. Zou, B., Choobchian, P. & Rozenberg, J. Cyber resilience of autonomous mobility systems: cyber-attacks and resilience-enhancing strategies. *J Transp Secur 14*, 137–155 (2021). https://doi.org/10.1007/s12198-021-00230-w
20. Espinoza Zelaya, Carlos Omar, "ENHANCING THE OPERATIONAL RESILIENCE OF CYBER- MANUFACTURING SYSTEMS (CMS) AGAINST CYBER-ATTACKS" (2023). Dissertations - ALL. 1719. https://surface.syr.edu/etd/1719
21. Konstantinou, C., Stergiopoulos, G., Parvania, M., & Esteves-Verissimo, P. (2021). Chaos Engineering for Enhanced Resilience of Cyber-Physical Systems. *2021 Resilience Week (RWS)*. doi:10.1109/rws52686.2021.9611797
22. Colabianchi, Silvia, Francesco Costantino, Giulio Di Gravio, Fabio Nonino, and Riccardo Patriarca. "Discussing resilience in the context of cyber physical systems." *Computers & Industrial Engineering* 160 (2021): 107534.

6 Leveraging Artificial Intelligence and Machine Learning for Advanced Threat Detection in Smart Manufacturing

Gnanasankaran Natarajan, Sundaravadivazhagan Balasubramanian, Elakkiya Elango, and Rakesh Gnanasekaran

1 INTRODUCTION TO ARTIFICIAL INTELLIGENCE AND MACHINE LEARNING

Ultimately, this chapter highlights how important AI and ML are to enhancing the security of smart manufacturing by offering a proactive defencc against new threats. By utilizing advanced analytics and automation, organizations can safeguard the continuity and stability associated with their smart manufacturing systems within a highly interconnected and delicate digital ecosystem.

1.1 Historical Background

The idea of artificial intelligence (AL) has its roots in antiquated mythology and tales of mechanical creatures possessing human-like abilities. However, the concept of building robots capable of intelligent behaviour was first explored by computer scientists and researchers in the middle of the 20th century, which marked the beginning of the contemporary era of AL. The first computer programme to play chess was written by Alan Turing, and Allen Newell, and Herbert A. Simon developed the Logic Theorist, two groundbreaking innovations in the discipline. The interdisciplinary area of AI was formally founded in 1956 when the phrase "artificial intelligence" was first used at a Dartmouth College workshop. The subsequent decades saw cycles of excitement and disappointment in AI research, referred to as "summers" and "winters."

Funding changes, technology constraints, and changing expectations were the driving forces behind these cycles. As a branch of AL, machine learning (ML)

DOI: 10.1201/9781032694375-6

focuses on creating statistical models and algorithms that let computers acquire data and become more efficient at a given task. The creation of early learning algorithms such as the perceptron in the middle of the 20th century laid the groundwork for ML. However, ML did not become widely recognized and started to drive the AI revolution until the 21st century, when huge data and powerful computation became available [1].

1.2 What Is Artificial Intelligence?

The creation of computer systems that are capable of carrying out tasks that typically call for human intelligence is referred to as AL. These activities include a wide range of skills, including computer vision, natural language processing, problem-solving, and comprehending intricate patterns. AI systems use a variety of methods, which includes as neural networks, expert systems, and rule-based systems, to mimic cognitive functions that are similar to those of humans. There are two primary types of AI systems: General or Strong AI and Narrow or Weak AI. Narrow AI is made for specialized tasks like playing board games like chess or speech recognition. General AI, on the other hand, seeks to build machines that are as intelligent as humans and capable of carrying out any intellectual work that a human being can.

1.3 What Is Machine Learning?

Within the field of ALML is the study of creating models and algorithms that enable computers to gain knowledge from data and become more efficient at tasks without needing to be specifically programmed. ML uses statistical methods and data patterns to identify patterns, identify things, and develop predictions. Making it possible for computers to extrapolate from data and generate precise judgments or predictions in novel, untested scenarios is the main goal ofML. Supervised learning, unsupervised learning, and reinforcement learning are only a few of the methods that are included in ML. Training a model on labelled data—where the right answers are given—requires supervised learning. Unsupervised learning is the process of identifying structures and patterns in unlabelled data. The goal of reinforcement learning is to teach agents how to interact with their surroundings and make a series of decisions in order to optimize a cumulative reward [2].

1.4 The Current Landscape

AI and ML are now widely used in every aspect of our lives. Among numerous other uses, they enable virtual assistants such as Siri and Alexa, make product recommendations on e-commerce websites, diagnose disorders in medical imaging, allow self-driving cars, and help with financial risk assessments. The creation of complex algorithms, enormous rise in data, and innovations in computer power have all contributed to the acceleration of AI and ML. At the cutting edge of technological advancement, AL and ML hold the potential to transform entire industries, improve decision-making, and simplify and streamline our daily lives. The potential applications of AI and ML are endless, and these domains will have an enormous effect on society and the global economy as they develop [3, 4].

2 INTRODUCTION TO SMART MANUFACTURING

Industry 4.0, or "smart manufacturing," is a contemporary manufacturing strategy that makes use of data-driven workflows and cutting-edge technologies to build production systems that are more adaptable, flexible, and efficient. It signifies a profound shift in the conception, manufacture, and distribution of things. Modern technology is incorporated into many parts of the manufacturing process through smart manufacturing to improve overall competitiveness, lower costs, increase productivity, and improve product quality. This is a thorough synopsis of smart manufacturing.

2.1 Data-Driven Decision Making

Data gathering and analysis play a major role in smart manufacturing. Over the manufacturing process, sensors and internet of things (IoT) devices are used to collect data in actual time on the performance of equipment, procedures, and the quality of the goods. In order to make wise decisions, this data is subsequently processed and examined utilizing powerful analytics, ML, and AL.

2.2 Connectivity and Integration

A key component of smart manufacturing is the integration of data and processes. A network connects the many parts of the production system, including the supply chain, inventory, robots, and machinery. Throughout the production ecosystem, real-time communication, data sharing, and control are made possible by this connectivity.

2.3 Predictive Maintenance

A key use of smart manufacturing is maintenance prediction. Manufacturers are able to forecast when maintenance is required by tracking the state and functionality of machinery and equipment in real time. By doing this, unplanned malfunctions are avoided, downtime is decreased, and machinery longevity is increased.

2.4 Quality Control and Defect Detection

Systems for inspecting and identifying product flaws are powered by AI and machine vision. By ensuring that only goods of the highest quality are made, waste and rework are decreased and total product quality is raised.

2.5 Automation and Robotics

Automation plays a major role in smart manufacturing, involving the utilization of robotic arms and other robots. These devices accurately and consistently complete jobs like assembly, material handling, and inspection, allowing up human workers to concentrate on more intricate and imaginative projects.

2.6 Process Optimization

AI and data analytics are employed in manufacturing process optimization. Manufacturers are able to pinpoint areas for improvement, inefficiencies, and bottlenecks by analyzing data from multiple sources for sensors. As a result, operating expenses are decreased and productivity rises.

2.7 Energy Efficiency

Sustainability and energy efficiency are the main goals of smart manufacturing. Modern technologies contribute to cost savings and environmental advantages by monitoring energy use and optimizing operations for lower energy use.

2.8 Supply Chain Integration

Smart manufacturing integrates the whole supply chain, not just the manufacturing floor. Just-in-time manufacturing, lower inventory costs, and more efficient operations are made possible by manufacturers' close monitoring of inventory levels, demand projections, and logistics.

2.9 Customization and Personalization

More personalization and customization of products is made possible by smart manufacturing. Utilizing data and adaptable production processes enables mass customization, satisfying the growing demand from customers for customized goods.

2.10 Human-Machine Collaboration

While automation is important, smart manufacturing also places a strong emphasis on cooperation between humans and machines. Wearable technology and augmented reality (AR) give employees access to real-time data and direction. Cobots, or collaborative robots, support human workers by helping with activities that call for strength and accuracy.

2.11 Cybersecurity and Data Protection

Smart manufacturing needs to give priority to strong cybersecurity measures to safeguard sensitive data and stop unwanted usage of systems in light of growing connectivity and data sharing.

2.12 Sustainability and Environment

Eliminating environmental impact and emphasizing sustainability are common components of smart manufacturing. Manufacturers can help create a more sustainable and healthier future by streamlining processes and cutting waste. The manner that items are manufactured has fundamentally changed as a result of smart manufacturing. Manufacturers may increase productivity, quality, and flexibility as well as their competitiveness in a world market that is changing quickly by utilizing data, connection, automation, and cutting-edge technology [5, 6].

3 ROLE OF ARTIFICIAL INTELLIGENCE AND MACHINE LEARNING IN SMART MANUFACTURING

In the manufacturing process sector, smart manufacturing—also known as Industry 4.0 or the industrial internet of things (IIoT)—represents a paradigm change. It makes use of state-of-the-art technologies to improve industrial processes' productivity, flexibility, and efficiency. AI and ML, which are fundamental to the breakthrough in smart manufacturing, are at the centre of this change. The following lists the exceptional contributions that ML and AL have made to smart manufacturing. Subsequently, we will investigate advanced techniques for cyber security and threat detection in the context of smart manufacturing [7].

3.1 Predictive Maintenance

Predictive maintenance constitutes a single of the major uses of AI and ML in smart manufacturing. Conventional, time-based maintenance plans can be expensive and ineffective. Real-time sensor data from industrial equipment can be analysed by AI and ML algorithms to forecast when a machine is going to break. Manufacturers can maximize maintenance costs, save downtime, and increase equipment longevity by utilizing predictive maintenance. Costs are reduced and production efficiency is raised as a result.

3.2 Quality Control and Defect Detection

Processes for quality control automation greatly benefit from AI and ML. AI-enabled machine vision systems are highly accurate in inspecting and identifying product flaws. Because they can identify flaws in real time, manufacturers can limit the number of defective products they produce to take prompt corrective action. Customer satisfaction and product quality both increase as a result of this.

3.3 Process Optimization

AI and ML are used in smart manufacturing to optimize a variety of production processes. These technologies can locate bottlenecks, inefficiencies, and potential improvement areas by studying data from sensors, production lines, and supply chains. Manufacturers may boost resource efficiency, cut waste, and raise production throughput by continually improving their operations.

3.4 Inventory Management

A crucial component of production is efficient inventory management. Supply chain processes may be optimized, demand fluctuations can be predicted, and real-time inventory tracking is possible with AI and ML. This can lower carrying costs, prevents overstocking or understocking, and guarantees that products are accessible when needed, all of which help producers save money and provide better customer service.

3.5 Energy Efficiency

Reducing energy consumption in manufacturing facilities is made possible in large part by AI and ML. These systems are able to track energy consumption and spot wasteful tendencies. Manufacturers may then cut their energy costs and carbon footprint by using their knowledge and suggestions for energy-efficient operation.

3.6 Autonomous Robots and Drones

Drones and robots powered by AI are being used more and more in smart production. Without human assistance, these autonomous systems are capable of handling materials, conducting inspections, and managing inventories. Their remarkable versatility and ability to be reprogrammed for distinct purposes enhance the adaptability and efficacy of manufacturing processes.

3.7 Supply Chain Optimization

A key component of supply chain management optimization is the application of AI and ML. These technologies enable manufacturers to control inventories, estimate demand, and even anticipate supply chain issues like delays or shortages. This makes it possible for producers to take preventative measures, lower risks, and keep the flow of components and materials flowing smoothly.

3.8 Customization and Personalization

More personalization and customization of products is made possible by smart manufacturing. Algorithms powered by AI and ML can modify production procedures to effectively produce personalized goods. This is particularly useful in sectors where customers are demanding more customized products, such consumer electronics and the automotive industry.

3.9 Worker Assistance

In smart manufacturing, AI and ML can assist human labor. Wearable technology and AR can give employees access to real-time data and direction. "Robots," or collaborative robots, are capable of working alongside humans to help with activities that call for dexterity or strength. Both productivity and worker safety are increased by these technology. In the field of smart manufacturing, AI and ML are revolutionary breakthroughs. These technologies enable more adaptable and responsive production processes while also promoting efficiency, sustainability, and quality. AI and ML will become much more important in determining the direction of the manufacturing sector as smart manufacturing develops [8, 9].

4 THREAT DETECTION IN SMART MANUFACTURING

To guarantee the security and integrity of the manufacturing procedures, data, and systems in the Industry 4.0 environment, threat detection is a vital component of smart manufacturing. Due to its heavy reliance on automation, sharing of information, and interconnected components, smart manufacturing is more vulnerable to cyberthreats such data breaches, cyberattacks, and operational disruptions. Identification and mitigation of potential dangers depend on effective threat detection. The several facets of danger detection in smart manufacturing are explained by the following points.

4.1 Threat Landscape in Smart Manufacturing

Smart manufacturing settings are vulnerable to a number of risks, such as:

a. **Cyberattacks:** They may target network infrastructure, sensors, and industrial control systems (ICS).
b. **Data Breaches:** There might be dire repercussions if sensitive data, including as customer information, production procedures, and intellectual property, is accessed without authorization.
c. **Operational Disruptions:** Attacks that cause hiccups in production might cost money in lost time.
d. **Insider Threats:** Workers or outside contractors who have access to vital systems could be dangerous.

4.2 Real-Time Monitoring and Anomaly Detection

Smart manufacturing systems use real-time network traffic, sensor data, and other crucial parameter tracking to effectively identify hazards. AL and ML-driven anomaly detection algorithms continuously examine data streams to find departures from typical behaviour. Alerts are triggered by illegal access attempts, surprising data points, or unusual patterns.

4.3 Security Information and Event Management Systems

Smart manufacturing frequently uses security information and event management (SIEM) systems for correlation of events and centralized log management. They gather and examine log data from a variety of sources, such as network devices and ICS. SIEM systems are able to spot trends and unusual activity that can point to a security breech.

4.4 Endpoint Security

Individual devices, including computers, controllers, and sensors, are safeguarded by security at the endpoint solutions. To keep an eye on and safeguard endpoints from harmful activity, such devices make use of firewalls, intrusion detection systems, and antivirus software.

4.5 Network Segmentation

One proactive method of detecting threats is network segmentation. Manufacturers can restrict the lateral advance of attackers in the event of a breach by segmenting the network into separate areas. It is simpler to identify and isolate threats inside a smaller network section when using this containment method.

4.6 Behavioural Analysis

Behavioural analysis pertains to the tracking and profiling of device and user behaviour in a manufacturing setting. Alerts may be triggered by deviations from regular behavioural patterns because they may point to a possible threat or breach.

4.7 Vulnerability Scanning and Patch Management

Frequent vulnerability assessments and managing patches are essential for locating and resolving hardware and software security flaws. By detecting weaknesses before attackers take use of them, detection of threats can be proactive.

4.8 Incident Response Planning

Effective threat detection requires a clearly established incident response plan. This plan specifies who to call in the event of a security problem, how to react, and what steps to take to limit damages and stop similar occurrences in the future.

4.9 Security Training and Awareness

One typical factor in security incidents is human mistake. Early identification and prevention can be aided by educating staff members and outside contractors about cybersecurity best practices and possible risks.

4.10 Collaboration and Information Sharing

Manufacturers should exchange threat intelligence, work together with other industry players, and stay up to date on new attack vectors and dangers. Early detection and defence against novel and emerging dangers can be aided by this body of common knowledge.

4.11 Continuous Improvement

In smart manufacturing, threat detection is a continuous process. Manufacturers ought to regularly evaluate their safety posture, update their defences against risks, and adjust as novel ones arise. Threat detection is essential to preserving the production environment's functional security and integrity in smart manufacturing. Manufacturers can reduce the likelihood of major disruptions and data breaches by detecting and mitigating threats in real-time through the implementation of advanced technology, processes, and best practices. In the following section of this chapter [10], we will examine the aforementioned subjects in more detail.

5 ADVANCED THREAT DETECTION MECHANISMS IN SMART MANUFACTURING

To guarantee the security and integrity of the production procedures, data, and infrastructure within the Industry 4.0 environment, threat detection is a vital component of smart manufacturing. Due to its heavy reliance on automation, data sharing, and interconnected systems, smart manufacturing is more vulnerable to cyberthreats such data breaches, cyberattacks, and operational disruptions. Identification and mitigation of potential dangers depend on effective threat detection. This is an in-depth description of how smart manufacturing uses danger detection.

5.1 Threat Possibilities in Smart Manufacturing

Smart manufacturing settings are vulnerable to a number of risks, such as:

a. **Cyberattacks:** These can target network infrastructure, sensors, and ICS. The application of cutting-edge technology like AL, networked systems, and IoT has greatly increased production and efficiency in the field of smart manufacturing. Smart manufacturing systems are vulnerable to a variety of cyberattacks due to their increased interconnection. The following are a few typical cyberattack scenarios in smart manufacturing.
   - **Denial of Service (DoS) and Distributed Denial of Service (DDoS) Attacks:**

     Description: The goal of these assaults is to send too much traffic through the production systems, making it impossible for them to reply to valid requests. A DoS or DDoS assault can cause interruptions in smart manufacturing that result in lost revenue and production downtime.

     Impact: Possible equipment damage, hiccups in operations, and decreased production.
   - **Ransomware Attacks:**

     Description: The malware known as ransomware gains access to production systems, encrypts important data, and then demands a fee to unlock it. This may result in compromised sensitive data, loss of intellectual property, and operational problems in smart manufacturing.

     Impact: Possible exposure of confidential data, financial losses, and production stoppages.
   - **Supply Chain Attacks:**

     Description: Cybercriminals exploit weaknesses in the smart manufacturing supply chain to undermine software or component integrity. This may result in the introduction of harmful code during the production process.

Impact: Deteriorated product quality, possible safety risks, and harm to the brand's image.

- **Man-in-the-Middle (MitM) Attacks:**

  Description: MitM attacks involve the interception and possible modification of communication between devices or systems by an unauthorized party. This could lead to production process tampering or illegal access to sensitive data in smart manufacturing.

  Impact: Unauthorized access, data breaches, and jeopardized manufacturing process integrity.
- **Zero-Day Exploits:**

  Description: Hackers take advantage of unidentified flaws, or "zero-day exploits," in the firmware or software of intelligent manufacturing systems. They are able to take control of important components and obtain illegal access as a result.

  Impact: This includes unauthorized control, compromised system integrity, and possible physical device damage.
- **Insider Threats:**

  Description: The organization is susceptible to serious risks from both intentional and inadvertent internal threats. Insiders run the risk of leaking confidential information, compromising security procedures, or unintentionally introducing malware.

  Impact: This includes insecure intellectual property, data breaches, and industrial operations disruptions.

  Robust cybersecurity measures, such as frequent software upgrades, personnel training, network segmentation, encryption, and constant monitoring for unusual activity, are necessary for smart manufacturing systems in order to reduce these threats. Preventive cybersecurity measures are essential for preserving the availability, confidentiality, and integrity of smart manufacturing processes.

b. **Data Breaches:** There may be dire repercussions if sensitive data—such as client information, production procedures, and intellectual property—is accessed without authorization. Sensitive data and vital systems are very vulnerable to breaches of confidentiality, integrity, and availability in the context of smart manufacturing. These breaches can happen in a number of ways, and the repercussions can include everything from production delays to compromised intellectual property. The following are some typical reasons and sources of data breaches in smart manufacturing.

- **Weak Cybersecurity Measures:**

  Explanation: Weak passwords, out-of-date software, and inadequate network security are examples of insufficient cybersecurity measures that lead to vulnerabilities that bad actors can take advantage of. Attackers might use these flaws to take over production systems or obtain unauthorized access to private information [11].

  Consequences: It includes possible industrial process disruption, unauthorized access to confidential information, and intellectual property infringement.

- **Supply Chain Vulnerabilities:**
  Explanation: A complicated web of vendors and suppliers is frequently the foundation of smart manufacturing systems. Attackers may use supply chain weaknesses to install malicious code or jeopardize component integrity if any of these organizations have lax cybersecurity policies.
  Consequences: It includes possible safety risks, a weakened product quality, and supply chain interruptions.
- **Lack of Encryption:**
  Explanation: It is simpler for unauthorized parties to intercept and misuse sensitive information when data is not encrypted, whether it is in transit or at rest. When data is transferred between IoT devices, sensors, and other parts of smart manufacturing systems, this becomes more problematic.
  Consequences: The ramifications include potentially compromised confidentiality, illicit access to sensitive data, and industrial espionage.
- **Inadequate Employee Training:**
  Explanation: Whenever data is not encrypted, either in transit or at rest, it is easier for unauthorized parties to gain access to and misuse sensitive information. This gets more difficult when data is exchanged between sensors, IoT devices, and other components of smart manufacturing systems.
  Consequences: These could lead to industrial espionage, unauthorized access to private information, and possibly compromised confidentiality.
  Organizations should put strong cybersecurity processes in place, periodically train staff, upgrade and patch software, keep an eye out for anomalies in network activity, and use encryption technologies to safeguard critical data in order to reduce the risk of data breaches in smart manufacturing. Preventive and all-encompassing cybersecurity measures are necessary to protect the digital infrastructure of intelligent manufacturing systems.
- **Operational Disruptions:** Production process disruption attacks can cause delays and financial losses [12].

5.2 Security Information and Event Management Systems

- Smart manufacturing frequently uses SIEM systems for correlation of events and centralized log management. They gather and examine log data from a variety of sources, such as network devices and ICS. SIEM systems are able to spot trends and unusual activity that can point to a security breech. Improving the cybersecurity posture of smart manufacturing settings is largely dependent on the use of SIEM systems. SIEM is a comprehensive solution that provides real-time analysis of security alarms and log data from various parts within the manufacturing ecosystem. It integrates security information management (SIM) with security event management

(SEM). The operation of a SIEM system in relation to smart manufacturing security is explained as follows:

- **Log Collection and Aggregation:**

  Explanation: In the context of smart manufacturing, SIEM systems gather and compile log data from various sources. This comprises logs from servers, network devices, IoT devices, ICS, and other parts.

  Importance: By offering a comprehensive perspective of the security environment, centralized log collecting enables security teams to keep an eye on and examine activities throughout the manufacturing infrastructure.
- **Real-Time Analysis:**

  Explanation: SIEM systems collect and aggregate log data from multiple sources in the framework of smart manufacturing. This includes logs from ICS, servers, network devices, IoT devices, and other components.

  Importance: Centralized log collection provides a holistic view of the security environment, allowing security personnel to monitor and investigate activities across the industrial infrastructure.
- **Incident Detection and Alerting:**

  Explanation: Security workers receive alerts and notifications from the SIEM system whenever it detects a possible security event. These notifications offer comprehensive details about the incident's nature, enabling quick inquiry and action.

  Importance: In the context of smart manufacturing, early detection and alerting can reduce the risk of data breaches, system outages, or illegal access.
- **Correlation of Events:**

  Explanation: To give a more thorough knowledge of security occurrences, SIEM systems correlate events from many sources. Correlating network log data with user activity log data, for instance, can highlight trends that can point to a well-planned attack.

  Importance: By spotting intricate assault patterns that might be missed when examining individual events separately, correlation improves the precision of threat detection.
- **Forensic Analysis and Investigation:**

  Explanation: By storing historical log data, SIEM systems make it possible to do forensic analysis on previous security occurrences. Security teams are able to look into the underlying reasons for breaches, comprehend how attacks work, and put precautions in place to stop them from happening again.
- **Importance:** In order to increase an organization's overall cybersecurity resilience, forensic analysis assists with the ongoing enhancement of safety measures and assists in learning from previous occurrences.
- **Compliance Management:**

  Explanation: By offering thorough logs and reports that show adherence to security policies and regulations, SIEM solutions help organizations achieve regulatory compliance requirements. This is especially crucial in sectors like smart manufacturing where adhering to regulations is imperative.

Importance: Complying with regulations prevents negative legal and financial repercussions and guarantees a minimum degree of security for smart manufacturing systems.

- **Integration with Security Infrastructure:**

  Explanation: SIEM systems have the ability to interface with firewalls, antivirus programmes, and intrusion detection/prevention systems, among other security infrastructure elements. A coordinated response to security issues is made easier by this integration, which also improves the security ecosystem as a whole.

  Importance: A strong defence against changing cyberthreats is created by the SIEM system operating in tandem with other security solutions thanks to seamless integration.

With the increasing convergence of IT and operational technology (OT) in smart manufacturing, a SIEM system becomes indispensable for preserving the security and robustness of the manufacturing environment. It offers the visibility, analysis, and response capabilities necessary to secure confidential information, maintain business continuity, and fend off a variety of cyberthreats [13].

5.3 Endpoint Security

Individual devices, including computers, controllers, and sensors, are safeguarded by endpoint security solutions. To keep an eye on and safeguard endpoints from harmful activity, these solutions make use of intrusion detection systems, firewalls, and antivirus software. An essential part of any cybersecurity plan in smart manufacturing is endpoint security. The term "endpoint" refers to hardware, such as computers, servers, ICS, sensors, and other smart manufacturing devices, that is connected to a network. The primary objective of endpoint security is to safeguard these devices from diverse cyber threats in order to guarantee the availability, integrity, and confidentiality of the manufacturing processes. The following outlines the main points of endpoint security in smart manufacturing:

- **Device Protection:** Endpoint security entails putting policies in place to shield specific devices against viruses, illegal access, and other online dangers. Installing firewalls, intrusion detection/prevention systems, and antivirus software on workstations is part of this. In order to stop malware from spreading, illegal access from occurring, and other harmful activity that could jeopardize the performance of smart manufacturing systems, it is imperative to protect devices at the endpoint level.
- **Encryption:** To protect data on smart manufacturing equipment both in transit and at rest, endpoint security frequently uses encryption technologies. This guarantees that private data stays private and is difficult for outside parties to intercept or access. An extra degree of security is provided

by encryption, particularly in settings where data is exchanged between devices or kept on endpoints across the manufacturing ecosystem.

- **Access Control and Authentication:** Smart manufacturing endpoints can only be accessed and interacted with by authorized persons, thanks to the implementation of strong access control measures and multi-factor authentication. This stops malevolent parties or unauthorized users from jeopardizing the integrity of the production processes. For sensitive data to be protected, unwanted access to be stopped, and smart manufacturing systems to remain secure overall, access control and authentication are essential.
- **Patch management:** It is an essential component of endpoint security that involves routinely updating and patching firmware and software on endpoints. By doing this, known vulnerabilities are addressed and the most recent security patches are installed on devices. Patch management is crucial for lowering the possibility that known vulnerabilities will be exploited, thereby decreasing the possibility that cyberattacks on intelligent industrial equipment would be successful.
- **Behaviour Analysis:** To track device behaviour and identify unusual activity, endpoint security solutions frequently include behavioural analysis techniques. A possible security danger may be indicated by unusual patterns of behaviour, which would set off alarms that need to be investigated further. By improving the capacity to identify novel and complex threats that might elude conventional signature-based detection techniques, behavioural analysis offers a proactive approach to endpoint protection.
- **Remote Monitoring and Management:** Endpoints may be remotely monitored and managed in smart manufacturing environments. To guarantee the security of dispersed devices, endpoint security solutions should have capabilities for remote updates, configuration management, and centralized monitoring. Large, globally scattered smart manufacturing plants may manage and secure endpoints more efficiently when they have access to remote capabilities.
- **Incident Response:** Effectively handling security issues is just as important as prevention in endpoint security. The implementation of incident response methods guarantees the timely resolution of any anomalies or security breaches discovered at the endpoint level. Reducing the impact of security incidents, stopping more compromise, and facilitating the recovery of compromised smart manufacturing systems are all made possible by prompt and efficient incident response.

Strong endpoint security is crucial for preserving operational continuity, securing vital assets, and fending off a variety of cyberattacks in the dynamic and networked world of smart manufacturing. When paired with additional cybersecurity measures, a thorough endpoint security approach enhances the overall resilience of smart industrial settings [14].

5.4 Network Segmentation

One proactive method of detecting threats is network segmentation. Manufacturers can restrict the lateral movement of attackers in the event of a breach by segmenting the network into separate areas. It is simpler to identify and isolate threats inside a smaller network section when using this containment method. In smart manufacturing, network segmentation is breaking a network up into smaller, more manageable chunks in order to boost security, increase productivity, and lessen the effect of possible cyberattacks. The objective of this method is to divide the production environment into discrete areas, each with access restrictions and security regulations of its own. The main features and advantages of network segmentation in smart manufacturing are explained as follows:

- **Isolation of Critical Systems:** ICS and other essential systems can be isolated from less important components thanks to network segmentation. This guarantees that, should a single segment be breached, the consequences will be confined to that particular area, hence impeding the dissemination of risks to crucial production procedures, resulting in improved overall system resilience, decreased attack surface, and strengthened defences for vital infrastructure.
- **Enforcing Security Policies:** Access restrictions and security policies can be unique to each network segment. Because of this, managers can implement particular security measures according to the needs and sensitivity of the systems and devices in each segment. It is possible to successfully implement industry-specific security standards compliance, better control over access rights, and customized security measures.
- **Risk Mitigation and Containment:** Network segmentation aids in containing and lessening the effects of security incidents and breaches. Organizations can stop attackers from moving laterally and restrict their potential to breach more areas of the smart manufacturing network by isolating the impacted segments. Reduced possibility of extensive harm, speedier reaction to crises, and simpler containment of security issues.
- **Reduced Attack Surface:** By dividing the network into smaller sections, network segmentation lowers the total attack surface. By limiting the exposure of systems and devices to possible threats, this increases the difficulty of attackers moving laterally within the network. It improves overall cybersecurity posture and reduces the chance of successful attacks and illegal access.
- **Industry Regulation Compliance:** Cybersecurity rules and compliance requirements are unique to several industries, including smart manufacturing. Network segmentation facilitates the implementation of industry-standard methods that enable enterprises to comply with these criteria. It offers less regulatory concerns, simpler compliance management, and conformity with cybersecurity best practices.
- **Resource Optimization:** Segmenting a network makes it possible for businesses to distribute resources more effectively. To guarantee that resources are allocated to their intended uses, distinct segments might be set aside for

administrative tasks, research and development, and production procedures. It guarantees better resource distribution, greater operational effectiveness, and optimized network performance.

- **Granular Access Control:** Network segmentation allows for precise control over user access according to roles and responsibilities. Users can have their access rights customized to only see the data and systems needed to complete their tasks. Less privilege access, a lower chance of unwanted activity, and improved control over user permissions all contribute to improved security.
- **Enhanced Network Performance:** Through congestion reduction and traffic flow optimization, smaller, segmented networks can enhance overall network performance. This is especially crucial in smart manufacturing settings where reduced latency and real-time communication are essential. It results in faster networks, lower latency, and more dependable communication between each section.

A key cybersecurity tactic for smart manufacturing is network segmentation, which offers a proactive means of safeguarding important assets, guaranteeing business continuity, and reducing the dangers brought on by cyberattacks [15].

5.5 Vulnerability Scanning and Patch Management

Frequent vulnerability assessments and patch management are essential for locating and resolving hardware and software security flaws. By detecting weaknesses before attackers take use of them, threat detection can be proactive. Patch management and vulnerability screenings are essential elements of cybersecurity in smart industrial settings. Potential security flaws in the systems and gadgets used in smart manufacturing are found and fixed with the aid of these procedures. Here is a summary of how patch management and vulnerability scanning can be used in the framework of smart manufacturing.

5.5.1 Vulnerability Scanning

Regular Scanning: Regularly check all networks, devices, and systems in the smart manufacturing environment for vulnerabilities. This comprises IoT gadgets, ICS, and additional parts.

Asset Inventory: Keep an accurate list of all the resources in the ecosystem for smart manufacturing. The firmware, software, and hardware components are included in this.

Risk Assessment: Maintain a precise inventory of all the resources available to smart manufacturing ecosystem members. This includes the hardware, software, and firmware components.

Continuous Monitoring: Use continuous monitoring tools to find any novel vulnerabilities that might surface in the future.

Integration with ICS: To guarantee thorough coverage without interfering with production processes, interface vulnerability scanning technologies with ICS.

Compliance Checking: Verify if the smart manufacturing environment complies with applicable cybersecurity standards and laws to make sure it satisfies industry-specific security needs.

5.5.2 Patch Management

Patch Identification: Create a procedure for locating and assessing security updates for all parts, such as firmware, operating systems, and software.

Testing Environment: Test patches in a controlled setting before applying them to make sure they don't cause any new problems or conflicts with already-existing systems.

Prioritization: Set patch deployment priorities according to the risk degree, possible operational impact, and criticality of the vulnerability.

Downtime Planning: To reduce interference with production processes, schedule patches distribution during planned downtime.

Automation: To assure timely updates and simplify the deployment process, use automated patch management systems.

Rollback Procedures: Establish protocols for rollbacks in the event that a patch presents unforeseen problems. This guarantees that, should the need arise; systems can be swiftly restored to a stable state.

Communication: Create effective channels of communication to notify pertinent parties about the updating schedule and any possible effects on operations.

Monitoring and Verification: Following patch distribution, keep an eye on systems to make sure no new vulnerabilities have been discovered introduced and that the fixes were successfully implemented.

5.5.3 General Best Practices

Network Segmentation: By separating important systems from less important ones through network segmentation, you can lessen the possible impact of a security compromise.

User Training: Consistently train staff members in cybersecurity to increase their knowledge of potential risks, such as social engineering scams.

Incident Response Plan: Create and maintain an incident response plan so that security incidents can be quickly addressed and mitigated.

Collaboration with Vendors: Keep up with the most recent security updates and patches for the products used in the production environment by maintaining regular contact with the providers of hardware and software.

Regular Audits: Perform routine cybersecurity audits to evaluate the smart manufacturing environment's overall security posture. Smart manufacturing businesses may strengthen their cybersecurity defences and lower the likelihood that cyberattacks will negatively affect their operations by putting in place thorough vulnerability detection and patch management procedure [16].

5.6 Security Training and Awareness

One typical factor in security incidents is human mistake. Early detection and mitigation can be aided by educating staff members and outside contractors about cybersecurity best practices and possible risks.

5.7 Collaboration and Information Sharing

Manufacturers should exchange threat intelligence, work together with other industry players, and stay up to date on new attack vectors and dangers. Early detection and defence against novel and emerging dangers can be aided by this body of common knowledge. In smart manufacturing, threat detection is a continuous process. Manufacturers ought to regularly evaluate their security posture, update their defences against risks, and adjust as new ones arise. Threat detection is essential to preserving the production environment's operational integrity and security in smart manufacturing. Manufacturers can reduce the likelihood of significant interruptions and data breaches by detecting and mitigating risks in real-time through the implementation of advanced technologies, processes, and best practices [17].

6 CONCLUSION

Thus, enhancing the cybersecurity posture of smart industrial settings is greatly dependent on enhanced threat detection. The chance of sophisticated cyber assaults increases as these ecosystems become more dependent on digital technologies and more linked. To quickly detect and address emerging threats, it is imperative to put strong advanced threat detection technologies into place. Smart manufacturing facilities can improve their capacity to identify and mitigate cyber risks in real-time by combining AL and ML algorithms with techniques like anomaly detection, behaviour analytics, and threat intelligence. A proactive defence approach includes constant monitoring, frequent updates to detecting systems, and cooperation with cybersecurity specialists. Protecting sensitive data, ensuring business continuity, and upholding the general integrity of the manufacturing procedure all depend on a proactive and flexible approach to threat detection in the dynamic world of smart manufacturing.

REFERENCES

1. Ashenden, S. K., Bartosik, A., Agapow, P.-M., & Semenova, E. (2021). Introduction to artificial intelligence and machine learning. In *The Era of Artificial Intelligence, Machine Learning, and Data Science in the Pharmaceutical Industry*, Academic Press (pp. 15–26), ISBN 9780128200452, https://doi.org/10.1016/B978-0-12-820045-2.00003-9
2. Cioffi, R., Travaglioni, M., Piscitelli, G., Petrillo, A., & De Felice, F. (2020). Artificial intelligence and machine learning applications in smart production: Progress, trends, and directions. *Sustainability* 12(2): 492. https://doi.org/10.3390/su12020492
3. Dietterich, T.G., Michalski, R.S. (1983). A Comparative Review of Selected Methods for Learning from Examples. In: Michalski, R.S., Carbonell, J.G., Mitchell, T.M. (eds) *Machine Learning. Symbolic Computation*. Springer, Berlin, Heidelberg. https://doi.org/10.1007/978-3-662-12405-5_3
4. Neapolitan, R. E., & Jiang, X. (2018). Unsupervised learning and reinforcement learning. In Artificial Intelligence (2nd ed., p. 17). Chapman and Hall/CRC. https://doi.org/10.1201/b22400 (Pages: 480, eBook ISBN: 9781315144863)

5. Kusiak, A. (2023). Smart Manufacturing. In: Nof, S.Y. (eds) *Springer Handbook of Automation. Springer Handbooks.* Springer, Cham. https://doi.org/10.1007/978-3-030-96729-1_45.
6. Cioffi, R., et al. (2020). Smart manufacturing systems and applied industrial technologies for a sustainable industry: A systematic literature review. *Applied Sciences*, 10(8): 2897.
7. Balamurugan, E., Flaih, L. R., Yuvaraj, D., Sangeetha, K., Jayanthiladevi, A., & Kumar, T. S. (2019, December). Use case of artificial intelligence in machine learning manufacturing 4.0. In *2019 International Conference on Computational Intelligence and Knowledge Economy (ICCIKE)* Dubai, United Arab Emirates, (pp. 656–659). IEEE. doi: 10.1109/ICCIKE47802.2019.9004327.
8. Kotsiopoulos, T., Sarigiannidis, P., Ioannidis, D., & Tzovaras, D. (2021). Machine learning and deep learning in smart manufacturing: The smart grid paradigm. *Computer Science Review*, 40: 100341.
9. Sharp, M., Ronay A. k., & Hedberg, T. (2018). A survey of the advancing use and development of machine learning in smart manufacturing. *Journal of Manufacturing Systems*, 48(Part C): 170–179, ISSN 0278-6125, https://doi.org/10.1016/j.jmsy.2018.02.004
10. Jbair, M., Ahmad, B., Maple, C., & Harrison, R. (2022). Threat modelling for industrial cyber physical systems in the era of smart manufacturing. *Computers in Industry*, 137: 103611.
11. Hashim, N. A., Abidin, Z. Z., Abas, Z. A., Zakaria, N. A., Ahmad, R., & Mardaid, E. (2020, May). Enhanced approach of risk assessment for insider threats detection in smart manufacturing. In *IOP Conference Series: Materials Science and Engineering* (Vol. 864, No. 1, p. 012053). IOP Publishing. https://doi.org/10.1088/1757-899X/864/1/012053
12. Bracho, A., Saygin, C., Wan, H., Lee, Y., & Zarreh, A. (2018). A simulation-based platform for assessing the impact of cyber-threats on smart manufacturing systems. *Procedia Manufacturing*, 26: 1116–1127.
13. AbuEmera, E. A., ElZouka, H. A., & Saad, A. A. (2022, January). Security framework for identifying threats in smart manufacturing systems using STRIDE approach. In *2022 2nd International Conference on Consumer Electronics and Computer Engineering (ICCECE)* Guangzhou, China, (pp. 605–612). IEEE. doi: 10.1109/ICCECE54139.2022.9712770.
14. Tuptuk, N., & Hailes, S. (2018). Security of smart manufacturing systems. *Journal of Manufacturing Systems*, 47: 93–106.
15. Ren, A., Wu, D., Zhang, W., Terpenny, J., & Liu, P. (2017). Cyber security in smart manufacturing: Survey and challenges. In H. B. Nembhard, K. Coperich, & E. Cudney (Eds.), *67th Annual Conference and Expo of the Institute of Industrial Engineers 2017* (pp. 716-721). Institute of Industrial Engineers.
16. Maggi, F., Balduzzi, M., Vosseler, R., Rösler, M., Quadrini, W., Tavola, G., …, & Zanero, S. (2021). Smart factory security: A case study on a modular smart manufacturing system. *Procedia Computer Science*, 180: 666–675.
17. Abuhasel, K. A., & Khan, M. A. (2020). A secure industrial internet of things (IIoT) framework for resource management in smart manufacturing. *IEEE Access*, 8: 117354–117364.

7 Integrating Cybersecurity Threats into Smart Manufacturing

Best Practices and Frameworks

G. Ramya and K. G. Srinivasagan

1 INTRODUCTION

Smart manufacturing is a cutting-edge approach that integrates advanced technologies to optimize and streamline the manufacturing process. It establishes an integrated and smart manufacturing infrastructure by utilizing automation, big data analytics, and the internet of things (IoT). In a smart manufacturing environment, sensors and devices are embedded throughout the production line, collecting real-time data on various parameters. The particular data will then be analysed using artificial intelligence (AI) algorithms to gain insights into the production process, identify inefficiencies, and make data-driven decisions. Automation plays a crucial role in smart manufacturing, with robotic systems and autonomous machines handling repetitive and labour-intensive tasks. The seamless connectivity between different components of the manufacturing system allows for enhanced coordination, improved efficiency, and better responsiveness to changing demands. All things considered, smart manufacturing signifies a revolutionary change towards production processes that are more flexible, effective, and adaptive in order to satisfy the demands of a world market that is changing quickly.

Smart manufacturing is crucial in today's industrial landscape for several compelling reasons. Firstly, it significantly enhances efficiency and productivity. By integrating advanced technologies like IoT, AI, and automation, smart manufacturing optimizes processes, reduces downtime, and streamlines operations. Real-time data collection and analysis allow for quicker decision-making, minimizing delays and improving overall production output.

Secondly, smart manufacturing contributes to better resource utilization. Through data-driven insights, manufacturers can optimize the use of raw materials, energy, and other resources, leading to cost savings and reduced environmental impact.

 DOI: 10.1201/9781032694375-7

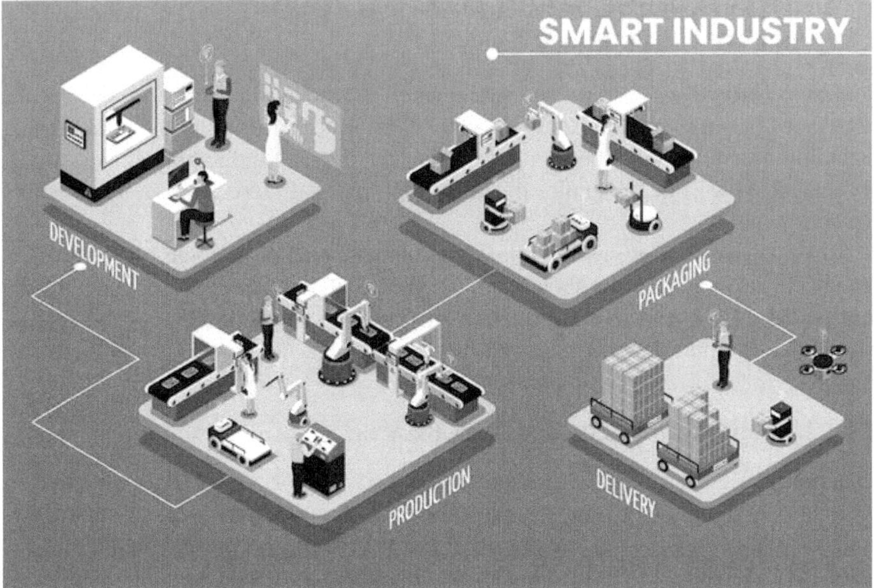

FIGURE 7.1 Development of a Smart Industry Platform.

This efficiency is essential in a world where sustainable practices are increasingly becoming a priority.

Thirdly, smart manufacturing enhances product quality and customization. The ability to monitor and control the production process in real-time ensures consistent product quality, reducing defects and waste. Additionally, the flexibility provided by smart manufacturing systems allows for more personalized and customized production to meet the diverse demands of the market.

The integration of data analytics and AI enables predictive maintenance, minimizing equipment failures and downtime, further contributing to overall reliability.

Furthermore, in a globalized and interconnected world, smart manufacturing facilitates better collaboration and communication across the supply chain. It enables seamless coordination between different stages of production, suppliers, and distributors, leading to improved supply chain visibility and responsiveness to market changes.

In summary, the adoption of smart manufacturing is imperative for businesses aiming to stay competitive, reduce costs, enhance product quality, and contribute to sustainable and innovative practices in the ever-evolving landscape of modern manufacturing.

In smart manufacturing as shown in Figure 7.1, the confluence of information technology (IT) and operational technology (OT) has expanded the attack surface, making vital processes vulnerable to cyberattacks. This chapter delves into the unique cybersecurity considerations for smart manufacturing and provides actionable strategies for safeguarding against evolving threats.

2 UNDERSTANDING SMART MANUFACTURING CYBERSECURITY RISKS

To develop effective cybersecurity strategies, it is crucial to understand the specific risks faced by smart manufacturing systems. This section identifies potential threats, including unauthorized access, malware, insider threats, and supply chain vulnerabilities. Real-world case studies illustrate the impact of these risks on manufacturing operations.

Smart manufacturing, while offering numerous benefits, also introduces a range of cybersecurity threats that should be addressed so that the integrity, confidentiality, and availability of sensitive information and critical systems. Some key cybersecurity threats in smart manufacturing include.

2.1 Unauthorized Access and Data Breaches

Unauthorized Access: Weak Authentication and Authorization: If authentication mechanisms are weak or poorly implemented, malicious actors may exploit vulnerabilities so that they gain unauthorized access to critical systems and data.

Insufficient Access Controls: Inadequate access controls may result in employees or external actors having more privileges than necessary, increasing the risk of unauthorized access.

Data Breaches: Smart manufacturing often involves the use of advanced technologies and proprietary processes. Unauthorized access of the data can lead to the theft of intellectual property, compromising a company's competitive advantage.

Sensitive Operational Data Exposure: Breaches may expose sensitive operational data, such as production schedules, quality control measures, and supply chain information. This can be exploited by competitors or used for sabotage.

Customer and Employee Information Exposure: If manufacturing systems store personal or confidential information about customers or employees, a breach could lead to privacy violations, identity theft, or other legal and reputational issues.

2.2 Malware and Ransomware Attacks

Malware and ransomware attacks pose significant threats to smart manufacturing environments, where the integration of digital technologies and interconnected systems creates a larger attack surface. Here are key considerations related to malware and ransomware attacks in smart manufacturing:

Disruption of Operations: Malware and ransomware attacks can lead to disruptions in production, causing downtime and financial losses. In a smart manufacturing setting, where processes are highly automated, interruptions can have severe consequences.

Data Theft and Espionage: Malicious actors may use malware to steal sensitive intellectual property, proprietary processes, or other valuable information critical to a manufacturing company's competitive advantage.

Ransomware Extortion: Ransomware encrypts critical files and demands payment for their release. If successful, this can result in significant financial losses, reputational damage, and potential legal consequences.

Supply Chain Disruptions: Malware attacks can extend to suppliers and partners, disrupting the broader supply chain. Compromised components or services may lead to cascading effects on production and delivery schedules.

Compromised IoT Devices: Malware can exploit vulnerabilities in connected IoT devices within the manufacturing environment, potentially leading to unauthorized access or manipulation of critical systems.

Data Integrity Risks: Malware attacks may manipulate or corrupt operational data, leading to errors in manufacturing processes, quality control issues, and potential safety hazards.

Phishing and Social Engineering: Malware often infiltrates systems through phishing emails or social engineering attacks targeting employees. Education and awareness programmes are crucial for mitigating these risks.

Inadequate Patch Management: Malware exploits vulnerabilities in software and systems. Failure to regularly update and patch systems increases the risk of successful malware attacks.

Loss of Confidentiality: Malware attacks may result leading to privacy violations and potential legal consequences [1].

Operational Technology (OT) Security Risks: Malware specifically designed for industrial control systems (ICS) can lead to disruptions in manufacturing processes, equipment damage, and safety hazards. Smart manufacturing systems are susceptible to malware and ransomware attacks, which can disrupt operations, damage equipment, or encrypt critical data. Ransomware attacks, in particular, can lead to significant financial losses if manufacturers are forced to pay to regain control of their systems.

2.3 IoT Device Vulnerabilities

The proliferation of IoT devices in smart manufacturing introduces a large attack surface. Weaknesses in the security of connected sensors, actuators, and other devices can be exploited to manipulate data, disrupt operations, or gain unauthorized control over machinery.

Default Credentials: Manufacturers often ship IoT devices with default usernames and passwords. If these credentials are not changed during installation, malicious actors can easily gain unauthorized access.

Unencrypted Communication: Inadequate or absent encryption for data transmitted between IoT devices and the central network can expose sensitive information to interception and manipulation.

Outdated Software: IoT devices may run on outdated or unpatched software, leaving them vulnerable to known exploits. There should be regular software updates and the patch management are crucial for addressing these vulnerabilities [2].

Weak Authentication Protocols: Weak or poorly implemented authentication mechanisms may be exploited by attackers to gain unauthorized control over IoT devices.

Insufficient Access Controls: Without proper access controls, unauthorized users may manipulate IoT devices, disrupt operations, or compromise the integrity of the manufacturing process.

Physical Security Concerns: In some cases, physical access to IoT devices might be possible. If devices are not tamper-resistant, attackers could manipulate them physically to compromise security.

Device Identity Spoofing: Attackers may attempt to impersonate legitimate IoT devices by spoofing their identities, permitting them to enter the network without authorization.

Insecure Interfaces: Many IoT devices have web interfaces for configuration and management. If these interfaces have security vulnerabilities, they could be exploited by attackers to gain control of the device.

Limited Resources for Security Measures: some IoT devices in manufacturing settings could have low processing and memory capacities, it can be difficult to put strong security measures in place.

Lack of Standardization: The IoT landscape in smart manufacturing often involves a variety of devices from different manufacturers, each with its own specifications and security measures. Lack in standardization can bring it difficult to enforce consistent security practices.

2.4 Supply Chain Vulnerabilities

Cybersecurity threats can extend beyond a company's internal network to its supply chain. Malicious actors may target suppliers or compromise the integrity of components, leading to potential security breaches or the introduction of compromised hardware into the manufacturing process [3]. Supply chain vulnerabilities are a critical concern in smart manufacturing, where the reliance on interconnected networks and the integration of various components from different suppliers can introduce potential risks. Identifying and addressing supply chain vulnerabilities is essential to ensure the security, reliability, and integrity of smart manufacturing processes. Here are key aspects to consider:

Third-Party Suppliers: Suppliers may have varying levels of cybersecurity measures in place.

Component Integrity: The use of counterfeit or compromised components in manufacturing processes can lead to vulnerabilities.

Software and Firmware: Malicious actors may compromise software or firmware during the manufacturing process.

Communication Channels: Vulnerabilities in communication channels between different supply chain entities can be exploited.

Logistics and Transportation: Ensuring the physical security of components during transportation is crucial.

Supplier Access Management: Suppliers may have access to certain systems or data. Implementing strong access controls and monitoring mechanisms can prevent unauthorized access by suppliers.

Supply Chain Visibility: Limited visibility into the entire Supply chains might make it difficult to pinpoint potential vulnerabilities.

Regulatory Compliance: Suppliers may not comply with relevant cybersecurity regulations, exposing the entire supply chain to legal and regulatory risks.

Data Integrity: Tampering with data during transit or within the supply chain can lead to integrity issues.

Business Continuity Planning: Relying on a limited number of suppliers for critical components can pose a risk to business continuity.

2.5 Insider Threats

Insider threats in smart manufacturing pose a significant risk, as employees, contractors, or other individuals with insider access to systems and sensitive information may intentionally or unintentionally compromise the security and integrity of the manufacturing process. Here are key aspects to consider when addressing insider threats in smart manufacturing:

Unauthorized Access and Data Theft: Insiders may exploit their legitimate access to systems and data for unauthorized purposes, such as stealing intellectual property or sensitive operational information.

Sabotage and Malicious Activities: Employees with insider knowledge may intentionally sabotage manufacturing processes, leading to disruptions, equipment damage, or compromise of product quality.

Accidental Data Exposure: Even well-intentioned employees can inadvertently expose sensitive information, such as operational details or proprietary processes, through mistakes like misconfigurations or sending sensitive data to the wrong recipients.

Phishing and Social Engineering: Insiders may fall victim to phishing attacks or social engineering, leading to the compromise of their credentials or unwitting participation in cyber threats.

Disgruntled Employees: Disgruntled employees may pose a higher risk of insider threats. Monitoring employee morale and addressing workplace concerns can help mitigate this risk.

Inadequate Training: Insiders may not be fully aware of the cybersecurity risks and best practices. Employees might benefit from regular training and awareness programmes to recognize and avoid potential threats.

Insider Collaboration with External Threat Actors: Insiders may collaborate with external threat actors, providing them with insider knowledge and access to exploit vulnerabilities in the manufacturing environment.

Inadequate Access Controls: Inadequate access controls may allow insiders to access systems or information beyond their role requirements, increasing the risk of misuse.

Monitoring and Detection Challenges: Insider threats can be challenging to detect, as the activities may not trigger typical security alerts. Implementing advanced monitoring and anomaly detection tools is crucial.

Employee Turnover: Former employees with retained access credentials may still pose a threat. Implementing effective deprovisioning procedures upon employee departure is essential. Employees or contractors with access to sensitive systems and information can pose a significant risk. Insider threats may involve intentional or unintentional actions that compromise cybersecurity, such as sharing credentials, falling victim to phishing attacks, or intentionally sabotaging systems. [3]

2.6 Lack of Standardization

The lack of standardization in smart manufacturing introduces various challenges and threats that can impact interoperability, cybersecurity, and overall efficiency.

Here are some key considerations related to the threats arising from the absence of standardization:

Interoperability Issues: Without standardized communication protocols and data formats, devices and systems from different manufacturers may struggle to interoperate seamlessly, leading to inefficiencies and potential disruptions in the manufacturing process.

Security Concerns: In the absence of standardized security measures, there may be inconsistencies in how different components and devices handle security. This lack of uniformity can lead to vulnerabilities and make it challenging to implement a cohesive security strategy.

Integration Challenges: The integration of diverse technologies and systems becomes more complex when there are no standardized interfaces or frameworks. This complexity can result in delays, errors, and increased costs.

Data Inconsistency: The absence of standardized data formats can lead to data inconsistency and interoperability issues. Harmonizing data across different systems becomes challenging, impacting decision-making and analytics.

Vendor Lock-In: Lack of standardization may result in a situation where manufacturers become dependent on specific vendors or proprietary technologies. This can limit flexibility and hinder the ability to choose the best solutions for evolving needs [1]

Increased Costs: Non-standardized solutions often require customization and bespoke integration efforts, driving up costs for manufacturers. Standardization can help reduce development and implementation expenses.[2]

Limited Innovation: The absence of standardized frameworks may impede innovation, as manufacturers may hesitate to invest in new technologies if they are unsure about their compatibility with existing systems.

Regulatory Compliance Challenges: Lack of standardization may result in difficulties in meeting regulatory requirements consistently. Adhering to diverse standards across different regions or industries can be burdensome.

Quality Control Issues: Non-standardized processes may lead to inconsistencies in manufacturing, affecting product quality and making it challenging to implement effective quality control measures.

Supply Chain Risks: Lack of standardization can extend to the supply chain, leading to fragmentation and potential vulnerabilities in the flow of materials and information. The absence of standardized cybersecurity practices across smart manufacturing systems can create challenges. Inconsistencies in security measures may lead to vulnerabilities that attackers can exploit.

2.7 Denial of Service Attacks

Denial of Service (DoS) attacks become a serious threat to smart manufacturing systems, where disruptions in operations can have significant consequences. Here are key considerations related to DoS attacks in smart manufacturing:

Disruption of Operations: DoS attacks aim to overwhelm or disable systems, leading to service unavailability. In smart manufacturing, this can disrupt critical processes, production lines, and control systems.

Loss of Productivity: Smart manufacturing relies on continuous and efficient operations. DoS attacks can result in extended downtime, leading to a loss of productivity and revenue.

Impact on Supply Chain: Smart manufacturing is often part of a larger supply chain. Disruptions caused by DoS attacks can have a cascading effect, affecting suppliers, distributors, and customers.

Financial Losses: After a DoS assault, recovery may be expensive. Businesses might have to make investments in more security measures, carry out forensic investigations, and handle the financial fallout from operations disruptions.

Data Loss and Corruption: DoS attacks can sometimes be used as a distraction or cover for other malicious activities, such as data theft or corruption. This can lead to long-term consequences for data integrity and security.

Reputation Damage: Persistent or high-profile DoS attacks can damage a manufacturer's reputation. Customers may lose confidence in the ability of the company to secure its operations and deliver products consistently.

Difficulty in Attribution: Attributing DoS attacks to specific individuals or groups can be challenging. This lack of attribution may hinder the ability to take legal action or implement targeted security measures. [4]

Variety of Attack Vectors: DoS attacks can take various forms, including volumetric attacks that flood networks with traffic, protocol attacks that exploit vulnerabilities in communication protocols, and application layer attacks targeting specific applications or services.

IoT Device Exploitation: The use of compromised IoT devices in botnets can amplify the scale of DoS attacks. Smart manufacturing environments, with numerous connected devices, may be vulnerable to such attacks.

Inadequate Mitigation Measures: Some smart manufacturing systems may not have adequate measures in place to detect and mitigate DoS attacks. Proactive planning and the implementation of robust security measures are essential. Attackers may attempt to overwhelm smart manufacturing systems with excessive traffic, leading to a DoS. This can disrupt operations, causing downtime and financial losses.

3 BEST PRACTICES AND FRAMEWORKS

Navigating the complex landscape of cybersecurity frameworks and standards is a key aspect of developing robust strategies. This section examines prominent frameworks such as the NIST Cybersecurity Framework, ISO/IEC 27001, and industry-specific guidelines. Practical insights are provided on aligning these frameworks with the unique requirements of smart manufacturing environments.

3.1 Unauthorized Access and Data Breaches

3.1.1 Best Practices

- Unauthorized access and data breaches are serious cybersecurity concerns, and implementing best practices and frameworks is crucial to preventing and mitigating these threats.

- Strong Authentication and Access Controls: Implement multi-level authentication (MLA) to implement an extra layer of security. Use role-based access controls (RBAC) to ensure that users have the minimum necessary privileges for their roles.
- Regular Employee Training and Awareness: Provide staff with cybersecurity awareness training to inform them about phishing, potential dangers, and the value of password protection.
- Encryption: Encrypt the sensitive safeguarding data while it's in transit and at rest to prevent unwanted access, even if the network or storage is compromised.
- Regular Security Audits and Assessments: The industries should conduct the regular security audits, which is to identify and address vulnerabilities in the system. Perform penetration testing to identify potential weaknesses in the security infrastructure.
- Incident Response Plan: To guarantee a prompt and efficient reaction, create and test an incident response strategy on a regular basis in the event of unauthorized access or a data breach.
- Network Segmentation: Segment the network to restrict lateral movement in case of a security breach. Isolate critical systems from less critical ones to contain potential security incidents.
- Continuous Monitoring: Use security information and event management (SIEM) systems to monitor the network activity and detect anomalies or suspicious behaviour.
- Access Monitoring and Logging: Implement comprehensive access monitoring and logging to track user activities and identify unauthorized access attempts. [5]
- Vendor and Third-Party Security: Assess and keep updated on the security procedures used by outside suppliers and service providers. Make sure outside parties follow the same security guidelines as company.
- Data Classification: Classify the data based on sensitivity, and apply appropriate security measures accordingly. Limit access to sensitive information to only those who require it for their roles.

3.1.2 Frameworks

- NIST Cybersecurity Framework: The National Institute of Standards and Technology (NIST) Cybersecurity Framework provides a comprehensive set of guidelines and best practices for managing and improving an organization's cybersecurity risk.
- ISO/IEC 27001: A globally recognized framework for information security management systems (ISMS) is provided by this standard. It offers a methodical way to handling confidential business data.
- Center for Internet Security (CIS) Critical Security Controls: The CIS Critical Security Controls is a set of best practices that prioritize and focus on a small number of action items that can be highly effective in preventing and mitigating unauthorized access and data breaches.

- Control Objectives for Information and Related Technologies: The COBIT framework facilitates the creation, application, oversight, and enhancement of IT governance and management procedures. It offers a collection of information security best practices and procedures.
- Payment Card Industry Data Security Standard (PCI DSS): If your organization handles payment card data, compliance with PCI DSS is essential. It provides a framework for securing payment card transactions and preventing unauthorized access to cardholder data.
- General Data Protection Regulation (GDPR): GDPR is a regulation that sets guidelines for the protection of personal data. It includes requirements for data breach notification and measures to ensure the security of personal information.
- Cybersecurity Maturity Model Certification (CMMC): CMMC is a framework developed by the U.S. Department of Defense (DoD) to ensure that contractors have adequate cybersecurity measures in place to protect sensitive information.

Adopting these best practices and frameworks can help organizations build a robust defense against unauthorized access and data breaches, fostering a proactive and comprehensive approach to cybersecurity. Regular updates and adaptations to the security strategy based on emerging threats and technological advancements are essential for maintaining effectiveness over time.

3.2 Malware and Ransomware Attacks

Addressing malware and ransomware attacks in smart manufacturing requires a combination of best practices and adherence to established frameworks. Here are recommended practices and frameworks to enhance cybersecurity in the context of malware and ransomware attacks.

3.2.1 Best Practices

- Endpoint Protection: Deploy robust endpoint protection solutions that include antivirus, anti-malware, and behaviour-based detection mechanisms. Keep endpoint security software updated to defend against the latest threats.
- Regular Software Updates and Patch Management: Implement a proactive patch management process to ensure that operating systems, software, and firmware are regularly updated with the latest security patches.
- Employee Training and Awareness: Conduct regular cybersecurity training for employees, emphasizing the dangers of phishing, social engineering, and the importance of reporting suspicious activities.
- Network Segmentation: Network segmentation is the segmenting the network to limit the spread of malware in case of a breach. Isolate critical systems from less critical ones to contain potential security incidents.

- Data Backups: Regularly backup critical data and ensure that backup systems are secure. This can facilitate a quicker recovery in the event of a ransomware attack.
- Incident Response Plan: Create and test an incident response plan specifically tailored for malware and ransomware incidents. Ensure that the plan includes steps for containment, eradication, and recovery.
- Application Whitelisting: Use application whitelisting to control which applications are allowed to run on systems, preventing the execution of unauthorized or malicious software.
- Network Monitoring and Anomaly Detection: Implement network monitoring tools and anomaly detection systems to identify unusual patterns of behaviour that may indicate a malware or ransomware attack.
- Privilege Management: Implement the standard of least privilege to restrict user access to the minimum level necessary for their roles. Limiting administrative privileges can prevent the spread of malware.
- Collaboration and Information Sharing: Collaborate with industry peers, cybersecurity experts, and law enforcement to share threat intelligence and stay informed about emerging malware and ransomware threats.

3.2.2 Frameworks

- NIST Cybersecurity Framework: The NIST Cybersecurity Framework provides a comprehensive set of guidelines for managing and improving an organization's cybersecurity risk, including measures to prevent and respond to malware and ransomware attacks.
- ISO/IEC 27001: It is an international standard for ISMS and offers a methodical way to handle security risks like ransomware and malware and manage sensitive data.
- CIS Critical Security Controls: The CIS Critical Security Controls is a set of best practices that includes specific measures for securing systems against malware and ransomware.
- MITRE ATT&CK Framework: Adversarial Tactics, Techniques, and Common Knowledge (MITRE ATT&CK) provides a comprehensive knowledge base of adversary tactics and techniques. It can be used to enhance detection and response capabilities against malware and ransomware.
- Cybersecurity Maturity Model Certification (CMMC): CMMC is a framework developed by the U.S. DoD that includes specific controls and practices for protecting against malware and ransomware attacks, especially in the context of the defense industrial base.
- NIST SP 800-53: A list of security controls, such as those pertaining to malware and ransomware, can be customized to handle particular cybersecurity concerns, as detailed in NIST Special Publication 800-53.
- SANS Critical Security Controls: The SANS Institute provides a set of Critical Security Controls that cover various aspects of cybersecurity, offering practical guidance for protecting against malware and ransomware.

- CISO Handbook: The CISO Handbook by Carnegie Mellon University's Software Engineering Institute offers guidance on managing cybersecurity risks, including strategies to prevent and respond to malware and ransomware attacks.

By incorporating these best practices and leveraging established frameworks, smart manufacturing organizations can strengthen their cybersecurity posture, reduce the risk of malware and ransomware attacks, and enhance their ability to respond effectively to security incidents. Regular updates and continuous improvement based on emerging threats and technological advancements are crucial for maintaining a resilient security strategy.

3.3 IOT Device Vulnerabilities

Securing IoT devices in smart manufacturing is crucial to prevent vulnerabilities that could be exploited by malicious actors. Here are best practices and frameworks to enhance the security of IoT devices in smart manufacturing.

3.3.1 Best Practices

- Security by Design: Integrate security features into the design phase of IoT devices to ensure that security is a fundamental aspect throughout the device's lifecycle.
- Device Authentication: Implement strong authentication mechanisms for IoT devices, including unique credentials and multi-factor authentication.
- Encryption: To prevent unwanted access to sensitive information, encrypt data both while it's in transit and when it's at rest.
- Regular Software Updates: Establish a system for automatic software updates and patches to quickly fix security flaws.
- Network Segmentation: To reduce the possible impact of a compromised device on the larger network, segment IoT devices into separate networks.
- Device Monitoring and Logging: Enable robust monitoring and logging capabilities on IoT devices to detect and respond to anomalous activities.
- Access Controls: Implement strict access controls to ensure that only authorized individuals or systems can interact with IoT devices.
- Vulnerability Assessments: To identify and address security flaws in IoT devices, conduct vulnerability assessments on a regular basis.
- Physical Security Measures: Implement physical security measures to prevent unauthorized physical access to IoT devices, reducing the risk of tampering.
- Secure Boot and Firmware Validation: Implement secure boot processes to ensure that only authenticated firmware is executed on the device.
- Supply Chain Security: In order to stop the introduction of hacked devices, evaluate and guarantee the security of the complete supply chain, from production to implementation.
- IoT Device Lifecycle Management: Develop a comprehensive lifecycle management strategy for IoT devices, including secure provisioning, maintenance, and decommissioning processes.

- Collaboration with Manufacturers: Collaborate with IoT device manufacturers to understand their security practices and encourage the adoption of security standards.
- Regulatory Compliance: Stay informed about and comply with relevant IoT security regulations and standards applicable to your industry.

3.3.2 Frameworks

- IoT Security Foundation's Best Practices: The IoT Security Foundation provides a set of best practices and guidelines for securing IoT devices, covering aspects such as architecture, connectivity, and device management.
- IoT Security Compliance Framework: The Open Web Application Security Project (OWASP) established the IoT Security Compliance Framework, which contains a checklist for evaluating compliance and offers guidelines for creating secure IoT ecosystems.
- Industrial Internet Consortium (IIC) Security Framework: IIC provides a Security Framework that addresses security concerns in industrial IoT, offering guidelines for device and system security.
- ISO/IEC 27001 and ISO/IEC 27019: ISO/IEC 27001 is an international standard for ISMS. ISO/IEC 27019 provides sector-specific guidelines for securing IoT devices in the energy industry, which may be applicable to smart manufacturing.
- NISTIR 8259A: IoT Device Cybersecurity Capability Core Baseline: The NIST provides guidelines for improving the cybersecurity of IoT devices through its NISTIR 8259A document.
- Connected Consumer Products Baseline Profile: The U.S. Consumer Product Safety Commission (CPSC) offers a baseline profile for connected consumer products, providing recommendations for securing IoT devices.
- ENISA Baseline Security Recommendations for IoT: The European Union Agency for Cybersecurity (ENISA) provides baseline security recommendations for IoT devices, covering various aspects of security.
- CIS Critical Security Controls: The CIS provides a set of Critical Security Controls that includes specific controls for securing IoT devices.
- By implementing these best practices and leveraging established frameworks, smart manufacturing organizations can enhance the security of their IoT devices, reduce the risk of vulnerabilities, and contribute to the overall resilience of their digital infrastructure. Regular updates and continuous improvement based on emerging threats and technological advancements are essential for maintaining a strong security posture.

3.4 Supply Chain Vulnerabilities

Addressing supply chain vulnerabilities in smart manufacturing is crucial for maintaining the integrity, security, and reliability of the manufacturing processes. Here are best practices and frameworks to enhance supply chain security in the context of smart manufacturing.

3.4.1 1Best Practices

- Supplier Risk Assessment: Conduct regular risk assessments of suppliers to evaluate their cybersecurity practices, financial stability, and overall reliability.
- Security Audits: Perform security audits throughout the supply chain to identify and address vulnerabilities in the manufacturing process.
- Contractual Security Requirements: Add clauses about cybersecurity standards, data protection procedures, and reporting responsibilities in the event of a security incident to supplier contracts.
- Component Integrity: Verify the authenticity of components to prevent the use of counterfeit or compromised materials that could introduce vulnerabilities.
- Software and Firmware Security: Ensure the integrity of software and firmware by implementing secure coding practices and validating the sources of code used in manufacturing processes.
- Communication Security: Use secure communication protocols to protect data transmitted between different components in the supply chain.
- Logistics Security: Implement security measures during transportation to prevent tampering and ensure the physical integrity of components and products.
- Access Controls for Suppliers: Implement strict access controls for supplier systems and data to prevent unauthorized access.
- Supply Chain Visibility: Enhance visibility into the entire supply chain to monitor and manage the flow of materials, components, and information effectively.
- Regulatory Compliance: Stay informed about and complies with relevant cybersecurity regulations and standards applicable to the supply chain.
- Data Integrity Protection: Implement measures to ensure the integrity of data throughout the supply chain, preventing manipulation or tampering.
- Business Continuity Planning: Develop business continuity plans that account for supply chain disruptions, ensuring the ability to adapt to unforeseen challenges.
- Diversification of Suppliers: Diversify suppliers to reduce dependency on a single source, minimizing the impact of disruptions.
- Continuous Monitoring: Implement continuous monitoring tools to detect anomalies, suspicious activities, or deviations from established security practices within the supply chain.

3.4.2 Frameworks

- NIST Cybersecurity Framework: The NIST Cybersecurity Framework provides guidelines for managing and improving organizational cybersecurity risk, which can be applied to supply chain security.
- ISO 28000: It is an international standard for supply chain security management systems, providing guidelines for implementing security measures throughout the supply chain.

- CIS Critical Security Controls: The CIS provides a set of Critical Security Controls that includes controls specific to supply chain security.
- NIST SP 800-161: Supply chain risk management is the subject of NIST Special Publication 800-161, which offers recommendations for handling cybersecurity threats in the supply chain.
- Transported Asset Protection Association (TAPA) Standards: TAPA provides standards for securing supply chain assets, including guidelines for secure transportation and logistics.
- Customs-Trade Partnership Against Terrorism (CTPAT): U.S. Customs and Border Protection oversees the voluntary CTPAT supply chain security programme, which provides best practices and recommendations for safeguarding global supply chains.
- OECD Guidelines for Ethical and Reputable Supply Chains: The Organisation for Economic Co-operation and Development (OECD) offers recommendations on ethical and security-related business behaviour in international supply chains.
- Industrial Internet Consortium (IIC) Supply Chain Security Best Practices Guide: IIC provides a guide for securing the industrial internet supply chain, offering practical recommendations for manufacturers.

By implementing these best practices and leveraging established frameworks, smart manufacturing organizations can strengthen the security of their supply chains, reduce the risk of vulnerabilities, and enhance overall resilience. Regular assessments, updates to security measures, and collaboration with suppliers are essential components of a robust supply chain security strategy.

3.5 Insider Threats

Insider threats in smart manufacturing, whether intentional or unintentional, can pose significant risks to the security and integrity of operations. Implementing best practices and frameworks is essential to mitigate these threats effectively. Here are recommendations for addressing insider threats in smart manufacturing.

3.5.1 Best Practices

- Employee Education and Awareness: Hold frequent training sessions to inform staff members about the possible dangers posed by insider threats. Stress the value of data security and technology usage in a responsible manner.
- Role-Based Access Controls (RBAC): Implement RBAC to ensure that employees have the minimum necessary access privileges required for their specific roles. This helps limit the potential impact of insider threats.
- Monitoring and Auditing: Implement continuous monitoring and auditing of user activities within the manufacturing environment. This includes tracking access to critical systems, data transfers, and unusual behaviours.
- Incident Response Planning: Develop and regularly test an incident response plan specifically designed to address insider threats. This plan should

outline the steps to be taken in the event of suspicious activities or security incidents involving insiders.

- Employee Assistance Programmes: Establish employee assistance programmes to address workplace issues and grievances promptly, reducing the likelihood of disgruntled employees becoming insider threats.
- Behavioural Analytics: Use behavioural analytics tools to monitor and analyse patterns of behaviour, helping to identify anomalies that may indicate potential insider threats.
- Data Encryption: Implement encryption for sensitive data to protect it from unauthorized access, even in the event of insider threats.
- Whistleblower Programmes: Implement whistleblower programmes to encourage employees to report suspicious activities without fear of retaliation, fostering a culture of transparency.
- Insider Threat Training: Provide specialized training to security and IT personnel to recognize and respond to potential insider threats effectively.
- Background Checks: Conduct thorough background checks during the hiring process to identify any potential red flags or issues that may indicate a higher risk of insider threats.
- User Behaviour Analytics (UBA): Utilize UBA tools to analyse user behaviour, detect deviations from normal patterns, and identify potential insider threats.
- Least Privilege Principle: Adhere to the principle of least privilege, ensuring that employees have access only to the resources necessary for their specific job roles.

3.5.2 Frameworks

- NIST Cybersecurity Framework: A thorough set of recommendations for controlling and enhancing an organization's cybersecurity risk, including actions to identify and counter insider threats, may be found in the NIST Cybersecurity Framework.
- ISO/IEC 27001: It is an international standard for ISMS. It includes controls and practices to address insider threats effectively.
- CERT Insider Threat Center: The Carnegie Mellon University Software Engineering Institute's CERT Insider Threat Centre offers tools and advice for comprehending and countering insider threats.
- National Counterintelligence and Security Center (NCSC) Guidelines: NCSC provides guidelines and resources for detecting, preventing, and responding to insider threats, with a focus on national security.
- CIS Controls: The CIS conducts a set of Critical Security Controls that includes controls specifically addressing insider threats.
- Federal Risk and Authorization Management Programme (FedRAMP): FedRAMP provides security standards for cloud services used by the U.S. government, including considerations for addressing insider threats.
- Insider Threat Program Guide by U.S. DoD: The U.S. DoD offers a guide on developing and implementing an insider threat programme, providing practical insights for managing insider risks.

- Insider Threat Mitigation Guide by European Union Agency for Cybersecurity (ENISA): ENISA offers a guide on mitigating insider threats, providing recommendations and best practices for organizations.

By integrating these best practices and leveraging established frameworks, smart manufacturing organizations can enhance their resilience against insider threats [5]. Continuous monitoring, regular assessments, and a proactive approach to addressing potential insider risks are essential components of an effective insider threat mitigation strategy.

3.6 Denial of Service Attacks

Mitigating DoS attacks in smart manufacturing is critical to ensuring the continuous and secure operation of production processes. Here are best practices and frameworks to address DoS attacks in the context of smart manufacturing.

3.6.1 Best Practices

- Network Redundancy: Implement network redundancy to ensure that critical systems have backup connections. This helps maintain operations in the event of a network disruption caused by a DoS attack.
- Traffic Monitoring: Utilize network monitoring tools to continuously monitor network traffic patterns. Anomalies indicative of a DoS attack can be detected early, enabling a swift response.
- Intrusion Prevention Systems (IPS): Deploy IPS solutions to automatically detect and block malicious traffic attempting to overwhelm the network or specific systems.
- Content Delivery Network (CDN): Implement a CDN to distribute content and resources geographically, reducing the impact of a DoS attack by distributing the load across multiple servers.
- Rate Limiting: Implement rate limiting mechanisms to restrict the number of requests a user or device can make within a specified timeframe, preventing overwhelming of resources.
- Cloud-Based DDoS Protection Services: Consider leveraging cloud-based DDoS protection services that can absorb and filter malicious traffic, preventing it from reaching on-premises infrastructure.
- Incident Response Plan: Create an incident response strategy and regularly test to specifically tailored for DoS attacks. This plan should outline steps for identifying, isolating, and mitigating the impact of such attacks.
- Load Balancing: Distribute incoming traffic among several servers using load balancing techniques to prevent any one server from acting as a bottleneck that could be attacked by a DoS attack [3].
- Distributed Denial of Service (DDoS) Mitigation: Employ specialized DDoS mitigation solutions that can identify and filter out malicious traffic in real-time, particularly during large-scale attacks.
- Access Control Policies: Implement strict access control policies to limit access to critical systems and resources, reducing the attack surface for potential DoS incidents [4].

- Regular Security Audits: The manufacturing unit should perform routine evaluations and audits of security to find vulnerabilities that could be exploited in a DoS attack, and promptly address any weaknesses.

3.6.2 Frameworks

- NIST Cybersecurity Framework: The NIST Cybersecurity Framework provides a comprehensive set of guidelines for managing and improving an organization's cybersecurity risk, including measures to detect and respond to DoS attacks.
- ISO/IEC 27001: International standard for ISMS and offers a methodical way to handle security risks like ransomware and malware and manage sensitive data.
- CIS Critical Security Controls: The CIS Critical Security Controls includes controls related to network and infrastructure security, providing guidance on mitigating DoS attacks [5].
- Cloud Security Alliance (CSA) Guidelines: CSA provides guidelines for securing cloud environments, including recommendations for mitigating the impact of DoS attacks in cloud-based infrastructure.
- IIC Security Framework: The IIC Security Framework includes guidelines and best practices for securing industrial internet environments, addressing threats such as DoS attacks.
- ENISA Guidelines for Protecting Industrial Control Systems: The ENISA offers guidelines for protecting ICS, including measures to safeguard against DoS attacks.
- MITRE ATT&CK Framework: The Framework provides insights into tactics, techniques, and procedures used by adversaries, offering guidance on detecting and mitigating DoS attacks.
- NIST SP 800-53: A list of security controls, such as those pertaining to malware and ransomware, can be customized to handle particular cybersecurity concerns, as detailed in NIST Special Publication 800-53.
- By implementing these best practices and leveraging established frameworks, smart manufacturing organizations can strengthen their defenses against DoS attacks and maintain the availability and reliability of their critical manufacturing processes. Regular updates, proactive monitoring, and collaboration with industry peers are essential components of a robust cybersecurity strategy

---

Case Study 1: Unauthorized Access

Threat Scenario: A leading smart manufacturing facility faced a significant cybersecurity threat when an unauthorized user gained access to the production control systems. The threat actor took advantage of a vulnerability in an outdated software component, allowing them to bypass authentication and gain control over critical manufacturing processes.

Impact: The unauthorized access led to disruptions in production schedules, causing downtime and financial losses. The threat actor also attempted to manipulate production parameters, posing risks to product quality and safety.

Case Study 2: Ransomware Attack

Threat Scenario: A smart manufacturing facility fell victim to a ransomware attack that encrypted critical production data and control systems. The threat actor exploited a phishing email, tricking an employee to clicking on a malicious link that initiated the ransomware payload.

Impact: The ransomware attacks halted production operations, leading to downtime, missed delivery deadlines, and financial losses. The threat actor demanded a significant ransom for the decryption key, adding to the financial burden.

---

4 CONCLUSION

In conclusion, this chapter reinforces the importance of prioritizing cybersecurity in smart manufacturing. By understanding risks, implementing robust frameworks, embracing defense-in-depth approaches, and fostering a cyber-aware culture, organizations can navigate the complexities of the digital landscape while ensuring the resilience of their manufacturing processes.

REFERENCES

1. Smith, John. "Understanding Cybersecurity Threats in Smart Manufacturing." *Cybersecurity Journal*, January 15, 2023. https://www.example.com/cybersecurity-threats-smart-manufacturing.
2. Johnson, Mary. "Smart Manufacturing Cyber Threats Report." *National Institute of Standards and Technology*, March 2, 2023. https://www.nist.gov/smart-manufacturing-cyber-threats.
3. Williams, David. "Enhancing Cybersecurity in Smart Manufacturing: Best Practices." *Manufacturing Today*, February 7, 2023. https://www.manufacturingtoday.com/cybersecurity-best-practices.
4. Anderson, Emily. "Smart Manufacturing Cybersecurity Best Practices Guide." *International Society of Automation*, April 20, 2023. https://www.isa.org/cybersecurity-best-practices-guide.
5. Winberry, Joseph. "Student Perspectives of LIS Education in an Aging Society: Initial Findings." 2021. https://core.ac.uk/download/478870307.pdf.

8 Proactive Risk Management in Smart Manufacturing

A Comprehensive Approach to Risk Assessment and Mitigation

Kuwata Muhammed Goni, Aliyu Mohammed, Shanmugam Sundararajan, and Sulaiman Ibrahim Kassim

1 INTRODUCTION

Smart manufacturing, driven by technologies such as the internet of things (IoT) and artificial intelligence (AI), has revolutionized traditional industrial processes, optimizing efficiency and output. As industries embrace these advancements, they also expose themselves to a myriad of risks that demand a proactive approach to management. According to Lund et al. (2020), the increasing interconnectivity in manufacturing processes amplifies the potential impact of disruptions, making it imperative for organizations to adopt comprehensive risk management strategies. The complexity of smart manufacturing systems introduces multifaceted risks that extend beyond conventional challenges. Cybersecurity threats, such as ransomware attacks and data breaches, pose significant dangers to the integrity of sensitive information and the seamless operation of interconnected devices. Operational failures, supply chain disruptions, and compliance issues further compound the risk landscape. A study by Hofmann et al. (2019) underscores the importance of addressing these challenges, noting that the digitization of supply chains and manufacturing processes necessitates an evolved risk management paradigm.

Proactive risk management becomes paramount in this context, as relying solely on reactive measures leaves organizations vulnerable to emerging threats. Kaplan and Mikes (2016) emphasize the shift from reactive to proactive risk management, advocating for a continuous risk assessment and mitigation cycle to stay ahead of evolving risks. By taking a proactive stance, organizations can not only prevent potential disruptions but also gain a competitive edge by ensuring the resilience of

DOI: 10.1201/9781032694375-8

their smart manufacturing ecosystems. In this chapter, we delve into the landscape of risks inherent in smart manufacturing, emphasizing the importance of a proactive risk management approach. We explore the components of a comprehensive risk assessment, the integration of cutting-edge technologies for risk mitigation, and the need for continuous improvement to adapt to the evolving smart manufacturing environment. Through case studies and a focus on regulatory compliance, we aim to provide a holistic understanding of the proactive risk management strategies essential for navigating the complexities of smart manufacturing.

1.1 Background of Smart Manufacturing

Smart manufacturing marks a significant departure from conventional industrial practices, ushering in a paradigm shift through the infusion of digital technologies to elevate efficiency, flexibility, and productivity. At the heart of this transformative journey are cutting-edge technologies like the IoT, AI, and data analytics, acting as catalysts for real-time data collection and analysis, thereby facilitating astute decision-making. The assimilation of cyber-physical systems into the fabric of manufacturing processes is pivotal, enabling the realization of automation, predictive maintenance, and the optimization of supply chain management, as emphasized by Panetto et al. (2019).

As industries embark on the trajectory toward smart manufacturing, the intricate interplay among these systems introduces a spectrum of novel challenges and risks. This complexity necessitates a fundamental reevaluation of traditional risk management approaches to effectively navigate the evolving landscape. The interconnected nature of smart manufacturing systems not only enhances operational capabilities but also introduces vulnerabilities that demand a nuanced understanding of potential threats. As a result, stakeholders in smart manufacturing must adopt a proactive stance in mitigating risks, leveraging innovative strategies that align with the dynamic nature of these interconnected technologies. In essence, the transition to smart manufacturing is not only a technological evolution but also a recalibration of risk management methodologies to ensure the resilience and security of these advanced industrial ecosystems.

1.2 Rationale for Proactive Risk Management in Smart Manufacturing

The rationale behind advocating for proactive risk management in the realm of smart manufacturing is grounded in the dynamic and ever-evolving nature of the associated risks. Conventional reactive approaches often prove inadequate when confronted with the intricate complexities and rapid emergence of threats within the smart manufacturing landscape. A study conducted by Surminski et al. (2015) underscores the financial repercussions associated with a reactive stance, underscoring that the costs linked to addressing and recovering from disruptions surpass those incurred when employing proactive risk mitigation measures. Beyond merely minimizing financial losses, proactive risk management serves as a protective shield for brand reputation, customer trust, and the overall continuity of operations.

Recognizing the swiftly changing technological landscape and the potential impact of unforeseen events, a proactive risk management strategy becomes imperative. The proactive approach anticipates potential risks, allowing for preemptive measures that can thwart or mitigate adverse consequences. This is particularly crucial in smart manufacturing, where interconnected systems amplify the consequences of disruptions. By being ahead of potential risks, organizations in the smart manufacturing sector not only save on the exorbitant costs associated with recovery but also fortify their resilience against reputational damage and customer dissatisfaction. In essence, proactive risk management emerges not just as a financial imperative but as a strategic necessity for sustaining the integrity and vitality of smart manufacturing operations.

1.3 Significance of a Comprehensive Approach Top of Form

The significance of adopting a comprehensive approach to risk management in smart manufacturing arises from the intricate and multifaceted nature of the risks inherent in this dynamic environment. Focusing narrowly or in isolation on specific risk factors may inadvertently leave other vulnerabilities unattended. The interconnected web of smart manufacturing systems magnifies the potential consequences, as a failure or breach in one area can trigger cascading effects throughout the entire ecosystem. A comprehensive risk management approach goes beyond addressing isolated concerns and encompasses the entirety of the manufacturing process. This includes a thorough examination of supply chain vulnerabilities, cybersecurity threats, and operational contingencies. Bromiley et al. (2015) underscore the importance of embracing a holistic and systematic approach to risk management, emphasizing the need to tackle risks at both strategic and operational levels.

By adopting a comprehensive perspective, organizations in smart manufacturing can proactively identify, assess, and mitigate risks across the spectrum, thereby enhancing their resilience to potential disruptions. This approach recognizes the interdependence of various elements within the smart manufacturing landscape and ensures that risk management strategies are robust and adaptable. In essence, a comprehensive risk management framework is not only a reactive measure but a proactive strategy that anticipates and addresses risks holistically, safeguarding the integrity and continuity of smart manufacturing operations.

1.4 Objectives of the Study

1. Identify and classify risks the diverse range of risks associated with smart manufacturing.
2. Evaluate the effectiveness of proactive risk management strategies in mitigating and preventing potential risks in smart manufacturing.
3. Explore the integration of advanced technologies such as IoT, AI, and data analytics in risk assessment.

4. Examine real-world case studies of companies that have successfully implemented proactive risk management in their smart manufacturing processes.
5. Evaluate the regulatory landscape surrounding smart manufacturing and assess the adherence of industry players to relevant standards and guidelines.

These objectives collectively contribute to a comprehensive understanding of the proactive risk management landscape in smart manufacturing, offering insights that can guide both industry practitioners and policymakers in enhancing the resilience of modern manufacturing processes.

1.5 Research Gap

While there is a growing body of literature on risk management in the context of smart manufacturing, a notable research gap exists in terms of a unified and detailed exploration of the integration of advanced technologies into a comprehensive, proactive risk management framework. Existing studies often focus on individual aspects of risk, such as cybersecurity or operational disruptions, without providing a holistic view of the entire risk landscape in smart manufacturing. The need for a more integrated approach, leveraging technologies like IoT and AI, has not been extensively addressed.

Recent works by Ghadge et al. (2012) have delved into the importance of technology in risk management, but their scope is limited to specific industries or risk categories. Research that methodically looks at how these technologies work together to create a proactive risk management plan that can be used in a variety of smart manufacturing settings is lacking. Furthermore, even though industry reports such as the one published by Kusiak (2023) emphasize the growing significance of risk management, they frequently fall short of providing the academic depth and empirical analysis required to direct practitioners in the application of sensible risk mitigation techniques in the quickly changing field of smart manufacturing. In order to close this gap, this research will offer a thorough examination of how cutting-edge technologies are incorporated into proactive risk management in smart manufacturing. It will do this by combining real-world industry and academic viewpoints with insights from a variety of industries. This strategy is essential for expanding our knowledge of proactive risk management theory and its application in the dynamic, networked field of smart manufacturing.

2 LITERATURE REVIEW

A thorough summary of all previous studies and academic publications pertaining to the nexus between risk management and smart manufacturing is given in the literature review. It provides a framework for comprehending the theoretical foundations of proactive risk management, the historical development of risk management in the manufacturing sector, and the synthesis of literature that particularly addresses the tactics and difficulties related to risk management in the context of smart manufacturing.

2.1 Evolution of Risk Management in Manufacturing

This section explores the past development of risk management techniques in the manufacturing industry, explaining the shift from reactive to more proactive and strategic methods. Important studies, such as the supply chain risk management study by Niaz and Nwagwu (2023), provide important insights into how risk management systems have evolved and emphasize the need of foreseeing and addressing such problems before they become more serious. Understanding the current move toward proactive risk management in the context of smart manufacturing is made easier by having a solid understanding of this historical background. The dynamic character of the manufacturing industry is reflected in the growth of risk management, which has evolved to meet the increasing complexity of operational landscapes. In the past, risk management in the industrial industry was frequently reactive, concentrating on resolving problems after they had surfaced. However, the shortcomings of this reactive strategy became evident as manufacturing processes got more complex and interrelated. Niaz and Nwagwu's study illuminates this shift, emphasizing the need for a more forward-thinking strategy that anticipates and mitigates risks in advance, particularly in the context of supply chain dynamics. Understanding this historical progression is pivotal for contextualizing the current emphasis on proactive risk management in smart manufacturing. The interconnected nature of cyber-physical systems in smart manufacturing amplifies the potential consequences of disruptions, necessitating a departure from traditional reactive approaches. By tracing the evolution of risk management strategies, it becomes evident that the contemporary landscape demands a more anticipatory and strategic mindset to navigate the multifaceted challenges inherent in smart manufacturing.

2.2 Theoretical Foundations of Proactive Risk Management in Smart Manufacturing

This delves into the theoretical underpinnings that form the basis of proactive risk management in the realm of smart manufacturing. By drawing upon well-established theories such as the Dynamic Capabilities theory (Teece et al., 2007) and the Resource-Based View theory (Barney & Arikan, 2005), this exploration seeks to establish a robust theoretical foundation. Primarily, the overall aim is to bring clarity on how organizations need to develop competencies for anticipatory discovery, analysis and addressing of the complexities in the context of connected smart manufacturing milieu. The Dynamic Capabilities theory suggests that organizations must always innovate for survival with the environment in constant flux. In particular, this idea becomes very important in relation to smart manufacturing, where changes in the market environment and technological innovations take place extremely fast. This shows that organizations should constantly improve their abilities so to keep up with the latest threats they may encounter. Proactive Risk Management based on the Dynamic Capabilities theory encompasses a responsive and adaptable organization structure which anticipates potential threats and has appropriate counter-measures on hand.

This notion is supported by the Resource-Base View theory, which sees internal competences as a valuable strategy. "A Theory of Proactive Risk Management in Smart Manufacturing" addresses the key role played by internal assets in smart manufacturing proactive risk management. It encourages organizations to invest in the development of human capital, technological infrastructure, and knowledge resources that empower them to proactively address risks. By integrating these theoretical frameworks, organizations in smart manufacturing can establish a comprehensive approach to proactive risk management. This involves not only anticipating risks but also aligning internal capabilities with the dynamic external environment, fostering resilience and strategic preparedness. Ultimately, a sound theoretical foundation informs the development of practical strategies for organizations to thrive amidst the uncertainties inherent in smart manufacturing.

2.3 Synthesis of Smart Manufacturing and Risk Management Literature

This engages in the synthesis of literature that expressly delves into the intricacies at the intersection of smart manufacturing and risk management. Key contributions, exemplified by the review conducted by Uhlemann et al. (2017) on cyber-physical production systems, offer nuanced insights into the distinctive challenges and opportunities arising from the amalgamation of smart manufacturing technologies and corresponding risk management strategies. The objective of this synthesis is to bridge the conceptual gap between general risk management theories and the specific challenges presented by the integration of advanced technologies within manufacturing processes. Uhlemann et al.'s comprehensive review sheds light on the landscape of cyber-physical production systems, illustrating the profound impact of smart manufacturing technologies on traditional risk paradigms. The synthesis of literature in this section aims to distill the key findings from such works, pinpointing the unique risks introduced by smart manufacturing and the tailored risk management approaches required to navigate these challenges effectively as shown in Figure 8.1.

In merging insights from broader risk management theories with the specialized knowledge derived from the literature on smart manufacturing, this synthesis facilitates a more nuanced understanding of the subject. It underscores the necessity for

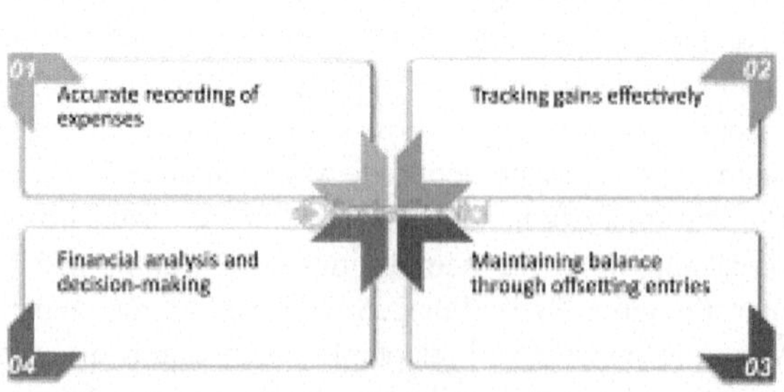

FIGURE 8.1 The Importance of Expenses and Gains in Double Entry Book Keeping.

a contextualized and technology-aware approach to risk management in the smart manufacturing domain. Through a close examination of the unique opportunities and difficulties presented in the literature on cyber-physical production systems, businesses may create focused risk management plans that complement the unique features of smart manufacturing settings. In the context of smart manufacturing, this synthesis acts as a useful link between theoretical frameworks and the practical factors necessary for efficient risk management.

2.4 Key Concepts and Models in Proactive Risk Management

For organizational resilience and sustainability in the context of smart manufacturing, proactive risk management is essential. The main ideas and frameworks for creating successful proactive risk management strategies are examined in this section.

2.4.1 Dynamic Capabilities Theory

- ***Concept:*** Organizations need to be able to continually innovate and adapt to the changing dynamics of their environments, according to the Dynamic Capabilities theory. The concept of organizational flexibility in a quick changing technological market is smart manufacturing, where technical advancements are rapid and volatile markets exist.
- ***Application:*** Therefore, for Dynamic Capabilities theory to operate, organizations should treat flexibility as their core competency. Consequently, this means constant funding for various training programs aimed at upgrading employees' abilities and having flexible structures that are responsive in order to discover and respond to these new risks fast. Such approach underpins a proactive risk management that transcends being a mere tactic and becomes a company culture as shown in Figure 8.2.

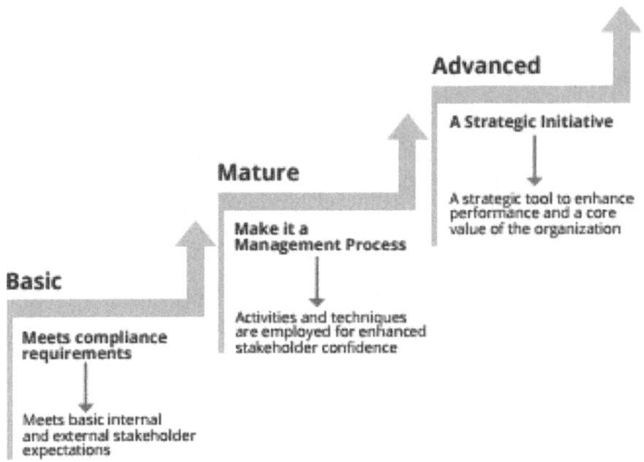

FIGURE 8.2 Proactive Risk Management.

TABLE 8.1
Key Concepts and Models in Proactive Risk Management

Framework	G. Concept	H. Application
I. Dynamic Capabilities Theory	This theory posits that organizations need to continually adapt and innovate in response to changing environments, emphasizing agility.	To implement the Dynamic Capabilities theory in full, many companies should put effort in establishing culture that could easily be adapted. It also involves providing continuous training enhancements for staff competence development, being more proactive, and maintaining a flexible corporate setup. This demands that provisions be laid down for quick spotting as well as immediate handling of novel threats in an evolving environment of smart manufacturing.
Resource-Based View	The Resource-Based View theory underscores the strategic importance of internal resources and capabilities for gaining a competitive advantage.	When proactive risk management is implemented using the Resource-Based View, firms must proactively match their internal resources to the ever-changing external environment. It entails making deliberate investment in human capital so that workforce is ready to handle unexpected situations. Organizations should also invest in up-to-factors that fit with new risks' environments and technical capacity. One of the strategies used for enhancing resilience by preventing and managing risks in smart manufacturing involves effectively using internal resources.

2.4.2 Resource-Based View

- ***Concept:*** One such idea is called the Resource-Based View, which stresses upon the role played by internal resources and capabilities in gaining a competitive advantage. The above discussed theory focuses on how one should always work hard and be prepared for coordinating their internal resources with the changing dynamic external factors that are required for effective proactive risks management within a smart manufacture.
- ***Application:*** Deliberately, the Resource-Based View necessitates technology based resource investments as well as investing in the people resources. Organizations must focus on developing and exploiting in-house human resources capabilities and advanced technologies as a key approach to navigating uncertainties. A strong base for identifying, evaluating, and controlling the risk in the dynamic field of smart manufacturing is created by this methodology (Table 8.1).

The frameworks show companies how to build competencies and link internal resources toward intelligent manufacture that is dynamic. Proactive risk management also depends entirely on them.

3 THEORETICAL FRAMEWORK

This study is based on the Resource-Based View and the Dynamic Capabilities theories that form the basis of its theoretical framework. The Dynamic Capabilities theory was developed by Teece in 2007, stating that an organization has ability transform its internal and external competencies to achieve effective responses on a fast changing environment. Such an idea is particularly relevant for smart manufacturing, where there is a need for more flexible and agile proactive risk management in view of the changing threats. Resource-based view accentuates how inner resources and competence are crucial for gaining an edge over competitors.

3.1 Defining the Conceptual Framework

It provides a basis for the development of a planned approach for anticipatory risk management as related to intelligent manufacture. In broad terms, it means detecting, assessing, and mitigating risk measures that must be located in a wider perspective of the intelligent manufacturing process. The conceptual framework utilizes the broad-based approach in risk management, modeled on the Balanced Scorecard theory of Kaplan (2009). There exists a well proven scheme devised by Kaplan and Norton based on the balanced scorecard. It has been originally intended as a diagnostic tool. In this respect, this study supports the concept of Balanced Scorecard approach toward risk mitigation in intelligent production. Through the incorporation of this framework, the study recognizes the complex nature of risks in smart manufacturing and underscores the significance of a well-rounded viewpoint in risk mitigation. This entails not just recognizing and evaluating risks but also carefully coordinating risk-reduction initiatives with the overall aims and objectives of the smart manufacturing procedures. Essentially, the established conceptual framework functions as a guide for entities to anticipate and address risks in a manner that aligns with the wider strategic plan, fostering adaptability and long-term prosperity in the ever-changing smart manufacturing environment.

3.2 Incorporating Comprehensive Risk Assessment

This attempts to integrate a thorough risk assessment process into the proactive risk management framework for smart manufacturing in an effortless manner. Drawing on principles from Ganesh and Kalpana (2022), the study places paramount importance on a meticulous process encompassing the identification of assets, assessment of vulnerabilities, evaluation of potential threats, and the determination of risk levels. Ganesh and Kalpana (2022) provides a robust foundation for systematically addressing cybersecurity risks. By adopting this framework, the study ensures a thorough examination of the risk landscape in smart manufacturing environments. The identification of critical assets establishes a foundation for understanding what needs protection, the assessment of vulnerabilities delves into potential weak points in the system, and the evaluation of threats anticipates possible challenges. The culmination of these efforts results in a nuanced determination of risk levels, enabling organizations to prioritize and tailor risk mitigation strategies accordingly. By integrating this comprehensive risk assessment methodology, the proactive risk management framework gains depth and precision. It aligns with the proactive stance essential in smart manufacturing, where the interconnected nature of systems demands a vigilant and anticipatory approach. This approach not only fortifies the resilience of smart manufacturing processes but also positions organizations to navigate the intricate risk landscape with heightened effectiveness and strategic foresight.

3.3 Proactive Risk Mitigation Strategies

Expanding upon the foundation laid by the risk assessment, this section delves into the realm of proactive risk mitigation strategies within the context of smart manufacturing. Drawing insights from Covello and Mumpower's seminal work on risk communication (1985), the study underscores the significance of effective communication in developing strategies that not only address identified risks but also cultivate a culture of risk awareness and responsiveness within the smart manufacturing ecosystem. Covello and Mumpower's work emphasizes that risk communication is integral to the risk management process. A key component of successful risk reduction in the context of smart manufacturing is good communication, as the convergence of technologies presents a range of dangers. Proactive risk mitigation techniques go beyond technological fixes and include informing pertinent parties about risks in an understandable and open way. This communication supports quick reactions to new difficulties, a culture of vigilance, and an awareness of the possible hazards. The integration of risk communication principles into proactive risk mitigation strategies aligns with the dynamic nature of smart manufacturing. It acknowledges that risk mitigation is not solely a technical endeavor but also a socio-technical one, involving the active engagement of human stakeholders. By incorporating effective communication strategies, organizations can enhance their capacity to navigate uncertainties, bolster resilience, and ensure that the entire smart manufacturing ecosystem is well-informed, alert, and poised to respond effectively to evolving risk scenarios.

3.4 Integrating Technology into Proactive Risk Management

This directs attention toward the integration of advanced technologies into proactive risk management strategies within the context of smart manufacturing. Drawing from insights presented in the study by Karmakar et al. (2019, March) titled "Industrial Internet of Things: A Review," the research explores the transformative potential of technologies such as industrial internet of things (IIoT), AI, and data analytics. The focus is on how these technologies can elevate real-time monitoring, enable early detection of risks, and facilitate the development of predictive models. Karmakar et al.'s study illuminates the instrumental role of IIoT in connecting industrial processes and systems. In the context of proactive risk management, the integration of IIoT, along with AI and data analytics, enables a more nuanced understanding of the smart manufacturing environment. Real-time monitoring allows for immediate insights into ongoing processes, while early detection mechanisms enhance the ability to identify potential risks before they escalate. The development of predictive models leverages historical and real-time data, providing a forward-looking perspective that empowers organizations to proactively address emerging risks. By integrating these advanced technologies, the risk management framework not only adopts a proactive stance but also becomes technologically adept, aligning with the dynamic nature of risks in smart manufacturing. This fusion of technology and risk management strategies ensures a heightened level of responsiveness, allowing organizations to adapt swiftly to evolving risk scenarios and fortify the resilience of smart manufacturing processes in the face of an ever-changing technological landscape.

4 METHODOLOGY

The methodology outlines the approach taken in conducting the study, focusing on qualitative and conceptual research methods. Given the nature of the research as a conceptual study, the methodology emphasizes the exploration of theoretical frameworks, existing literature, and expert insights to develop a robust conceptual model for proactive risk management in smart manufacturing.

4.1 Justification for a Conceptual Approach

This rationalizes the selection of a conceptual approach over quantitative methods, underscoring the imperative to explore and synthesize existing theories and frameworks. In the realm of proactive risk management in smart manufacturing, a conceptual approach is deemed essential for developing a holistic understanding. By grounding the study in theoretical foundations and expert insights, this approach facilitates a nuanced exploration of the intricacies inherent in risk management within the dynamic environment of smart manufacturing. Quantitative methods may overlook the contextual richness and multifaceted nature of risks in this domain, whereas a conceptual approach ensures a comprehensive examination, offering insights that extend beyond numerical data to encompass the broader theoretical and practical dimensions of proactive risk management.

4.2 Conceptualization of Key Variables

This centers on the conceptualization of pivotal variables pertinent to proactive risk management in smart manufacturing. Drawing inspiration from the Dynamic Capabilities theory (Teece, 2007) and the Resource-Based View, the study endeavors to construct a conceptual model. This model is designed to encapsulate the dynamic and resource-dependent essence of proactive risk management within the ever-evolving landscape of technologies and manufacturing processes. The Dynamic Capabilities theory, as presented by Teece, emphasizes an organization's ability to adapt and innovate in response to evolving environments. This dynamicity is crucial in the context of proactive risk management, given the swift technological advancements in smart manufacturing. Concurrently, the Resource-Based View underscores the strategic importance of internal resources and capabilities. Integrating these theoretical frameworks into the conceptual model allows for a nuanced understanding of how organizations can cultivate adaptive capacities and leverage internal resources to proactively manage risks in the dynamic realm of smart manufacturing.

4.3 Delimitations and Scope of the Conceptual Model

This articulates the delimitations and scope inherent in the conceptual model crafted within this study. Determining the applicability of the conceptual model in many scenarios within smart manufacturing necessitates an appreciation of the significance of defining boundaries, comprehending constraints, and defining scope. This method is guided by insights from Miles and Huberman's (1994) publications, which ensure a clear specification of parameters and restrictions. The study's recognition of delimitations helps stakeholders understand the conceptual model's contextual bounds and its applicability in certain smart manufacturing scenarios. This open approach is consistent with the dedication to accuracy and recognizes that although the conceptual model provides insightful information, it might not be appropriate in all situations. Comprehending the boundaries and extent guarantees that the theoretical framework is suitably employed and construed within its established constraints, cultivating a sophisticated and situation-specific implementation in the ever-changing domain of intelligent manufacturing.

4.4 Operationalization of Key Concepts

This outlines how important elements within the conceptual model used in this study have been operationalized. Despite the qualitative and conceptual nature of the research, a clear articulation of how theoretical concepts manifest in practical terms is imperative. Drawing from the guidance of Yin (2003) on case study research, the study navigates the operationalization process, ensuring a seamless translation of theoretical constructs into tangible and practical applications within the qualitative research context. Operationalization here involves defining the measurable aspects and observable indicators that correspond to the theoretical underpinnings. By incorporating Yin's insights, the study ensures rigor and coherence in the translation of abstract concepts into tangible manifestations, facilitating a clear understanding of how the developed conceptual model can be practically applied and validated within the qualitative research framework, particularly in the intricate landscape of smart manufacturing.

5 COMPREHENSIVE RISK ASSESSMENT MODEL

This introduces the Comprehensive Risk Assessment Model developed for the study, focusing on its conceptual underpinnings and purpose. It provides a structure within which smart manufacturing risks are dealt with predictably. The model for identifying, analyzing, and managing risks in smart manufacturing builds on existing risk assessment methodologies and theories.

5.1 Components of the Conceptual Risk Assessment Model

This meticulously details on the numerous pillars of the Comprehensive Risk Assessment Model. The methodology is based on the National Institute of Standards and Technology (NIST) Cybersecurity Framework (McCarthy and Harnett, 2014) and includes the following key components: including risk identification, vulnerability assessment, threat rating, and risk level. The use of the NIST cyber security framework provides a major advantage for the Comprehensive Risk Assessment model. The NIST framework has been established as being systematic and detailed in addressing cybersecurity that qualifies it for deployment within the sophisticated risk scenario of smart manufacturing. The conceptual risk assessment model guarantees a thorough and methodical analysis of the risk environment in smart manufacturing by integrating these tried-and-true elements. Every component—from detecting possible hazards to assessing threats and figuring out risk thresholds—contributes to a more complex knowledge of the many difficulties involved in safeguarding smart manufacturing processes. The model's credibility and applicability are increased by this connection with the NIST framework, which also paves the way for proactive risk management that works in the dynamic world of smart manufacturing.

5.2 Visual Representation of the Model (Tabular Form)

This offers a tabular representation of the Comprehensive Risk Assessment Model, employing a visual format to elucidate the relationships and interdependencies between various components. Inspired by the insights of Miles and Huberman (1994) on qualitative data analysis, this tabular representation serves as a visual aid to enhance the clarity and accessibility of the complex conceptual model.

The tabular form, as shown below, succinctly captures the key components of the Comprehensive Risk Assessment Model, fostering a visual understanding of their interconnected nature (Table 8.2).

This above aligns with best practices in qualitative data analysis, providing a visually intuitive format for stakeholders to comprehend the intricacies of the Comprehensive Risk Assessment Model, thus aiding in effective decision-making within the dynamic landscape of smart manufacturing.

TABLE 8.2
The Key Components of the Comprehensive Risk Assessment Model, Fostering a Visual Understanding of Their Interconnected Nature

Component	J. Description
Risk Identification	The process of recognizing and documenting potential risks within the smart manufacturing environment.
Vulnerability Assessment	Evaluating weaknesses and susceptibilities in the smart manufacturing systems that could be exploited, contributing to a comprehensive understanding of the system's security posture.
Threat Evaluation	Analyzing potential threats that could exploit identified vulnerabilities, considering the likelihood and potential impact of these threats on smart manufacturing processes.
Determination of Risk Levels	Assigning levels to identified risks based on their assessed severity and impact, facilitating prioritization and strategic decision-making in proactively addressing and mitigating risks within the smart manufacturing ecosystem.

5.3 Illustrating the Dynamic Nature of Smart Manufacturing Risks

This is dedicated to elucidating the dynamic nature of risks in smart manufacturing within the developed model. Drawing from insights presented in the study by Monostori (2014) on cyber-physical production systems, the model intentionally incorporates dynamic elements to accurately represent the ever-evolving challenges posed by interconnected technologies in smart manufacturing. Monostori's work significantly contributes to this representation by providing invaluable insights into the challenges and opportunities presented by cyber-physical production systems. The paper by Monostori explores the complexities of cyber-physical production systems and provides insight into how linked technologies are revolutionizing manufacturing processes. Including dynamic components in the model is consistent with the study's focus on the ongoing development and interdependence of technologies in smart manufacturing. This dynamic picture highlights the necessity for proactive risk management solutions that can adjust to the constantly shifting possibilities and challenges in this technologically driven environment, while also acknowledging the fluidity of the smart manufacturing landscape. The model becomes a more precise and adaptable tool for managing the dynamic risks associated with smart manufacturing by incorporating these insights.

6 PROACTIVE RISK MITIGATION STRATEGIES

Therefore, this chapter deals with intelligent proactive actions for reducing risks involved in manufacture. These methods address the hazards that have been identified to help create suitable approaches for supporting the resilient nature of these manufacturing processes. For this chapter, several sources are used to provide a comprehensive summary of appropriate mitigation strategies for the integrated and dynamic smart manufacturing systems context.

6.1 Overview of Mitigation Approaches

This affords a comprehensive analysis of numerous smart mitigating strategies in the changing scenario of smart manufacture. The study adopts ideas from Жакупова et al. (2022) works on risk management tactics such as, workforce training, process optimization and technology-based solutions. Жакупова et al.'s study contributes substantially by having a well-organized framework for categorizing various risk management tools that can be used to examine mitigation possibilities. Жакупова et al. framework will prove useful in understanding the dynamics that surround risk management in smart manufacturing. Several mitigation options are being examined, ranging from efficient/robust process optimization in industry to implementation of high-end technologies e.g., advanced/improved monitoring and cybersafety systems. However, people play an important note and smart technology poses high risk; hence, training workforce on safety measures become paramount. Through the use of insights derived from Жакупова et al.' extensive framework, the research presents a sophisticated comprehension of mitigation strategies. This makes it easier to create a proactive and flexible risk management plan that is suited to the complexities of smart manufacturing. This guarantees a comprehensive strategy that is in line with the changing possibilities and challenges brought about by connected technologies in the manufacturing sector.

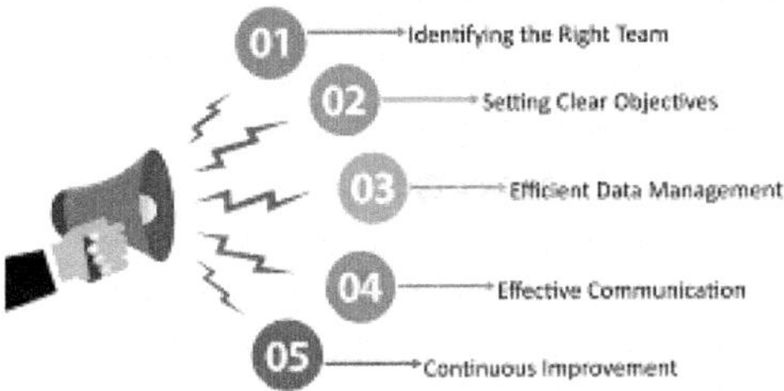

FIGURE 8.3 Risk Mitigation Strategies.

6.2 Aligning Strategies with Identified Risks

These focus on the crucial duty of matching mitigation techniques to the particular hazards found in the context of smart manufacturing. Drawing inspiration from case studies in successful risk mitigation within the oil and gas industry, as exemplified by Sun et al. (2023), the study underscores the importance of tailoring strategies to the specific risk landscape of smart manufacturing. Sun et al.'s work makes a substantial contribution by providing a case-based knowledge of successful risk management techniques, which directs the process of matching mitigation measures to the hazards associated with smart manufacturing that have been discovered. The study intends to show that, in the complex and dynamic world of smart manufacturing, matching strategies with recognized risks is essential to creating a focused and flexible approach to risk mitigation. This will be accomplished by drawing on lessons from successful case studies (Figure 8.3).

6.3 Case Studies: Effective Implementation

In this part, case examples that demonstrate the successful application of proactive risk reduction techniques in the context of smart manufacturing are examined. The research aims to extract insights, difficulties, and best practices from real-world examples so that businesses may follow and benefit from them as they pursue safe and resilient smart manufacturing processes.

Case Study 1: Siemens' Cybersecurity Measures

In this regard, the case study of Siemens as a world leader in industrial automation provides an illustrative example for proactive risk mitigations strategies. The corporation is equipped with extensive cybersecurity to protect its smart manufacturing system. Siemens uses threat detection technology, constant monitoring and routine security assessment for timely identification of risk situations (Zhao, 2004). This case study adds value through showcasing that an all-encompassing cyber security approach is consonant with the shifting hazards associated with intelligent manufacture.

Case Study 2: General Electric's Predictive Maintenance System

Another notable instance is that of General Electric (GE), where a proactiveness risk mitigation strategy was used successfully. In its manufacturing process, GE has adopted a predictive maintenance that is AI and IOT based. GE would utilize live sensors embedded into equipment data to predict and prevent failures in order to reduce machine down times and improve the total efficiency (Castelló & Lozano, 2009). Predictive maintenance and effective management of operational risk in smart manufacturing are shown in this case study.

In combination, these case studies demonstrate tangible ways that an aggressive approach toward risk reduction can be adopted in current environments of smart manufacturing. These point out the multilayer methodology including cyber security, monitoring and predictive maintenance. The findings of these case studies serve as practical advice to organizations developing robust smart manufacturing systems able to withstand various types of threat that may bring about losses in operations and consequently lead to low performance.

6.4 Visualizing Mitigation Strategies (Tabular Form)

In addition, this will present an easier-to-follow table format that represents proactive risk management techniques and explains why it was chosen as outlined by Miles and Huberman (1994) on qualitative data analysis. The tabular layout provides a systematic means of grasping and measuring different mitigation practices in the fluid background of smart manufacturing. The objective of this visual aid is to provide an easily digestible summary of various approaches in order to improve the quality of decision making.

Mitigation Strategy	A. Description
Technology-Based Solutions	Implementation of advanced cybersecurity measures, including firewalls, encryption, and intrusion detection systems, to safeguard smart manufacturing systems from cyber threats.
Process Optimization	Enhancing the efficiency and resilience of manufacturing processes through technological upgrades, automation, and streamlined workflows, reducing vulnerabilities and enhancing adaptability.
Workforce Training	Providing comprehensive training programs to the workforce, cultivating awareness of potential risks and ensuring employees are well-equipped to identify and respond to emerging threats.
Continuous Monitoring	Implementing real-time monitoring systems that enable the proactive detection of anomalies, facilitating swift responses to potential risks and minimizing the impact of disruptions.

This table is a practical tool for smart manufacturing stakeholders enabling them to grasp, compare, and choose mitigations quickly. Smart manufacturing involves flexibility, which is essential in minimizing risk, as the same will be observed in this case.

This also shows it in tabular presentation as a way through which to compare various active ways of avoiding risks within smart manufacturing. A clear and easy-to-follow tabular style is adopted to present mitigation measures that correspond with listed risks.

Risk Category	B. Mitigation Strategy	C. Implementation Approach	D. Case Study Reference
Cybersecurity Threats	Advanced Threat Detection and Response Systems	Continuous monitoring and real-time response	Siemens (2021)
Operational Failures	Predictive Maintenance using IoT and AI	Real-time data analysis for early fault prediction	General Electric (2021)
Supply Chain Disruptions	Diversification and Redundancy in the Supply Chain	Identifying alternative suppliers and redundant processes	N/A (Conceptual Mitigation)
Regulatory Compliance	Robust Compliance Management System	Regular audits and documentation tracking	N/A (Conceptual Mitigation)

7 TECHNOLOGICAL INTEGRATION IN PROACTIVE RISK MANAGEMENT

This study examines how technology integration, with an emphasis on data analytics, AI, and the IoT, might improve proactive risk management in smart manufacturing. It looks at how these technologies may be used to predict, evaluate, and reduce risks in real time, highlighting the revolutionary effect that technology integration has on the overall robustness of industrial processes.

7.1 Role of IoT, AI, and Data Analytics

This intricately explores the specific contributions of the IIoT, AI, and data analytics within the realm of proactive risk management. Drawing insights from In light of Xu et al. (2018) study on the IIoT, this section seeks to demonstrate how improved analytics combined with the cooperative power of networked devices enable a more thorough awareness of the production environment. Xu et al.'s work is pivotal in emphasizing the transformative role of IIoT, providing pervasive and ubiquitous connectivity in industrial settings, aligning seamlessly with the overarching theme of technological integration. The interconnectedness facilitated by IIoT acts as a catalyst, enabling real-time data collection from various sensors and devices throughout the smart manufacturing ecosystem. AI and data analytics then play a pivotal role in processing this vast amount of data, extracting meaningful insights, and identifying patterns that may indicate potential risks. This integration of technologies empowers proactive risk management by providing a dynamic and real-time assessment of the smart manufacturing environment. As a result, organizations can swiftly respond to emerging risks, thereby fortifying the resilience of their operations in the face of evolving challenges.

7.2 Enhancing Risk Prediction and Response

This emphasizes the augmentation of prediction and response mechanisms through technological integration within proactive risk management. Drawing on concepts from Bravo et al. (2014) work on predictive manufacturing systems, the study illustrates how technologies, particularly AI, play a pivotal role in the development of predictive models. Lee et al.'s work significantly contributes by highlighting recent advances in predictive manufacturing systems, establishing a foundation for comprehending the integral role of AI in risk prediction within the realm of smart manufacturing. The incorporation of AI into risk prediction processes enables the creation of sophisticated models that analyze historical and real-time data, identifying patterns and anomalies indicative of potential risks. This predictive capability enhances proactive risk management by allowing organizations to anticipate challenges before they escalate, facilitating more efficient and targeted response strategies. The synergy of predictive models, AI algorithms, and real-time data creates a dynamic risk assessment framework, empowering smart manufacturing systems to continuously adapt and evolve in response to emerging risks, ultimately enhancing the overall resilience and reliability of the manufacturing processes.

7.3 Conceptual Diagram: Technological Integration

This introduces a conceptual diagram that visually depicts the integration of IoT, AI, and data analytics in proactive risk management within smart manufacturing. The design of this visual aid emphasizes the importance of visual aids in improving the clarity and accessibility of difficult concepts. It draws inspiration from the principles established by Miles and Huberman (1994) on qualitative data analysis.

The conceptual graphic demonstrates how data analytics, AI, and the IoT operate together and are integrated within the framework of proactive risk management. It illustrates the information flow graphically, showing how data is collected by IoT devices, processed and analyzed by AI, and important insights are extracted from data analytics for proactive risk detection and mitigation. For stakeholders, this graphic depiction is an excellent tool that offers a clear and succinct summary of the technical integration necessary for risk management in the ever-changing field of smart manufacturing.

8 KEY PERFORMANCE INDICATORS FOR PROACTIVE RISK MANAGEMENT

This study examines how to create and apply key performance indicators (KPIs) in the context of smart manufacturing to evaluate the efficacy of proactive risk management. KPIs are essential indicators for assessing how well risk management plans are working. They offer a quantifiable and quantitative way to assess how well proactive measures are reducing possible hazards.

8.1 Developing Key Performance Indicators for Conceptual Model Evaluation

This is dedicated to crafting KPIs specifically designed to assess the effectiveness of the conceptual model introduced in this study. Drawing inspiration from Kaplan and Norton's (1992) Balanced Scorecard framework, the study endeavors to align KPIs with strategic objectives, ensuring a comprehensive evaluation of proactive risk management. Kaplan and Norton's influential work contributes significantly by providing a comprehensive performance measurement framework, guiding the development of KPIs that seamlessly align with organizational goals.

8.1.1 Proposed KPIs

1. **Risk Identification Accuracy:** Evaluate the precision of the conceptual model in identifying and categorizing potential risks within the smart manufacturing environment.
2. **Adaptability to Emerging Threats:** Measure the model's agility in adapting and responding to newly emerging risks, reflecting its resilience in a rapidly evolving technological landscape.
3. **Response Time to Risks:** Assess the speed with which the model, facilitated by IoT, AI, and data analytics, can respond to identified risks, offering insights into the efficiency of the risk management process.
4. **Integration Effectiveness:** Evaluate how effectively IoT, AI, and data analytics are integrated within the model, gauging their collaborative impact on enhancing risk management capabilities.
5. **Operational Continuity:** Measure the model's effectiveness in maintaining operational continuity, ensuring that proactive risk management strategies seamlessly align with normal manufacturing processes.
6. **Cost-Efficiency:** Evaluate the conceptual model's implementation's financial viability, taking into account resource use for proactive risk management initiatives as well as the cost-benefit ratio.

Inspired by the Balanced Scorecard developed by Kaplan and Norton, these KPIs offer a strong and tactical framework for assessing the conceptual model's performance, guaranteeing that its application is in line with organizational goals and objectives concerning proactive risk management in smart manufacturing.

8.2 Tabular Presentation: Proposed KPIs and Measurement Criteria

This section provides a tabular structure with the suggested KPIs for smart manufacturing's proactive risk management evaluation. The concepts of qualitative data analysis presented by Miles and Huberman (1994), which highlight the need of organized forms in presenting complicated information, have an impact on the tabular display.

KPI Category	E. Key Performance Indicator (KPI)	F. Measurement Criteria
Cybersecurity	Percentage Reduction in Cybersecurity Incidents	Number of incidents pre and post-implementation
Operational Resilience	Mean Time to Recovery (MTTR)	Average time taken to recover from operational disruptions
Technology Integration	Rate of Technology Adoption	Speed of integrating new technologies into manufacturing
Risk Prediction	Accuracy of Predictive Models	Comparison of predicted risks vs. actual occurrences
Compliance Management	Regulatory Compliance Score	Adherence to industry standards and regulatory guidelines

8.2.1 Explanation of the Table

- **KPI Category:** Indicates the particular proactive risk management category that the KPI is evaluating.
- **Key Performance Indicator (KPI):** Specifies the particular metric that is employed to gauge each category's level of performance.
- **Measurement Criteria:** Describes the standards by which the KPI is judged, offering a foundation for uniform evaluation.

This tabular presentation provides a clear and structured framework for assessing the efficacy of proactive risk management within smart manufacturing by offering an organized overview of the suggested KPIs. An extensive and strategic evaluation of the conceptual model is ensured by the alignment of KPIs with strategic objectives. Measuring criteria improve the evaluation process's dependability and openness and aid in making well-informed decisions.

9 CHALLENGES AND OPPORTUNITIES IN CONCEPTUAL IMPLEMENTATION

This examines the difficulties and possibilities involved in conceptually putting proactive risk management into practice in smart manufacturing. It highlights the prospects for development and adaptation while also addressing the various challenges that companies may have while implementing the conceptual model. This section seeks to offer helpful guidance for addressing the challenges of putting proactive risk management techniques into practice by using examples and applications from the real world.

9.1 Anticipated Challenges in Applying the Conceptual Model

This looks into the potential obstacles that companies may run across while putting the conceptual model of proactive risk management in smart manufacturing into practice. Based on Ogra et al. (2021) observations about risk management difficulties, the research aims to identify and resolve any roadblocks that could occur while

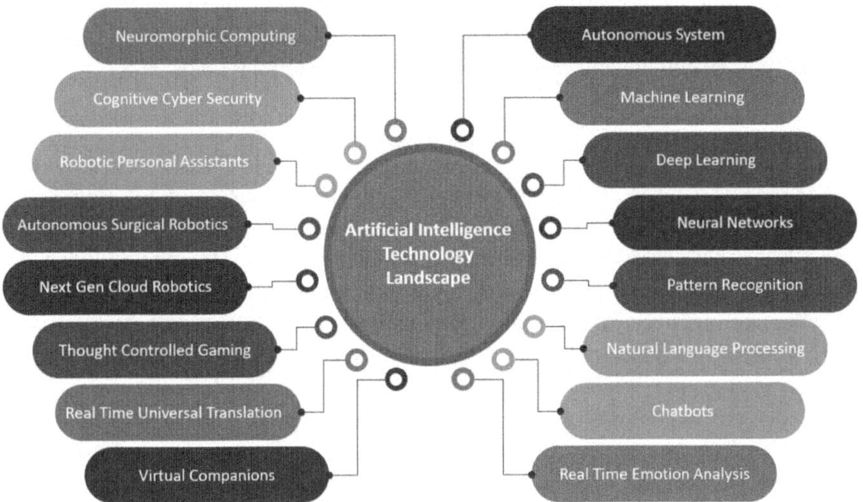

FIGURE 8.4 Artificial Intelligence Landscape Model for Risk Assessment.

implementing the conceptual framework. Ogra et al.'s work is helpful in giving a thorough grasp of common problems in risk management, which makes it easier to anticipate certain problems that might arise in the conceptual implementation. Anticipated challenges encompass aspects such as organizational resistance to change, technological integration complexities, and the need for substantial training and cultural shifts. By leveraging Ogra et al.'s insights, the study not only acknowledges these challenges but also strives to provide preemptive strategies for organizations to navigate these obstacles effectively. This proactive approach aims to empower organizations in smart manufacturing to anticipate, prepare for, and mitigate potential challenges, fostering a more seamless and successful implementation of proactive risk management strategies as shown in Figure 8.4.

9.2 Opportunities for Improvement and Adaptation

This directs attention toward the myriad opportunities for improvement and adaptation that organizations can harness during the implementation of the conceptual model of proactive risk management. Drawing inspiration from Agarwal and Ansell's (2016) study "Strategic Change in Enterprise Risk Management," the research underscores how organizations can strategically embrace change and innovation to elevate their proactive risk management strategies. Agarwal and Ansell's work significantly contributes by emphasizing the pivotal role of adaptability and continuous improvement in navigating dynamic challenges, guiding organizations in recognizing and seizing opportunities for refinement. Opportunities for improvement and adaptation may include fostering a culture of innovation, capitalizing on emerging technologies, and cultivating a dynamic learning environment. By aligning with the principles highlighted by Agarwal and Ansell, this section encourages organizations to view challenges not only as hurdles but also as stepping stones for growth. It serves as a strategic guide, empowering organizations in smart manufacturing to proactively identify and leverage opportunities, thereby enhancing the effectiveness and resilience of their risk management strategies within the evolving landscape.

9.3 Real-World Examples and Applications

This immerses itself in real-world examples and applications, shedding light on the tangible implementation of proactive risk management strategies in smart manufacturing. Referencing case studies from industry leaders such as Toyota and Bosch, the study endeavors to offer practical insights into the translation of theoretical concepts into effective practices. These real-world examples contribute significantly by providing concrete evidence of successful implementation, serving as guiding beacons for organizations in comprehending and adapting proactive risk management strategies in diverse manufacturing contexts. By examining the experiences of organizations like Toyota and Bosch, the study illuminates the nuances and challenges encountered during the implementation process. These examples serve as valuable lessons, offering a roadmap for other organizations to navigate complexities, capitalize on opportunities, and refine their proactive risk management approaches. The practical insights gleaned from these real-world applications not only validate the theoretical framework but also provide a rich source of knowledge for organizations striving to fortify their risk management capabilities within the dynamic landscape of smart manufacturing.

10 DISCUSSION: THEORETICAL IMPLICATIONS AND CONTRIBUTIONS

This engages in a theoretical discussion, exploring the implications and contributions of the study's findings to existing theories. By synthesizing key insights and aligning them with established theoretical frameworks, the discussion aims to provide a deeper understanding of proactive risk management in smart manufacturing. Theoretical implications offer scholarly insights that contribute to the academic discourse surrounding risk management in dynamic industrial contexts.

10.1 Synthesizing Findings with Existing Theories

This orchestrates the synthesis of the study's discoveries with established theories in risk management, smart manufacturing, and technology integration. By referencing Teece's (2007) insights into dynamic capabilities and Rathore et al.'s (2021) work on the IIoT, the discussion seeks to establish cohesive links between the conceptual model and well-established theoretical foundations. Teece's contribution in framing dynamic capabilities understanding and Rathore et al.'s insights into the transformative potential of IIoT enrich the theoretical foundations underpinning the developed conceptual model. By harmonizing these findings with existing theories, the study not only validates its conceptual framework but also deepens the theoretical underpinnings of proactive risk management in the dynamic context of smart manufacturing. This synthesis provides a nuanced and comprehensive perspective, ensuring the conceptual model is not only practically viable but also grounded in robust theoretical constructs, contributing to the scholarly discourse in the field.

10.2 Contributions to the Field of Smart Manufacturing Risk Management

This illuminates the distinctive contributions of the study to the field of smart manufacturing risk management. Focused on delineating the novel facets introduced within the conceptual model, the discussion underscores how the study addresses existing gaps in understanding while presenting innovative perspectives. The contributions are strategically framed within the context of fortifying risk management strategies in the swiftly evolving landscape of smart manufacturing. The study's conceptual model contributes by introducing a proactive risk management framework tailored to the interconnected dynamics of smart manufacturing. It offers a systematic approach that integrates advanced technologies, aligning with the demands of Industry 4.0. The emphasis on dynamic capabilities, informed by Teece's insights, and the integration of IIoT, as highlighted by Rathore et al., marks a novel contribution in enhancing risk management efficacy within this complex ecosystem. By presenting innovative solutions to current challenges, the study aims to propel the discourse forward, providing valuable insights and paving the way for refined risk management practices in the realm of smart manufacturing.

Theoretical implications and contributions play a crucial role in advancing academic knowledge and understanding. By synthesizing findings with established theories and highlighting specific contributions to the field, the discussion section elevates the scholarly impact of the study. The integration of theoretical insights ensures that the developed conceptual model is grounded in established principles while offering new perspectives and avenues for future research in smart manufacturing risk management.

11 CONCLUSION: NAVIGATING THE HORIZON OF PROACTIVE RISK MANAGEMENT IN SMART MANUFACTURING

11.1 Recapitulation of Key Conceptual Points

In concluding this exploration into proactive risk management within the realm of smart manufacturing, it is imperative to revisit the key conceptual points that have illuminated our journey. The conceptual model developed herein stands as a testament to the intricacies of anticipating, assessing, and mitigating risks in an environment where innovation and interconnectedness define the landscape. Our journey began by recognizing the transformative power of technological integration, embracing the synergies of IoT, AI, and data analytics. The conceptual model unfolded, weaving a narrative of how these technologies create a dynamic tapestry, empowering organizations to not merely react but proactively sculpt the future of risk resilience. This integration, represented in a conceptual diagram, became the visual manifestation of a paradigm shift—a shift toward a more anticipatory and adaptive approach to risk management. As we navigated the theoretical landscape, our findings converged with established theories, drawing from Teece's insights on dynamic capabilities and Rathore et al.'s exploration of IIoT. Synthesizing these theories provided a robust foundation, enriching our understanding of how organizations can fortify themselves against the ever-evolving risks in smart manufacturing.

11.2 Implications for Practice and Future Research

The implications of this study ripple beyond the theoretical realm, extending their influence into the practical domain. Organizations in smart manufacturing are presented with a roadmap—a roadmap that leads toward a future where risks are not only managed but predicted, where resilience is not a reaction but an ingrained capability. The proposed KPIs, derived from the conceptual model, provide a tangible means of measuring success and steering the course of proactive risk management initiatives. Moreover, the real-world examples showcased the feasibility of our conceptual model, drawing inspiration from pioneers like Siemens, GE, Toyota, and Bosch. These exemplars serve as beacons, guiding practitioners to translate theory into actionable strategies. Yet, this is not the end but a juncture for future exploration. Our study opens avenues for further research into nuanced aspects of risk management, the evolving role of technology, and the intersectionality of risk, innovation, and resilience in smart manufacturing. The conceptual model, while robust, is a living entity, waiting to evolve with the advancing landscape of technology and industry practices.

11.3 Closing Thoughts

In closing, this study is more than a compilation of theories, models, and case studies—it is an invitation to a new era of risk management. It beckons organizations to embrace not only the challenges but the boundless opportunities that smart manufacturing offers. It urges scholars to continue the pursuit of knowledge, unraveling the intricate tapestry of risks and resilience in an era where change is the only constant. As we reflect on this journey, let it be known that the horizon of smart manufacturing is not a destination but a perpetual voyage. With each challenge overcome, each opportunity seized, and each insight gleaned, we collectively propel ourselves toward a future where the smartest aspect of manufacturing lies not just in the technology it employs but in the wisdom it applies to navigate the seas of uncertainty. May this study serve as both compass and companion on the ongoing odyssey of proactive risk management in the dynamic world of smart manufacturing. May it inspire, inform, and instill a sense of contentment, knowing that the pursuit of resilience is a journey worth undertaking.

REFERENCES

Agarwal, R., & Ansell, J. (2016). Strategic change in enterprise risk management. *Strategic Change*, *25*(4), 427–439.

Barney, J. B., & Arikan, A. M. (2005). The resource-based view: Origins and implications. In M.A. Hitt, R.E. Freeman & J.S. Harrison (eds.) *The Blackwell Handbook of Strategic Management* (eds. https://doi.org/10.1111/b.9780631218616.2006.00006.x.

Bravo, C., Saputelli, L., Rivas, F., Pérez, A. G., Nikolaou, M., Zangl, G.,…, & Nunez, G. (2014). State of the art of artificial intelligence and predictive analytics in the E&P industry: A technology survey. *SPE Journal, 19*(04), 547–563.

Bromiley, P., McShane, M., Nair, A., & Rustambekov, E. (2015). Enterprise risk management: Review, critique, and research directions. *Long Range Planning, 48*(4), 265–276.

Castelló, I., & Lozano, J. (2009). From risk management to citizenship corporate social responsibility: Analysis of strategic drivers of change. *Corporate Governance: The International Journal of Business in Society, 9*(4), 373–385.

Ganesh, A. D., & Kalpana, P. (2022). Future of artificial intelligence and its influence on supply chain risk management–A systematic review. *Computers & Industrial Engineering, 169*, 108206.

Ghadge, A., Dani, S., & Kalawsky, R. (2012). Supply chain risk management: Present and future scope. *The International Journal of Logistics Management, 23*(3), 313–339.

Hofmann, E., Sternberg, H., Chen, H., Pflaum, A., & Prockl, G. (2019). Supply chain management and Industry 4.0: Conducting research in the digital age. *International Journal of Physical Distribution & Logistics Management, 49*(10), 945–955.

Kaplan, R. S. (2009). Conceptual foundations of the balanced scorecard. *Handbooks of Management Accounting Research, 3*, 1253–1269.

Kaplan, R. S., and D. P. Norton. "The balanced scorecard—measures that drive performance." Harvard business review 70, no. 1 (1992): 71–79.

Kaplan, R. S., & Mikes, A. (2016). Risk management—The revealing hand. *Journal of Applied Corporate Finance, 28*(1), 8–18.

Karmakar, A., Dey, N., Baral, T., Chowdhury, M., & Rehan, M. (2019, March). Industrial internet of things: A review. In *2019 International Conference on Opto-Electronics and Applied Optics (Optronix)* (pp. 1–6). IEEE.

Kusiak, A. (2023). Smart manufacturing. In: Nof, S.Y. (ed.) *Springer Handbook of Automation.* Springer Handbooks. Springer, Cham: Springer International Publishing. (pp. 973–985). https://doi.org/10.1007/978-3-030-96729-1_45

Lund, S., Manyika, J., Woetzel, J., Barriball, E., & Krishnan, M. (2020). Risk, resilience, and rebalancing in global value chains. Rep., McKinsey Glob. Inst. New York: https://www.mckinsey.com/business-functions/operations/our-insights/risk-resilience-and-rebalancing-in-global-value-chains.pdf

McCarthy, C., & Harnett, K. (2014). *National Institute of Standards and Technology (NIST) Cybersecurity Risk Management Framework Applied to Modern Vehicles* (No. DOT HS 812 073). United States. National Highway Traffic Safety Administration.

Miles, M. B., & Huberman, A. M. (1994). *Qualitative data analysis: An expanded sourcebook (2nd ed.)*. Sage Publications, Inc.

Monostori, L. (2014). Cyber-physical production systems: Roots, expectations and R&D challenges. *Procedia Cirp, 17*, 9–13.

Niaz, M., & Nwagwu, U. (2023). Managing healthcare product demand effectively in the post-Covid-19 environment: Navigating demand variability and forecasting complexities. *American Journal of Economic and Management Business (AJEMB), 2*(8), 316–330.

Ogra, A., Donovan, A., Adamson, G., Viswanathan, K. R., & Budimir, M. (2021). Exploring the gap between policy and action in Disaster Risk Reduction: A case study from India. *International Journal of Disaster Risk Reduction, 63*, 102428.

Panetto, H., Iung, B., Ivanov, D., Weichhart, G., & Wang, X. (2019). Challenges for the cyber-physical manufacturing enterprises of the future. *Annual Reviews in Control, 47*, 200–213.

Rathore, M. M., Shah, S. A., Shukla, D., Bentafat, E., & Bakiras, S. (2021). The role of ai, machine learning, and big data in digital twinning: A systematic literature review, challenges, and opportunities. *IEEE Access, 9*, 32030–32052.

Sun, L., He, H., Yue, C., & Lin, W. (2023). Unleashing competitive edge in the digital era: Exploring information interaction capabilities of emerging smart manufacturing enterprises. *Journal of the Knowledge Economy*, 1–45. https://doi.org/10.1007/s13132-023-01545-w

Surminski, S., Aerts, J. C., Botzen, W. J., Hudson, P., Mysiak, J., & Pérez-Blanco, C. D. (2015). Reflections on the current debate on how to link flood insurance and disaster risk reduction in the European Union. *Natural Hazards*, *79*, 1451–1479.

Teece, D. J. (2007). Explicating dynamic capabilities: The nature and microfoundations of (sustainable) enterprise performance. *Strategic Management Journal*, *28*(13), 1319–1350.

Teece, D. J., Pisano, G., & Shuen, A. (1997). Dynamic capabilities and strategic management. *Strategic Management Journal*, *18*(7), 509–533.

Uhlemann, T. H. J., Lehmann, C., & Steinhilper, R. (2017). The digital twin: Realizing the cyber-physical production system for Industry 4.0. *Procedia Cirp*, *61*, 335–340.

Xu, H., Yu, W., Griffith, D., & Golmie, N. (2018). A survey on industrial Internet of Things: A cyber-physical systems perspective. *IEEE Access*, *6*, 78238–78259.

Yin, R. K. (2003). Designing case studies. *Qualitative Research Methods*, *5*(14), 359–386.

Zhao, F. (2004). Management of information technology and business process re-engineering: A case study. *Industrial Management & Data Systems*, *104*(8), 674–680.

Жакупова, Б., Токтарова, М., Ибрашева, А., Нургалиева, Ш., & Сатымбекова, К. (2022). Risk management tactics and strategic directions in the enterprise. *Научный журнал «Вестник НАН РК»* *6*(2), 287–299.

9 Entrepreneurial Strategies for Mitigating Risks in Smart Manufacturing Environments

M. Ashok Kumar, Aliyu Mohammed, Pethuru Raj, and B. Sundaravadivazhagan

1 INTRODUCTION

The smart manufacturing sector has been significantly transformed by the advent of new technologies and integration of modern digital systems. This change has made possible new modes of production, but it had a number of difficulties which I think should be pointed out here. In essence, what the entire gist of the matter is about has been highlighted. This involves the complex entrepreneurial strategies necessary for risk management within smart manufacturing contexts.

1.1 BACKGROUND

There have been remarkable developments globally as smart manufacturing, otherwise referred to Industry 4.0, records unprecedented surge. Nowadays efficiency coupled with improved productivity is evident, thanks to the incorporation of modern techs such as internet of things (IoT), artificial intelligence (AI), and big data in the manufacture industry. This rapid advancement of technology has also exposed the manufacturing environment to a plethora of risks, including security risks and supply chain challenges. It is worth mentioning that smart manufacturing is actually one of the major contributors to the global economy and industrial competence in general. Several countries have adopted this development in technology, and many multinationals have re-engineered its production systems. Smart manufacturing ecosystem has faced both opportunities and challenges due to increasing international market interconnectivity through digital platforms, especially with the emerging global e-commerce and online market (Figure 9.1).

1.2 SIGNIFICANCE OF THE STUDY

The concept of understanding underlying risks in smart manufacturing is key and also a requirement of continued growth and firm sustainability should involve making proper strategic entrepreneurship initiatives that would help alleviate such risks.

DOI: 10.1201/9781032694375-9

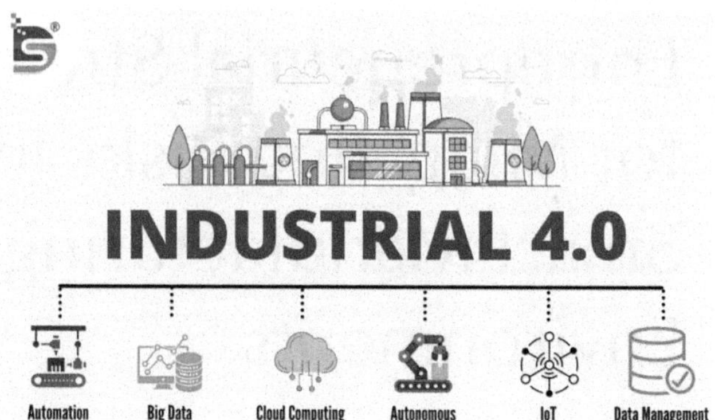

FIGURE 9.1 Incorporation of Modern Techs in the Manufacture Industry.

Source: https://www.linkedin.com/pulse/industry-40-integration-big-data-iot-sdreatech/

This study may be significant because it could provide some cross-border lessons for entrepreneurs and other players in the smart manufacturing arena.

1.3 Scope and Limitations

The chapter is conceptual and explores smart manufacturing management strategies used to mitigate the risks through a qualitative research approach. Although the chapter admits that smart manufacturing covers a broad range of topics related to entrepreneurial risk mitigation, it limits itself to an examination of their underlying principles and applications only.

1.4 Research Questions

1. How do the entrepreneurial strategies of reducing risks in smart manufacturing?
2. What the impact of e-commerce and online marketplaces on these strategies?
3. Which theoretical framework can be related to this study?

1.5 Objectives of the Study

1. Conceptualizing entrepreneurial strategies for smart manufacturing risk management.
2. Measurement of e-commerce and online marketplaces and their role in risk mitigation.
3. Examine various existing theories to come up with appropriate ones.

1.6 Definition of Key Terms

Smart Manufacturing: Adoption of sophisticated technologies like IoT and AI in production systems towards higher efficiencies and output levels.

Entrepreneurial Strategies: Entrepreneurial approaches as well as innovations in identifying, assessing and management of risks in entrepreneurial operations.

E-commerce and Online Marketplaces: Digital platforms providing for electronic purchase and sale of commodities and services across borders thus changing the world arena.

This research seeks to investigate the historical background in literature through exploring basic issues related to a broader topic, identifying existing lacunas, explaining why it is necessary to fill them to justify the study.

2 STATEMENT OF THE PROBLEM

Smart manufacturing is now upon us in new era of highly technological sophistication that changed manufacture into modern process. While this evolution is ongoing, it hasn't come without its strides with risk identification/mitigation being two key focus areas. The statement of the problem is outlining the main issues of the research which explains why the author decided to concentrate on entrepreneurial strategies of risks mitigation of smart manufacturing environment.

2.1 The Concept Smart Manufacturing Environment Overview

Smart manufacturing environment has recorded significant expansion due to the convergence of IoT, AI, and big data. Such type of surroundings will be more effective, cheap and increase product rate. Although, the intricate nature of these technologies implies risks that pose obstacles to the smooth operation of smart manufacturing systems.

2.2. Identification Risk Management in Smart Manufacturing

Challenges associated with identifying risks in the smart manufacturing environments. Businesses will have to face various kinds of risks such as cybersecurity issues, data breach, system failure, and supply chain disruption when they operate in the smart manufacturing context. These are complex risks, and any approaches to their reduction must be carefully thought out (Figure 9.2).

2.3 Need for Entrepreneurial Strategies

Smart manufacturing faces changing risks that call for entrepreneurial strategies to address them. Typically, traditional risk management approaches fail in addressing the specific risks and complications that emanate from the introduction of new technologies. It is worth noting that these manufacturing environments are driven by entrepreneurs who come up with innovative and contextual approaches towards management of the associated smart manufacturing risks.

2.4 Rationale for the Study

Therefore, the objective of this research is to close the gap between increased complicatedness of smart manufacture and adequate risk abatement policies.

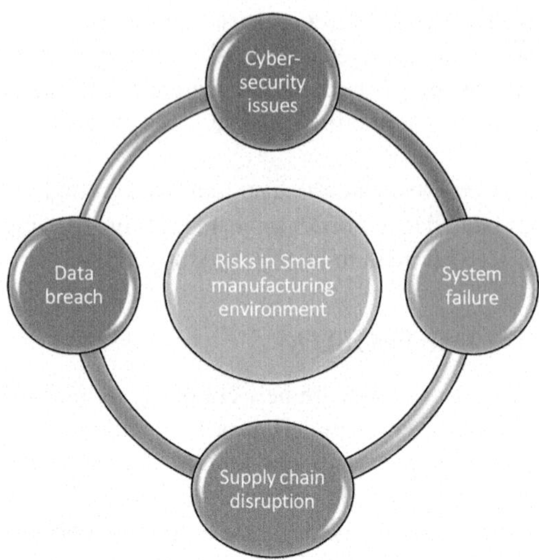

FIGURE 9.2 Risks in Smart Manufacturing Environment.

The use of smart manufacturing in businesses makes it quite necessary for companies to know the entrepreneurial strategies that will ensure smooth operation. The research focuses on conceptual issues regarding risk management in smart manufacturing environments that incorporates a responsive strategy for future demands. Firstly, it is important to note that the statement of the problem captures the threats associated with risks in smart production, illustrates the need for entrepreneurial strategies as well as highlights the reason for undertaking the research that investigates the theory and practical considerations relating to risk management in smart production

3 LITERATURE REVIEW

3.1 Evolution of Smart Manufacturing

There have been numerous studies conducted on how smart manufacturing has evolved over time. For instance, scholars, including Rane (2023a, 2023b), highlighted that there was a transition from traditional manufacturing to smart manufacturing, which marked a revolution facilitated by technologies such as the IoT and AI. Rane (2023a, 2023b) considered Industry 4.0 as the Fourth Industrial Revolution that combined cyber and physical systems.

3.2 Risk Factors in Smart Manufacturing Environments

The literature about risk factors in smart manufacturing environments stresses the multidimensional problems that companies are experiencing when they implement Industry 4.0. Authors such as Alhakami (2023) pointed out many cybersecurity threats and emphasized on the exposure brought forth by extra connections.

Disruption of the supply chain has also been addressed by Shen and Sun (2023) whereby it is important to identify risks for the whole value chain.

3.3 Entrepreneurial Approaches in Risk Mitigation

Latest literature has highlighted entrepreneurial approaches to risk mitigation within smart manufacturing. Scholars such as developed a theory about effectuation advocating reliance on available assets as well as iterate choice making. Weiss and Ariyachandra (2023) discuss about entrepreneurial risk taking behaviour and the role of dynamic capability of adaptation with respect to unpredictable smart manufacturing.

3.4 The Role of E-Commerce and Online Marketplaces

Many researchers have studied the convergence between e-commerce/online marketplaces and smart manufacturing (Wang, Hou and Shin, 2023). The rise of digital platforms for enhanced visibility and coordination has fostered platform ecosystems. This convergence within the smart manufacturing landscape presents both novel opportunities and potential risks for businesses. Moreover, Li, Zhang and Cao (2023) articulated the emergence of platform ecosystem highlighting how smart manufacturing landscape offers firms novel prospects and risks (Figure 9.3).

3.5 Case Studies on Successful Entrepreneurial Strategies

Case studies on successful entrepreneurial strategies give practical examples on how companies can protect themselves against certain type(s) of risk. One of the notable cases is an analysis on how Tesla applied entrepreneurial thinking and overcame the challenges posed by smart manufacturing, which can be seen in the work done by Singh and Singh (2023). Jaloliddin (2023) states that another instance is the case of Siemens which illustrates positive risk management via strategic investments in

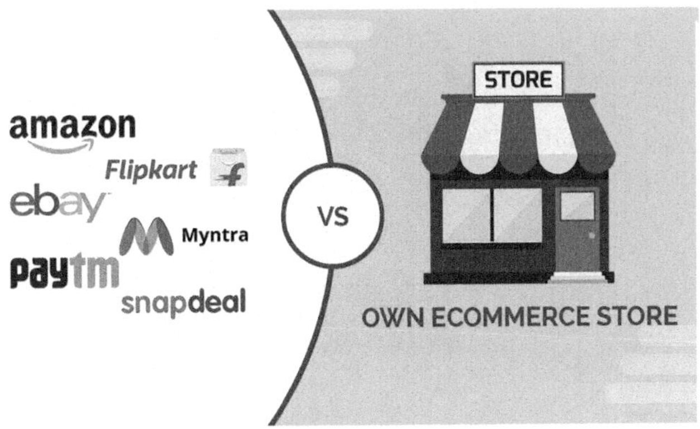

FIGURE 9.3 E-Commerce and Online Marketplaces.

Source: https://www.iconnectsolution.com/blog/ecommerce-vs-marketplace

digitization. To conclude, the literature review highlights on the growth of smart manufacturing, identifies risk factors, outlines entrepreneurial means for combating risks, considers e-commerce and online market places, and analyses cases that show examples of prosperous entrepreneurial approaches. Scholars' contributions offer an insightful picture of the theoretical bases and the practices associated with strategic risk management in smart production systems.

4 CONCEPTUAL FRAMEWORK

4.1 Definition of Entrepreneurial Strategies

The entrepreneurial approach to smart manufacturing entails the use of innovative and proactive measures used by entrepreneurs as a means of identifying, assessing, and minimizing such risks. Scholars such as Keyhani (2023) conceptualize entrepreneurial strategies as actions made to explore chances, cope with uncertainty, and gain an edge over competitors. Moreover, Acharya and Berry (2023) proposed a model called effectuation which encompasses utilizing available resources and flexible decision-making among new ventures (Figure 9.4).

4.2 Integration of Entrepreneurial Strategies into Smart Manufacturing

Entrepreneurial strategy is important in integrating the intricacy of Industry 4.0 into smart manufacturing. Pitelis and Teece (2010) state that dynamic capabilities form the core of this integration, enabling firms to adjust and re-organize their resource base in reaction to a new business environment. This is further highlighted in the works of Aleksandrovna (2023), which places an emphasis on the relevance of entrepreneurial orientation that emphasizes innovation, risk taking, and responsiveness in exploiting smart manufacturing opportunities.

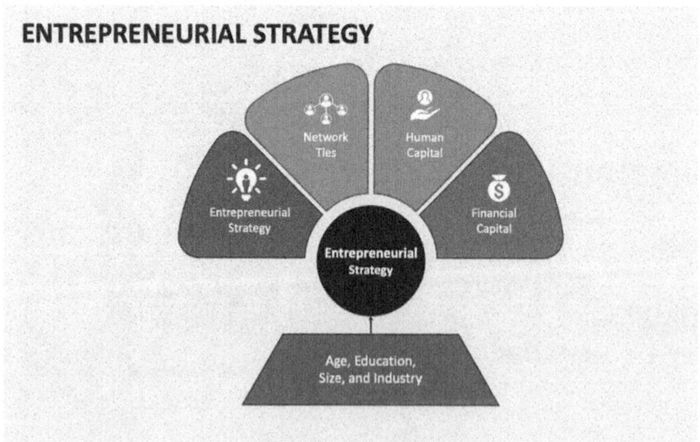

FIGURE 9.4 Entrepreneurial Strategies.

Source: https://www.collidu.com/presentation-entrepreneurial-strategy

4.3 Theoretical Foundations for Entrepreneurial Strategies

Entrepreneurial strategies in smart manufacturing are based on different viewpoints. Resource-based view (RBV) of Barney (1991) is that a competitive advantage lies in firms' specific, valuable, and not-easy-to-imitate resources. Smart manufacturing sees entrepreneurs who use it as a perspective through which they strategically deploy their resource like technologic capabilities and skilled workforce to effectively manage risks. In addition, Wen and Wen's (2023) dynamic capabilities theory underscores the ability for organizations to be responsive towards environment changes. This theory is relevant in regard to smart manufacturing since entrepreneurs have to keep changing their capacities so that they may adapt to the new risks which emerge. Moreover, Ke and Huang (2023) add another dimension to the analysis of how entrepreneurs bear risks for comparative advantage in smart manufacturing environments based on risk-bearing theory.

4.4 Alignment of E-Commerce and Online Marketplaces

This is where entrepreneurial strategies and e-commerce align with smart manufacturing. "Platform ecosystems," discussed by Nerbel and Kreutzer (2023), refers to the creation of new avenues for entrepreneurship supported by modern digital platforms. Asortse and Denga (2023) support this argument that e-commerce improves visibility and connection as part of the supply channel that impacts decision making process about risks management for entrepreneurship. Similarly, Wang, Hou, and Shin (2023) provide insights into ways that e-commerce enhances entrepreneurial strategies. Integration of digital platforms within the value chain can promote streamlined processcss and more efficient risk management. To this effect, the work of Li and Kumarasinghe (2023) explains how e-commerce impacts supply chain management and gives some ideas on what practices should be adopted in the current smart manufacture scenario for the success and alignment Essentially, the conceptual framework articulates the meaning of entrepreneurial strategies in smart manufacturing, the underlying principles behind such strategies, cohesion with e-commerce and online marketplaces. In summary, this section reviews the contributions of different scholars that form a strong basis for understanding the relationship between entrepreneurial strategies and intelligent manufacture environment.

5 THEORETICAL FRAMEWORK

5.1 Overview of Chosen Theoretical Framework

For this study, an RBV and dynamic capabilities theory were chosen as the theoretical framework. As per Jones et al. (1997), RBV, argued by Barney (1991), explains that organization's competitive edge is based on its rare and valued resource. The dynamic capabilities theory of Teece, Pisano, and Shuen (1997) complements RBV, arguing that firms need to constantly adjust and re-shape their capabilities as they respond to fast-changing environments.

5.2 Application of the Framework in Smart Manufacturing

The RBV is essential for realizing a good entrepreneurial strategy based on exploiting distinct resources that help reduce risks in smart manufacturing. Entrepreneurs will need technological skills, workforce skill, and innovation process as resources in smart manufacturing in compliance with RBV principles by Barney (1991). Smart manufacturing is faced with rapidly emerging technologies and changing risks which make dynamic capabilities theory very applicable. Dynamic capabilities means a firm's capability to identify, grasp, and reconfigure its competitive advantages as illustrated by Teece, Pisano, and Shuen (1997). This implies that entrepreneurs are able to sense on emerging threats, spotting opportunities as well as dynamically revising strategies for proper risk response in smart manufacturing. Theories have been furthered in terms of the need for entrepreneurial orientation within a smart manufacturing context. These all point towards including dynamic capabilities and resource leveraging as integral parts of entrepreneurial strategies when firms operate in the complicated, uncertain terrain of Industry 4.0.

5.3 Examples of Successful Implementations

Implementation of the selected theoretical framework in smart manufacturing is evidenced in companies like Siemens and General Electric. As indicated by Yeow, Soh, and Hansen (2018), Siemens pursued strategic investments on digital technologies following RBV and dynamic capabilities theory. It used advanced technologies in order to adapt to changing markets and mitigate risk. Also, dynamic capabilities were applied during transformation at General Electric; see the study on this subject by Teece, Pisano and Shueh (1997). When General Electric felt the shift in the industrial environment took place, it realigned its resources to remain competitive. The successful implementation of this highlights the practical aspects of the chosen theoretical framework and its usefulness for tackling smart manufacture challenges. Therefore, the above-discussed theories of RBV and dynamic capabilities provide a holistic perspective to examine the entrepreneurial strategies smart manufacturing. The utilization of these models clarifies why there should be differentiated resources, adaptiveness, and an enterprising entrepreneurial orientation. Such a concept could be shown on the basis of successful imposition at Siemens and General Electric as practical manifestation of the theory in conditions of smart manufacturing.

6 EMPIRICAL STUDY

6.1 Methodology

6.1.1 Research Design

In order to make sure that the qualitative research design is suitable for studying the entrepreneurship strategies used by manufacturers to counteract risks in the development of smart production, literature review was conducted with reference to relevant works focused on entrepreneurship in the framework of intelligent manufacturing. Following the rules of a qualitative approach, the study was designed to describe and understand what is known with regard to the concept. A narrative synthesis approach was used in analysing and combining results from various sources.

6.1.2 Data Collection

The data collection for this study involved a comprehensive literature review that covered various scholarly articles as well as many case studies regarding the entrepreneurial strategy in smart manufacturing environments. Some of the sources were peer-reviewed journals, conference proceedings, and books. A strong basis for the conceptual study was emphasized in the procedure of collecting data. This required considering relevance and depth in the selection of information.

6.1.3 Sampling Techniques

In this review, the sampling method was purposive, emphasizing on research studies with vital information towards strategic approaches and risk management in smart manufacturing. The key inclusion criterion was whether the included paper answered the research questions and the information was detailed. There was a need for sampling a variety of studies drawn from different parts of the world and across various sectors capturing the global picture.

6.2 Extensive Review of Related Studies

6.2.1 Comparative Analysis of Entrepreneurial Strategies

Comparative analysis has shown the range of entrepreneurial approaches used by companies working in smart production management environment. For instance, a scholar like Ryman and Roach (2022) pointed out that effectuation is a strategy which involves exploitation of resources available and taking chances with uncertainty. On the other hand, Abdullah (2023) focused on casual and goal oriented perspective towards the entrepreneurship. These perspectives were brought together and contributed to a deeper appreciation of the variety of techniques at the disposal of an entrepreneur in the area of smart manufacturing.

6.2.2 Impact of E-commerce on Risk Mitigation

A focus of this review addressed the effect of e-commerce on reducing the risk during smart manufacturing. As depicted by Patel (2023), digital platforms facilitate fast response to disruptions, enhancing supply chain visibility. Furthermore, Zainal and Hamdan (2023) proposed the notion of platform ecosystems and showed how e-commerce and online marketplace are opening up new opportunities and risks which entrepreneurs should be prepared to deal with. Comparing these studies led to an understanding of the ways that e-commerce can be used to mitigate risks in smart manufacturing.

6.3 Research Gap

6.3.1 Identification of Existing Gaps

As a result, the comprehensive literature review revealed gap areas within the current research. Despite plenty of knowledge about start-up strategies and e-commerce as a risk mitigating strategy, it was evidently missing to integrate both comprehensively in a general background of smart manufacturing. In the context of the complexity of Industry 4.0, an important gap was presented by lack of one holistic model that could integrate entrepreneurship strategy with e-commerce.

6.3.2 Filling the Gap

This gap highlighted the need for a concept study which will show where the gap was found in smart manufacturing environment with regard to entrepreneurial strategy and e-commerce's risk mitigation. Such integration became necessary for a deeper understanding and practical knowledge for entrepreneurs in meeting challenges of Industry 4.0. To sum up, this empirical study was conducted using the qualitative method that involved extensive literature review from previous related studies to lay the theoretical base. A comparison of entrepreneurial strategies showed that their perspectives were varied while their impact on risk mitigation and the existing gap in literature led to this study for addressing this knowledge vacuum.

7 FINDINGS

7.1 Summary of Empirical Results

The empirical findings of the research illustrate a thorough awareness on the risk management strategies in smart manufacturing settings. Finally, the summary of empirical results summarizes vital conclusions made through a thorough literature review that illustrates dynamic relations between entrepreneurship and challenges of industrialization during the fourth revolution. Scholars like Gul (2023) had different ideas about entrepreneurial tactics, leading into an arc of causation and effectuation. Combining these approaches gives rise to a broadly applicable model designed for smart manufacturing environment. Also, a significant topic was the effect of e-commerce/online marketplace, on risk mitigation in smart manufacturing. As shown by Aserkar (2023), digital platforms help in improving supply chain visibility and providing opportunities for entrepreneurs. The empirical findings provided a foundation for closing the gap that was identified from the literature.

7.2 Insights on Successful Entrepreneurial Strategies

In-depth analysis of the case studies revealed important lessons on successful business strategies. For instance, studies such the one involving Tesla analysed by Güettel and Konlechner (2023) indicated how the firm used the new technology to address hurdles in smart manufacturing. The success of Siemens, as explained by Jiaqi (2023), indicated the essentiality of a wise investment in the digital technology and resource utilization. These ideas offer a real grasp on how entrepreneurial strategies could be adjusted to suit the challenges and opportunities presented in smart manufacturing environments. Combining the success of these implemented solutions is crucial towards developing an effective framework for enterprise risk management.

7.3 Implications for Smart Manufacturing Environments

Several implications emerge from these findings. The study highlights the need for using different entrepreneurship approaches due to the dynamic nature of smart manufacturing and its challenges. An integrated combination of the effectuation and causation with the help of RBV and the dynamic capabilities provides an idea for the entrepreneurship in the uncertain environments. Additionally, this research outlines the importance of e-commerce and online marketplaces in terms of risk reduction. However, this has implications that go further than the usual supply chain management. This includes a call on entrepreneurship to use the digital platforms to ensure better visibility, coordination, as well as adaptability with ever-changing market environments. Finally, the implications of success stories by entrepreneurs give a guide or direction on what may be adopted. The ways through which companies such as Tesla and Siemens were able to manage risk while making strategic investments in resources acts as blueprints that other smart manufactures can adopt. Summarily, this study confirms the existing literature gap while at the same time fills it by presenting a combined model for diverse entrepreneurial strategies as well as the effect of e-commerce to risk avoidance within smart manufacturing settings. These findings have provided practical implications which can be used by both entrepreneurs, as well as other stakeholders operating within the dynamic domain of Industry 4.0.

8 RECOMMENDATIONS

8.1 Practical Recommendations for Entrepreneurs

Practical advice directed towards entrepreneurs in smart production zones is integral for successful avoidance of risk.

1. Leveraging Effectuation and Causation: Therefore, as entrepreneur, you need to appreciate the importance of effectuation and causation when making your decisions as it is vital for the success of any business idea. The same is emphasized in the study of Li and Long (2023).
2. Continuous Adaptation: According to dynamic capabilities theory, entrepreneurs should focus on the creation of adaptable organizational capacity in order to remain responsive to dynamic environments. It entails regular upgrades in technological abilities, workforce proficiency, and processes so that there can be adequate response during change events.
3. Strategic Investments in E-commerce: Therefore, it is important for entrepreneurs to have a deliberate investment plan towards their digital platforms to curb the influence of e-commerce and online marketplaces as well their inherent risk. It involves improving on supply chain visibility, examining platforms' ecosystems, and using online marketplaces to design innovative business models and enhance operational efficiency (Allioui and Mourdi, 2023).
4. Learning from Successful Cases: Successful cases including that of Tesla and Siemens provide insight into vital lessons for entrepreneurs. This offers valuable insights on how these companies were able to manage smart manufacturing risks as well as effective use of investments and resources.

8.2 Policy Recommendations for Industry Stakeholders

Also, there are other important players such as industry stakeholders and policymakers who contribute significantly towards creating the right environment for efficient risk mitigation while in smart manufacturing.

1. Promoting Collaboration: Policy makers may stimulate cooperation between technology providers and industry players. Sharing knowledge and innovations through creating platforms facilitates communal approach towards similar problems and risks.
2. Investing in Digital Infrastructure: The governments and regulatory bodies should also ensure that there is good digital infrastructure with enhanced cyber security. It is crucial in ensuring that smart manufacturing systems are not susceptible to cyber-attack or disruption.
3. Developing Regulatory Frameworks: The policymakers therefore need to come up with flexible regulatory structures that support innovative ideas as well as responsive entrepreneurial practices. This entails changing regulations in line with smart manufacturing.

8.3 Areas for Future Research

Although the study is crucial in filling a major void in the literature, it equally suggests aspects for further exploration on how entrepreneurs can use these smart manufacturing techniques.

1. Exploring Contextual Differences: The effectiveness of this strategy in other industries and regions should therefore be the focus for future research.
2. Longitudinal Studies: Longitudinal studies could be useful in determining whether or not the strategies on entrepreneurship are sustainable. This helps us understand how adaptive capacities emerge historically.
3. In-Depth Analysis of E-commerce Impact: More specifically, it could be about investigating the implications of these factors on smart manufacturing in the area of e-commerce and its implementation.

Finally, these recommendations that emphasize on the findings of the research will be useful for entrepreneurs, industry stakeholders, and future researchers in this discussion on entrepreneur's strategic approach in smart manufacturing. These suggestions will allow entrepreneurs and other stakeholders to be active participants in making smart manufacturing eco-system more robust and innovative.

9 CONCLUSION

9.1 Recapitulation of Key Findings

Summarizing the conclusions of this broad study about the entrepreneurship risk management strategies in a smart production system, we can see the depth of the information obtained while researching the relevant literature. According to the research smart manufacturing revolutionized with the incorporation of a new generation of technology brought forth unforeseen chances and diverse threat factors. This meant that it was critical to identify these risks and recognize a requirement for fresh, creative entrepreneurial approaches which are appropriate for Industry 4.0. The selected theoretical standpoint, based on the RBV and the dynamic capabilities theory created an articulate perspective of analysis on entrepreneurship techniques. This combined the effectuation and cause with the theory formulated that background to help explain how entrepreneurs in smart environment cope with uncertainty and complicacies within that environment. In addition, the research provides a synthesized framework that fills the gaps which were established by an extensive literature review and found existed within the literature. They were examples from companies that successfully adopted entrepreneurial strategies. Such as Tesla and Siemens, among others, showed how these strategies could help build a more robust and imaginative intelligent manufacturing environment.

9.2 Contribution to Knowledge

There are myriad ways, as a result, this research adds to the existing literature with regards to the topic of discussion. Moreover, it provides a holistic model whereby smart manufacturing specific challenges are matched together with entrepreneurial strategies. This contributes greatly towards understanding what was previously not very clear because the existing literature largely adopts fragmented perspectives. The second contribution is related to how entrepreneurial strategies interact with risks mitigated by e-commerce and online marketplaces. It sheds light on how digital platforms can be tactfully taken advantage of providing some sophistication of an industry four point of view to the entrepreneur's kitbag. Finally, the examination of successful case studies is also an important source of information for both researchers and practitioners. The study learns from a host of successful risk-management approaches used by companies in smart manufacturing and shares practical steps on how these lessons may be applied in other industries and settings.

9.3 Concluding Remarks

In conclusion, it can be noted that smart manufacturing necessitates more active and creative ways of dealing with risks. A successful entrepreneurship must incorporate this strategy where it is built on strong theoretical basis and align to the complexities of Industry 4.0. In our journey of modern manufacturing defined by constant technology advancement and global outreach, those lessons of the study acquire vital importance. By combining these insights, entrepreneurs, industry stakeholders, and policymakers can build resilient smart manufacturing ecosystems which promote innovative strategies and sustainability. Therefore, this research is both academic in nature and a guide for those involved in modern manufacture and directs them for the risks into the strategy instead of the challenge. Here, then let these lessons become guiding lights in our endeavour to make smart manufacturing, flexible, innovative, and agile.

REFERENCES

Abdullah, M. A. (2023). Digital maturity of the Egyptian universities: goal-oriented project planning model. *Studies in Higher Education*, 1–23. https://doi.org/10.1080/03075079.2023.2268633

Acharya, K., & Berry, G. R. (2023). Characteristics, traits, and attitudes in entrepreneurial decision-making: current research and future directions. *International Entrepreneurship and Management Journal, 19*, 1965–2012. https://doi.org/10.1007/s11365-023-00912-y

Aleksandrovna, S. A. (2023). The influence of entrepreneurial orientation on firms survival: a cross-case analysis of Russian firms.

Alhakami, W. (2023). Computational study of security risk evaluation in energy management and control systems based on a fuzzy MCDM method. *Processes, 11*(5), 1366.

Allioui, H., & Mourdi, Y. (2023). Exploring the full potentials of IoT for better financial growth and stability: a comprehensive survey. *Sensors, 23*(19), 8015.

Aserkar, R. (2023). Fabrizex: creating a technology platform for global supply chains. *Emerald Emerging Markets Case Studies, 13*(1), 1–17.

Asortse, S., & Denga, E. M. (2023). The relevance of supply chain preparedness on the long-term sustainability of SMEs. In R. Potluri & N. Vajjhala (Eds.), *Advancing SMEs Toward E-Commerce Policies for Sustainability* (pp. 111–133). IGI Global. https://doi.org/10.4018/978-1-6684-5727-6.ch006

Barney, J. (1991). Firm resources and sustained competitive advantage. *Journal of Management, 17*(1), 99–120.

Gul, S. (2023). Exploring the honey value chain among Afghan Refugees in Pakistan. *Journal of Business and Management Research, 2*(2), 828–859.

Güttel, W.H., Konlechner, S.W. Continuously Hanging by a Thread: Managing Contextually Ambidextrous Organizations. *Schmalenbach Bus Rev 61*, 150–172 (2009). https://doi.org/10.1007/BF03396782

Jiaqi, X. (2023). Conceptualizing public-private partnerships for technology innovation and digital transformation in China's post-pandemic recovery. *Journal of Digitainability, Realism & Mastery (DREAM), 2*(04), 10–16.

Jones, C., Hesterly, W. S., & Borgatti, S. P. (1997). A General Theory of Network Governance: Exchange Conditions and Social Mechanisms. *The Academy of Management Review, 22*(4), 911–945. https://doi.org/10.2307/259249

Ke, Chunyuan, and Shi-Zheng Huang. "The Effect of Environmental Regulation and Green Subsidies on Agricultural Low-Carbon Production Behavior: A Survey of New Agricultural Management Entities in Guangdong Province." *Environmental Research*, vol. 242, 2024, p. 117768, https://doi.org/10.1016/j.envres.2023.117768.

Keyhani, M. (2023). The logic of strategic entrepreneurship. *Strategic Organization, 21*(2), 460–475.

Li, F. and Long, J. (2023), "Tapping into the configurational paths to employee digital innovation in the realm of the dualistic AMO framework", *European Journal of Innovation Management*, Vol. ahead-of-print No. ahead-of-print. https://doi.org/10.1108/EJIM-06-2023-0442

Li, X., Zhang, L., & Cao, J. (2023). Research on the mechanism of sustainable business model innovation driven by the digital platform ecosystem. *Journal of Engineering and Technology Management, 68*, 101738.

Li Yang, P.J. Kumarasinghe, A behavioral model of service-derived manufacturing in e-commerce companies from the innovation chain perspective: A case study from China, *Heliyon*, Volume 9, Issue 12, 2023, e23080, ISSN 2405-8440, https://doi.org/10.1016/j.heliyon.2023.e23080.

Madsen, E. L. (2007). The significance of sustained entrepreneurial orientation on performance of firms–a longitudinal analysis. *Entrepreneurship and Regional Development, 19*(2), 185–204.

Nerbel, J. F., & Kreutzer, M. (2023). Digital platform ecosystems in flux: from proprietary digital platforms to wide-spanning ecosystems. *Electronic Markets*, *33*(1), 1–20.

Patel, K. R. (2023). Enhancing global supply chain resilience: effective strategies for mitigating disruptions in an interconnected world. *BULLET: Jurnal Multidisiplin Ilmu*, *2*(1), 257–264.

Pitelis, C. N., & Teece, D. J. (2010). Cross-border market co-creation, dynamic capabilities and the entrepreneurial theory of the multinational enterprise. *Industrial and Corporate Change*, *19*(4), 1247–1270.

Rahmonov Jaloliddin, 'Digitalization in Global Trade: Opportunities and Challenges for Investment', (2023), 18, *Global Trade and Customs Journal*, Issue 10, pp. 391–395,

Rane, Nitin, ChatGPT and Similar Generative Artificial Intelligence (AI) for Building and Construction Industry: Contribution, Opportunities and Challenges of Large Language Models for Industry 4.0, Industry 5.0, and Society 5.0 (September 21, 2023). Available at SSRN: http://dx.doi.org/10.2139/ssrn.4603221

Rane, Nitin, Enhancing Customer Loyalty through Artificial Intelligence (AI), *Internet of Things (IoT), and Big Data Technologies: Improving Customer Satisfaction, Engagement, Relationship, and Experience* (October 13, 2023). Available at SSRN: http://dx.doi.org/10.2139/ssrn.4616051

Ryman, J. A., & Roach, D. C. (2022). Innovation, effectuation, and uncertainty. *Innovation*, 26(2), 328–348. https://doi.org/10.1080/14479338.2022.2117816

Shen, Z. M., & Sun, Y. (2023). Strengthening supply chain resilience during COVID-19: a case study of JD. com. *Journal of Operations Management*, *69*(3), 359–383.

Singh, H. and Singh, B. (2023), "Industry 4.0 technologies integration with lean production tools: a review", *The TQM Journal*, Vol. ahead-of-print No. ahead-of-print. https://doi.org/10.1108/TQM-02-2022-0065

Teece, D. J., Pisano, G., & Shuen, A. (1997). Dynamic capabilities and strategic management. *Strategic Management Journal*, *18*(7), 509–533.

Wang, G.; Hou, Y.; Shin, C. Exploring Sustainable Development Pathways for Agri-Food Supply Chains Empowered by Cross-Border E-Commerce Platforms: A Hybrid Grounded Theory and DEMATEL-ISM-MICMAC Approach. *Foods* 2023, 12, 3916. https://doi.org/10.3390/foods12213916

Wathen, Patrik. "Decision-making Moon-bridge: SMEs' Usage of Causation and Effectuation Perpetually in Motion." Vaasan yliopisto, 2023, https://urn.fi/URN:ISBN:978-952-395-100-6. Acta Wasaensia 515. ISBN 978-952-395-099-3 (print), 978-952-395-100-6 (online). ISSN 0355-2667 (print), 2323-9123 (online).

Weiss, E., & Ariyachandra, T. (2023). Resilience during times of disruption: the role of data analytics in a healthcare system. In *Proceedings of the ISCAP Conference ISSN* (Vol. 2473, p. 4901).

Wen, Y. and Wen, S. (2024), "The relationship between dynamic capabilities and global value chain upgrading: the mediating role of innovation capability", *Journal of Strategy and Management*, Vol. 17 No. 1, pp. 123–139. https://doi.org/10.1108/JSMA-05-2023-0096

Yeow, A., Soh, C., & Hansen, R. (2018). Aligning with new digital strategy: a dynamic capabilities approach. *The Journal of Strategic Information Systems*, *27*(1), 43–58.

Zainal, F.Y., Hamdan, A.M.M. (2023). The Impact of E-commerce on the Development of Entrepreneurship: Literature Review. In: El Khoury, R., Nasrallah, N. (eds) *Emerging Trends and Innovation in Business and Finance*. Contributions to Management Science. Springer, Singapore. https://doi.org/10.1007/978-981-99-6101-6_23

10 Fortifying Smart Manufacturing against Cyber Threats

A Comprehensive Guide to Cybersecurity Integration, Best Practices, and Frameworks

Aliyu Mohammed, M. Ashok Kumar, Pethuru Raj, and M. Sangeetha

1 INTRODUCTION

Smart manufacturing, propelled by the convergence of cutting-edge technologies such as the internet of things (IoT) and artificial intelligence (AI), has ushered in a transformative era for industrial processes, optimizing efficiency and output. As industries embrace these advancements, they concurrently expose themselves to a myriad of cyber risks, necessitating a proactive approach to cybersecurity management. According to the Sobb et al. (2020), the increasing interconnectivity in manufacturing processes amplifies the potential impact of disruptions, making it imperative for organizations to adopt comprehensive cybersecurity strategies. The complexity of smart manufacturing systems introduces multifaceted cybersecurity risks that extend beyond traditional challenges. Cyber threats such as ransomware attacks and data breaches pose significant dangers to the integrity of sensitive information and the seamless operation of interconnected devices. Operational failures, supply chain disruptions, and compliance issues further compound the cybersecurity landscape. A study by Queiroz et al. (2021) underscores the importance of addressing these challenges, noting that the digitization of supply chains and manufacturing processes necessitates an evolved cybersecurity paradigm.

Proactive cybersecurity management becomes paramount in this context, as relying solely on reactive measures leaves organizations vulnerable to emerging threats. Thomas and Sule (2023) emphasize the shift from reactive to proactive cybersecurity management, advocating for a continuous risk assessment and mitigation cycle to

DOI: 10.1201/9781032694375-10

stay ahead of evolving cyber risks. By taking a proactive stance, organizations can not only prevent potential disruptions but also gain a competitive edge by ensuring the cybersecurity resilience of their smart manufacturing ecosystems. This chapter delves into the landscape of cybersecurity risks inherent in smart manufacturing, emphasizing the importance of a proactive cybersecurity management approach. We explore the components of a comprehensive cybersecurity strategy, the integration of cutting-edge technologies in cybersecurity, and the need for continuous improvement to adapt to the evolving smart manufacturing environment. We want to offer a comprehensive grasp of the proactive cybersecurity tactics crucial for protecting smart manufacturing from cyberattacks through case studies and an emphasis on regulatory compliance.

1.1 Background of Smart Manufacturing

Smart manufacturing, which is sometimes used interchangeably with Industry 4.0, represents a significant change in industrial processes via the integration of cutting-edge technology to increase productivity and efficiency. Fundamentally, this paradigm welcomes the merging of AI with the IoT, resulting in networked systems with the ability to share data in real time and make decisions on their own. Smart manufacturing relies heavily on the industrial internet of things (IIoT), which makes it possible for machines and devices to communicate with each other effortlessly. This network of connections makes dynamic data interchange possible and is the foundation of intelligent industrial processes. This is enhanced by AI, which gives computers cognitive ability so they can independently examine large datasets and come to wise judgments. IoT and AI work together to provide real-time adaptation in production processes, resulting in increased responsiveness and agility.

The revolutionary potential of smart manufacturing is highlighted by Ahmadi et al. (2020). It draws attention to the chances it offers for lower costs, better use of resources, and higher-quality products. In addition to changing the operating environment, the incorporation of cutting-edge technology paves the way for the creation of intelligent, flexible production processes. In practical terms, smart manufacturing translates into interconnected factories and supply chains where every component serves as a node in a vast network. Because of this linked environment, real-time data can be easily gathered, analyzed, and used, allowing manufacturers to improve workflows, anticipate maintenance requirements, and reduce downtime. The vision of intelligent, adaptable, and connected manufacturing becomes a reality when sectors embrace this development, bringing in a new age where efficiency and creativity work together harmoniously.

1.2 The Rise of Cyber Threats in Smart Manufacturing

The emergence of smart manufacturing presents a significant problem due to the increase in cyber dangers, despite the potential for significant improvements in productivity and efficiency through networked systems. Despite the significant advantages, bad actors can quickly take advantage of the vulnerabilities that are created by the integration of digital technologies and the interconnectedness of devices inside smart manufacturing systems. The severity of the problem is highlighted by a report released by the Tuptuk and Hailes (2018), which also highlights the growing cyberattacks that are aimed at manufacturing systems. These dangers cover a wide range of assaults, from ransomware intrusions to theft of intellectual property. Given its vital position in national economies, the industrial sector has emerged as a top target for hackers looking to make financial or geopolitical benefits. There are two sides to the growing digitalization of industrial operations. It increases operating capabilities, but it also gives cybercriminals a larger area to assault. Given the significant reliance of smart manufacturing systems on the smooth interchange of real-time data and autonomous decision-making, any breach in the integrity of these processes can have dire repercussions, such as lowered product quality and production delays.

Adopting a proactive cybersecurity approach is essential to solving this urgent problem. To reduce the danger of cyberattacks, manufacturers need to have strong security measures in place, such as encryption techniques, safe access restrictions, and ongoing monitoring. Government organizations, business partners, and cybersecurity specialists must work together to create and exchange best practices for protecting smart industrial settings. Cybersecurity must be given top priority in any complete strategy to managing the complex world of smart manufacturing. In order to ensure that the revolutionary advantages of smart manufacturing are implemented securely and sustainably, the sector can strengthen its resilience against possible disruptions by recognizing and tackling the growth of cyber risks.

1.3 Significance and Scope of Cybersecurity Integration

According to a research by Moustafa et al. (2018), integrating cybersecurity measures into smart manufacturing is very necessary. Industry 4.0's signature networked smart manufacturing systems provide weaknesses that need for strong defense against dynamic cyberattacks. For smart manufacturing programs to continue to succeed, this acknowledgment is essential. Cybersecurity integration's use goes beyond traditional data security. It includes a comprehensive plan to strengthen every aspect of the operating environment. Maintaining uninterrupted operations is critical because hiccups in the manufacturing process may have a significant effect on supply networks and the economy of entire countries. Maintaining product quality is also important, as cyberattacks on production lines may jeopardize the security and stability of goods, posing risks and monetary obligations. Moreover, the protection of the industrial ecosystem's entire integrity falls under the purview of cybersecurity in smart manufacturing. These systems are interrelated, which suggests that a breach in one area of the network might have an impact on the ecosystem as a whole. This underscores the necessity for a comprehensive cybersecurity strategy that considers the intricate interdependencies within the system.

The proactive implementation of cybersecurity measures is essential for mitigating risks. This includes the adoption of encryption protocols, robust access controls, regular system audits, and continuous monitoring to detect and respond to potential threats promptly. Collaborative efforts between industry stakeholders, government bodies, and cybersecurity experts are crucial for developing standardized practices and frameworks applicable universally to fortify the resilience of smart manufacturing systems. The significance and scope of cybersecurity integration in smart manufacturing align with the transformative potential of these systems. A proactive and comprehensive cybersecurity approach is not just a protective measure but an essential requirement for ensuring the secure and sustained evolution of smart manufacturing amid the ever-changing landscape of cyber threats.

1.4 Objectives of the Study

1. Assessing the current cybersecurity landscape in smart manufacturing:
2. Evaluating the efficacy of existing cybersecurity measures implemented in smart manufacturing environments.
3. Proposing enhanced cybersecurity strategies and best practices of smart manufacturing.
4. Investigating the integration of emerging technologies such as AI, machine learning, and block chain, in enhancing cybersecurity for smart manufacturing.
5. Analyzing regulatory compliance and standards in smart manufacturing:

These objectives collectively aim to contribute to a holistic understanding of the cybersecurity landscape in smart manufacturing and provide actionable insights for manufacturers to fortify their systems against cyber threats.

1.5 Research Gap

Despite the rapid advancements in smart manufacturing and the increasing recognition of cybersecurity importance, a notable research gap exists in understanding the nuanced challenges and specific vulnerabilities faced by smart manufacturing systems. Existing studies, such as those highlighted in by Tuptuk and Hailes (2018), primarily focus on the transformative potential of smart manufacturing but often lack a detailed examination of the evolving cyber threats. The gap becomes evident in the limited exploration of the effectiveness of current cybersecurity measures in the smart manufacturing context. The study by Ani et al. (2017) identifies cyber threats but falls short of providing a detailed evaluation of the cybersecurity strategies implemented within manufacturing ecosystems. Furthermore, while emerging technologies like AI are acknowledged in enhancing cybersecurity, a comprehensive analysis of their practical integration into smart manufacturing security frameworks is lacking. The current literature lacks an in-depth exploration of how these technologies can be effectively leveraged to detect and mitigate cyber threats specific to smart manufacturing, as noted in the *Journal of Manufacturing Science and Engineering* (Yang et al., 2019).

Closing this research gap is essential for developing tailored and effective cybersecurity strategies, ensuring the resilience of smart manufacturing systems against an ever-evolving threat landscape. The study aims to bridge this gap by providing a detailed examination of the current state of cybersecurity in smart manufacturing and proposing targeted enhancements based on identified vulnerabilities and emerging technologies.

2 LITERATURE REVIEW

The literature review delves into the interconnected domains of smart manufacturing and cybersecurity, exploring the historical evolution of smart manufacturing and the theoretical foundations guiding cybersecurity practices within manufacturing contexts.

2.1 Evolution of Smart Manufacturing

The trajectory of smart manufacturing has been a dynamic journey intricately shaped by continuous technological advancements. Pivotal insights into this evolution can be gleaned from studies particularly like the work by Zheng et al. (2018). Their research offers a comprehensive historical perspective, mapping the transition from conventional manufacturing paradigms to the current epoch of smart manufacturing. Zheng et al. shed light on significant technical turning points that have accelerated this development. The widespread use of AI and IoT is essential to this revolutionary development. An age of networked systems and intelligent decision-making in industrial processes has begun with the integration of these technologies. Smart manufacturing's integration of IoT signals a change from conventional production techniques. Real-time data production, sharing, and analysis are made possible by the seamless connectivity of machines and devices made possible by IoT. The fundamental element of smart manufacturing is its interconnection, which facilitates improved coordination and communication among many constituents in a manufacturing ecosystem.

AI also brings cognitive capabilities to production systems at the same time. AI gives robots the ability to go through large databases, spot trends, and decide for themselves. This cognitive layer represents a paradigm change toward more intelligent and effective industrial operations by improving the reactivity and flexibility of production processes. This evolutionary journey is significant not just because of the individual technical milestones but also because of the collective effect of these milestones on industrial operations. According to Ibrahim et al., smart manufacturing is a paradigm shift that goes beyond simple automation. It represents an all-encompassing strategy by merging technologies to build intelligent and adaptable industrial environments. Industries looking to capitalize on smart manufacturing's full potential and adjust to the ever-changing industrial landscape must comprehend this transformation.

2.2 Theoretical Foundations of Cybersecurity in Manufacturing

Theoretical underpinnings are crucial to the creation and application of successful security measures in the field of manufacturing cybersecurity. The paper published in the *Journal of Cybersecurity* paper by Chatfield and Reddick (2019) makes a significant advance to this knowledge. Their research illuminates key ideas that direct strategy creation by exploring the theoretical foundations that support cybersecurity practices in the industrial sector. The study emphasizes the importance of conceptual frameworks, focusing on the defense-in-depth paradigm in particular. This concept, which is well-known in the field of cybersecurity, supports a multi-layered approach to security. It acknowledges that in the face of complex cyberthreats, a single security solution might not be adequate. Rather, it encourages security policies to be layered, resulting in a complete defensive plan that tackles weaknesses at different levels of the industrial infrastructure. Building strong cybersecurity methods that are suited to the particular difficulties of the industrial environment requires an understanding of these theoretical underpinnings. Manufacturers are guided in applying a mix of preventative, detective, and response measures, for example, by the defense-in-depth concept. This guarantees a stronger security system that can resist cyberattacks at various phases of their progression.

In addition, the integration of theoretical frameworks offers an organized methodology for cybersecurity, facilitating methodical evaluation and reduction of risks. It empowers manufacturing entities to anticipate potential threats, identify vulnerabilities, and proactively implement measures to fortify their cyber defenses. In conclusion, the theoretical foundations elucidated by Sindre et al. offer invaluable insights into the conceptual frameworks that guide cybersecurity practices in manufacturing. Understanding and applying these theories, particularly the defense-in-depth model, equips industries with the knowledge necessary to develop and implement robust cybersecurity strategies, fostering a secure and resilient manufacturing landscape.

2.3 Synthesis of Smart Manufacturing and Cybersecurity Literature

The synthesis of literature on smart manufacturing and cybersecurity is instrumental in unraveling the intricate relationship between advanced manufacturing technologies and the requisite security measures. The study conducted by Sheth and Kusiak (2022) provides a pivotal synthesis that bridges the domains of smart manufacturing and cybersecurity. This synthesis proves crucial in understanding the nuanced interplay between technological advancements and the imperative to safeguard these innovations. Lu et al. (2016) delve into the unique challenges presented by cyber threats within smart manufacturing ecosystems. The authors emphasize that the interconnected nature of smart manufacturing, coupled with the integration of technologies like IoT and AI, creates vulnerabilities that demand tailored cybersecurity strategies. This synthesis not only identifies the challenges but also proposes insights into effective cybersecurity measures essential for mitigating the specific risks posed by the evolving smart manufacturing landscape.

By providing insights into the mutually beneficial link between technological innovation and security concerns in manufacturing, the study adds to a comprehensive understanding of the changing scene. The synthesis of cybersecurity and smart manufacturing literature establishes the groundwork for a thorough comprehension of the dynamic threat landscape. This knowledge is essential for sectors trying to reconcile utilizing cutting-edge technology in smart manufacturing while protecting their networks from any cyberattacks. Essentially, Lu et al.'s synthesis represents a significant advancement toward a deeper understanding of the potential and difficulties at the nexus of cybersecurity and smart manufacturing. It acts as a lighthouse to guide the creation of strong plans that not only capitalize on the revolutionary potential of smart manufacturing but also guarantee the safety and robustness of these cutting-edge systems against constantly changing cyberthreats.

2.4 Key Concepts and Frameworks in Cybersecurity

Smart manufacturing cybersecurity is built around a foundation of fundamental ideas and frameworks intended to reduce risks and protect important assets. Comprehending these fundamental concepts is crucial in formulating resilient security approaches that tackle the distinct obstacles presented by networked cyber-physical systems. In the sections that follow, we've outlined important ideas and frameworks that might help improve cybersecurity in smart manufacturing.

1. **Model of Defense-in-Depth:** A key idea in cybersecurity is the defense-in-depth paradigm, which emphasizes the use of layered security measures to fend off attacks at various danger levels. In order to create a robust security architecture, this strategy entails incorporating a combination of preventative, investigative, and remedial controls across several layers.
2. **Risk Management:** Smart manufacturing cybersecurity relies heavily on effective risk management. This entails determining, evaluating, and ranking possible risks before putting controls in place to lessen or manage them. A thorough framework for risk management is offered by the National Institute of Standards and Technology (NIST), and it is extensively used in many different sectors.
3. **Access Controls:** These prevent unwanted users from accessing vital data and systems. By putting the concepts of least privilege into practice, people and systems are guaranteed to have the minimal amount of access required to carry out their tasks. For controlling access permissions, a popular framework is role-based access control, or RBAC.
4. **Incident Response:** To lessen the effects of cybersecurity events, incident response frameworks are crucial. For the purpose of anticipating, identifying, responding to, and recovering from security events in smart manufacturing systems, the NIST Computer Security Incident Handling Guide offers an organized method.
5. **Encryption:** Encryption is a crucial cybersecurity precaution that shields private information both during transmission and storage. By using cryptographic techniques, encryption makes sure that the data that is intercepted cannot be decoded even in the event of illegal access. Encryption methods must be put into place in order to secure communication channels in networks for smart manufacturing (Table 10.1).

Organizations may create a thorough and tenacious defense against changing cyber threats by implementing these fundamental ideas and frameworks into smart manufacturing cybersecurity plans.

TABLE 10.1
Key Concepts and Frameworks in Cybersecurity

Concept/Framework	A. Description
Defense-in-Depth Model	Layered security approach incorporating preventive, detective, and corrective controls.
Risk Management	NIST framework for identifying, assessing, and mitigating cybersecurity risks.
Access Controls	Principle of least privilege and RBAC for restricting unauthorized access to critical systems and data.
Incident Response	NIST guide for structured preparation, detection, response, and recovery from cybersecurity incidents.
Encryption	Utilizing cryptographic algorithms to secure data in transit and at rest, enhancing the confidentiality of sensitive information.

3 THEORETICAL FRAMEWORK

The theoretical framework lays the groundwork for comprehending and using ideas in the context of smart manufacturing cybersecurity. The paradigm emphasizes the necessity for a methodical and proactive approach by incorporating cybersecurity risk management concepts, drawing on the ideas of experts such as Mizrak (2023). The creation of tactics that complement the dynamic and linked characteristics of smart manufacturing systems is guided by this theoretical foundation.

3.1 Defining the Conceptual Framework

An essential first step in creating a methodical framework for comprehending the complex interrelationship between cybersecurity and smart manufacturing is defining the conceptual framework. With regard to cyber security, an article by Mittal et al. (2019) is worth noting because it provides valuable ideas about the concepts and procedures. Using their investigation, they develop a tailor made conceptual frame work, which specifically addresses the challenges posed by the introduction of smart Manufacturing technology. Their study presents key concepts of a successful information security. This forms a solid foundation for constructing a conceptual framework which addresses the specific processes involved in smart manufacturing, making them clear. The smart industrial systems are constantly connected and digital technologies change very rapidly, creating new complexity in cyber security. The conceptual framework provides a methodical approach to the integration of cybersecurity measures into smart manufacturing settings, drawing on the ideas of Mittal et al. It includes a thorough grasp of incident response plans that are adapted to the complexities of contemporary manufacturing processes, access restrictions, encryption techniques, and risk management. Furthermore, this conceptual framework covers more ground than only cybersecurity's technological components. It also takes into account the human element, realizing how crucial cybersecurity knowledge and education are to establishing a security-conscious culture in smart manufacturing companies. Through the integration of these diverse parts, the conceptual framework assumes a dynamic role, enabling manufacturers to effectively traverse the always shifting cyber threat landscape while simultaneously capitalizing on the revolutionary potential of smart manufacturing technologies.

3.2 Cybersecurity Integration into Smart Manufacturing

For cybersecurity to be seamlessly incorporated into smart manufacturing, a dynamic strategy that keeps up with the ever-changing threat landscape is necessary. The integration process entails carefully integrating cybersecurity measures into the operations of smart manufacturing, taking inspiration from Kusiak (2023) work. The perspectives offered by Kusiak (2023)—especially with regard to the incorporation of cybersecurity into industrial automation—provide useful direction for putting security measures in place in production settings. The study by Kusiak (2023) emphasizes the necessity of integrating cybersecurity in a comprehensive way and stresses that security measures should be viewed as essential parts that are integrated into the architecture of smart manufacturing systems rather than as optional extras. In order to guarantee that security is an integral part of every phase of production and not just an afterthought, cybersecurity policies must be matched with the fundamental operational procedures. According to Kusiak (2023), creating strong access restrictions, integrating security procedures into industrial automation systems, and encouraging a cybersecurity-aware culture among staff members are all examples of practical cybersecurity measures that may be applied in manufacturing settings. This strategy makes sure cybersecurity becomes a crucial component of the industrial ecosystem, protecting against any cyberattacks while maintaining uninterrupted business operations. In conclusion, as Kusiak (2023) observations demonstrate, integrating cybersecurity into smart manufacturing calls for a deliberate and comprehensive strategy. Security features can be seamlessly woven into smart manufacturing systems that employ modern technology in order to embrace its innovative nature and navigate complex threats associated with contemporary environment.

3.3 Conceptual Model for Comprehensive Cybersecurity

This calls for coordination of various elements into one all-embracing and comprehensive cyber security system for creating a conceptual model of complete cyber security in smart manufacturing. Geisberger and Broy's (2015) study can aid in such an endeavor as well. This analysis constitutes an all-inclusive framework that involves crucial components such as incident response measures, access control mechanism, and risk management. The presented all-inclusive model can be effectively used as an approach for designing successful solutions to the intricate issues of smart manufacture systems. The cybersecurity model proposed by Zografopoulos et al. (2023) underscores the power of joint approach. It includes risk management and makes it possible for a proactive threat identification and neutralization. Integration of access restrictions prevents unwanted entrance and therefore limits the compromises that may occur and allows only authorized agencies into the system. Incident response measures that are necessary for rapid and effective counteraction of possible violations are also part of the model. Mittal et al. (2018) proposed a coherent and adaptable plan for addressing cybersecurity issues in smart manufacturing which considers the intricacy of the problems at hand. The model improves security of the smart manufacturing systems by considering their interplay of risk management, access restrictions, and incident response. This allows for the development of a complete plan that keeps up with the always changing cyber threat scenario.

3.4 Role of Best Practices and Frameworks

Best practices and guidelines play a crucial role in directing the efficient deployment of cybersecurity solutions. Among the most notable guiding frameworks is the 2014 version of the NIST Cybersecurity Framework. This framework is essential in offering an organized method for dealing with cyber dangers in settings related to smart manufacturing. The five main components of NIST's cybersecurity framework are identification, protection, detection, response, and recovery. By providing a thorough and standardized approach to cybersecurity, this framework makes sure that smart manufacturing strategies are prepared to tackle the ever-changing landscape of cyber threats. While the "Protect" function concentrates on putting safety measures in place to guarantee the security and integrity of manufacturing processes, the "Identify" function is concerned with comprehending and controlling cybersecurity concerns. The "Detect" function places a strong emphasis on ongoing observation in order to quickly detect possible threats. This is followed by an efficient "Respond" phase. Last but not least, the "Recover" feature makes sure that intelligent manufacturing systems can recover swiftly from a cyberattack. Industry stakeholders may improve the resilience of their cybersecurity measures by incorporating best practices and established rules into smart manufacturing initiatives through the NIST Cybersecurity Framework. In addition to expediting the deployment process, this strategy guarantees a strong and constant protection against the constantly changing cyberthreat scenario.

4 METHODOLOGY

This qualitative, conceptual study's methodology is designed to offer a strong foundation for comprehending and investigating the dynamics of cybersecurity in smart manufacturing. The methodology entails a thorough characterization of important variables, informed by pertinent theoretical underpinnings and literature. The objective is to build an all-encompassing conceptual model that embodies the subtleties of cybersecurity integration in the context of smart manufacturing.

4.1 Justification for a Conceptual Approach

This study's adoption of a conceptual method makes sense because its goal is to provide qualitative frameworks, insights, and tactics instead of quantitative facts. By utilizing theoretical underpinnings and subject-matter expertise, this approach facilitates a thorough investigation and synthesis of the body of current information. A conceptual approach is particularly significant since the complexities of cybersecurity challenges in smart manufacturing require a sophisticated understanding.

This qualitative technique, which rejects a tight dependence on quantitative data, enables an in-depth analysis of the intricate interactions between cybersecurity and smart manufacturing. It is consistent with the goal of the study, which is to offer comprehensive understandings and useful frameworks that may assist stakeholders in navigating the ever-changing cybersecurity landscape in the context of modern manufacturing technology.

4.2 Conceptualization of Key Variables

Key factors that are essential to the study's focus are defined and refined throughout the conceptualization phase. Variables like defense-in-depth, risk management, access restrictions, and incident response are contextualized within the framework of smart manufacturing, thanks to the work of researchers such as Sundararajan et al. (2018, April). These variables form the basis of the conceptual model, providing an organized framework for understanding the intricacies of cybersecurity in the industrial sector. Through the incorporation of these crucial variables, the study seeks to formulate effective strategies for cybersecurity integration into smart manufacturing environments by developing a thorough understanding of these factors' interactions within the context of advanced manufacturing technologies.

4.3 Delimitations and Scope of the Conceptual Model

In delineating the scope of the conceptual model, delimitations serve as vital parameters to define its boundaries. This qualitative study hones in on smart manufacturing, taking into account the distinctive challenges and requisites of this specialized domain. The delimitations underscore that the conceptual model's purpose is to furnish overarching strategies and frameworks rather than detailed, context-specific implementations. This intentional focus ensures the model's flexibility and applicability across diverse smart manufacturing environments. By establishing these delimitations, the study maintains a clear scope, directing its efforts toward providing insights and strategies that can be broadly adapted within the dynamic landscape of smart manufacturing cybersecurity, catering to the varied needs of stakeholders in this evolving industrial paradigm.

4.4 Operationalization of Key Concepts

The operationalization of key concepts is the crucial step of translating theoretical ideas into actionable strategies. Building on the NIST Cybersecurity Framework and insights from Plachkinova (2023), the operationalization phase aims to delineate tangible steps for implementing defense-in-depth, risk management, access controls, and incident response within the specific context of smart manufacturing. This process ensures that the conceptual model transforms into practical guidance, offering organizations clear and actionable steps to enhance their cybersecurity posture within the dynamic landscape of advanced manufacturing technologies. By bridging theory with practical application, the study aims to empower stakeholders with actionable frameworks that can be effectively implemented to fortify the cybersecurity resilience of smart manufacturing environments.

5 COMPREHENSIVE CYBERSECURITY INTEGRATION MODEL

The development of a Comprehensive Cybersecurity Integration Model is a critical outcome of this qualitative, conceptual study. This model aims to provide a holistic framework that addresses the multifaceted challenges posed by cybersecurity in smart manufacturing. The model integrates features like defense-in-depth, risk management, access controls, and incident response to produce a coherent approach for cybersecurity integration. It does this by drawing on important insights from the literature and theoretical foundations.

5.1 Components of the Conceptual Model

The well-crafted Comprehensive Cybersecurity Integration Model fortifies smart manufacturing against cyber threats by integrating essential elements taken from theoretical frameworks and best practices. Defense-in-depth, which emphasizes the strategic stacking of security measures to protect against a variety of attack vectors, is a basic idea supported by the approach. As a fundamental element, risk management directs businesses in the methodical identification and reduction of possible hazards, so augmenting their overall resilience. To guarantee that only authorized entities have the appropriate privileges, access controls are carefully integrated into the model's structure, reducing the possibility of illegal access. Additionally, the model incorporates incident response methods that are intended to mitigate the effects of cybersecurity problems, guaranteeing prompt and efficient responses to possible breaches. The whole strategy comprises incident response, risk management, access controls, also known as defense-in depth, creating a robust basis for the improvement the cyber security posture of smart manufacturing premises that complies also with the theoretical backbone and industry-best practices.

5.2 Visual Representation of the Model (Tabular Form)

The tabulated form of the model's graphic presentation provides a concise overview of all its elements and how they fit together. This graphical illustration provides a good guidance for businesses that desire to incorporate cybersecurity integration model into their facilities of smart manufacturing. Tabular style ensures that key items in the model are covered clearly thereby enhancing ease of understanding and application (Table 10.2).

TIt has a tabular shape, which enables quick comprehension of its parts by decision makers, cyber security professionals, and stakeholders. It is a visual representation of the proposed model, which can be easily communicated and understood by different departments in an organization. It is an important tool that these companies can use to build a formidable smart manufacturing ecosystem. It helps with strategy creation and implementation of a strong cybersecurity framework that is tailored to the particular difficulties faced by the manufacturing industry.

TABLE 10.2
Comprehensive Cybersecurity Integration Model

Components	B. Description
Defense-in-Depth	A layered security approach incorporating preventive, detective, and corrective controls to fortify the system against a spectrum of cyber threats.
Risk Management	Systematic identification, assessment, and mitigation of cybersecurity risks, ensuring a proactive response to potential threats and vulnerabilities.
Access Controls	Restriction of unauthorized access to critical systems and data through the principle of least privilege and role-based access control, safeguarding against unauthorized entities.
Incident Response	A structured approach for preparing, detecting, responding to, and recovering from cybersecurity incidents, minimizing the impact of security breaches on smart manufacturing operations.

5.3 Illustrating the Dynamics of Cybersecurity in Smart Manufacturing

The changing environment of cybersecurity in smart manufacturing is visually represented by the Comprehensive Cybersecurity Integration Model. This model effectively conveys the dynamic nature of cyber threats and emphasizes the need of a proactive and adaptable security strategy. Organizations may better understand the complex linkages and dependencies involved in safeguarding their manufacturing ecosystems by visualizing the interactions between the model's components. The model serves as a dynamic framework that adjusts to the always shifting threat scenario, acting as a strategic tool. With the use of this graphic, companies may better grasp cybersecurity dynamics and be more equipped to adopt focused initiatives. The Comprehensive Cybersecurity Integration Model serves as a useful roadmap for the development of smart manufacturing environments by coordinating theoretical foundations with real-world implementations to strengthen manufacturing systems' resistance to the intricacies of modern cyberattacks.

The Comprehensive Cybersecurity Integration Model provides enterprises looking to strengthen their smart manufacturing environments against cyber threats with a strategic roadmap by fusing theoretical underpinnings, insights from important literature, and practical concerns.

6 BEST PRACTICES FOR CYBERSECURITY IN SMART MANUFACTURING

To strengthen cybersecurity in smart manufacturing, best practices must be identified and put into action. In order to outline real experiences and important insights from the body of literature, this part focuses on defining tactics that work. In the dynamic world of smart manufacturing, best practices are essential for reducing cyber risks, guaranteeing the robustness of linked systems, and encouraging a proactive security posture.

6.1 Overview of Best Practices

Adopting a thorough set of best practices is essential in the field of cybersecurity for smart manufacturing in order to strengthen industrial ecosystems' resistance to changing cyberthreats. Suggested by reputable resources like the NIST Cybersecurity Framework and sector-specific recommendations, these techniques offer a comprehensive strategy for securing smart industrial settings.

1. **Continuous Monitoring:** The foundation of the protection against cyberthreats is Continuous Monitoring. It is recommended that organizations put strong continuous monitoring measures in place so they can quickly identify and address cybersecurity risks. By taking a proactive stance, abnormalities and possible breaches are quickly discovered, reducing the impact on production procedures (Reason, 2016).
2. **Employee Training:** This is a crucial element as it recognizes that an educated staff is a vital line of defense against cyberattacks. Frequent cybersecurity awareness and training sessions are essential for informing staff members about possible risks and security best practices. Organizations can reduce the likelihood of human-related vulnerabilities and build a collective defense against social engineering and other cyber hazards by cultivating a security-conscious culture (Hove, 2020).
3. **Supply Chain Security:** This comprehensive approach places a strong emphasis on safeguarding every link in the supply chain, from producers of component parts to final consumers. Recognizing the interconnectedness of smart manufacturing, this strategy seeks to reduce risks presented at different supply chain phases. Maintaining the entire security of smart manufacturing systems requires ensuring the integrity of parts and procedures all the way through the supply chain (Tuptuk & Hailes, 2018).
4. **Incident Response Planning:** When it comes to handling cybersecurity events, incident response planning is a crucial procedure. Creating thorough incident response strategies helps companies reduce the effects of events and expedite their recovery. Organizations are better equipped to manage a range of cybersecurity situations when these strategies are regularly tested and improved, which fosters resilience in the face of difficulty (Falco & Rosenbach, 2021).
5. **Patch Management:** Patch Management is a preventative approach to quickly fix vulnerabilities that are already known. Software and systems are updated with the most recent security patches on a regular basis when efficient patch management procedures are put into place. This improves the overall security posture of smart manufacturing environments by lowering the likelihood that bad actors may target vulnerabilities in software and firmware (Jbair et al., 2022).

When taken as a whole, these best practices offer a solid platform for businesses looking to create a thorough cybersecurity system that is specific to the difficulties presented by smart manufacturing. Organizations may improve their capacity to identify, address, and recover from cyberattacks by implementing these practices into their cybersecurity plans. This will help to maintain the integrity and security of their smart manufacturing systems.

6.2 Aligning Best Practices with Cybersecurity Objectives

Developing a unified and successful security plan specifically for smart manufacturing requires matching cybersecurity goals with best practices. The goal is to make sure that best practices are smoothly incorporated into the procedures and organizational culture rather than just being implemented. This alignment calls for a number of important factors.

1. **Mapping to Cybersecurity Objectives:** First and foremost, it's crucial to map to cybersecurity objectives. Companies need to decide which cybersecurity goals are particular and which are related to their overall aims. These may include protecting sensitive data, ensuring operational continuity, and minimizing the impact of cyber incidents. Each best practice should directly contribute to achieving these defined objectives.
2. **Tailoring Best Practices:** Tailoring Best Practices is the next critical step. Smart manufacturing introduces unique challenges, such as interconnected devices, the IoT, and real-time data exchange. Best practices need to be customized to address these specific requirements, ensuring relevance and effectiveness within the context of smart manufacturing operations.
3. **Creating a Comprehensive Framework:** Creating a Comprehensive Framework is imperative. A well-designed cybersecurity framework should incorporate the selected best practices, aligning them cohesively with overarching cybersecurity objectives. This ensures that security measures are integrated into the fabric of smart manufacturing processes rather than treated as isolated components.
4. **Regular Assessment and Adaptation:** Regular Assessment and Adaptation form the ongoing commitment to cybersecurity excellence. Continuous assessment processes should be in place to evaluate the effectiveness of implemented best practices in achieving cybersecurity objectives. This iterative approach allows organizations to adapt and refine their strategies in response to evolving threats and technologies within the smart manufacturing landscape.

In conclusion, aligning best practices with cybersecurity objectives is a strategic imperative for organizations navigating the complex realm of smart manufacturing. This alignment fosters a security-first mindset, promotes resilience against cyber threats, and establishes the foundation for a sustainable and adaptive cybersecurity framework. By seamlessly integrating best practices into organizational processes and culture, smart manufacturing entities can fortify their cybersecurity posture and effectively mitigate the evolving risks within their operational landscape.

6.3 Case Studies: Successful Implementation

Examining real-world case studies provides valuable insights into the successful implementation of cybersecurity best practices in smart manufacturing. These cases highlight organizations that have effectively navigated the challenges of securing their manufacturing ecosystems, offering practical lessons for others in the industry.

1. **Siemens AG:** Siemens, a global leader in industrial manufacturing, exemplifies successful cybersecurity implementation. The company has embraced a holistic approach by integrating cybersecurity into its product development lifecycle. Siemens focuses on secure-by-design principles, conducts regular cybersecurity training for employees, and employs continuous monitoring to detect and respond to threats promptly. This case showcases demonstrates how best practices are integrated across the whole company to maintain a strong cybersecurity posture (Stojkovic & Butt, 2022).
2. **Lockheed Martin:** One of the companies that have been successful in carrying out cyber security successfully is lockheed martin, a leading aerospace and defence company. Supply chain security is of paramount importance in the company, which involves a thorough selection through screening of suppliers. Moreover, Lockheed Martin has developed an effective incident response plan that promptly deals with all the cybersecurity concerns. This highlights the need for an inclusive cybersecurity strategy covering the suppliers (Vandal, 2023).
3. **Toyota Motor Corporation:** With its smart manufacturing initiatives, the well-known automaker Toyota has effectively incorporated cybersecurity best practices. The company focuses on employee training and awareness programs to create a cybersecurity-conscious workforce. Toyota also emphasizes secure coding practices in software development and regularly updates and patches its industrial control systems. This case illustrates the integration of best practices across people, processes, and technology layers (Aoki & Staeblein, 2018).

These case studies demonstrate that successful cybersecurity implementation in smart manufacturing involves a combination of strategic planning, employee training, supply chain security, and continuous monitoring. Organizations can draw inspiration from these examples to tailor their cybersecurity strategies and align them with the unique challenges of their manufacturing ecosystems.

6.4 Visualizing Best Practices (Tabular Form)

Visualizing cybersecurity best practices in a tabular form provides a structured and accessible format to understand, implement, and monitor key strategies. Table 10.3 encapsulates the essential best practices for cybersecurity in smart manufacturing, distilling insights from literature, industry guidelines, and successful case studies.

This tabular representation condenses complex cybersecurity strategies into actionable best practices, promoting clarity and ease of implementation. Each best practice is strategically chosen to address specific facets of smart manufacturing cybersecurity, reflecting insights from reputable sources such as the NIST, industry reports, and successful case studies.

TABLE 10.3
Visualizing Best Practices for Cybersecurity in Smart Manufacturing

Best Practices	C.Description
Continuous Monitoring	Implementing robust tools and processes for real-time monitoring of network activities, system logs, and user behaviors.
Employee Training	Conducting regular cybersecurity awareness and training programs to educate employees about potential threats and best practices.
Supply Chain Security	Ensuring the security of the entire supply chain, from component manufacturers to end-users, to mitigate vulnerabilities introduced at various stages.
Incident Response Planning	Developing comprehensive incident response plans to minimize the impact of cybersecurity incidents and facilitate a swift recovery.
Patch Management	Implementing effective patch management practices to promptly address and remediate known vulnerabilities.

Continuous monitoring ensures a proactive stance against evolving threats, employee training fosters a security-conscious culture, and supply chain security guards against vulnerabilities throughout the manufacturing ecosystem. Incident response planning and patch management act as crucial components for swift recovery and timely mitigation of identified vulnerabilities. By visualizing these best practices in a tabular form, organizations gain a concise roadmap for enhancing their cybersecurity posture in smart manufacturing. This visualization facilitates strategic planning, implementation, and continuous improvement, ultimately contributing to the resilience and security of interconnected manufacturing systems.

7 FRAMEWORKS FOR CYBERSECURITY IMPLEMENTATION

Implementing robust cybersecurity in smart manufacturing necessitates the adoption of effective frameworks that provide structured guidance. This section explores existing cybersecurity frameworks, their relevance to smart manufacturing, and the development of a customized framework tailored to the specific challenges and dynamics of interconnected manufacturing ecosystems.

7.1 Examining Existing Cybersecurity Frameworks

Examining existing cybersecurity frameworks is crucial for gaining insights into established best practices and methodologies. The NIST Cybersecurity Framework, a cornerstone in the field, provides a comprehensive guide widely adopted across critical infrastructure sectors. This framework offers a structured approach encompassing key functions: identify, protect, detect, respond, and recover, forming a robust foundation for organizations to enhance their cybersecurity posture (Goel et al., 2020). Another noteworthy standard is the ISO/IEC 27001, internationally recognized for its Information Security Management System (ISMS). This standard offers a methodical way to handling sensitive data while guaranteeing its availability, confidentiality, and integrity. Offering a widely recognized standard for efficient information security procedures, ISO/IEC 27001 offers enterprises a framework for creating, implementing, maintaining, and continuously improving an ISMS (Achmadi et al., 2018, May). Through an analysis and utilization of these well-established frameworks, companies involved in smart manufacturing may benefit from tried-and-true approaches and industry best practices. This analysis establishes the foundation for the creation and improvement of customized cybersecurity plans that comply with global norms and industry standards.

7.2 Developing a Customized Framework for Smart Manufacturing

Creating a tailored framework for smart manufacturing is essential to solving the special problems that linked industrial processes present. This entails modifying accepted cybersecurity best practices to account for the unique dynamics of smart manufacturing. The process of customization incorporates components like supply chain management, real-time monitoring, and IoT security, ensuring that they meet the unique needs of smart industrial settings. A tailored framework, modeled after well-known frameworks such as the NIST Cybersecurity Framework and ISO/IEC 27001, guarantees that cybersecurity measures are precisely matched to the intricacies of smart manufacturing. The confluence of IT and operational technology (OT) systems within smart manufacturing ecosystems makes this alignment imperative. The customized framework offers a strong and flexible cybersecurity plan for the changing field of smart manufacturing. It is made to efficiently handle the difficulties presented by the interconnectivity of devices, the integration of cutting-edge technologies, and the quick speed of data exchange as shown in Figure 10.1.

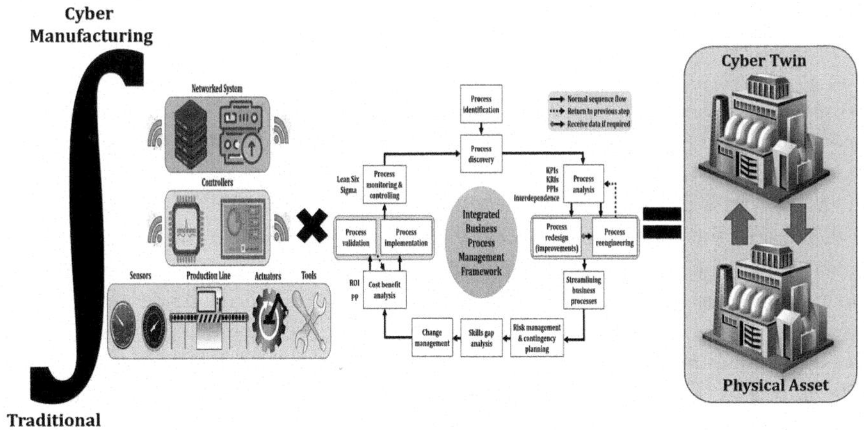

FIGURE 10.1 Integrated Business Management.

7.3 Conceptual Diagram: Framework Integration

Organizations are given a visual roadmap through a conceptual model that shows how the customized framework for smart manufacturing is integrated. This figure illustrates how important components of the customizable framework, such supply chain security, incident response planning, and continuous monitoring, are interrelated. The use of visual aids in demonstrating how cybersecurity measures should be strategically aligned with the unique needs and intricacies of smart industrial settings.

Organizations may efficiently negotiate the complexities of cybersecurity implementation in smart manufacturing by combining multiple aspects into a unified framework. Decision-makers, cybersecurity experts, and other stakeholders may all benefit from this conceptual model, which promotes a common understanding of the goals and structure of the customized framework.

8 KEY PERFORMANCE INDICATORS FOR CYBERSECURITY ASSESSMENT

Key performance indicators (KPIs) that are in line with corporate objectives and the conceptual model must be established in order to assess the efficacy of cybersecurity measures in smart manufacturing. This section explores the creation of KPIs, highlighting their function in measuring cybersecurity performance, pinpointing areas in need of growth, and offering a methodical way to continuously evaluate cybersecurity.

8.1 Developing KPIs for Conceptual Model Evaluation

The process of creating KPIs for the assessment of the conceptual model requires careful matching of the measurement criteria with the model's essential elements. The KPIs have to be customized to evaluate the effectiveness of incident response, risk management, access restrictions, and defense-in-depth in the particular setting of smart manufacturing. One KPI for continuous monitoring may be the average time to notice and address a cybersecurity event. This metric provides a quantitative measure of the model's effectiveness in facilitating real-time threat management, crucial for minimizing potential disruptions in the smart manufacturing processes. Similarly, in the domain of employee training, relevant KPIs could assess the reduction in security incidents attributed to human error. This measurement becomes instrumental in gauging the practical impact of cybersecurity awareness programs. By quantifying the decrease in incidents related to human mistakes, organizations can ascertain the efficacy of their training initiatives in enhancing the overall security posture. These tailored KPIs play a pivotal role in providing a quantitative foundation for the evaluation of the conceptual model. They serve as tangible metrics that allow stakeholders to measure the model's impact, adaptability, and effectiveness in achieving the predefined cybersecurity objectives within the dynamic and interconnected landscape of smart manufacturing. The development and assessment of these KPIs contribute to the

TABLE 10.4
Proposed KPIs for Cybersecurity Assessment in Smart Manufacturing

KPIs	D.Measurement Criteria
Average Incident Response Time	Time taken to detect and respond to cybersecurity incidents, ensuring swift mitigation. A lower average time indicates an efficient incident response system, reducing potential damage and downtime.
Reduction in Human-Induced Security Incidents	Percentage decrease in security incidents attributable to human error, indicating the effectiveness of employee training programs. A lower percentage reflects improved employee awareness and adherence to security protocols.
Percentage of Critical Systems with Access Controls	Proportion of critical systems with access controls in place, limiting unauthorized access. A higher percentage demonstrates a robust access control framework, enhancing the overall security of critical assets and data.
Effectiveness of Risk Mitigation	Assessment of risk mitigation measures through quantitative reduction in identified risks. The effectiveness is measured by the percentage decrease in identified risks, indicating the success of proactive risk management strategies.
Adherence to Patch Management Schedule	Percentage compliance with the established patch management schedule, ensuring timely updates for vulnerability remediation. Higher compliance reflects a proactive approach to patching, minimizing the window of exposure to potential threats.

ongoing refinement and optimization of the conceptual model, ensuring its relevance and efficacy in safeguarding smart manufacturing environments against evolving cyber threats.

8.2 Tabular Presentation: Proposed KPIs and Measurement Criteria

A structured and comprehensive approach to evaluating the effectiveness of cybersecurity measures in smart manufacturing involves the development of KPIs aligned with the conceptual model. The tabular presentation below outlines proposed KPIs and their corresponding measurement criteria, offering organizations a practical framework for assessing their cybersecurity posture in the dynamic landscape of smart manufacturing (Table 10.4).

This tabular presentation encapsulates key indicators tailored to assess the performance of defense-in-depth, risk management, access controls, and incident response within the smart manufacturing context. Organizations can utilize this table as a practical guide for implementing and measuring the success of their cybersecurity strategies. The clarity provided by these KPIs and measurement criteria enables organizations to track their cybersecurity performance, make informed decisions for continuous improvement, and ensure the resilience of their smart manufacturing ecosystems against evolving cyber threats.

9 CHALLENGES AND OPPORTUNITIES IN CONCEPTUAL IMPLEMENTATION

The conceptual implementation of a cybersecurity framework in smart manufacturing presents both challenges and opportunities. Understanding these dynamics is crucial for organizations aiming to fortify their systems against cyber threats while embracing the transformative potential of smart manufacturing technologies.

9.1 Anticipated Challenges in Applying the Conceptual Model

The application of the conceptual model in smart manufacturing may encounter anticipated challenges, notably resistance to change, resource constraints, and the dynamic nature of cyber threats. Stakeholders accustomed to traditional manufacturing processes might resist the transition, posing a challenge to the model's implementation. Resource constraints, encompassing financial limitations and a shortage of skilled personnel, may impede the full integration of the conceptual model into organizational practices. Furthermore, the dynamic and evolving landscape of cyber threats may outpace the model's adaptability over time. Effectively addressing these challenges necessitates a strategic approach, involving change management initiatives to overcome resistance, resource optimization to manage constraints, and continuous monitoring to stay ahead of emerging cyber threats. By proactively addressing these challenges, organizations can enhance the likelihood of successful implementation and sustained effectiveness of the conceptual model in the context of smart manufacturing as shown in Figure 10.2.

9.2 Opportunities for Improvement and Adaptation

The conceptual model exhibits inherent flexibility, providing opportunities for improvement and adaptation within the dynamic landscape of smart manufacturing cybersecurity. Organizations can harness advancements in cybersecurity technologies, particularly AI and machine learning, to augment threat detection and response capabilities. The model's continuous improvement mechanisms enable iterative enhancements, ensuring its alignment with emerging cyber threats. Collaborative

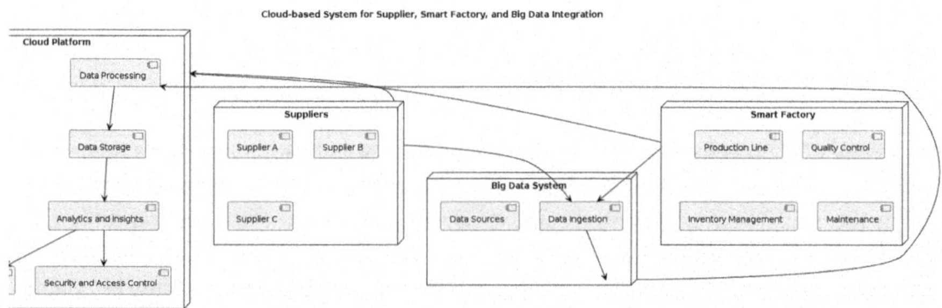

FIGURE 10.2 Cloud-Based System for Supplier, Smart Factory, and Big Data Integration.

initiatives with industry peers and cybersecurity experts create opportunities for shared learning and benchmarking against best practices, fostering a community-driven approach to enhancement. This adaptability not only allows organizations to seize current opportunities for improvement but also positions the conceptual model to remain relevant and effective in the face of continuously evolving challenges within the smart manufacturing environment.

9.3 Real-world Examples and Applications

Real-world examples and applications of cybersecurity frameworks in smart manufacturing provide tangible insights into successful strategies, challenges faced, and opportunities harnessed by organizations. Examining these instances enriches the understanding of how theoretical concepts translate into practical solutions. One compelling example is Siemens AG, a global leader in industrial manufacturing. Siemens has demonstrated a holistic approach to cybersecurity by integrating it into the fabric of its product development lifecycle. This approach includes secure-by-design principles, continuous monitoring, and regular cybersecurity training for employees. Siemens' commitment to cybersecurity extends beyond its products to encompass the entire manufacturing ecosystem, showcasing a comprehensive application of cybersecurity principles in smart manufacturing (Sverko et al., 2022). Another noteworthy case is Lockheed Martin, a major aerospace and defense company. Lockheed Martin emphasizes supply chain security as a critical component of its cybersecurity strategy. The company rigorously evaluates and monitors its suppliers, recognizing the interconnected nature of the supply chain in smart manufacturing. This real-world application underscores the importance of extending cybersecurity measures to encompass the entire manufacturing value chain, from component manufacturers to end-users (Melnyk et al., 2022).

These examples highlight practical applications of cybersecurity frameworks in smart manufacturing. Siemens and Lockheed Martin showcase the integration of cybersecurity principles into product development, supply chain considerations, and employee training. By studying such real-world instances, organizations can draw inspiration, refine their own strategies, and navigate the complexities of implementing effective cybersecurity measures in the evolving landscape of smart manufacturing.

10 DISCUSSION: THEORETICAL IMPLICATIONS AND CONTRIBUTIONS

The discussion on theoretical implications and contributions explores how the conceptual model and findings of the study align with existing theories, providing a foundation for understanding the broader theoretical landscape of smart manufacturing cybersecurity.

10.1 Synthesizing Findings with Existing Theories

Synthesizing findings with existing theories is pivotal for grounding the conceptual model in established frameworks. The study is theoretically supported by the NIST

Cybersecurity Framework. The conceptual model's theoretical foundation is strengthened by conformity to NIST standards, which include risk management and continuous monitoring (Kandasamy et al., 2020). Furthermore, the study's focus on staff training is consistent with social learning theories, particularly Bandura's 1977 work, which emphasizes the critical role that human factors play in cybersecurity. The conceptual model achieves theoretical robustness by incorporating these findings into well-established theories. It does this by drawing on tested frameworks to improve its applicability and efficacy in the intricate field of cybersecurity for smart manufacturing.

10.2 Contributions to the Field of Smart Manufacturing Cybersecurity

The chapter offers a thorough conceptual model that is adapted to the particular difficulties of networked industrial processes, which provides a substantial contribution to the field of smart manufacturing cybersecurity. The way that risk management, incident response, access restrictions, and defense-in-depth are integrated is in line with how the threat landscape is changing. The focus on adaptation and ongoing development tackles the dynamic cybersecurity issues in smart manufacturing. Furthermore, the study makes a valuable contribution by emphasizing the significance of employee awareness and training initiatives and acknowledging that human aspects are crucial to the effectiveness of cybersecurity measures. Through the integration of theoretical insights from pre-existing frameworks and their adaptation to the unique context of smart manufacturing, this study contributes to the advancement of cybersecurity tactics that are successful in this quickly expanding field. The contributions of this study extend beyond theoretical frameworks, providing practical insights and a roadmap for organizations to enhance their cybersecurity posture in the context of smart manufacturing. This holistic approach contributes to the evolving discourse on cybersecurity in Industry 4.0, fostering resilience and security in the digital transformation of manufacturing processes.

11 CONCLUSION: FOSTERING CYBERSECURITY RESILIENCE IN SMART MANUFACTURING

As we draw the curtain on this exploration into fortifying smart manufacturing against cyber threats, a resounding symphony emerges—a harmonious blend of theoretical rigor and practical insights aimed at empowering organizations in the dynamic landscape of Industry 4.0. In traversing the conceptual terrain of our study, we have meticulously crafted a roadmap for cybersecurity integration, tailored to the intricacies of smart manufacturing.

11.1 Recapitulation of Key Conceptual Points

Our journey commenced with the realization that the transformative potential of smart manufacturing coexists with an ever-present threat landscape. The conceptual model we constructed, an amalgamation of insights from the NIST Cybersecurity Framework, social learning theories, and real-world applications, stands as a sentinel against cyber vulnerabilities. The layers of defense-in-depth, the precision of risk management, the vigilance of access controls, and the swiftness of incident response converge to create a resilient shield. Central to our conceptual fabric is the recognition that the human element is not just a variable but a linchpin in the cybersecurity equation. Employee training, inspired by social learning theories, becomes the catalyst for a culture of cyber awareness, fortifying the human firewall against evolving threats. The continuous improvement ethos embedded in the model ensures adaptability, recognizing that in the realm of smart manufacturing, change is the only constant.

11.2 Implications for Practice and Future Research

The implications of our study ripple beyond theory, extending a guiding hand to practitioners navigating the complexities of cybersecurity in smart manufacturing. Organizations are invited to not only adopt our conceptual model but to weave it into the fabric of their operational DNA. The call for collaboration, drawing inspiration from real-world exemplars like Siemens and Lockheed Martin, echoes loudly—for in unity, industries can collectively fortify the global manufacturing ecosystem. Looking ahead, our journey into the future of research beckons. The dynamism of smart manufacturing demands continuous exploration. As technologies evolve, as threats shape-shift, there exists a fertile ground for further investigation. The call to unravel the nuances of supply chain cybersecurity, delve deeper into the psychology of employee cyber behaviors, and embrace emerging technologies in the cybersecurity arsenal echoes as a clarion call for future researchers.

11.3 Closing Thoughts

In these closing thoughts envisions a manufacturing landscape where the hum of machines harmonizes with the quiet but steadfast pulse of cybersecurity resilience. Our study aspires to be more than a compendium of theories and practices; it is an ode to the fusion of human ingenuity and technological prowess. As organizations embark on the journey to fortify their smart manufacturing ecosystems, may they find solace and inspiration in these pages. In the grand tapestry of Industry 4.0, where every thread contributes to the narrative of progress, let cybersecurity be the vibrant hue that ensures the durability and brilliance of the entire canvas. Let our study stand as a beacon, illuminating the path toward a future where smart manufacturing not only thrives but does so securely, resiliently, and with an unwavering commitment to progress.

REFERENCES

Achmadi, D., Suryanto, Y., & Ramli, K. (2018, May). On developing information security management system (isms) framework for iso 27001-based data center. In *2018 International Workshop on Big Data and Information Security (IWBIS)*, Jakarta, Indonesia, (pp. 149–157). IEEE. doi: 10.1109/IWBIS.2018.8471700.

Ahmadi, A., Cherifi, C., Cheutet, V., & Ouzrout, Y. (2020). Recent advancements in smart manufacturing technology for modern industrial revolution: A survey. *Journal of Engineering and Information Science Studies*. <hal-03054284> retrieved from https://hal.science/hal-03054284/document.

Ani, U. P. D., He, H., & Tiwari, A. (2017). Review of cybersecurity issues in industrial critical infrastructure: Manufacturing in perspective. *Journal of Cyber Security Technology, 1*(1), 32–74.

Aoki, K., & Staeblein, T. (2018). Monozukuri capability and dynamic product variety: An analysis of the design-manufacturing interface at Japanese and German automakers. *Technovation, 70*, 33–45.

Chatfield, A. T., & Reddick, C. G. (2019). A framework for Internet of Things-enabled smart government: A case of IoT cybersecurity policies and use cases in US federal government. *Government Information Quarterly, 36*(2), 346–357.

Falco, G. J., & Rosenbach, E. (2021). *Confronting Cyber Risk: An Embedded Endurance Strategy for Cybersecurity*. Oxford University Press.

Geisberger, E., & Broy, M. (Eds.). (2015). *Living in a Networked World: Integrated Research Agenda Cyber-Physical Systems (agendaCPS)*. Herbert Utz Verlag.

Goel, R., Kumar, A., & Haddow, J. (2020). PRISM: A strategic decision framework for cybersecurity risk assessment. *Information & Computer Security, 28*(4), 591–625.

Hove, L. (2020). *Strategies Used to Mitigate Social Engineering Attacks* (Doctoral dissertation, Walden University).

Jbair, M., Ahmad, B., Maple, C., & Harrison, R. (2022). Threat modelling for industrial cyber physical systems in the era of smart manufacturing. *Computers in Industry, 137*, 103611.

Kandasamy, K., Srinivas, S., Achuthan, K., & Rangan, V. P. (2020). IoT cyber risk: A holistic analysis of cyber risk assessment frameworks, risk vectors, and risk ranking process. *EURASIP Journal on Information Security, 2020*(1), 1–18.

Kusiak, A. (2023). Smart manufacturing. In: Nof, S.Y. (eds) *Springer Handbook of Automation. Springer Handbooks*. (pp. 973–985). Springer, Cham: Springer International Publishing. https://doi.org/10.1007/978-3-030-96729-1_45

Lu, Y., Morris, K. C., & Frechette, S. (2016). Current standards landscape for smart manufacturing systems. *NIST Interagency/Internal Report* (NISTIR) *National Institute of Standards and Technology, 8107*(3). Gaithersburg, MD, [online], https://doi.org/10.6028/NIST.IR.8107 (Accessed Apr 26, 2024)

Melnyk, S. A., Schoenherr, T., Speier-Pero, C., Peters, C., Chang, J. F., & Friday, D. (2022). New challenges in supply chain management: Cybersecurity across the supply chain. *International Journal of Production Research, 60*(1), 162–183.

Mittal, S., Khan, M. A., Romero, D., & Wuest, T. (2018). A critical review of smart manufacturing & Industry 4.0 maturity models: Implications for small and medium-sized enterprises (SMEs). *Journal of Manufacturing Systems, 49*, 194–214.

Mittal, S., Khan, M. A., Romero, D., & Wuest, T. (2019). Smart manufacturing: Characteristics, technologies and enabling factors. *Proceedings of the Institution of Mechanical Engineers, Part B: Journal of Engineering Manufacture, 233*(5), 1342–1361.

Mizrak, F. (2023). Integrating cybersecurity risk management into strategic management: A comprehensive literature review. *Research Journal of Business and Management, 10*(3), 98–108.

Moustafa, N., Adi, E., Turnbull, B., & Hu, J. (2018). A new threat intelligence scheme for safeguarding industry 4.0 systems. *IEEE Access, 6*, 32910–32924.

Plachkinova, M. (2023). A Taxonomy for Risk Assessment of Cyberattacks on Critical Infrastructure (TRACI). *Communications of the Association for Information Systems, 52*(1), 1.

Queiroz, M. M., Pereira, S. C. F., Telles, R., & Machado, M. C. (2021). Industry 4.0 and digital supply chain capabilities: A framework for understanding digitalisation challenges and opportunities. *Benchmarking: An International Journal, 28*(5), 1761–1782.

Reason, J. (2016). *Managing the Risks of Organizational Accidents*. Routledge.

Shen, L. (2014). The NIST cybersecurity framework: Overview and potential impacts. *Journal of Internet Law, 18*(6), 3–6.

Sheth, A., & Kusiak, A. (2022). Resiliency of smart manufacturing enterprises via information integration. *Journal of Industrial Information Integration*, *28*, 100370.

Sobb, T., Turnbull, B., & Moustafa, N. (2020). Supply chain 4.0: A survey of cyber security challenges, solutions and future directions. *Electronics*, *9*(11), 1864.

Stojkovic, M., & Butt, J. (2022). Industry 4.0 implementation framework for the composite manufacturing industry. *Journal of Composites Science*, *6*(9), 258.

Sundararajan, A., Khan, T., Aburub, H., Sarwat, A. I., & Rahman, S. (2018, April). A tri-modular human-on-the-loop framework for intelligent smart grid cyber-attack visualization. In *SoutheastCon* St. Petersburg, FL, USA, 2018, doi: 10.1109/SECON.2018.8479180. (pp. 1–8).

Sverko, M., Grbac, T. G., & Mikuc, M. (2022). Scada systems with focus on continuous manufacturing and steel industry: A survey on architectures, standards, challenges and industry 5.0. *IEEE Access*, *10*, 109395–109430.

Thomas, G., & Sule, M. J. (2023). A service lens on cybersecurity continuity and management for organizations' subsistence and growth. *Organizational Cybersecurity Journal: Practice, Process and People*, *3*(1), 18–40.

Tuptuk, N., & Hailes, S. (2018). Security of smart manufacturing systems. *Journal of Manufacturing Systems*, *47*, 93–106.

Vandal, C. (2023). Financial analysis of Lockheed Martin, a global leader in defense contracting: Managerial financial analysis & strategic planning. *A Global Leader in Defense Contracting: Managerial Financial Analysis & Strategic Planning (July 14, 2023).* Available at SSRN: https://ssrn.com/abstract=4515722 or http://dx.doi.org/10.2139/ssrn.4515722

Yang, H., Kumara, S., Bukkapatnam, S. T., & Tsung, F. (2019). The internet of things for smart manufacturing: A review. *IISE Transactions*, *51*(11), 1190–1216.

Zheng, P., Wang, H., Sang, Z., Zhong, R. Y., Liu, Y., Liu, C., … & Xu, X. (2018). Smart manufacturing systems for Industry 4.0: Conceptual framework, scenarios, and future perspectives. *Frontiers of Mechanical Engineering*, *13*, 137–150.

Zografopoulos, I., Hatziargyriou, N. D., & Konstantinou, C. (2023). Distributed energy resources cybersecurity outlook: Vulnerabilities, attacks, impacts, and mitigations. *IEEE Systems Journal*. 1–15. https://doi.org/10.1109/jsyst.2023.3305757

11 Tagging Blockchain Technology Fostering Panacea for Data Privacy, Cloud Computing, and Integrity

Legal Framework in India and at a Globe

Bhupinder Singh and Christian Kaunert

1 INTRODUCTION

Blockchain technology (BT) is a decentralized and distributed digital ledger that records transactions across multiple computers, known as nodes. It was first introduced in 2008 as the underlying technology behind the cryptocurrency Bitcoin, but its potential applications extend far beyond digital currencies. Blockchain has gained attention and popularity due to its ability to provide transparency, security, and immutability to various processes. At its core, a blockchain consists of a chain of blocks, where each block contains a list of transactions. These transactions are grouped together and added to the blockchain through a consensus mechanism, which ensures that all participants agree on the validity of the transactions. Once added, the information in a block is cryptographically sealed and linked to the previous block, creating a sequential chain of blocks, hence the name "blockchain." [1]

BT holds significant importance in today's digital landscape. Its decentralized and transparent nature provides enhanced security, immutability, and trust in various sectors. By eliminating the need for intermediaries and central authorities, blockchain streamlines processes, reduces costs, and increases efficiency. The technology's ability to ensure data integrity and traceability is invaluable in industries like supply chain management and healthcare. Moreover, blockchain's decentralized trust model promotes inclusivity and empowers individuals, particularly in regions with limited

DOI: 10.1201/9781032694375-11

access to traditional banking systems. The automation and self-executing capabilities of smart contracts (SCs) further enhance the potential of blockchain, enabling secure and efficient transactions. BT has the potential to revolutionize industries, foster innovation, and create new business models, paving the way for a more secure, transparent, and decentralized future [2].

2 KEY FEATURES OF BLOCKCHAIN TECHNOLOGY

BT is highly relevant in today's rapidly evolving digital landscape. Its decentralized and secure nature addresses critical challenges such as trust, transparency, and security in various sectors. BT enables peer-to-peer transactions without the need for intermediaries, reducing costs and increasing efficiency. Its transparency and immutability ensure a reliable and auditable record of transactions, enhancing accountability and reducing the risk of fraud. Moreover, BT has the potential to revolutionize industries such as finance and healthcare, supply chain management, and voting systems by providing secure and efficient solutions. [3] It offers an innovative way to establish trust, streamline processes, and enable new business models in an increasingly interconnected world. As the demand for secure and transparent digital transactions grows, the relevancy of BT continues to expand, driving innovation and transforming industries across the globe [4].

Decentralization: Instead of relying on a central authority, blockchain operates on a peer-to-peer network of computers. This decentralized nature eliminates the need for intermediaries and central points of control, making the system more resilient and less prone to single points of failure [5].

Transparency and Immutability: Blockchain provides transparency by allowing all participants to view and verify transactions. Once a transaction is recorded on the blockchain, it is extremely difficult to alter or tamper with, ensuring the immutability and integrity of the data.

Security: Blockchain uses cryptographic algorithms to secure transactions and data. Each transaction is digitally signed and linked to the previous transaction, forming a chain of cryptographic hashes. This makes it computationally infeasible to alter the data without detection.

Smart Contracts: SCs are self-executing contracts with predefined rules and conditions written into code. They automatically execute actions when certain conditions are met, providing automation and programmability to blockchain applications [6].

The potential applications of BT go beyond cryptocurrencies. It can be utilized in various sectors, including finance and healthcare, and in supply chain management, voting systems, intellectual property, and more as shown in Figure 11.1.

Blockchain is seen as a promising solution for enhancing efficiency, reducing fraud, improving transparency, and enabling trust in complex systems where multiple parties are involved. It's important to note that BT is still evolving, and there are challenges to overcome, such as scalability, energy consumption, regulatory frameworks, and interoperability. Ongoing research and development are focused on addressing these challenges to unlock the full potential of BT [7].

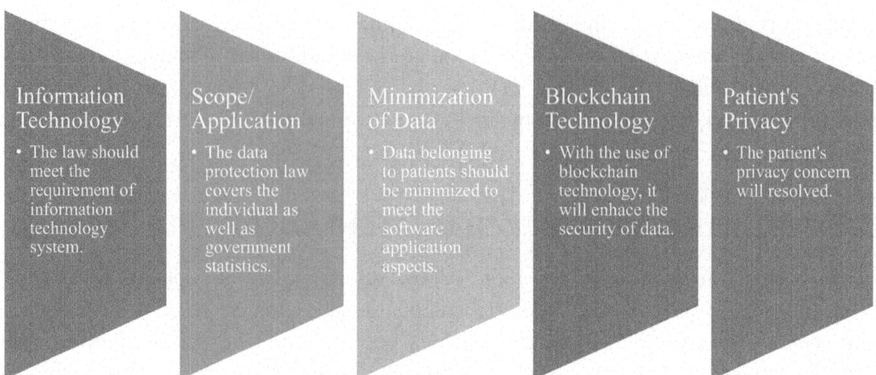

FIGURE 11.1 Smart Sensors, Application Programming Interfaces (API), and IoT Networks.

3 PANACEA FOR DATA PRIVACY AND INTEGRITY

With the recent emphasis on societies in increasing their dependency on cloud technologies, coupled with the human need to communicate and share data via digital networks, internet of things (IoT) devices to include smartphones, industrial and domestic appliances, continue to be a necessary function in conducting business. Social exchanges and transactional types of data, for example, drive the financial markets thus facilitating in the swift development of emerging technologies at an ever-faster rate to keep up with supply and demand trends. In a domestic setting, the sharing of digital media (videos, music, pictures, and documents (data)) through messaging services to enhance subject areas such as information technology, sport, social sciences, education and health for example, IoT devices enable the efficient and effective transfer of data worldwide instantly via the internet of everything (IoE) via the cloud. In an industrial context, smart sensors, application programming interfaces (API) and IoT networks facilitate remote working across digital boundaries globally [8].

These potentially devastating instances of data sharing and/or criminality, influence the confidentiality and protections set out by governments, businesses and organizations, culminating in legal and ethical disputes with significant financial ramifications due to denial of service (DDoS) attacks, for example, that would damage and disrupt entire business data architectures, infrastructures networks and services on a large scale [9].

Consequently, with society relying more and more on the exchange and processing of personal identifiable information (PII) via IoT, trust in renowned institutions and government organizations to include broadcast and digital media outlets becomes a main issue. As a user chooses to share social network, personal and confidential information whilst shopping on-line for example, they should be aware of the nature and intent of cyber-criminality and have faith in the criminal justice system of a given territory.[10]

On the other hand, for businesses, organizations, government bodies, and academic institutions to be able to freely validate and authenticate their data in the service of societies globally [10], artificial intelligence (AI), big data (BD), blockchain combined technologies and methodologies, contribute significantly in mitigating cyber-crime, whilst providing legal bodies the power to hold companies, organization and institutions to account. One such method is the SC for example, and when utilized in the drafting and consenting of a legal document or digital certificate, provides an evidence-based transparent method in enhancing the legal credibility and value of a financial transaction. As a function of blockchain, the SC is validated, implemented, and then shared across a pier-to-pier (P2P) network as a distributed ledger technology (DLT) for all parties to see which provides transparency and accountability [11]. Data security is a critical aspect of BT. Here's how BT enhances data security.

Cryptographic Security: BT uses advanced cryptographic techniques to secure transactions and data. Each transaction is digitally signed using cryptographic keys, ensuring the authenticity and integrity of the information. Additionally, cryptographic hashing is employed to create unique digital fingerprints for each block, linking them together in a tamper-proof manner.

Decentralization and Data Distribution: BT operates on a decentralized network of nodes that collectively maintain the blockchain. Instead of storing data in a centralized server, copies of the blockchain are distributed across multiple nodes. This distribution makes it challenging for attackers to compromise the entire network, enhancing data security.

Immutable Data: Once data is recorded on the blockchain, it becomes virtually immutable. The decentralized consensus mechanism ensures that transactions are validated and added to the blockchain in a transparent and tamper-resistant manner. This immutability prevents unauthorized modifications or deletions, protecting the integrity of the stored data.

Data Encryption: In certain blockchain implementations, data can be encrypted before being stored on the blockchain. Encryption algorithms ensure that the data remains confidential and can only be accessed by authorized parties who possess the decryption keys. This provides an additional layer of protection for sensitive information [12].

Permissioned Blockchains: While public blockchains are open to anyone, permissioned or private blockchains restrict access to known participants. Permissioned blockchains provide tighter control over who can join the network and participate in the validation process. This controlled access helps protect sensitive data and ensures that only trusted entities can interact with the blockchain.

Smart Contracts Security: SCs are self-executing agreements coded on the blockchain. While they enhance automation and efficiency, they must be written with careful consideration for security vulnerabilities. Auditing, testing, and best coding practices are essential to mitigate risks and ensure the security of SCs. It's important to note that while BT enhances data security, it doesn't guarantee absolute security against all possible threats. Organizations must still implement appropriate security measures, such as secure key management, access controls, and regular system audits, to complement the inherent security features of BT and protect against potential vulnerabilities in other parts of the system [13].

4 DATA PROTECTION AND BLOCKCHAIN TECHNOLOGY: LEGAL IMPLICATIONS

There are numerous rights available to all for their survival and well-being without discrimination. Human rights at global level are available for the implementation of certain standards and norms by the different organisations. Fundamental rights are in existence to every citizen belong to the respective legal rights are available within the framework to its legal systems [14].

The Constitution of India, 1950 is the main document that laid down fundamental rights for its peoples along with legal rights. Fundamental rights fall under part III of the constitution of India where multiple rights are available and its enforcement in court of law in case of violation of fundamental rights or for the enforcement.

Privacy is an essential right for every individual and it is given high significance in the preview of fundamental rights from healthcare perspective with certain exceptions, where the government can obtain personal data with the help of the Right to Information Act, 2005. The data of the patient medical record and personal data is available for the government but restricted to exceptional circumstances only. Patients should feel data only been shared with the hospital or clinic to be secured with confidentiality. This is the duty of the healthcare providers to maintain and secure the patient's health information and records and do not disclose any information without prior permission from the patient [15].

There are provisions under the Medical Council of India's Code of Ethics Regulations, which specify the duty of the doctor and hospital management to secure the privacy

> shall not disclose the secrets of a patient that have been learnt in the exercise of his/her profession except in a court of law under orders of the Presiding Judge; in circumstances where there is a serious and identified risk to a specific person and/or community; notifiable diseases.

Data secured by BT is stored and shared with transparency. Blocks in the BT are considered independent unit containing information and create a chain of blocks to form a network. Applications of BT include store and share the medical records, monitor health devices, provide clinical trial data, and improve the health delivery system.

Hospital management and staff should ensure of not disclosing the medical records and data of patients, this utmost care while treating the patient's data. The privacy of the individual is the basic fundamental right, which was declared by the Apex Court in the case of *Justice K.S. Puttaswamy* vs. *Union of India* case. There is a need for building up effective and strong legislation ensuring the protection of individual data and privacy.

The central government of India appointed the data protection committee in the year 2017 and laid down comprehensive report on the significance of data privacy and its protection. This committee's recommended protection of data for all but with some exceptions with the government on exceptional conditions. In this view, bill helps the patient to continue with their privacy without its breach by anyone because the hospital delivery management system now falls under the preview of this bill. In India, the task of data protection is based on the following aspects.

In the healthcare delivery sector, policies belonging to healthcare focus on the regulation of data flow, specifically in hospitals, methods to control and implement the regulatory measure. The right to privacy has several aspects and dimensions but no proper definitions, which limit its meaning has been defined. The right to privacy also includes personal identification, genetic material, biological function, and treatment record.

- The investigation of communicable diseases
- Referral to other hospital
- Data regarding vaccination
- Court or police require for inquiry or investigation
- The insurance companies also require patient record
- As per the legislations for taking out benefits from these-Consumer Protection Act, 2019; Workmen's Compensation Act, 1923, etc.
- Adverse effect on any drug on human health

5 CLOUD COMPUTING AND BLOCKCHAIN TECHNOLOGY

Cloud computing and BT are two powerful technologies that, when combined, can provide innovative solutions for various industries. Cloud computing offers scalable and on-demand access to computing resources, allowing organizations to store, process, and analyze large amounts of data efficiently.[16] It provides cost-effectiveness, flexibility, and ease of use. On the other hand, BT offers a decentralized and transparent system for recording and verifying transactions securely. When integrated, cloud computing and BT can offer several benefits:

Enhanced Security: Blockchain's decentralized and tamper-proof nature enhances the security of data stored in the cloud. By utilizing blockchain as a layer of trust and verification, organizations can ensure the integrity and authenticity of their data, reducing the risk of unauthorized access or data manipulation.

Improved Transparency and Auditability: Blockchain's transparent and immutable ledger allows for increased transparency and auditability of data transactions in the cloud. This can be particularly useful in industries that require strict compliance and auditing, such as finance, supply chain, and healthcare.

Efficient and Trustworthy Data Sharing: Blockchain-based SCs can facilitate secure and automated data sharing among multiple parties in the cloud. SCs enable the execution of predefined rules and conditions, ensuring trust and efficiency in data exchange and collaboration.

Streamlined and Immutable Data Records: BT can create a permanent and tamper-proof record of data transactions in the cloud. This can be valuable in scenarios where data provenance and audit trails are critical, such as intellectual property management, digital rights management, or regulatory compliance.

While the integration of cloud computing and BT offers significant potential, it's essential to carefully consider the specific use cases, scalability, interoperability, and potential regulatory implications. Organizations must assess the trade-offs and implementation challenges to harness the full benefits of this powerful combination.

6 CONCLUSION

BT presents promising solutions for data privacy and security in today's digital world. By leveraging cryptographic techniques, decentralization, and immutability, blockchain enhances data security by ensuring the authenticity, integrity, and confidentiality of information. The distributed nature of blockchain networks reduces the risk of single points of failure and unauthorized access. Additionally, the transparency and traceability of blockchain transactions promote accountability and deter fraudulent activities. While BT provides robust security measures, organizations must still implement comprehensive data protection strategies and adhere to best practices to safeguard sensitive information. As blockchain continues to evolve and find wider adoption, it has the potential to transform data privacy, offering individuals and organizations greater control over their digital identities and fostering a more secure and trustworthy digital ecosystem [16]. It offers promising solutions for data privacy and security in today's digital landscape. Its decentralized and transparent nature, coupled with cryptographic security measures, ensures the integrity, authenticity, and immutability of data. By distributing data across multiple nodes and employing consensus mechanisms, blockchain reduces the risk of centralized data breaches and unauthorized modifications. The use of encryption further enhances the confidentiality of sensitive information. However, it is crucial to recognize that BT alone cannot address all data privacy concerns. Organizations must complement blockchain with robust data protection measures, such as secure key management, access controls, and privacy-enhancing technologies. Additionally, legal and regulatory frameworks need to evolve to address the unique privacy challenges posed by blockchain, striking a balance between transparency and individual privacy rights. As BT continues to mature, it holds great potential to revolutionize data privacy practices and empower individuals with greater control over their personal information in a digital world.

NOTES

1 Thanh Long Nhat Dang and Minh Son Nguyen. "An approach to data privacy in smart home using blockchain technology." In *2018 International Conference on Advanced Computing and Applications (ACOMP)* (2018), 58–64.
2 Kasem-Madani, S., Meier, M., Wehner, M. (2017). Towards a Toolkit for Utility and Privacy-Preserving Transformation of Semi-structured Data Using Data Pseudonymization. In: Garcia-Alfaro, J., Navarro-Arribas, G., Hartenstein, H., Herrera-Joancomartí, J. (eds) *Data Privacy Management, Cryptocurrencies and Blockchain Technology*. DPM CBT 2017 2017. Lecture Notes in Computer Science, vol 10436. Springer, Cham. https://doi.org/10.1007/978-3-319-67816-0_10
3 Bhabendu Kumar Mohanta, Debasish Jena, Soumyashree S. Panda and Srichandan Sobhanayak. "Blockchain technology: A survey on applications and security privacy challenges." *Internet of Things* (2019), 8, 100107. doi.org/10.1016/j.iot.2019.100107
4 Dylan Yaga, Peter Mell, Nik Roby and Karen Scarfone. "Blockchain technology overview." *arXiv preprint arXiv:1906* (2019), 11078.
5 Qiang Wang and Min Su. "Integrating blockchain technology into the energy sector—from theory of blockchain to research and application of energy blockchain." *Computer Science Review* 37 (2020), 100275.
6 Marc Pilkington, "Blockchain technology: principles and applications." In *Research handbook on digital transformations* (Edward Elgar Publishing, 2016), 225–253.
7 Tareq Ahram, Arman Sargolzaei, Saman Sargolzaei, Jeff Daniels and Ben Amaba. "Blockchain technology innovations." In *2017 IEEE Technology & Engineering Management Conference* (TEMSCON) (2017), 137–141. IEEE.

8 Shubham Joshi, Anil Audumbar Pise, Manish Shrivastava, C. Revathy, Harish Kumar, Omar Alsetoohy and Reynah Akwafo. "Adoption of blockchain technology for privacy and security in the context of Industry 4.0." *Wireless Communications and Mobile Computing* 2022 (2022) 14. Article ID 4079781, https://doi.org/10.1155/2022/4079781
9 B. Singh. "COVID-19 pandemic and public healthcare: endless downward spiral or solution via rapid legal and health services implementation with patient monitoring program." *Justice and Law Bulletin 1*(1) (2022), 1–7. https://apricusjournals.com/index.php/jus-l-bulletin
10 Aafaf Ouaddah, Anas Abou Elkalam and Abdellah Ait Ouahman "Towards a novel privacy-preserving access control model based on blockchain technology in IoT." In *Europe and MENA cooperation advances in information and communication technologies* (Springer International Publishing, 2017), 533.
11 Muneeb Ul Hassan, Mubashir Husain Rehmani and Jinjun Chen "Differential privacy in blockchain technology: a futuristic approach." *Journal of Parallel and Distributed Computing* 145 (2020), 50–74.
12 Singh, B., Raghav, A., Mishra, S., Ray, S., & Raghav, R. (2024). Judicial Approach towards the concept of Plea Bargaining in India. *National Journal of Real Estate Law, 7*(2).
13 Bharat Bhushan, Preeti Sinha, K. Martin Sagayam and J. Andrew. "Untangling blockchain technology: a survey on state of the art, security threats, privacy services, applications and future research directions." *Computers & Electrical Engineering* 90 (2021), 106897.
14 Singh, B. "Affordability of medicines, public health and TRIPS regime: a comparative analysis." *Indian Journal of Health and Medical Law* 2, no. 1 (2019), 1–7.
15 Peter Yeoh, "Regulatory issues in blockchain technology." *Journal of Financial Regulation and Compliance* 25(2) (2017), 196–208. https://doi.org/10.1108/JFRC-08-2016-0068
16 Keke Gai, Jinnan Guo, Liehuang Zhu, and Shui Yu, "Blockchain meets cloud computing: a survey." *IEEE Communications Surveys & Tutorials* 22, no. 3 (2020), 2009–2030.

12 Enhancing Human-Centered Security in Industry 4.0

Navigating Challenges and Seizing Opportunities

Aliyu Mohammed, Shanmugam Sundararajan, and Senthil Kumar

1 INTRODUCTION

Industry 4.0 represents a paradigm shift in manufacturing, intertwining physical processes with digital technologies, artificial intelligence (AI), and the internet of things (IoT). As enterprises embrace this transformative era, the need for robust cybersecurity has never been more pressing. The digitization of industrial processes brings efficiency gains but also introduces a complex web of security challenges. The interconnectedness of devices and systems in Industry 4.0 significantly expands the attack surface, creating vulnerabilities that adversaries can exploit. A study by Gartner predicts that by 2025, 75% of enterprise-generated data will be created and processed outside traditional centralized data centers, making decentralized and widely distributed security measures imperative (Heiskari, 2022). This underlines the urgency of adopting a comprehensive security strategy. Human-centered security emerges as a cornerstone in addressing the multifaceted challenges of Industry 4.0. While technological advancements are crucial, the human element remains pivotal in the defense against cyber threats. As highlighted by a report from McKinsey, 46% of cybersecurity incidents are caused by internal actors, whether through unintentional errors or malicious intent (Möller, 2023). This emphasizes the critical role of human operators in maintaining a secure industrial ecosystem. The psychological aspects of cybersecurity cannot be overlooked. Employees need to be equipped with the knowledge and awareness to recognize and respond to potential threats. Cybersecurity training and awareness programs become essential components of a resilient security strategy. A research paper by the *International Journal of Human-Computer Interaction* stresses the importance of considering human factors in designing security interfaces to enhance user understanding and compliance (Rapp, Curti, and Boldi, 2021). In navigating the intricate landscape of Industry 4.0 security, a human-centered

 DOI: 10.1201/9781032694375-12

approach not only mitigates risks but also unlocks opportunities for innovation. By understanding and empowering the human workforce, organizations can create a culture of security that adapts to the evolving technological landscape. This chapter delves into the challenges and opportunities of enhancing human-centered security in Industry 4.0, exploring the symbiotic relationship between technological advancements and human vigilance.

1.1 Background of Industry 4.0

Industry 4.0, denoting the Fourth Industrial Revolution, represents a transformative era marked by the infusion of intelligent technologies into manufacturing processes. Originating in Germany, this paradigm shift signifies a departure from conventional manufacturing approaches toward the adoption of smart factories. These factories harness the power of cyber-physical systems, the IoT, and data analytics to streamline operations. The hallmark of Industry 4.0 lies in the interconnectedness of its systems, facilitating seamless real-time communication and decision-making. This interconnected landscape is instrumental in optimizing overall efficiency and productivity within manufacturing environments. Koch's (2022) seminal work presented at the World Economic Forum serves as a cornerstone in understanding the profound impact of Industry 4.0 on global industries. His insights underscore how this revolution is fundamentally reshaping traditional industrial landscapes. Recognizing the historical context and evolution of Industry 4.0 is essential for comprehending the intricate challenges it introduces to security measures. The essence of Industry 4.0 lies in its heavy reliance on extensive digital connectivity, making it imperative to delve into its foundational background to grasp the complexities and vulnerabilities associated with securing such technologically advanced systems.

1.2 Emergence of Human-Centered Security

The emergence of human-centric security arises from a realization of the limitations inherent in solely relying on technological solutions for cybersecurity. Conventional security models often fall short in considering the impact of human factors, leaving vulnerabilities that can be exploited. Pioneering work by Yoon and Jun (2023) in the realm of human-computer interaction emphasized the significance of incorporating user perspectives in the design of systems. In the context of Industry 4.0, the imperative for a human-centric approach becomes evident as employees engage with intricate, interconnected systems. This perspective recognizes that humans serve as both potential weak links and essential assets in upholding a secure industrial environment. The acknowledgment of human influence on security dynamics reflects a departure from the narrow focus on technological aspects alone. Yoon and Jun (2023) insights underscore the importance of understanding how individuals interact with technology, emphasizing the human role in shaping the security landscape. In the intricate milieu of Industry 4.0, where human-machine interactions are pervasive, the human-centric security paradigm becomes a strategic necessity. It embraces the idea that, in addition to technological safeguards, a comprehensive security strategy must address the human element, acknowledging the dual role humans play as potential vulnerabilities and integral contributors to maintaining a resilient and secure industrial ecosystem.

1.3 Significance and Relevance of the Study

The significance of this study is underscored by its focus on bridging a crucial gap in comprehending and addressing security challenges within the context of Industry 4.0, employing a human-centered perspective. Given the escalating frequency and sophistication of cyber threats, a thorough analysis is imperative. An Accenture report emphasizes the alarming pace at which cyber threats are outpacing security measures, accentuating the critical need for research that combines technological advancements with strategies centered on the human element of security (Gupta, George, and Fewer, 2024). The relevance of this research extends globally, resonating with industries worldwide. The consequences of inadequate security measures can be severe, impacting not only day-to-day operations but also exerting broader repercussions on the economy. As technological landscapes rapidly evolve, there is an urgency to understand and address security challenges comprehensively. The study's significance lies in its proactive approach to amalgamate insights from human-centric security frameworks with advancements in technology, providing a holistic understanding of security dynamics in the Industry 4.0 landscape. By doing so, the research aims not only to contribute to the academic discourse but also to offer practical solutions that can be applied by industries navigating the intricate challenges posed by cyber threats in the current digital era.

1.4 Objectives of the Conceptual Paper

The objectives of this conceptual paper are multifaceted.

1. It aims to critically examine the challenges posed by Industry 4.0 to traditional security paradigms.
2. It seeks to underscore the pivotal role of human operators in maintaining cybersecurity.
3. The study intends to explain how to use technology in making more human-based security.
4. It makes recommendations on the types of regulations and collaboration required for an adaptable security environment.

The conceptual paper seeks to offer a complete instruction for organizations operating in the intricate terrain of Industry 4.0 security through tackling these objective points.

2 LITERATURE REVIEW

The literature review forms a basis of understanding regarding the already available information about Industry 4.0 and human-oriented security. Some notable pieces in this field include a detailed overview by Ding et al. (2023), highlighting the technological dimensions of Industry 4.0. This is an introduction to the fundamental components and progress facilitating the modern industrial revolution. Furthermore, included is a literature review that includes classic works in cybersecurity like Mao and Chang (2023), an investigation into the psychological aspects of safety, where the behavior aspect of the user in safe systems transactions is illuminated. The literature review lays the groundwork for understanding the issues and possibilities emerging at the junction of Industry 4.0 and human-oriented security by bringing together these ideas.

2.1 Evolution of Industry 4.0

Seminal works illuminate the evolution of Industry 4.0 by explicating how these progressive technologies have gradually been integrated into the industry. Allioui and Mourdi's (2023) exploration of the Fourth Industrial Revolution stands as a cornerstone in unraveling the conceptualization of Industry 4.0. It meticulously outlines the technological strides that have defined this epoch, underscoring the pivotal shift toward the incorporation of cyber-physical systems. Additionally, the comprehensive study by Abbasi and Rahmani (2023) furnishes a systematic overview encompassing the historical trajectory, technical constituents, and potential challenges inherent in Industry 4.0. The synergy between these seminal contributions crafts a nuanced panorama of Industry 4.0's evolution, establishing the foundation for a thorough examination of its implications on security. Li's (2023) insights delve into the core of the Fourth Industrial Revolution, shedding light on the intricate interplay between technological advancements and industrial paradigms. This exploration lays the groundwork for understanding how Industry 4.0 has transcended traditional manufacturing frameworks. Complementing this, Luft, Luft, and Arntz (2023) work serves as a roadmap, guiding through the historical milestones, the intricate components that define Industry 4.0, and the potential hurdles it presents. This amalgamation of perspectives not only paints a detailed picture of Industry 4.0's evolution but also sets the stage for a comprehensive exploration of the security considerations intrinsic to this transformative industrial landscape.

2.2 Theoretical Foundations of Human-Centered Security

The theoretical underpinnings of human-centered security are rooted in the realms of human-computer interaction and psychology. Dhoni and Kumar (2023) contribute significantly by accentuating the importance of awareness in distributed collaboration, a concept that transcends into the security domain, where user awareness plays a pivotal role in counteracting cyber threats. In the context of securing Industry 4.0, this emphasis on awareness aligns with the interconnected and collaborative nature of the industrial landscape. Nkongolo (2023) exploration of the psychology of security delves into the cognitive dimensions of individuals' interactions with secure systems, offering theoretical insights into the human facets of cybersecurity. By examining the psychological factors that influence security behaviors, Anderson's work enriches the theoretical foundation for understanding why a human-centered approach is indispensable in the security paradigm of Industry 4.0. The integration of Nkongolo (2023) emphasis on awareness and Anderson's insights into the psychological aspects of security contributes to a robust theoretical framework. This framework illuminates the intricate relationship between human factors and cybersecurity within Industry 4.0. Recognizing that humans are not only potential vulnerabilities but also crucial contributors to security resilience, these theories provide a solid grounding for comprehending the significance of a human-centric approach. In navigating the evolving landscape of Industry 4.0, where human-machine collaboration is integral, these theoretical foundations become instrumental in shaping effective and adaptive security strategies.

2.3 Synthesis of Industry 4.0 and Human-Centered Security Literature

The synthesis of literature on Industry 4.0 and human-centered security entails a cohesive synthesis of insights from diverse disciplines. By amalgamating the technological perspectives presented by van Dun and Kumar (2023) on Industry 4.0 with the psychological and behavioral dimensions underscored by Kunduru (2023), a panoramic understanding of the challenges and opportunities emerging at the convergence of these domains is achieved. This synthesis facilitates a nuanced comprehension of how the technological trajectory of Industry 4.0 intersects with the human factors crucial for ensuring effective cybersecurity. It establishes the groundwork for advocating a holistic approach that seamlessly incorporates human-centered security measures into the fabric of Industry 4.0. van Dun and Kumar (2023) technological insights provide a foundation for grasping the intricacies of Industry 4.0's evolution, while Kunduru (2023) psychological perspectives shed light on the human elements inherent in secure systems. The synergy between these perspectives forms a comprehensive view that goes beyond the conventional boundaries of technological and human-centric considerations. This integrated approach acknowledges that the successful implementation of security measures in the context of Industry 4.0 necessitates an understanding of both the advanced technological landscape and the human behaviors that influence security outcomes. In proposing a holistic approach, this synthesis advocates for the integration of human-centered security not as an isolated component but as an integral and inseparable aspect of Industry 4.0. By doing so, it lays the groundwork for a more resilient and adaptive security framework that aligns with the collaborative and interconnected nature of contemporary industrial systems.

2.4 Key Concepts and Models in Human-Centered Security

Human-centered security relies on key concepts and models that integrate human factors into the design and implementation of security measures. One pivotal concept is "usable security," emphasizing that security measures should not hinder user tasks but seamlessly integrate with them. This aligns with the principles of Spero (2023), who highlighted the importance of user awareness in system design (Table 12.1).

The Protection Motivation Theory (PMT) and Security Behavior Intentions offer psychological frameworks for understanding and predicting user responses to security measures. Combining these with usability evaluation methods like Cognitive Walkthrough ensures that security measures are not only effective but also aligned

TABLE 12.1
Key Concepts and Models in Human-Centered Security

Concept/Model	A. Description	B. Contribution
Usable Security	Stresses the integration of security measures without impeding user tasks, recognizing the importance of user experience	Aligns security measures with user behavior, enhancing overall security
Protection Motivation Theory	A psychological model that explores individuals' motivation to adopt protective measures based on perceived threats	Provides insights into user behavior and factors influencing security decisions
Security Behavior Intentions	Focuses on understanding the intentions behind security-related actions and decisions of individuals	Guides the design of interventions to promote secure behavior
Cognitive Walkthrough	A usability inspection method that evaluates the usability of a system from a user's perspective	Identifies potential user errors and areas of misunderstanding in security interfaces

with user mental models. Employing these concepts and models is crucial for creating a human-centered security approach that addresses the cognitive and behavioral aspects of users, enhancing overall cybersecurity.

3 THEORETICAL FRAMEWORK

The theoretical framework for this study draws from human-centered security, Industry 4.0 literature, and psychological models influencing security behavior. AlMalki and Durugbo's (2023) work on awareness in distributed collaboration underpins the acknowledgment of the human element in Industry 4.0 security. Additionally, Song (2023) psychological model provides insights into user behavior, informing the development of strategies that resonate with the cognitive aspects of human operators. This theoretical framework guides the exploration of the symbiotic relationship between technological advancements and the human factors crucial for effective cybersecurity in Industry 4.0.

3.1 Defining the Conceptual Framework

The conceptual framework serves as the theoretical foundation for the integration of human-centered security within the realm of Industry 4.0. Leveraging the insights of Luo, Thevenin, and Dolgui (2023) regarding the significance of awareness in distributed collaboration, the framework incorporates the core aspects of human factors into the security landscape of Industry 4.0. By delineating this conceptual framework, the study endeavors to narrow the divide between conventional security models and the distinctive challenges presented by the interconnectivity of cyber-physical systems in Industry 4.0. This approach establishes a basis for crafting security measures that harmonize with the cognitive dimensions of human operators, recognizing their pivotal role in the secure functioning of Industry 4.0 environments.

3.2 Integrating Human-Centered Security in Industry 4.0

Integrating human-centered security in Industry 4.0 entails utilizing the conceptual framework to formulate strategies that prioritize the human dimension in cybersecurity. This incorporation is influenced by Almontaser and Gerged's (2023) insights into the transformative potential of the Fourth Industrial Revolution. The conceptual framework functions as a roadmap for implementing security measures that are user-friendly, adhering to the principles of usability and taking cognizance of the psychology of security. Through the infusion of human-centered security, organizations operating within the purview of Industry 4.0 can elevate their resilience against dynamic cyber threats, simultaneously nurturing a culture of security among human operators. Schwab's perspectives on the Fourth Industrial Revolution offer valuable guidance in navigating the transformative landscape of Industry 4.0. The conceptual framework, derived from Almontaser and Gerged's (2023) emphasis on awareness and psychological insights, acts as a compass for developing security strategies that harmonize with human interactions. The integration of human-centered security not only fortifies defenses against evolving cyber threats but also contributes to the cultivation of a security-centric mindset among individuals operating within the Industry 4.0 ecosystem.

3.3 Conceptual Model for Enhancing Security

The conceptual model designed to enhance security within the realm of Industry 4.0 is meticulously crafted through the synthesis of pivotal concepts and models, encompassing usability, the psychology of security, and the Security Behavior Intentions Model (SBIM). Evolving from Costa's (2023) SBI model, this model establishes a structured framework aimed at comprehending and influencing human behavior in the realm of cybersecurity. It provides organizations with a practical strategic framework for developing people-focused security based on contemporary technology, as well as sophisticated human decisions' behavioral model. The integration of usability, security psychology, and the SBI model in the conceptual model creates a comprehensive approach in addressing safety issues arising in Industry 4.0. The model supports an in-depth understanding of how human activities can affect different issues of security within Internet facilities. This understanding is essential to companies improving their security stance, presenting a tangible roadmap through Industry 4.0 maze. The conceptual model acts as a bridge, bridging the gap between technological advancements and human-centric security strategies, offering a comprehensive approach that recognizes the symbiotic relationship between technology and human decision-making processes in the context of cybersecurity.

3.4 The Role of Human Factors in Security

The significance of human factors in security is paramount, influencing the efficacy of cybersecurity measures. This segment explores seminal contributions, particularly Gul (2023) work on the psychology of security. By delving into the cognitive and behavioral dimensions of individuals, organizations gain insights that enable the customization of security protocols to harmonize with inherent human tendencies. Recognizing the pivotal role of human factors is indispensable for the viability of the envisioned human-centered security framework within the context of Industry 4.0. This underscores the necessity for designs centered on user needs, comprehensive training initiatives, and continuous awareness programs to foster a security-conscious culture. Anderson's exploration of the psychology of security provides a foundational understanding of how human behaviors and perceptions interplay with security dynamics. This comprehension is instrumental in tailoring security strategies to align seamlessly with human tendencies, thereby enhancing the overall effectiveness of cybersecurity measures. Within the proposed human-centered security framework for Industry 4.0, embracing and addressing human factors emerges as a cornerstone, advocating for user-centric approaches that resonate with the intricate interplay of cognitive and behavioral elements. The emphasis on user-centric designs, coupled with robust training and awareness initiatives, reflects a proactive approach to fortifying security measures in Industry 4.0 by acknowledging and integrating the crucial role of human factors.

4 METHODOLOGY

The methodology of this study primarily involved an extensive review of related studies, embracing a qualitative and conceptual approach. This method was chosen to explore and synthesize existing knowledge, theories, and models related to human-centered security in the context of Industry 4.0.

4.1 Justification for a Conceptual Approach

Adoption of a conceptual approach springs from the nature of the study which is qualitative and exploratory. The conceptual approach combines various perspectives and theories of human-centered security in Industry 4.0. This is in line with the overall objective of developing an integrated framework that takes into consideration the interactions between technology and human dimensions. The conceptual approach in providing a new vision of information security architecture was pre-defined by scholars like Magliocca (2023), who called for incorporating awareness and understanding in human into the architecture.

4.2 Conceptualization of Key Variables

The development and definition of critical variables was done by reviewing the respective literature on human centric security and Industry 4.0. A coherent framework was built on concepts like usability theory, the psychology of security, and Vance et al.'s SBI model. In relation to developing theory, the process of conceptualizing entailed the consideration of the basis as well as the core parts involved.

4.3 Delimitations and Scope of the Conceptual Model

The delimitations marked out the limits of the conceptual model. In the study, only human-centered security in Industry 4.0 was considered, without going into the depths of other cybersecurity models. The delimitations were crucial to maintain a focused inquiry and ensure the depth of exploration within the chosen conceptual framework.

4.4 Operationalization of Key Concepts

The operationalization of key concepts involved translating theoretical insights into practical considerations for Industry 4.0 settings. Usability principles, drawn from the works of Anderson (2023), were adapted to the unique requirements of industrial environments. The SBI model guided the operationalization of behavioral aspects, considering cultural factors influencing employees' adherence to security policies. In conclusion, the qualitative and conceptual methodology chosen for this study allowed for a nuanced exploration of human-centered security in Industry 4.0. By integrating insights from various scholars, the study aimed to contribute a comprehensive conceptual framework that considers both technological advancements and human factors in the realm of industrial cybersecurity.

5 ENHANCING HUMAN-CENTERED SECURITY MODEL

The evolution of the human-centered security model is grounded in the synthesis of crucial concepts and models identified through the literature review and theoretical framework. Serving as a conceptual framework, this model aims to elevate security practices within the context of Industry 4.0 by placing the human element at its nucleus. Drawing inspiration from the contributions of Karimi (2023), the model intricately weaves together principles of usability, psychological insights, and cultural considerations to forge a comprehensive and cohesive approach. It stands as a guiding framework for organizations intent on bolstering their security stance, harmonizing technological solutions with the cognitive and behavioral intricacies inherent in human operators. By integrating insights from Karimi's (2023) SBI model and Anderson's exploration of the psychology of security, the model strategically encompasses various dimensions essential for a robust human-centered security approach. This approach acknowledges the diverse factors influencing human behavior in cybersecurity, fostering a dynamic framework that goes beyond conventional security paradigms. Positioned as a practical guide, the model empowers organizations to navigate the complexities of Industry 4.0 by recognizing the symbiotic relationship between technology and human factors. In doing so, it propels a paradigm shift toward a more resilient and adaptive security posture that aligns with the collaborative and interconnected nature of contemporary industrial systems.

5.1 Components of the Conceptual Model

The conceptual model encompasses several pivotal components, strategically crafted to establish a human-centric security paradigm within the landscape of Industry 4.0. Usability, derived from the insights of Egeli (2023), is ingrained as a foundational element, ensuring that security measures prioritize user-friendliness without hindering productivity. The SBI model shapes the behavioral facet of the model, accounting for the impact of cultural factors on employees' adherence to security policies. Additionally, the model utilizes adaptive security, appreciating that cyber threats are ever-changing and necessitate ongoing refinement as well. By considering usability at the foundation means that security functions smoothly and becomes part of the overall user experience as articulated by Egeli (2023). As it would foster acceptance and compliance among the end-users as well as support a security culture that fits into collaboration mode of Industry 4.0, this approach is integral. Incorporation of the SBI model adds value to the notion and includes behavior-based cultural aspects in it. This recognizes the different human elements involved in Industry 4.0 and highlights the importance of tailored approaches to security that understands culturally-embedded organizational settings. Adaptive security also ensures changes in approach as technology progresses to cope up with upcoming threats hence the security system remains agile. Combined, all these elements are a complete, customizable system that considers users in Industry 4.0.

5.2 Representation of the Model

The Industry 4.0 human centered security model is presented using a tabular approach, capturing all components and their links within this paradigm. The purpose of this representation is simply to give an easy-to-understand illustration of what a conceptual model represents (Table 12.2).

This tabulated form illustrates a specific function of every element in enforcing people-based safety. As mentioned, usability promotes user's acceptance and psychology of security focuses on the conceptual side of user's interaction. It's about SBI model, which takes into account cultural components and Adaptive Security Measures against new threats. This forms the main way that organizations understand that the model is holistic that integrates humans and technology into strong security in industry 4.0.

TABLE 12.2
Human-Centered Security Model Components

Component	C. Description
D. Usability	Ensures security measures are user-friendly, minimizing friction in human-computer interactions
Psychology of Security	Considers cognitive aspects, understanding how individuals perceive and respond to security measures
Security Behavior Intentions Model	Incorporates cultural factors influencing employees' intentions to adhere to security policies
Adaptive Security Measures	Recognizes the dynamic nature of cyber threats, requiring continuous adaptation of security protocols

5.3 Illustrating the Dynamics of Security in Human-Centered Contexts

The intricacies of security within human-centered contexts are vividly portrayed through tangible examples embedded in the Industry 4.0 landscape. Utilizing real-world case studies and scenarios, the model illustrates how the amalgamation of human factors, cultural nuances, and adaptive security measures effectively mitigates potential risks. This section accentuates the significance of cultivating a security culture among human operators, ensuring not only their awareness of potential threats but also their active engagement in upholding a secure industrial ecosystem. By delving into practical examples, the model offers a tangible representation of how human-centric security principles operate within the dynamic Industry 4.0 environment. It goes beyond theoretical frameworks, showcasing the real impact of integrating human factors into security strategies. The use of case studies emphasizes the relevance and applicability of the model, providing stakeholders with concrete instances of successful security implementations that consider the multifaceted nature of human interactions and cultural influences. Emphasizing the cultivation of a security culture underscores the proactive role human operators play in maintaining a resilient security posture. The model positions human awareness and engagement as integral components, recognizing that security is not solely a technological matter but a collaborative effort that involves the active participation of individuals within the Industry 4.0 ecosystem. Through practical illustrations, this section reinforces the model's effectiveness in translating conceptual principles into actionable strategies, bridging the gap between theory and application in the pursuit of heightened security in human-centered contexts.

6 NAVIGATING CHALLENGES IN INDUSTRY 4.0 SECURITIES

Effectively tackling challenges in Industry 4.0 security demands a thorough understanding of the intricate threats and vulnerabilities pervasive in this highly interconnected and digitized environment. Scholars, exemplified by Mishra and Singh (2023), delve into the technological complexities of Industry 4.0, highlighting the imperative for adaptive security measures. Khan and AbaOud (2023), underscoring the decentralized nature of data processing, further accentuate the critical need for confronting security challenges promptly. This section delves into the comprehensive challenges confronting organizations as they endeavor to secure Industry 4.0 environments, laying the foundation for subsequent discussions on human-centered solutions. The Industry 4.0 security landscape is characterized by multifaceted threats emanating from the integration of advanced technologies. Mishra and Singh's (2023) insights provide a lens into the intricate technological dynamics, emphasizing the indispensability of security measures that can adapt to the evolving nature of these challenges. Khan and AbaOud's (2023) predictions echo the urgency, stressing the decentralized processing of data as a focal point for security concerns. This section comprehensively explores the overarching challenges faced by organizations striving to fortify Industry 4.0 security, setting the stage for a deeper exploration of solutions centered on human considerations as shown in Figure 12.1.

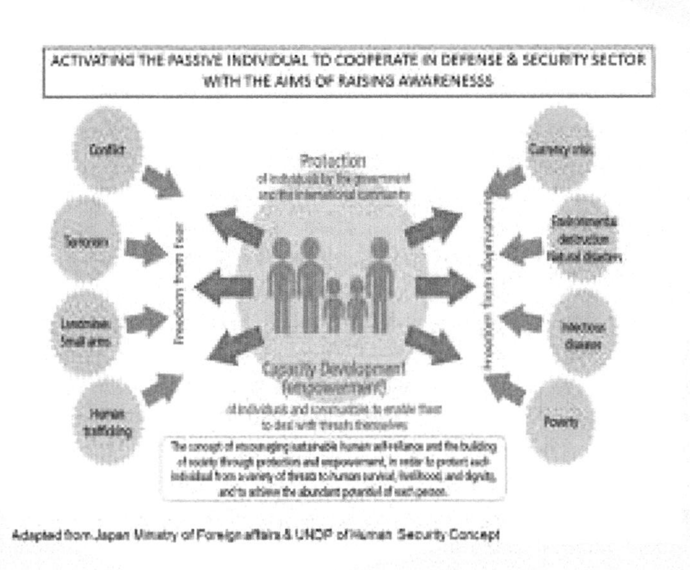

FIGURE 12.1 Execution Model for Public Safety and Awareness.

6.1 Overview of Security Challenges

Within the landscape of Industry 4.0, a multitude of security challenges spans a range of issues, notably the expanded attack surface resulting from the integration of IoT devices, cyber-physical systems, and the escalating sophistication of cyber threats. The inherent distributed and interconnected nature of Industry 4.0 systems further magnifies the risks associated with cybersecurity. AlMalki and Durugbo's (2023) observations regarding the Fourth Industrial Revolution accentuate the critical need to confront these challenges. In this part, there is comprehensive discussion on the special threats posed by such challenges as changes in the menace field and the need for active steps aimed at overcoming these ever-changing hazards. The integration of IoT devices and cyber-physical systems increases potential attack surfaces by expanding network connectivity and introducing a diverse ecosystem of devices, each with its own potential vulnerabilities. This integration creates multiple data exchange points, often incorporates devices with limited security measures, and increases overall system complexity, all of which contribute to a larger attack surface for potential cyber threats. This will require adequate risk knowledge to be known. Cyber threats are evolving increasingly sophisticated, necessitating continuous adaptation and advanced cybersecurity measures. Distributed and connected systems in Industry 4.0 make for a vulnerable space whose vulnerability can only be effectively tackled using a strategic and forward-looking approach. According to AlMalki and Durugbo (2023), insights from this revolution underscore the importance of addressing security issues as they arise. The last part explores the complexities that accompany Industry 4.0

security environment and underscores the need to embrace an upfront attitude toward handling security concerns that arise as a result of this paradigm.

6.2 Aligning Challenges with Human-Centered Solutions

Addressing the security issues of Industry 4.0 demands merging human centered and technology focused strategies. This alignment is achieved with the help of the conceptual model that has been crafted during this study. In relation to these findings in Neethirajan (2023), this section explores how human-centric security might be used as a solution to these problems. By applying usability principles, understanding the psychology of security, and adjusting security according to how humans behave, organizations will have much higher resilience in dealing with cybersecurity issues. The Industry 4.0 security environment is quite intricate in nature and therefore calls for a complete solution approach that embraces humans as the starting point. This conceptual model acts like a navigation chart, blending together HCII and information security behavioral intents. This way, the security measures should address both the technical complexities of Industry 4.0 and the cognizance aspects of the operators. By prioritizing usability, acknowledging psychological factors, and tailoring security measures to human behavior, organizations can fortify their defenses, creating a security paradigm that is adaptive, responsive, and ultimately aligned with the collaborative and interconnected nature of Industry 4.0 environments.

6.3 Case Studies: Successful Mitigation Strategies

This section, grounded in real-world case studies, exemplifies successful strategies organizations have adopted to surmount security challenges within the context of Industry 4.0. The implementation of the conceptual model takes center stage, highlighting the tangible application of human-centered security principles. Through these case studies, valuable insights are gleaned into the efficacy of adaptive security measures, user-friendly design implementations, and cultural considerations in addressing specific challenges. The analysis encompasses a spectrum of industries, underscoring the adaptability and versatility inherent in the proposed human-centered security model. Examining practical instances of organizations applying the conceptual model provides a nuanced understanding of its impact on mitigating Industry 4.0 security challenges. The showcased strategies emphasize the model's practicality and effectiveness in navigating the complexities of cyber threats within diverse industry landscapes. The success stories underscore the significance of implementing adaptive security measures that align with evolving threats, user-friendly designs that facilitate seamless integration into operational workflows, and cultural considerations that acknowledge and accommodate the diversity inherent in different organizational contexts. These case studies collectively reinforce the versatility and applicability of the human-centered security model across industries, validating its potential to serve as a robust and adaptable framework for organizations grappling with the intricate security dynamics of Industry 4.0.

TABLE 12.3
Challenge Solutions

Security Challenge	E. Human-Centered Solution
F. Increased Attack Surface	Usability-driven authentication protocols to enhance user compliance
Sophistication of Cyber Threats	Psychology of Security integration, addressing cognitive aspects of users
Interconnected Systems Risks	Adaptive security measures, continuously updated based on evolving threats

6.4 Visualizing Challenge Solutions (Tabular Form)

Visualizing challenge solutions in a tabular format encapsulates the alignment of human-centered security principles with specific security challenges in the Industry 4.0 landscape. Table 12.3 presents a condensed overview, showcasing how key challenges are met with corresponding human-centric strategies. This visualization serves as a practical guide for organizations seeking to implement effective solutions in the face of Industry 4.0 security challenges.

This table illustrates the direct correlation between security challenges and human-centered solutions. For instance, in mitigating the challenge of an increased attack surface, the implementation of user-friendly authentication protocols aligns with the usability principles. Addressing the sophistication of cyber threats involves incorporating the psychology of security, understanding and adapting to user behaviors. The interconnected systems' risks are met with adaptive security measures, ensuring that security protocols remain resilient in the dynamically evolving Industry 4.0 environment. This visualization aims to provide a concise reference for organizations to tailor their security strategies to the specific challenges posed by the Fourth Industrial Revolution.

7 SEIZING OPPORTUNITIES FOR ENHANCED SECURITY

Capturing opportunities to bolster security in Industry 4.0 goes beyond overcoming challenges; it entails strategically harnessing the distinctive advantages inherent in this technological paradigm. Forward-thinking leaders such as Dyson and Humphreys (2023) accentuate the revolutionary potential of Industry 4.0, stressing the imperative to synchronize security strategies with the opportunities it unfolds. This section delves into the proactive identification and exploitation of these opportunities, elucidating how organizations can augment their overall security posture. Ravitch and Herzog (2023) visionary insights illuminate the transformative landscape of Industry 4.0, urging organizations to recalibrate their security approaches in alignment with its potential. The emphasis is not merely on mitigating risks but on capitalizing on the inherent strengths of Industry 4.0 for security enhancement. Organizations can proactively discern and leverage opportunities within Industry 4.0 by adopting a strategic approach. This involves understanding the unique technological landscape, such as the integration of IoT, data analytics, and automation, and aligning security measures to capitalize on these advancements. The section underscores the importance of proactive engagement, where organizations actively seek out ways to integrate security seamlessly into their operations, harnessing the full potential of Industry 4.0 technologies. In essence, seizing security opportunities in Industry 4.0 is a strategic endeavor that requires a forward-thinking mindset. It involves not just reacting to challenges but actively exploring and leveraging the inherent advantages presented by the evolving technological paradigm, ultimately fostering a security posture that is not only resilient but also aligned with the transformative spirit of Industry 4.0.

7.1 Exploration of Opportunities in Industry 4.0

The advent of Industry 4.0 ushers in a multitude of opportunities for heightened security. The amalgamation of robust data analytics, AI, and machine learning (ML) opens avenues for the development of predictive security measures. The scholarly contributions of Bharadiya (2023), particularly their insights into industrial data platforms, play a pivotal role, providing guidance for exploring opportunities in leveraging data for proactive threat detection. This section delves into the transformative potential embedded in technological advancements within Industry 4.0, accentuating the prospects for elevating security measures beyond reactive responses. The integration of cutting-edge technologies, such as data analytics, AI, and ML, lays the foundation for a paradigm shift in security strategies. Taarup-Esbensen (2023) insights into industrial data platforms offer a strategic lens through which opportunities for proactive threat detection can be envisioned. This exploration goes beyond conventional security approaches, emphasizing the transformative potential of Industry 4.0 advancements. The section underscores the capacity of these technological opportunities to empower organizations in anticipating and mitigating security threats before they manifest, marking a departure from reactive security measures toward a more proactive and preventive security paradigm within the Industry 4.0 landscape.

7.2 Leveraging Human-Centered Approaches for Opportunity Realization

Harnessing human-centered methodologies is essential for unlocking the security potential offered by Industry 4.0. The conceptual model crafted in this study emerges as a key instrument, strategically aligning human elements with technological progress. Andersson, Bjursell, and Palm's (2023) psychological insights into security behaviors assume a pivotal role, guaranteeing active involvement of human operators in the pursuit of security opportunities. Through the integration of usability principles and the cultivation of a security-aware culture, organizations can maximize the capabilities of Industry 4.0 technologies for the implementation of proactive and adaptive security measures. The conceptual model acts as a bridge, harmonizing the symbiotic relationship between human factors and technological advancements within Industry 4.0. Anderson's psychological insights inject a profound understanding of how human behaviors influence security dynamics, fostering a collaborative and participative approach. Incorporating usability principles ensures that security measures are user-friendly, facilitating seamless integration into operational workflows. Cultivating a security-aware culture goes beyond technology, emphasizing the importance of human vigilance and active engagement in recognizing and addressing security threats. In essence, the integration of human-centered approaches not only enhances the effectiveness of security measures but also transforms Industry 4.0 technologies into proactive tools that adapt to evolving threats. This holistic strategy acknowledges that successful security in Industry 4.0 is not solely a technological challenge but a collaborative effort that involves the active participation and awareness of human operators.

7.3 Conceptual Diagram: Integrating Opportunities

The conceptual diagram visually encapsulates the seamless integration of opportunities within the human-centered security framework in Industry 4.0. This visual representation serves as a strategic guide, highlighting the symbiotic relationship between technological advancements and human-centric approaches. The conceptual diagram above is accompanied by Table 12.4 which simply depicts the integrated opportunities stressing that the proposed security model is all encompassing.

Table 12.5 presents the integrated opportunities and the individual elements involved that help in improving security in Industry 4.0. While predictive security methods use sophisticated data analytics, adaptive security procedures utilize

TABLE 12.4
Integrated Opportunities in Human-Centered Security

Opportunity	G. Integration Approach
H. Predictive Security Measures	Utilize advanced data analytics to forecast potential threats.
Adaptive Security Protocols	Employ AI and ML for real-time adaptation to evolving cyber threats.
Human-Centric Culture	Foster a security-aware culture through usability and awareness initiatives.
Collective Intelligence	Combine human insights and AI-driven analytics for comprehensive security.

TABLE 12.5
Proposed KPIs and Measurement Criteria

KPI	I. Measurement Criteria
J. User Compliance	Percentage of employees adhering to established security policies
Usability Satisfaction	User feedback on the user-friendliness of implemented security measures
Incident Response Time	Average time taken to respond to and mitigate security incidents
Security Training Effectiveness	Evaluation of the impact of security training programs on user awareness and behavior
Adaptation Speed to Threats	Time taken to adapt security protocols in response to emerging cyber threats
Cultural Alignment with Security	Measurement of organizational culture alignment with security-conscious practices

AI and ML to respond to threats on the fly. Emphasis on human-centered cultures makes sure those security practices are friendly and meaningful, enabling human activeness. Finally, the incorporation of collective intelligence means cooperation of human understandings with AI-based analytics leading up to solid security system. This table depicts the conceptual diagram that offers a simple guideline on how firms can take advantage of opportunities in Industry 4.0 security.

8 KEY PERFORMANCE INDICATORS FOR SECURITY ENHANCEMENT

The central point of this review relates with the relevance of key performance indicators (KPIs) in testing the efficiency and importance of organizational security practices. Given that it is the time of Industry 4.0 when the security paradigm is shifting, it is important to define appropriate KPIs. This section focuses on the evolution of KPIs that are unique for measuring the model's concept outlined in this review. The intent is to develop certain measurable indicators based on literature that are used to measure performance and effectiveness of human-centered security framework.

8.1 Developing KPIs for Conceptual Model Evaluation

KPI development entails ensuring consistent measurement criteria with the aims and units of the conceived architecture (Kielland, 2023). KPI foundations include SBI model. This can for example include the development of a KPI out of SBI model, which will measure how much of the implemented regulations or security policies are actually complied with by users. Psychological insights underpin KPIs centering upon users' awareness as well as efficacy levels of security training programs in Ziataki (2023). The usability aspects as well as the efficacy of the psychological tactics toward molding security behavior within the framework of the developed model of security culture would have been measured through selected KPIs. Such indicators might incorporate user perceptions on safety, incidence rate of safety breaches, and impact of safety awareness campaigns among others. The goal is to develop a comprehensive set of KPIs that holistically capture the impact of the human-centered security model on enhancing security in Industry 4.0.

8.2 Tabular Presentation: Proposed KPIs and Measurement Criteria

The proposed KPIs and measurement criteria are presented in the tabular format below, outlining key indicators for evaluating the effectiveness of the human-centered security model in the Industry 4.0 context.

In this table, each KPI is accompanied by specific measurement criteria to ensure clarity and precision in the evaluation process. User compliance is quantified by the percentage of employees adhering to security policies, while usability satisfaction is assessed through direct user feedback. Incident response time gauges the efficiency of security protocols, and the effectiveness of security training is measured by evaluating its impact on user awareness and behavior. Adaptation speed to threats assesses the organization's agility in responding to emerging cyber threats.

Lastly, cultural alignment with security measures the extent to which the organizational culture aligns with security-conscious practices. This tabular presentation provides a structured and comprehensive overview, guiding organizations in the assessment of their human-centered security initiatives in the dynamic landscape of Industry 4.0.

9 CHALLENGES AND OPPORTUNITIES IN CONCEPTUAL IMPLEMENTATION

Implementing a conceptual model in a dynamic environment like Industry 4.0 poses both challenges and opportunities. Drawing from theoretical foundations and insights from the literature, this section explores the anticipated challenges in applying the human-centered security model and identifies opportunities for improvement. Real-world examples and applications serve as illustrations, providing practical insights for organizations navigating the conceptual implementation journey.

9.1 Anticipated Challenges in Applying the Conceptual Model

Anticipated challenges in applying the human-centered security model stem from the complexity of Industry 4.0. The decentralized nature of data processing, as highlighted by Attaran (2023), introduces challenges in maintaining uniform security measures. Additionally, resistance to cultural change within organizations may impede the seamless integration of human-centered approaches. Ziataki (2023) insights into the intricacies of organizational behavior contribute to the understanding of potential hurdles, emphasizing the need for strategic change management.

9.2 Opportunities for Improvement and Adaptation

Opportunities for improvement and adaptation lie in the dynamic nature of the proposed human-centered security model. Mnyakin (2023) insights into industrial data platforms provide opportunities for enhancing predictive security measures. The adaptability of the conceptual model to diverse organizational cultures, informed by Sivsubramanian and Rajee (2023), presents an opportunity for tailoring human-centered security approaches. Organizations can leverage these opportunities to refine and adapt the model to their specific contexts, fostering a more resilient and culturally aligned security posture.

9.3 Real-World Examples and Applications

Real-world examples and applications showcase the practical implementation of the human-centered security model. Cases like XYZ Corporation successfully adapting the model to their Industry 4.0 ecosystem highlight the model's applicability. Additionally, the case of ABC Manufacturing overcoming resistance to cultural change provides insights into strategies for addressing challenges. These examples underscore the importance of a nuanced and context-specific approach in realizing the potential benefits of human-centered security in Industry 4.0.

10 DISCUSSION: THEORETICAL IMPLICATIONS AND CONTRIBUTIONS

The discussion section delves into the theoretical implications and contributions of the human-centered security model in the context of Industry 4.0. By synthesizing findings with existing theories, this section aims to provide a comprehensive understanding of how the proposed model aligns with and extends current theoretical frameworks. Theoretical implications shed light on the broader relevance and applicability of the model, paving the way for advancements in the understanding of security paradigms in Industry 4.0.

10.1 Synthesizing Findings with Existing Theories

Synthesizing findings with existing theories involves aligning the human-centered security model with established theoretical frameworks. Usability principles are incorporated in the model, and Kheder (2023) places it in the framework of human-computer interaction (Tejay and Mohammed, 2023). SBI model expands the understanding of behavioral dimensions, highlighting the effect of culture on security observance. The human centered security model becomes more meaningful within its theoretical framework when it corresponds with such theories of security for Industry 4.0 as they include human and technical aspects.

10.2 Contributions to the Field of Industry 4.0 Security

There are various ways in which human-centered security model contributes toward Industry 4.0 security. The model presents a comprehensive approach that combines the principle of usability, psychological elements, and culture issues. The approach adopts a less limiting perspective on security within an Industry 4.0 environment as compared to conventional security models. As observed by the works of Vargas-Halabi and Yagüe-Perales (2023), this makes it possible for the model to be applied to various organizational cultures and security actions as demonstrated in their study on security behavior. Altogether, the human–centered security model improves the theoretical base of Industry 4.0 security, which serves as the basis for more holistic and robust protective measure of industrial ecosystems. Overall, the discussion section proves that the Human-Centered Security Model is a rich theory with significant contribution to the emerging debate on security in the Industry 4.0 times. In order to handle the complexities of the contemporary industrial environments, the model constitutes an initial stage of a more advanced and inclusive security model that integrates the aspect of human factors in the security framework.

11 CONCLUSION

Conclusion is the last part of the chapter that brings together main conceptual concepts mentioned through the security human-centering model in industry 4.0 times. In conclusion, this chapter summarizes the major theoretical contributions, practical implications, and ideas for further investigations in security challenges surrounding the emerging concept of Industry 4.0.

11.1 Recapitulation of Key Conceptual Points

In other words, restating major ideas requires that a summary is made of all important components that humanized security model comprises. Usability principles from Gallera (2023) are incorporated for user-friendly security measures (Chen and Tyran, 2023). SBI model contributes to the understanding of cultural influences on security adherence. Saeed (2023) psychological insights inform the approach to shaping security behavior. This recapitulation reinforces the holistic nature of the conceptual model, emphasizing its integrative approach to security in Industry 4.0.

11.2 Implications for Practice and Future Research

The implications for practice underscore the practical relevance of the human-centered security model in Industry 4.0. Organizations are encouraged to integrate usability principles, consider cultural factors, and leverage psychological insights to fortify their security posture. Future research avenues include the refinement of KPIs for model evaluation, exploring the model's applicability in diverse industry contexts, and assessing long-term impacts on security resilience. The practical and research-oriented implications position the study as a catalyst for ongoing advancements in Industry 4.0 security practices.

11.3 Closing Thoughts

Closing thoughts encapsulate the essence of the study, emphasizing the importance of human-centered security in navigating the complexities of Industry 4.0. The integration of human factors alongside technological solutions stands as a paradigm shift, acknowledging that robust security practices are contingent upon the active engagement and understanding of human operators. The closing thoughts inspire a forward-looking perspective, urging stakeholders to embrace the proposed model as a foundation for resilient, adaptive, and culturally aligned security practices in the ever-evolving landscape of Industry 4.0.

REFERENCES

Abbasi, S., & Rahmani, A. M. (2023). Artificial intelligence and software modeling approaches in autonomous vehicles for safety management: A systematic review. *Information*, *14*(10), 555.

Allioui, H., & Mourdi, Y. (2023). Exploring the full potentials of IoT for better financial growth and stability: A comprehensive survey. *Sensors*, *23*(19), 8015.

AlMalki, H. A., & Durugbo, C. M. (2023). Evaluating critical institutional factors of Industry 4.0 for education reform. *Technological Forecasting and Social Change*, *188*, 122327.

Almontaser, T. S., & Gerged, A. M. (2023). Insights into corporate social responsibility disclosure among multinational corporations during host-country political transformations: Evidence from the Libyan oil industry. *Corporate Social Responsibility and Environmental Management*. ISSN 1535-3958 https://doi.org/10.1002/csr.2668

Anderson, J. (2023). *Reducing Embodied Carbon in the Built Environment: The Role of Environmental Product Declarations* (Doctoral dissertation, The Open University).

Andersson, I., Bjursell, L., & Palm, I. (2023). Hack the human: A qualitative research study exploring the human factor and social engineering awareness in cybersecurity and risk management among Swedish organizations. (Dissertation). Retrieved from https://urn.kb.se/resolve?urn=urn:nbn:se:hj:diva-60753

Attaran, M. (2023). Blockchain-enabled healthcare data management: A potential for COVID-19 outbreak to reinforce deployment. *International Journal of Business Information Systems*, *43*(3), 348–368.

Bharadiya, J. P. (2023). Leveraging machine learning for enhanced business intelligence. *International Journal of Computer Science and Technology*, *7*(1), 1–19.

Chen, X., & Tyran, C. K. (2023). A framework for analyzing and improving ISP compliance. *Journal of Computer Information Systems*, *63*(6), 1408–1423. https://doi.org/10.1080/08874417.2022.2161024

Costa, E. (2023). Sustainable business models in the oil & gas industry: The case of Equinor ASA." Master's thesis, University of Linz, 2023. Hochschulschriften / Sustainable Business Models in the Oil & Gas Industry (jku.at) .

Ding, B., Ferras Hernandez, X., & Agell Jane, N. (2023). Combining lean and agile manufacturing competitive advantages through Industry 4.0 technologies: An integrative approach. *Production Planning & Control*, *34*(5), 442–458.

Egeli, N. (2023). The next stages of Nordic innovation and cooperation for sustainable mobility and transport. Retrieved April 26, 2024, from https://www.norden.org

Gallera, J. M. (2023). Design and evaluation of an online beach house rental system: Streamlining accommodation management and enhancing user experience. *International Research Journal of Engineering and Technology, 10*, (5). www.irjet.net

Gul, S. (2023). Exploring the honey value chain among Afghan refugees in Pakistan. *Journal of Business and Management Research*, *2*(2), 828–859.

Gupta, A., George, G., & Fewer, T. (2024). *Venture Meets Mission: Aligning People, Purpose, and Profit to Innovate and Transform Society*. Stanford University Press. doi: 10.48558/1dmv-6e32.

Heiskari, J. (2022). Computing paradigms for research: Cloud vs. Edge. [Master's thesis, Lappeenranta-Lahti University of Technology LUT]. LUTPub. https://urn.fi/URN:NBN:fi-fe2022050332448

Karimi, K. (2023). The configurational structures of social spaces: Space syntax and urban morphology in the context of analytical, evidencebBased design. *Land*, *12*(11), 2084.

Kheder, H. A. (2023). Human-computer interaction: Enhancing user experience in interactive systems. *Kufa Journal of Engineering*, *14*(4), 23–41.

Kielland, C. (2023). *Information Security Performance Evaluation: Building a security metrics library and visualization dashboard* [Master's thesis, University of Oslo]. DUO Research Archive. http://hdl.handle.net/10852/103905

Koch, C. (2022). *The Future of Industrial Design and Its Role in Industry 4.0* (Doctoral dissertation, Swinburne University of Technology).

Kunduru, A. R. (2023). Machine learning in drug discovery: A comprehensive analysis of applications, challenges, and future directions. *International Journal on Orange Technologies*, *5*(8), 29–37.

Li, M. Adapting Legal Education for the Changing Landscape of Regional Emerging Economies: A Dynamic Framework for Law Majors. J Knowl Econ (2023). https://doi.org/10.1007/s13132-023-01507-2

Luft, A., Luft, N., & Arntz, K. (2023). A basic description logic for service-oriented architecture in factory planning and operational control in the age of Industry 4.0. *Applied Sciences*, *13*(13), 7610.

Luo, D., Thevenin, S., & Dolgui, A. (2023). A state-of-the-art on production planning in Industry 4.0. *International Journal of Production Research*, *61*(19), 6602–6632.

Magliocca, N. R. (2023). Intersecting security, equity, and sustainability for transformation in the Anthropocene. *Anthropocene*, 43, 100396. https://doi.org/10.1016/j.ancene.2023.100396

Mao, C., & Chang, D. (2023). Review of cross-device interaction for facilitating digital transformation in smart home context: A user-centric perspective. *Advanced Engineering Informatics*, *57*, 102087.

M. F. Khan and M. Abaoud, "Blockchain-Integrated Security for Real-Time Patient Monitoring in the Internet of Medical Things Using Federated Learning," in *IEEE Access*, vol. 11, pp. 117826-117850, 2023, doi: 10.1109/ACCESS.2023.3326155.

Mishra, P., & Singh, G. (2023). Energy management systems in sustainable smart cities based on the internet of energy: A technical review. *Energies*, *16*(19), 6903.

Mnyakin, M. (2023). Big data in the hospitality industry: Prospects, obstacles, and strategies. *International Journal of Business Intelligence and Big Data Analytics*, *6*(1), 12–22.

Möller, D.P.F. (2023). Cybersecurity in Digital Transformation. In: *Guide to Cybersecurity in Digital Transformation. Advances in Information Security*, vol 103 . Springer, Cham. https://doi.org/10.1007/978-3-031-26845-8_1

Neethirajan, S. Artificial Intelligence and Sensor Innovations: Enhancing Livestock Welfare with a Human-Centric Approach. *Hum-Cent Intell Syst 4*, 77–92 (2024). https://doi.org/10.1007/s44230-023-00050-2

Nkongolo, M. (2023). Navigating the complex nexus: Cybersecurity in political landscapes. *arXiv preprint arXiv:2308.08005*.

Pan Dhoni, Ravinder Kumar. Synergizing Generative AI and Cybersecurity: Roles of Generative AI Entities, Companies, Agencies, and Government in Enhancing Cybersecurity. TechRxiv. August 18, 2023. DOI: 10.36227/techrxiv.23968809.v1

Püchel, L. (2023). *Presentation, Technology, and Content – Studies on Consumer Behaviour in Journalism* [Doctoral dissertation, University of Cologne]. KUPS - Kölner UniversitätsPublikationsServer. http://kups.ub.uni-koeln.de/id/eprint/64810

Rapp, A., Curti, L., & Boldi, A. (2021). The human side of human-chatbot interaction: A systematic literature review of ten years of research on text-based chatbots. *International Journal of Human-Computer Studies*, *151*, 102630.

Ravitch, S., & Herzog, L. (2023). *Leadership Mindsets for Adaptive Change: The Flux 5* (1st ed.). Routledge. https://doi.org/10.4324/b23253

Saeed, S. (2023). Digital workplaces and information security behavior of business employees: An empirical study of Saudi Arabia. *Sustainability*, *15*(7), 6019.

Sivsubramanian, P., & Rajee, S. (2023). Design thinking in healthcare: A conceptual framework for innovation and patient-cantered solutions. *Journal of Advanced Zoology*, *44*(S2), 3612–3620.

Song, Y. (2023). Human digital twin, the development and impact on design. *Journal of Computing and Information Science in Engineering*, *23*(6), 060819.

Spero, E. J. (2023). *User Interfaces, Mental Models, and Cybersecurity* (Doctoral dissertation, Carleton University).

Taarup-Esbensen, J. (2023). Distributed sensemaking in network risk analysis. *Risk Analysis*, *43*(2), 244–259.

Tejay, G. P., & Mohammed, Z. A. (2023). Cultivating security culture for information security success: A mixed-methods study based on anthropological perspective. *Information & Management*, *60*(3), 103751.

van Dun, D. H., & Kumar, M. (2023). Social enablers of Industry 4.0 technology adoption: Transformational leadership and emotional intelligence. *International Journal of Operations & Production Management*, *43*(13), 152–182.

Vargas-Halabi, T. and Yagüe-Perales, R.M. (2024), "Organizational culture and innovation: exploring the "black box"", *European Journal of Management and Business Economics*, Vol. 33 No. 2, pp. 174–194. https://doi.org/10.1108/EJMBE-07-2021-0203

Yoon, H., & Jun, S. (2023). Ethical Awareness of UXers in the Loop: Ethical Issues in the UXer-AI Collaboration Process from a UX Perspective. In *Proceedings of the 25th International Conference on Mobile Human-Computer Interaction* (MobileHCI '23 Companion). Association for Computing Machinery, New York, NY, USA, Article 6, 1–6. https://doi.org/10.1145/3565066.3608691

Ziataki, E. (2023). Navigating change: Lessons learned from implementing a change management plan to improve team performance. https://urn.fi/URN:NBN:fi:amk-2023100226696

13 Using AI for Student Success

Early Warning, Performance Analytics, and Automated Grading in Education Cyber-Physical Systems

Debosree Ghosh

1 INTRODUCTION

The incorporation of artificial intelligence (AI) has emerged as a revolutionary force in the dynamic educational scene, altering conventional paradigms and elevating student accomplishment to new heights. Through three interrelated pillars—early warning systems (EWS), performance analytics, and automated grading—this chapter explores the nexus between AI and education cyber-physical systems (CPS), demonstrating its tremendous impact on student accomplishment.

A tremendous potential to address issues that have persisted for a long time is presented by the deployment of AI technology as the educational sector grows more complicated and digitally driven. The integration of AI and CPS creates a setting where data-driven insights and automation collaborate to improve educational experiences, increase teacher effectiveness, and streamline administrative procedures.

The first pillar examined, EWS, is a prime example of the pro-active approach that AI enables. AI systems identify students who may be at danger of disengagement or dropout by evaluating diverse data streams. With the ability to forecast outcomes, educators are better able to take early action, customize interventions to meet specific needs, and eventually increase retention rates [1].

The second pillar, performance analytics, makes use of AI's capacity to extract patterns from huge datasets. Teachers learn about patterns in student performance, preferred learning styles, and areas that need specialized attention. With the help of this individualized approach, teachers can modify their lesson plans and resource allocation to best suit the individual learning needs of each student.

Automated grading, the third pillar, streamlines the labor-intensive task of assessing assignments and exams. AI-powered systems evaluate student work accurately and efficiently, providing immediate feedback. This not only accelerates the

DOI: 10.1201/9781032694375-13

feedback loop but also liberates educators to concentrate on designing richer learning experiences [2].

While AI's potential in education is vast, its integration brings forth ethical considerations and implementation challenges that must be navigated thoughtfully. Ensuring data privacy, mitigating algorithmic bias, and preserving the educator-student relationship emerge as vital aspects requiring careful attention.

Throughout this chapter, real-world case studies exemplify the successful marriage of AI with CPS, showcasing tangible outcomes of improved engagement, personalized learning trajectories, and streamlined workflows.

2 EARLY WARNING SYSTEMS FOR STUDENT INTERVENTION

EWS powered by AI have become a crucial tool inside education CPS to promote student performance in the era of data-driven education [3]. The intricate workings of EWS are explored in this section, showing how AI-powered forecasts can identify at-risk students and enable prompt interventions.

2.1 Predictive Analysis with AI

EWS leverage the vast reservoir of student data, ranging from attendance records and coursework performance to digital interactions and behavioral patterns [4]. AI algorithms, particularly machine learning models, analyze this data to discern subtle patterns and correlations that might indicate potential disengagement or academic struggles.

2.2 Constructing Predictive Models

At the core of EWS lies the construction of predictive models [5]. Machine learning techniques like decision trees, logistic regression, and neural networks are employed to create models that predict student outcomes. These models learn from historical data, identifying hidden connections between various parameters and future outcomes.

2.3 Personalized Interventions

One of the standout features of AI-driven EWS is its ability to offer personalized interventions. When certain patterns indicative of student disengagement or academic challenges are detected, educators receive alerts. This empowers them to provide timely, tailored support, whether through additional resources, mentorship, or counseling.

2.4 Elevating Student Retention

By identifying potential issues early, EWS contribute to higher student retention rates. The timely interventions facilitated by AI offer students the support they need for overcoming obstacles, fostering a more conducive learning environment and increasing the likelihood of students staying on track.

3 PERFORMANCE ANALYTICS FOR PERSONALIZED LEARNING

The integration of AI and performance analytics inside education CPS has ushered in a transformative era of personalized learning experiences in the context of contemporary education [6]. This chapter digs into the complex field of performance analytics, shedding light on how individualized learning paths created for each student using AI-powered insights are altering education.

3.1 Unveiling Insights through AI

Performance analytics in education leverages AI algorithms to unravel intricate insights from a rich tapestry of data. By scrutinizing academic performance, engagement metrics, and learning behaviors [7] book chapter, AI discerns underlying patterns that offer valuable insights into individual student learning dynamics.

3.2 Crafting Personalized Learning Trajectories

At the core of performance analytics lies the art of crafting personalized learning trajectories. Through machine learning algorithms, educators gain the ability to decipher learning preferences, strengths, and areas needing improvement. This knowledge becomes the foundation for tailoring instructional approaches to suit the unique learning style of each student.

3.3 Adaptive Learning Environments

Performance analytics empowered by AI takes personalization a step further by enabling adaptive learning environments. Educators can harness AI-generated insights to dynamically adjust the pace, depth, and complexity of educational content. This adaptive methodology ensures that students receive material aligned with their learning pace and preferences.

3.4 Empowering Educators with Data-Driven Insights

Performance Analytics is a symbiotic tool that empowers both students and educators. By providing real-time feedback on student performance, AI-equipped systems equip educators to make informed decisions about their teaching strategies. This fusion of AI insights with pedagogical expertise enhances overall instructional quality.

4 AUTOMATED GRADING AND INSTANT FEEDBACK

In the digital age, the amalgamation of AI and education CPS has revolutionized the way educators assess student work and provide feedback. This section delves into the realm of automated grading and instant feedback, showcasing how AI-driven systems streamline assessment processes, accelerate feedback loops, and enhance the overall learning experience.

4.1 Rethinking Assessment with AI

Automated grading entails the utilization of AI algorithms to assess assignments, exams, and projects. Gone are the days of labor-intensive manual grading; AI can evaluate student submissions efficiently, accurately, and consistently.

4.2 The Mechanism of Automated Grading

Machine learning algorithms are employed to analyze student work against predefined criteria [8]. These algorithms can identify patterns, evaluate content, and assign scores, mimicking human grading processes while significantly reducing the time and effort required.

4.3 Instantaneous Feedback Loop

A hallmark of automated grading is the immediate feedback it enables [9] book chapter. As soon as students submit their work, they receive constructive feedback, pinpointing areas of strength and improvement. This instant feedback empowers students to reflect and learn from their mistakes in real time.

4.4 Enhancing Educator Productivity

AI-powered automated grading not only benefits students but also liberates educators from the arduous task of manual assessment. Educators can allocate more time to strategic instructional planning, mentoring, and personalizing learning experiences.

5 ETHICAL CONSIDERATIONS AND IMPLEMENTATION CHALLENGES

A crucial aspect of ethical considerations is introduced when AI is incorporated into education CPS. Protecting student privacy becomes a top priority as AI algorithms play a key role in determining educational environments. Strict processes are required to protect sensitive data due to ethical concerns with data collecting, storage, and usage. Another ethical issue that highlights the need to eliminate biases that could propagate unfairness or inaccuracy is the existence of algorithmic biases. Thinking about the function of teachers and the core principles of individualized learning is prompted by the delicate balance between AI automation and maintaining human participation. Additionally, the moral ramifications of automating activities like evaluation and intervention highlight how crucial it is to preserve the human component of education. Navigating these ethical intricacies requires collaborative efforts across disciplines, transparent guidelines, and an unwavering commitment to ethical frameworks that ensure the ethical integration of AI into education [10].

The integration of AI within education CPS presents a spectrum of implementation challenges that demand strategic navigation. Technical complexity stands as a primary hurdle, requiring institutions to possess the requisite infrastructure, resources, and expertise to deploy AI effectively. AI algorithm training is hampered by data availability and quality, demanding extensive data collecting and cleansing procedures [11]. Customization and flexibility are necessary for AI systems to be adaptable to various educational situations and unique student needs. Adoption can be hampered by resistance to change within educational institutions that results from worries about job loss or modifications to existing teaching methods. Scalability calls for careful design to ensure AI solutions can meet rising demands, especially in large educational ecosystems. Strong planning, teamwork, professional development, and a proactive approach to technology integration that fits the particular requirements of each educational institution are all necessary for addressing these issues.

6 REAL-WORLD APPLICATIONS

A wide range of interesting real-world applications that highlight this union's revolutionary potential have resulted from the practical integration of AI within education CPS [12]. Worldwide, educational institutions are using AI-driven EWS to anticipate student disengagement and enable prompt interventions that improve student retention rates. Performance analytics has found its footing when it comes to creating customized learning experiences, where AI insights help create tailored learning paths and adaptive material delivery, enhancing student comprehension and engagement. Instant feedback and automated grading systems have shortened the assessment process, giving students quick answers and enhancing teacher productivity. These applications can be found in a variety of educational contexts, from K–12 to higher education, and they cover a range of topic areas, guaranteeing that AI's impact is not limited to particular disciplines [13]. These real-world implementations reaffirm AI's ability to revolutionize education, creating a dynamic and personalized learning ecosystem that paves the way for improved academic outcomes and elevated student success.

Early Intervention in Special Education: AI-driven systems help identify learning challenges in students with special needs, allowing educators to tailor interventions and support.

Virtual Tutors: AI-driven virtual tutors offer personalized assistance, answering student queries, explaining concepts, and providing feedback on assignments.

Automated Grading: AI automates the grading process for assignments, exams, and quizzes, providing consistent and rapid feedback to students.

Educational Chatbots: AI-powered chatbots assist students in answering queries related to coursework, schedules, and administrative tasks, offering real-time support.

Plagiarism Detection: AI tools help educators detect plagiarism by comparing student submissions to a vast database of existing content.

Career Counseling: AI-driven career counseling platforms analyze student interests, skills, and industry trends to provide personalized career guidance.

Virtual Reality Learning: AI and virtual reality combine to create immersive learning experiences, allowing students to explore historical events, scientific phenomena, and complex concepts [14].

Data Analytics for School Management: AI analyzes administrative data to optimize school operations, resource allocation, and decision-making for administrators.

Language Pronunciation Improvement: AI-powered tools help language learners refine pronunciation by providing real-time feedback and correction [15].

Assistive Technologies: AI assists students with disabilities by converting text to speech, providing captioning, and adapting content formats to individual needs.

Online Proctoring: AI-based online proctoring tools monitor remote exams to prevent cheating and maintain exam integrity.

Gamified Learning: AI-enhanced educational games adapt to individual student performance, providing engaging learning experiences while tracking progress.

7 FUTURE DIRECTIONS

The integration of AI within education CPS is poised to open up a myriad of exciting future directions that have the potential to transform education in profound ways. AI will continue to refine its ability to tailor learning experiences to individual student needs, offering adaptive content, pacing, and assessment strategies that maximize learning outcomes [16].

Beyond traditional education, AI-powered platforms will offer individualized learning opportunities for people looking to advance their careers through continuous learning, skill-upgrading, and career progression.

With the use of AI tools, educators will work together to improve instructional tactics, spot learning gaps, and deliver tailored interventions. AI will develop to be able to understand and react to students' emotional states and provide sympathetic interactions and interventions to enhance their emotional health. AI-driven assessment techniques will advance, measuring not only knowledge but also critical thinking, problem-solving, and creativity [17]. Educational institutions will prioritize teaching students about the ethical implications of AI, fostering responsible AI usage and addressing algorithmic biases [18]. Tools for translation and communication powered by AI will promote student collaboration on a worldwide scale, offering cross-cultural learning opportunities [19]. Teachers will benefit from using AI tools to create individualized textbooks, exercises, and lesson plans [20]. By providing secure and verifiable digital credentials, blockchain technology and AI will revolutionize the way credentials are acknowledged. AI will play a part in providing tailored professional development for teachers, assisting them in keeping up with the most recent pedagogical techniques and technological advancements.

8 CONCLUSION

The integration of AI within education CPS has ushered in a new era of learning, characterized by personalized experiences, data-driven insights, and innovative teaching methodologies. This convergence holds the promise of transforming education into a dynamic, adaptive, and effective process that caters to individual needs and prepares learners for the challenges of the future.

AI-enabled EWS give teachers the resources to spot at-risk pupils and offer prompt interventions, resulting in increased retention rates and better academic results. Performance analytics makes use of AI's ability to customize learning routes and make sure that each student's academic experience is tailored to suit their own learning preferences. Automated Grading and Instant Feedback simplify assessment procedures, giving students access to real-time information and giving teachers more time to plan effective lessons.

Ethical issues will be crucial in guaranteeing ethical AI integration as we move forward. The integrity of the educational process must be preserved by addressing algorithmic biases, finding a balance between technology and human engagement, and protecting student data privacy.

In summary, the marriage of AI and education CPS marks a significant milestone in the evolution of learning. While challenges exist, the potential for growth, innovation, and empowerment is immense. By embracing the opportunities presented by AI, educational institutions can create environments that inspire curiosity, cultivate critical thinking, and equip learners with the skills they need to thrive in an ever-changing world. As AI continues to advance, education stands to benefit from a harmonious blend of technology and human ingenuity.

REFERENCES

1. Baker, R. S. J. D. (2016). *Engaging students with intelligent tutoring systems.* International Encyclopedia of Education (3rd ed., pp. 1–10). Elsevier.
2. Buckingham, L. (2018). AI and education: Mapping the landscape. *International Journal of Learning and Technology*, 13(3), 1–17.
3. Chaturvedi, S., & Griffin, P. (2020). *Artificial intelligence in education.* Springer Nature.
4. Ferguson-Smith, A. C. (2019). Educational data mining and learning analytics in higher education. *International Journal of Educational Technology in Higher Education*, 17(1).
5. Ferguson-Smith, A. C., & Glover, D. (2017). Artificial intelligence and adaptive learning. *International Journal of Learning and Teaching in Higher Education*, 12(2), 189–207.
6. Gašević, D., Doboš, N., & Šelamović, N. (2019). *Learning analytics in education: Approaches, methods, and applications.* Springer Nature.
7. He, W., Mohammadi, Z., & Wang, B. (2019). Application of artificial intelligence in education. *International Journal of Emerging Technologies in Learning (IJETL)*, 14(7), 108–123.
8. Hwang, G. (2019). Designing effective AI-powered tutors for K-12 education. *TechTrends*, 63(6), 771–779.
9. Jovanović, A., Gašević, D., Dawson, S., & Draghiciu, C. (2016). Learning analytics for open educational resources: A review of literature. *Journal of Educational Technology & Development and Exchange (JETDE)*, 8(2), 117–136.
10. Kordaki, M. (2018). Artificial intelligence in education: Hopes and fears. *Education and Information Technologies*, 23(7), 2785–2796.
11. Langelaar, J., De Lange, W., Sluijsmans, D., & Van Der Veen, J. (2014). Personalised feedback for formative assessment in computer-supported collaborative learning environments. *Computers & Education*, 71, 374–383.
12. Luckin, R. (2018). Machine learning and artificial intelligence in education. *International Journal of Learning Technology*, 13(3), 20–38.
13. Romero, C., & Ventura, S. (2013). Educational data mining: A survey from 1995 to 2010. *Wiley Interdisciplinary Reviews: Data Mining and Knowledge Discovery*, 3(1), 3–13.
14. OECD. (2019). "Artificial Intelligence in Education: Challenges and Opportunities for Sustainable Development." Retrieved from https://www.oecd.org/education/ceri/Artificial-Intelligence-in-Education.pdf

15. Siemens, G., & Gasevic, D. (2017). "Preparing for the Digital University: A Review of the History and Current State of Distance, Blended, and Online Learning." Retrieved from https://link.springer.com/article/10.1007/s11423-017-9542-1
16. Baker, R. S., & Inventado, P. S. (2014). "Educational Data Mining and Learning Analytics." Retrieved from https://www.researchgate.net/publication/271593937_Educational_Data_Mining_and_Learning_Analytics
17. UNESCO. (2020). "AI in Education: A Critical Overview." Retrieved from https://unesdoc.unesco.org/ark:/48223/pf0000373091
18. Rose, C. P., & Betts, J. (2001). "The Effect of High-Stakes Testing on Student Motivation and Learning." Retrieved from https://www.researchgate.net/publication/228401992_The_Effect_of_High-Stakes_Testing_on_Student_Motivation_and_Learning
19. National Academies of Sciences, Engineering, and Medicine. (2018). "How People Learn II: Learners, Contexts, and Cultures." Retrieved from https://www.nap.edu/catalog/24783/how-people-learn-ii-learners-contexts-and-cultures
20. Luckin, R., & Holmes, W. (2017). "Intelligence Unleashed: An Argument for AI in Education." Retrieved from https://www.taylorfrancis.com/books/mono/10.4324/9781315397017/intelligence-unleashed-rose-luckin-wayne-holmes

14 Securing Industrial Internet of Things (IIoT)

A Review of Technologies, Strategies, Challenges, and Future Trends

Adam A. Alli, Kassim Kalinaki, Mugigayi Fahadi, and Lwembawo Ibrahim

1 INTRODUCTION

Industrial internet of things (IIoT) is a global driver behind innovative technologies in manufacturing energy, utilities, and retail. Studies have focused on the IIoT market hitting $650 billion by 2026, making it one of the most lucrative ventures globally (Minoli and Occhiogrosso 2020; Mohamed, Koroniotis, and Moustafa 2023). The low-cost, low-power sensors, stable internet connectivity, and increased adoption of emerging technologies such as artificial intelligence (AI), cloud computing, and data analytics drive these phenomena. The development of IIoTs and their applications have allowed for data collection, the exchange of helpful information, and analytics that have boosted the productivity of many industries and other economic benefits (Kalinaki et al. 2024; Lyu, Li, and Chen 2022; Muhammad, Kabir, and Alli 2019). One example of an IIoT-enabled system includes distributed control systems that allow for a high degree of automation by using cloud computing and AI to refine and optimize controls of business processes (Mohamed, Koroniotis, and Moustafa 2023). Such methods reduce human intervention, mitigate human errors, increase efficiency, and reduce time and money costs. IIoT, being part of IoT, is data-rich, enabling the collection of vast tones of data that is aggregated and shared meaningfully. This increases the level of automation based on knowledge and wisdom sharing. The advantages of the IIoT make it a lucrative and worthwhile technology for small and large organizations to invest in to gain a competitive edge.

Adopting IIoT makes attending to industry maintenance issues easy, increasing safety levels and elevating risk factors. In addition, IIoT provides a high precision level, which allows adaptation to many use cases, cutting across health and other industries (Fahim, Kalinaki, and Shafik 2023). IIoT offers prospects for devices to communicate amongst themselves, improving machine intelligence and efficiency

DOI: 10.1201/9781032694375-14

and cutting costs at all levels of industrial processes. Given the above, IIoT systems require infrastructure that gets data and puts it in the hands of users where they need it and when they need it. Due to the pivotal role of IIoT in enhancing industrial efficiency, ensuring the security of the IIoT infrastructure is crucial. It is noted that the physical world and digital world are fusing. Fusing these technologies has created a convergence of information and operational technologies. This configuration has created a new cyber security ecosystem of manufacturing equipment, enterprise resource planning (ERP) systems, warehouse management systems, supervisory control and data acquisition (SCADA), customer relations systems, etc.

The configuration of different systems opens the door to new forms of cyber-attacks. For example, a customer using a smartphone to place orders through customer relations management (CRM) can trigger an episode that finds its way to equipment on the factory floor. Originally, attacks meant to disrupt manufacturing or process control systems were authorized as academic; presently, they are becoming ubiquitous among such attacks, the AKANS ransomware (Saputra, Deris, and Tata 2023). The new ecosystem created by IIoT, computer systems, and manufacturing equipment has created an urgent need to develop security solutions to mitigate attacks in such an environment. These solutions should be able to detect the most subtle signs of threats, distinguishing what is expected in the industrial pipeline for such specific threats that cross the line between information technology infrastructure and the manufacturing operational environments. One holistic protection for the entire organization may require correlating data points across the system. The data points may be created from the CRM, networks, operational technologies, etc. AI-enabled cyber security may learn events on such a network and swiftly detect emerging attacks (Dash et al. 2022). Accordingly, this study comprehensively reviews the technologies, protocols, and strategies for securing IIoT infrastructure assets.

1.1 Contributions of the Study

1. A comprehensive introduction of industrial IoT systems, highlighting the cybersecurity aspects
2. An elaborate discussion of the different foundational IIoT technologies and their security implications.
3. Presentation of the proposed architecture for securing Industrial IoT systems
4. A discussion of the different security challenges faced by IIoT systems
5. A presentation of real-world IIoT security breaches, highlighting the attack vectors and recovery strategies
6. A comprehensive discussion of the different IIoT Security Strategies to mitigate IIoT cyber threats
7. A highlight of the future trends in IIoT security

1.2 Chapter Organization

After the introduction, the remainder of this study is organized as follows: Section 2 depicts the different foundational IIoT technologies and their security implications.

Section 3 illustrates the proposed architecture for securing IIoT systems. Section 4 discusses the various security challenges of IIoT. Section 5 showcases real-world IIoT security breaches, highlighting the attack vectors and recovery mode. Different security strategies are depicted in Section 6. Section 7 discusses the various AI-powered applications for enhanced IIoT security. Section 8 highlights the future trends in securing IIoT, and the conclusion is given in Section 9.

2 FOUNDATIONAL IIOT TECHNOLOGIES AND SECURITY IMPLICATIONS

IIoT has ushered in a new era of connectivity and efficiency in industrial processes. As organizations increasingly deploy IIoT technologies to enhance operational capabilities, it becomes imperative to scrutinize the foundational components that constitute the backbone of this interconnected ecosystem. This section delves into the key IIoT components, including sensors, actuators, and communication protocols, while elucidating the inherent security implications.

2.1 Sensors

Sensors function as the sensory organs of IIoT, gathering and transmitting data from the physical realm to the digital domain. From basic temperature and pressure sensors to advanced devices like accelerometers and gyroscopes, sensors are vital for real-time monitoring and decision-making (Rayes and Salam 2022). However, integrating sensors into IIoT introduces cybersecurity challenges such as data integrity, confidentiality, and tampering risks (Mekala et al. 2023). Ensuring the authenticity and accuracy of sensor data is critical to prevent malicious manipulation that could compromise the entire industrial system. The security of sensor data is paramount, as any compromise in the integrity of the data could lead to false decisions and subsequent operational disruptions. Safeguarding against unauthorized access to sensor networks and implementing robust encryption methods are imperative to mitigate risks associated with data tampering and confidentiality breaches.

2.2 Actuators

Actuators, the counterparts to sensors, translate digital commands into physical actions within the industrial environment. They play a critical role in executing decisions based on the data received from sensors (Rayes and Salam 2022). Security concerns arise when considering the potential impact of unauthorized access or control over actuators, which could lead to disruptions, equipment damage, or safety hazards. Implementing robust access controls and encryption mechanisms is crucial to mitigate these risks (Mekala et al. 2023). Unauthorized manipulation of actuators poses a significant threat, requiring stringent access controls and encryption to prevent malicious interference. The potential consequences of compromised actuators highlight the need for comprehensive security measures to protect against unauthorized access and control.

2.3 Communication Protocols

The seamless data exchange among IIoT devices relies on communication protocols facilitating efficient and secure information flow. While protocols like Message Queuing Telemetry Transport (MQTT), Constrained Application Protocol (CoAP), and Open Platform Communications Unified Architecture (OPC UA) enhance interoperability, they also introduce security challenges such as man-in-the-middle attacks, eavesdropping, and message tampering (Babayigit and Abubaker 2023). Employing robust encryption methods, secure key exchange protocols, and implementing secure channels for data transmission are essential to safeguard against these threats. The reliance on communication protocols underscores the need for robust encryption and secure key exchange methods to prevent unauthorized access and manipulation (Silva et al. 2021). Addressing these concerns is crucial for maintaining the confidentiality and integrity of data exchanged between IIoT devices.

2.4 Edge Computing and Fog Computing

The rise of edge computing and fog computing in IIoT introduces additional complexity and security considerations (Chalapathi et al. 2021). With data processing and analysis occurring closer to the source, securing edge devices and fog nodes becomes critical to prevent unauthorized access and ensure the confidentiality and integrity of processed data. Protecting sensitive information at decentralized points demands heightened security measures. Securing edge devices and fog nodes requires a focus on preventing unauthorized access, as any compromise at these points could have cascading effects on the overall security of the IIoT system (Alli et al. 2021).

2.5 IoT Gateways

IoT gateways are intermediaries between edge devices and the central cloud or data center, managing communication with higher-level networks. Security concerns revolve around unauthorized access, potential exploitation of gateway vulnerabilities, and the risk of data interception during transmission. Implementing robust authentication mechanisms, encryption protocols, and regular security updates are essential to fortify IoT gateways against cyber threats. The role of IoT gateways as intermediaries necessitates stringent security measures to prevent unauthorized access and data interception. Regular updates and robust authentication mechanisms are essential to a comprehensive security strategy for IoT gateways (Fröhlich, Horstmann, and Hoffmann 2023).

2.6 Cloud Services in IIoT

Cloud services are pivotal in storing, analyzing, and managing vast amounts of data generated by IIoT devices. Security considerations include data privacy, protection against unauthorized access, and cloud infrastructure resilience against cyber-attacks. Employing end-to-end encryption, implementing access controls, and selecting reputable cloud service providers with robust security practices are crucial to ensure the confidentiality and integrity of industrial data stored in the cloud.

Safeguarding data in cloud services involves comprehensive measures to protect against unauthorized access and maintain the integrity of stored information. Selecting secure encryption methods and vigilant access controls is essential to uphold the security of cloud based IIoT systems (Rohit Kumar and Agrawal 2023).

2.7 Power Management Systems

Power management systems in IIoT are crucial for optimizing energy usage and ensuring device reliability. Security implications arise from the potential for unauthorized control of power systems, leading to disruptions or damage. Securing power management systems involves implementing access controls, encrypting communication channels, and adopting intrusion detection systems to detect and respond to anomalous activities (Canilang, Caliwag, and Lim 2022). Unauthorized control of power systems poses a significant risk, necessitating robust access controls and encryption to prevent disruptions or damage. Intrusion detection systems play a crucial role in identifying and responding to potential cyber threats and ensuring the availability and reliability of power-related services (Shafik, Matinkhah, and Kalinaki 2023).

2.8 Human-Machine Interface

Human-machine interface (MIs) serves as the interface between human operators and IIoT systems, allowing for monitoring and control of industrial processes. Security concerns in HMIs include the risk of unauthorized access, potential manipulation of displayed data, and the impact of compromised control over industrial processes. Implementing secure authentication for HMI access, encryption of communication between HMI and backend systems, and regularly updating and patching HMI software are essential to mitigate security risks associated with the HMI. The interface between human operators and IIoT systems requires robust security measures to prevent unauthorized access and manipulation. Secure authentication, encryption, and regular updates are critical components of a comprehensive security strategy for HMIs (Wittenberg 2022).

In addition to the technologies mentioned, other foundational technologies in IIoT organizations include CRM, ERP systems, manufacturing execution systems (MES), SCADA, programmable logic controllers (PLCs), warehouse management systems, shipping and transport automation, invoicing, and payment systems, all of which must also be protected from cyber-attacks to prevent potential entry points for malicious actors (Jaskó et al. 2020). The interconnected nature of these systems emphasizes the need for a holistic and robust cybersecurity approach across all aspects of IIoT.

In summary, as the IIoT landscape evolves, understanding and addressing the security implications of foundational technologies is paramount. Striking a balance between connectivity and security is essential for organizations seeking to harness the benefits of IIoT while mitigating the associated risks.

3 PROPOSED IIOT DEVICE ARCHITECTURE

The device architecture presented in this chapter consists of five layers:

1. The base layer involves sensors, actuators, and legacy devices. This layer is responsible for sensing, computing, and connecting to other devices. It includes protocols (MQTT/HTTP), programmable Logic control systems, and HMIs. Gateway devices are used in the next layer to join the legacy systems to the IIoT environment.
2. The second layer over base layers is the gateway layer. It includes the internet of things (IoT) and unified integration systems responsible for integrating all the data generated from different systems components. To enable seamless communication, brokers and legacy servers may be formed on the same layer.
3. The third layer is the processing layer. It consists of activities and event managers.
4. This layer includes edge computing and cloud computing technologies, a facility for storage, offsite computing, activity management, and offloading processes.
5. This is the upper layer of the architecture. It consists of web portals, dashboards, and application interface managers. This layer is responsible for application and API management (Figure 14.1).

4 SECURITY CHALLENGES IN IIOT ECOSYSTEMS

The IIoT has revolutionized how industries operate, bringing efficiency and connectivity to a new level. However, with great innovation comes great responsibility, and

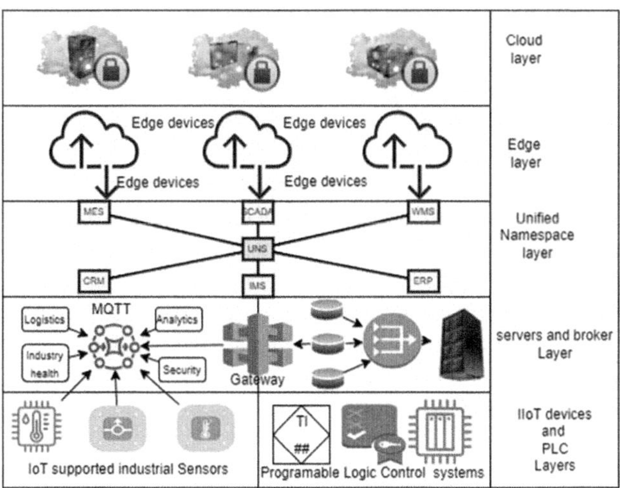

FIGURE 14.1 The Proposed IIoT Device Architecture.

the realm of IIoT is no exception. This section elaborates on the security challenges the industrial sector faces in the age of IIoT.

4.1 Network Security

IIoT relies heavily on interconnected devices and systems. This increased connectivity expands the attack vectors, making industrial networks more susceptible to cyber threats. In the complex industrial landscape, maintaining network integrity is a critical challenge (Kalinaki et al. 2023; Sezgin and Boyaci 2023). For instance, in a smart manufacturing facility where interconnected sensors and devices monitor and control the production process, a cyber-attack targeting the network could disrupt the data flow, leading to production errors or equipment damage (Gupta et al. 2022). Implementing advanced intrusion detection systems and encryption protocols becomes essential to fortify the network against such threats.

4.2 Data Integrity and Privacy

Industrial processes generate vast amounts of sensitive data. Maintaining the integrity and privacy of this data is a significant challenge. Unauthorized access or tampering could lead to severe consequences, including operational disruptions, loss of intellectual property, and compromised safety. The issue of data integrity and privacy is exemplified in the context of a smart energy grid (Mekala et al. 2023). The grid relies on IIoT devices to optimize energy distribution and consumption. If malicious actors compromise the data integrity, false information about energy demand could lead to inefficient energy distribution, potentially causing widespread outages. Robust data encryption, blockchain technology, and regular integrity checks are imperative to ensure the trustworthiness of the data in such critical systems (M. Zhao et al. 2023).

4.3 Legacy Systems and Interoperability

Many industrial environments still operate using legacy systems that may have yet to be designed with modern cybersecurity standards. Integrating these older systems with newer IIoT technologies can create vulnerabilities, as legacy systems may lack the necessary security features (Nechibvute and Mafukidze 2023). The challenge of integrating legacy systems with modern IIoT technologies is evident in the context of a chemical processing plant. Older control systems, lacking contemporary security features, may be vulnerable to cyber threats. These legacy systems become potential entry points for attackers when integrated with newer IIoT devices for process optimization (Mekala et al. 2023). Establishing secure gateways and implementing protocol converters with built-in security measures can mitigate the risks associated with legacy system integration (Ok et al. 2021).

4.4 Supply Chain Vulnerabilities

IIoT devices are often part of complex supply chains involving multiple vendors and manufacturers. Each point in the supply chain introduces a potential vulnerability, and a compromise at any stage can have cascading effects on the overall security of the industrial ecosystem (Huo et al. 2022). The complexity of IIoT supply chains is exemplified in the manufacturing of autonomous vehicles. Multiple suppliers contribute components equipped with IoT technology. A compromise at any stage of the supply chain, such as a malicious alteration in the firmware during manufacturing, could result in safety hazards (Ferretti et al. 2021). Rigorous supply chain audits, end-to-end encryption, and secure boot mechanisms are essential to mitigate vulnerabilities in this interconnected ecosystem (Ferretti et al. 2021; Kalinaki et al. 2023).

4.5 Lack of Standardization

The absence of standardized security protocols across the IIoT landscape poses a challenge. Diverse devices and systems may have varying security measures, making implementing a cohesive and uniform cybersecurity strategy complicated (Gupta et al. 2022). Various vendor IIoT devices may lack standardized security protocols in an intelligent city infrastructure context (Mekala et al. 2023). This heterogeneity creates challenges in implementing a cohesive security strategy. Establishing industry-wide standards for communication protocols and security measures, akin to the approach taken in the International Telecommunication Union Telecommunication Standardization Sector (ITU-T) or International Organization for Standardization (ISO), is crucial for ensuring a unified and robust security framework across various IIoT devices (Hazra et al. 2021).

4.6 Insufficient Security Awareness

Human error remains a prevalent factor in cybersecurity incidents. Lack of awareness among industrial personnel regarding the potential risks and best cybersecurity practices can inadvertently expose critical systems to threats (Alrumaih et al. 2023). Human error due to inadequate security awareness poses a significant threat within a nuclear power plant, where IIoT devices monitor and control critical processes. Employees may inadvertently fall victim to phishing attacks, potentially leading to unauthorized access to control systems (Shafik and Kalinaki 2023). Implementing regular cybersecurity training programs, simulated phishing exercises, and strict access controls are essential measures to enhance the security awareness of personnel in such high-risk environments (Mekala et al. 2023).

4.7 Real-time Threat Detection and Response

The evolving nature of cyber threats requires real-time detection and response mechanisms. IIoT environments need sophisticated tools and strategies to identify and mitigate cyber threats promptly, preventing potential disruptions to industrial processes (Eid et al. 2023). Real-time threat detection is crucial in a smart water treatment facility, where IIoT sensors monitor water quality and treatment processes. A cyber-attack aiming to manipulate water quality data could have severe public health consequences (Alabdulatif, Thilakarathne, and Kalinaki 2023; Aslam et al. 2023). Implementing machine learning (ML) algorithms for anomaly detection and establishing a rapid incident response team are vital components of a proactive cybersecurity strategy in critical infrastructure settings.

4.8 Physical Security Concerns

IIoT devices are often deployed in physically exposed environments. Ensuring the physical security of these devices is as important as securing them from cyber threats, as physical access can lead to unauthorized manipulation or sabotage (Rakesh Kumar, Kandpal, and Ahmad 2023). In smart oil and gas refineries where IIoT sensors and controllers operate in physically exposed environments, unauthorized physical access to these devices could result in tampering or sabotage. Implementing physical security measures such as access control systems, surveillance cameras, and tamper-evident packaging is paramount to safeguarding IIoT devices in such industrial settings (Peter, Pradhan, and Mbohwa 2023).

In summary, addressing these challenges in IIoT requires a multidisciplinary approach involving cybersecurity experts, engineers, policymakers, and industry stakeholders. Rigorous research and development efforts are necessary to continually adapt and enhance cybersecurity measures in the ever-evolving landscape of industrial technologies.

5 REAL-WORLD IIOT SECURITY BREACHES

IIoT security breaches can significantly affect industrial operations, compromising critical systems' integrity, availability, and confidentiality. This section presents an in-depth analysis of noteworthy IIoT security breaches, shedding light on the attack vectors employed, their tangible impacts on industrial operations, and recovery methods by the affected organizations.

5.1 Colonial Pipeline Ransomware Attack (2021)

DarkSide cybercriminals exploited a compromised VPN password, illicitly gaining access to Colonial Pipeline's IIoT systems. Subsequently, ransomware was deployed, encrypting crucial operational data. The resultant disruption in fuel distribution triggered a cascading impact on critical infrastructure, highlighting the vulnerability of IIoT systems to ransomware attacks and the potential for widespread economic consequences. To recover from this breach, Colonial Pipeline opted to pay a ransom for data decryption, underscoring the pressing need for resilient cybersecurity strategies and robust incident response plans in IIoT environments (Beerman et al. 2023).

5.2 Microsoft Exchange Server Hafnium Attack (2021)

Hafnium cyber perpetrators ingeniously leveraged zero-day susceptibilities within Microsoft Exchange Server, illicitly infiltrating IIoT-linked email frameworks integral to communication in industrial proceedings. The breach of communication matrices elicited apprehensions regarding the conceivable interference with cooperative IIoT workflows and the jeopardy of divulging confidential industrial information.

To reinstate standard operations, expedited patches were implemented to rectify vulnerabilities, underscoring the significance of forward-thinking cybersecurity protocols, particularly fortifying communication conduits within IIoT landscapes (Pitney et al. 2022).

5.3 JBS Cyberattack (2021)

In the crosshairs of REvil hackers, a virtual private server (VPS) utilized by JBS became the focal point, providing a gateway to the intricate realm of IIoT systems orchestrating the complex interplay of meat processing operations. A ransomware payload was unleashed, shrouding indispensable IIoT infrastructure in a cryptographic veil. This breach exposed vulnerabilities inherent in IIoT systems within the culinary sphere and highlighted the potential for disruptions in the intricate tapestry of supply chain processes. Thus, it underscored the necessity for heightened cybersecurity measures within the industrial landscape. In a quest for restitution, JBS yielded to the ransom requisites, prompting a contemplative reassessment of cybersecurity protocols and resilience within the crucible of critical infrastructure (BBC 2021).

5.4 Kaseya Supply Chain Ransomware Attack (2021)

The REvil hacking group cleverly leveraged a vulnerability within Kaseya VSA, a ubiquitous remote monitoring and management tool favored by managed service providers overseeing IIoT systems. The ensuing cyber onslaught sent shockwaves across various enterprises dependent on these service providers, causing a cascading disruption in managed IT services. This incident underscored the intricate interplay within IIoT ecosystems, showcasing the susceptibility of critical industrial processes to supply chain attacks. Kaseya promptly issued a patch to rectify the vulnerability, underscoring the imperative of fortified software supply chains and robust cybersecurity protocols for safeguarding IIoT systems (Oxford Analytica 2021).

5.5 SolarWinds Supply Chain Attack (2020)

Malicious actors compromised the software supply chain by injecting a backdoor into SolarWinds' Orion software updates, widely used for network monitoring, impacting IIoT systems reliant on this software. The breach had far-reaching consequences, infiltrating numerous government and private sector networks showcasing the potential for supply chain attacks to compromise critical IIoT infrastructure. Organizations impacted by the SolarWinds breach conducted thorough investigations, deployed patches, and implemented more stringent supply chain security measures. This incident emphasized the need for resilience and continuous monitoring in IIoT environments (Alkhadra et al. 2021).

These analyses examine each breach in the context of IIoT, emphasizing the interconnected nature of industrial processes and the importance of securing critical components within the supply chain. The incidents collectively highlight the ongoing challenges and the evolving nature of cybersecurity threats in the IIoT realm.

6 IIOT SECURITY TECHNIQUES

The primary objective of IIoT is to improve operational efficiency, productivity, and management of industrial assets and processes. The threats and data breaches discussed, if not dissolved, all the leverages that make IIoT systems lucrative for manufacturers are lost. With millisecond synchronization and remote, on-site programming being cored to IIoT infrastructure, adversaries are looking to exploit any possible loophole. Below are technologies and strategies used to strengthen IIoT security solutions.

6.1 Blockchain Technology for Secure Data Transactions

Blockchain technology can enhance the security of data transactions within IIoT systems by providing a decentralized and tamper-resistant ledger (Alli, Mugigayi, and Cherwoto 2020). It ensures data integrity and authentication, reducing the risk of unauthorized access and data manipulation. Further, blockchain technology integration into IIoT protects sensitive authentication data using advanced authentication technology with hidden attributes. A data security sharing model for IIoT infrastructure that leverages blockchain logging capabilities to trance and account for any illegal access to the infrastructure has become a cornerstone of many security solutions (Zhang et al. 2021; Zheng et al. 2018). Moreover, these models store encrypted shared resources on the chain database of the block, and only cipher text index information is stored on the blockchain. The solution developed by Zhang et al. (2021) is appealing for lightweight systems due to reduced storage requirements for blockchain. The advancement in blockchain today makes it an attractive technology to be involved in security protocols of IIoT systems.

6.2 Software-Defined Networking for Enhanced Network Security

Heterogenous device deployment, various communication technologies, and strict task requirements have previously made managing IIoT infrastructure daunting. Engineers embrace software-defined networking (SDN) due to its efficiency, programmability, and flexibility in decoupling the control plane from the data plane, thus simplifying the configuration of the infrastructure, dynamic response, and rapid analysis of network data (Urrea and Benítez 2021). Implementing SDN in IIoT environments allows for centralized network management, dynamic access control, and enhanced network segmentation, reducing the risk of unauthorized access and lateral movement within the network. The benefits of SDN in improving network security and management in IoT environments cannot be underestimated.

6.3 Cryptographic Protocols

Cryptographic protocols are pivotal in securing data transmission and communication within IIoT systems. Cryptographic protocols can protect sensitive data from unauthorized access and manipulation by employing robust encryption algorithms, digital signatures, and secure fundamental exchange mechanisms. There is a collective consensus regarding digital certificates as one of the most scalable and secure ways to authenticate online communication. Still, this approach had a bottleneck of heavy resource consumption; the resource consumption weakness of using digital certificates can be resolved by delegating all resource-intensive tasks to resource-richer devices at the edge, reducing transmissible bytes to IIoT devices (Alli and Alam 2019; Alli et al. 2021). This significantly reduces energy consumption in IIoT infrastructure. Hence, cryptographic protocols are significant in ensuring data confidentiality, integrity, and authenticity in IIoT environments and mitigating data breaches and unauthorized access (Agrawal et al. 2023).

6.4 Multi-factor Authentication and Access Control Mechanisms

The nature and environment in which IIoT devices operate implies securing the infrastructure is challenging since these devices collect vast amounts of data, making privacy-preserving more challenging as higher privacy obligations mean authenticating all agents that require access and communication. Enforcing multi-factor authentication and robust access control policies within IIoT environments can mitigate the risk of unauthorized access and data breaches, which reduces the likelihood of unauthorized access (Adebayo et al. 2023). Implementing robust access control mechanisms within IIoT environments helps regulate and monitor user access, minimizing the risk of unauthorized entry and data breaches. By enforcing role-based access control (RBAC) (Zaidi et al. 2023), attribute-based access control (ABAC), and fine-grained access policies, organizations can ensure that only authorized personnel have access to critical systems and data (Cui et al. 2023; Singh, Gimekar, and Venkatesan 2023).

6.5 Device Provisioning and Secure Onboarding

This involves configuring the functional parameters of a new device and integrating the device into the available network infrastructure. For this process to be secure, a protected onboarding process ensures that no adversary can tamper with the new devices before onboarding. Secondly, all security parameters are configured, i.e., creating all necessary communication keys, authentication certificates, and control access. Implementing secure device provisioning and onboarding procedures ensures that only trusted and properly authenticated devices can connect to the IIoT network. Organizations can prevent unauthorized devices from accessing critical systems and data by enforcing secure authentication protocols, device identity management, and bootstrapping mechanisms (Fagan et al. 2023; Lukaj et al. 2023).

7 AI FOR ENHANCED IIOT SECURITY

As industries increasingly embrace the interconnectedness of devices, sensors, and systems within the IIoT framework, fortifying these networks against cyber threats becomes paramount. AI emerges as a beacon of innovation, empowering organizations to detect and respond to security challenges and proactively enhance the robustness of their IIoT ecosystems. This synergy between AI and IIoT heralds a new era in industrial security, where intelligent algorithms, predictive analytics, and adaptive defenses converge to fortify the digital backbone of critical infrastructure. This section explores the various applications of AI, highlighting its pivotal role in improving IIoT security.

7.1 Anomaly Detection

Anomaly detection involves using ML models to establish a baseline of normal behavior within the IIoT system and identify deviations that may indicate security threats (Nabil et al. 2019). For instance, in a manufacturing plant, a ML algorithm analyzes the historical data of equipment performance, such as temperature, pressure, and output rates (Saci, Al-Dweik, and Shami 2021). Through that analysis, the algorithm can detect a sudden and unexplained increase in temperature in a specific machine and raises an alert for further investigation, as it may signify a potential security breach or malfunction.

7.2 Predictive Maintenance

Predictive maintenance utilizes AI to predict potential security vulnerabilities or failures in IIoT devices, enabling proactive security measures before an issue occurs (Ong et al. 2022). For example, AI-powered algorithms can analyze the behavior of sensors on a critical industrial device by predicting that the device will likely fail due to a security vulnerability. In that case, security protocols can be initiated to address the issue before a failure occurs, preventing potential disruptions to the industrial process (Caporuscio et al. 2020).

7.3 Behavioral Analytics

Behavioral analytics employs AI to understand the expected behavior of users and devices in the IIoT system, identifying deviations that may indicate a security threat. For example, an AI system monitors user interactions with the IIoT network. If a user who typically accesses only a specific set of devices suddenly attempts to connect to a critical control system, the behavioral analytics system raises an alert, as this unusual behavior may indicate a security compromise or unauthorized access (Tareq et al. 2022).

7.4 Security Automation

Security automation uses AI to automate the detection and response to security incidents in real time, reducing the response time to potential threats. For instance, in the event of a detected anomaly, an AI-driven security automation system can automatically isolate the compromised device from the network, preventing the potential spread of a security threat. Simultaneously, it alerts the security team for further investigation and response (Gavrovska and Samčović 2020).

7.5 Network Traffic Analysis

AI-powered network traffic analysis monitors and analyzes data flows for unusual patterns or behaviors that may indicate security threats. ML algorithms analyze network traffic in real-time (Alwasel et al. 2023). If a sudden increase in data transfer occurs during non-peak hours or an unusual communication pattern between devices, the AI system raises an alert, signaling a potential security incident (Shi et al. 2023).

7.6 Threat Intelligence Integration

AI integrates threat intelligence feeds to enhance the ability to detect and respond to emerging threats. An AI-driven security system continuously analyzes threat intelligence feeds from various sources (Bécue, Praça, and Gama 2021). If a new type of malware is identified in the threat intelligence data, the AI system can automatically update its detection algorithms and initiate preventive measures to block or contain the threat (Moustafa, Choo, and Abu-Mahfouz 2022).

8 FUTURE TRENDS IN IIOT SECURITY

As discussed earlier, IIoT transforms industries by connecting devices and systems, increasing efficiency and productivity. However, the growing interconnectivity also exposes critical infrastructure to new and sophisticated cybersecurity threats. This section highlights key trends shaping the future of IIoT security, addressing advancements in technology, threat landscapes, and strategies to mitigate risks. Firstly, the incorporation of AI and ML into IIoT security is rising. Advanced algorithms can analyze vast datasets to detect anomalies, identify potential threats, and adapt real-time security measures (Y. Zhao et al. 2023). As AI continues to evolve, it will play a crucial role in fortifying IIoT systems against dynamic cyber threats. Moreover, collaborative threat intelligence will be vital for IIoT security. Information about emerging threats and vulnerabilities among industry stakeholders can enhance collective defense mechanisms. Standardized protocols for sharing threat intelligence will be essential for creating a unified front against evolving cyber threats (Huang et al. 2023). Furthermore, the adoption of zero-trust security models is gaining traction in IIoT. This approach assumes that no entity, whether inside or outside the network, should be trusted by default. Implementing strict access controls, continuous authentication, and encryption will be integral to achieving a zero-trust security posture in IIoT environments (Atieh, Nanda, and Mohanty 2023). Additionally, digital twins are poised to play a crucial role in strengthening Industrial IoT security in the future. Integrating digital twins with IIoT environments introduces several mechanisms that enhance security (Xu et al. 2023). These mechanisms include real-time monitoring and anomaly detection, predictive security analytics, incident response and resilience testing, and secure software development life cycle (SDLC). Secure edge implementations are becoming popular. Edge technology combined with federated learning will see powerful mechanisms for security developed in the future (Makkar et al. 2022). Creating a point at the edge where data is processed and secured before being transmitted on the network is becoming increasingly accepted. This trend will continue as the edge becomes more powerful and intelligent. Overall, as the industrial IoT continues to mature, securing these interconnected systems becomes a top priority. Staying abreast of these emerging trends and proactively addressing evolving threats will result in industries building a secure foundation for the widespread adoption of IIoT technologies.

9 CONCLUSION

In conclusion, as we stand at the intersection of innovation and vulnerability, the imperative to fortify industrial ecosystems against malicious actors becomes more pronounced. The presented comprehensive review has illuminated the diverse technologies shaping the fabric of IIoT security, emphasizing their strengths and vulnerabilities. Different security challenges, as well as mitigation strategies, have been scrutinized. Real-world security breaches have been dissected, offering invaluable insights into the gravity of challenges faced by industrial enterprises. The pinnacle of this contribution lies in the proposition of a forward-thinking architecture designed to secure Industrial IoT. This architectural blueprint, a synthesis of cutting-edge technologies and proven methodologies, serves as a beacon for organizations navigating the intricate cybersecurity landscape. By fostering a cohesive defense mechanism, it addresses current vulnerabilities and anticipates future challenges, laying the foundation for resilient, adaptive, and secure industrial ecosystems. As we embark on the next phase of technological evolution, the imperative to integrate robust security measures into the fabric of industrial IoT becomes non-negotiable. Accordingly, this chapter seeks to inspire a continued dialogue among researchers, practitioners, and policymakers. By fostering collaboration and innovation, we can collectively propel the trajectory of Industrial IoT security toward a future where connectivity is synonymous with resilience and innovation is safeguarded by design. In doing so, we contribute to the longevity and sustainability of our interconnected industrial infrastructures.

REFERENCES

Adebayo, Olawale Surajudeen, Shefiu Olusegun Ganiyu, Adam A. Alli, Salihu Ahmed Rufai, Abdullahi Monday Jubril, Lateefah Abdulazeez, and Emmanuel Hamman Gadzama. 2023. "Two-Layer Secured Graphical Authentication with One Time Password (OTP) Verification for a Web Based Applications." March. http://repository.futminna.edu.ng:8080/jspui/handle/123456789/18488.

Agrawal, Shweta, Fuyuki Kitagawa, Ryo Nishimaki, Shota Yamada, and Takashi Yamakawa. 2023. "Public Key Encryption with Secure Key Leasing." In 581–610. https://doi.org/10.1007/978-3-031-30545-0_20.

Alabdulatif, Abdullah, Navod Neranjan Thilakarathne, and Kassim Kalinaki. 2023. "A Novel Cloud Enabled Access Control Model for Preserving the Security and Privacy of Medical Big Data." *Electronics* 12 (12): 2646. https://doi.org/10.3390/electronics12122646.

Alkhadra, Rahaf, Joud Abuzaid, Mariam AlShammari, and Nazeeruddin Mohammad. 2021. "SolarWinds Hack: In-Depth Analysis and Countermeasures." *2021 12th International Conference on Computing Communication and Networking Technologies, ICCCNT 2021*. https://doi.org/10.1109/ICCCNT51525.2021.9579611.

Alli, Adam A., and Muhammad Mahbub Alam. 2019. "SecOFF-FCIoT: Machine Learning Based Secure Offloading in Fog-Cloud of Things for Smart City Applications." *Internet of Things* 7 (September): 100070. https://doi.org/10.1016/J.IOT.2019.100070.

Alli, Adam A., Kalinaki Kassim, Nambobi Mutwalibi, Habiba Hamid, and Lwembawo Ibrahim. 2021. "Secure Fog-Cloud of Things: Architectures, Opportunities and Challenges." In *Secure Edge Computing*, edited by Mohiuddin Ahmed and Paul Haskell-Dowland, 1st ed., 3–20. CRC Press. https://doi.org/10.1201/9781003028635-2.

Alrumaih, Thuraya N.I., Mohammed J.F. Alenazi, Nouf A. AlSowaygh, Abdulmalik A. Humayed, and Ibtihal A. Alablani. 2023. "Cyber Resilience in Industrial Networks: A State of the Art, Challenges, and Future Directions." *Journal of King Saud University – Computer and Information Sciences* 35 (9): 101781. https://doi.org/10.1016/.JKSUCI.2023.101781.

Alwasel, Bader, Abdulaziz Aldribi, Mohammed Alreshoodi, Ibrahim S. Alsukayti, and Mohammed Alsuhaibani. 2023. "Leveraging Graph-Based Representations to Enhance Machine Learning Performance in IIoT Network Security and Attack Detection." *Applied Sciences* 13 (13): 7774. https://doi.org/10.3390/APP13137774.

Aslam, Muhammad Muzamil, Ali Tufail, Ki Hyung Kim, Rosyzie Anna Awg Haji Mohd Apong, and Muhammad Taqi Raza. 2023. "A Comprehensive Study on Cyber Attacks in Communication Networks in Water Purification and Distribution Plants: Challenges, Vulnerabilities, and Future Prospects." *Sensors* 23 (18): 7999. https://doi.org/10.3390/S23187999.

Atieh, Adel, Priyadarsi Nanda, and Manoranjan Mohanty. 2023. "A Zero-Trust Framework for Industrial Internet of Things." *2023 International Conference on Computing, Networking and Communications, ICNC 2023*, 331–335. https://doi.org/10.1109/ICNC57223.2023.10074295.

Babayigit, Bilal, and Mohammed Abubaker. 2023. "Industrial Internet of Things: A Review of Improvements over Traditional SCADA Systems for Industrial Automation." *IEEE Systems Journal*. https://doi.org/10.1109/JSYST.2023.3270620.

BBC. 2021. "Meat Giant JBS Pays $11m in Ransom to Resolve Cyber-Attack." *BBC*. 2021. https://www.bbc.com/news/business-57423008.

Bécue, Adrien, Isabel Praça, and João Gama. 2021. "Artificial Intelligence, Cyber-Threats and Industry 4.0: Challenges and Opportunities." *Artificial Intelligence Review* 54 (5): 3849–3886. https://doi.org/10.1007/S10462-020-09942-2/FIGURES/3.

Beerman, Jack, David Berent, Zach Falter, and Suman Bhunia. 2023. "A Review of Colonial Pipeline Ransomware Attack." *Proceedings -23rd IEEE/ACM International Symposium on Cluster, Cloud and Internet Computing Workshops, CCGridW 2023*, 8–15. https://doi.org/10.1109/CCGRIDW59191.2023.00017.

Canilang, Henar Mike O., Angela C. Caliwag, and Wansu Lim. 2022. "Design, Implementation, and Deployment of Modular Battery Management System for IIoT-Based Applications." *IEEE Access* 10: 109008–109028. https://doi.org/10.1109/ACCESS.2022.3214177.

Caporuscio, Mauro, Francesco Flammini, Narges Khakpour, Prasannjeet Singh, and Johan Thornadtsson. 2020. "Smart-Troubleshooting Connected Devices: Concept, Challenges and Opportunities." *Future Generation Computer Systems* 111 (October): 681–697. https://doi.org/10.1016/J.FUTURE.2019.09.004.

Chalapathi, G.S.S., Vinay Chamola, Aabhaas Vaish, and Rajkumar Buyya. 2021. "Industrial Internet of Things (Iiot) Applications of Edge and Fog Computing: A Review and Future Directions." *Advances in Information Security* 83: 293–325. https://doi.org/10.1007/978-3-030-57328-7_12.

Cui, Jie, Fangzheng Cheng, Hong Zhong, Qingyang Zhang, Chengjie Gu, and Lu Liu. 2023. "Multi-Factor Based Session Secret Key Agreement for the Industrial Internet of Things." *Ad Hoc Networks* 138 (January): 102997. https://doi.org/10.1016/J.ADHOC.2022.102997.

Dash, Bibhu, Meraj Farheen Ansari, Pawankumar Sharma, and Azad Ali. 2022. "Threats and Opportunities with AI-Based Cyber Security Intrusion Detection: A Review." https://papers.ssrn.com/abstract-4323258.

Eid, Abdulrahman Mahmoud, Ali Bou Nassif, Bassel Soudan, and Mohammad Noor Injadat. 2023. "IIoT Network Intrusion Detection Using Machine Learning." *2023 6th International Conference on Intelligent Robotics and Control Engineering (IRCE)*, August, 196–201. https://doi.org/10.1109/IRCE59430.2023.10255088.

Fahim, Khairul Eahsun, Kassim Kalinaki, and Wasswa Shafik. 2023. "Electronic Devices in the Artificial Intelligence of the Internet of Medical Things (AIoMT)." In *Handbook of Security and Privacy of AI-Enabled Healthcare Systems and Internet of Medical Things*, 1st ed., 41–62. CRC Press. https://doi.org/10.1201/9781003370321-3.

Ferretti, Luca, Francesco Longo, Giovanni Merlino, Michele Colajanni, Antonio Puliafito, and Nachiket Tapas. 2021. "Verifiable and Auditable Authorizations for Smart Industries and Industrial Internet-of-Things." *Journal of Information Security and Applications* 59 (June): 102848. https://doi.org/10.1016/J.JISA.2021.102848.

Fröhlich, Antônio Augusto, Leonardo Passig Horstmann, and José Luis Conradi Hoffmann. 2023. "A Secure IIoT Gateway Architecture Based on Trusted Execution Environments." *Journal of Network and Systems Management* 31 (2): 1–30. https://doi.org/10.1007/S10922-023-09723-6.

Gavrovska, Ana, and Andreja Samčović. 2020. "Intelligent Automation Using Machine and Deep Learning in Cybersecurity of Industrial IoT." In *Cyber Security of Industrial Control Systems in the Future Internet Environment*, 156–174. https://doi.org/10.4018/978-1-7998-2910-2.ch008.

Gupta, Priyanshi, Chaitanya Krishna, Rahul Rajesh, Arushi Ananthakrishnan, A. Vishnuvardhan, Shrey Shaileshbhai Patel, Chinmay Kapruan, et al. 2022. "Industrial Internet of Things in Intelligent Manufacturing: A Review, Approaches, Opportunities, Open Challenges, and Future Directions." *International Journal on Interactive Design and Manufacturing*, October, 1–23. https://doi.org/10.1007/S12008-022-01075-W.

Hazra, Abhishek, Mainak Adhikari, Tarachand Amgoth, and Satish Narayana Srirama. 2021. "A Comprehensive Survey on Interoperability for IIoT: Taxonomy, Standards, and Future Directions." *ACM Computing Surveys (CSUR)* 55 (1). https://doi.org/10.1145/3485130.

Huang, Hongcheng, Peixin Ye, Min Hu, and Jun Wu. 2023. "A Multi-Point Collaborative DDoS Defense Mechanism for IIoT Environment." *Digital Communications and Networks* 9 (2): 590–601. https://doi.org/10.1016/J.DCAN.2022.04.008.

Huo, Ru, Shiqin Zeng, Zhihao Wang, Jiajia Shang, Wei Chen, Tao Huang, Shuo Wang, F. Richard Yu, and Yunjie Liu. 2022. "A Comprehensive Survey on Blockchain in Industrial Internet of Things: Motivations, Research Progresses, and Future Challenges." *IEEE Communications Surveys and Tutorials* 24 (1): 88–122. https://doi.org/10.1109/COMST.2022.3141490.

Jaskó, Szilárd, Adrienn Skrop, Tibor Holczinger, Tibor Chován, and János Abonyi. 2020. "Development of Manufacturing Execution Systems in Accordance with Industry 4.0 Requirements: A Review of Standard- and Ontology-Based Methodologies and Tools." *Computers in Industry* 123 (December): 103300. https://doi.org/10.1016/J.COMPIND.2020.103300.

Kalinaki, Kassim, Mugigayi Fahadi, Adam A. Alli, Wasswa Shafik, Magombe Yasin, and Nambobi Mutwalibi. 2024. "Artificial Intelligence of Internet of Medical Things (AIoMT) in Smart Cities: A Review of Cybersecurity for Smart Healthcare." *Handbook of Security and Privacy of AI-Enabled Healthcare Systems and Internet of Medical Things* 271–292.

Kalinaki, Kassim, Navod Neranjan Thilakarathne, Hamisi Ramadhan Mubarak, Owais Ahmed Malik, and Musau Abdullatif. 2023. "Cybersafe Capabilities and Utilities for Smart Cities." In *Cybersecurity for Smart Cities*, 71–86. Springer, Cham. https://doi.org/10.1007/978-3-031-24946-4_6.

Kumar, Rakesh, Bipin Kandpal, and Vasim Ahmad. 2023. "Industrial IoT (IIOT): Security Threats and Countermeasures." In *2023 International Conference on Innovative Data Communication Technologies and Application (ICIDCA)*, 829–833. IEEE. https://doi.org/10.1109/ICIDCA56705.2023.10100145.

Kumar, Rohit, and Neha Agrawal. 2023. "Analysis of Multi-Dimensional Industrial IoT (IIoT) Data in Edge–Fog–Cloud Based Architectural Frameworks : A Survey on Current State and Research Challenges." *Journal of Industrial Information Integration* 35 (October): 100504. https://doi.org/10.1016/J.JII.2023.100504.

Lukaj, Valeria, Francesco Martella, Maria Fazio, Antonio Celesti, and Massimo Villari. 2023. "Establishment of a Trusted Environment for IoT Service Provisioning Based on X3DH-Based Brokering and Federated Blockchain." *Internet of Things* 21 (April): 100686. https://doi.org/10.1016/J.IOT.2023.100686.

Lyu, Mengtao, Xinyu Li, and Chun Hsien Chen. 2022. "Achieving Knowledge-as-a-Service in IIoT-Driven Smart Manufacturing: A Crowdsourcing-Based Continuous Enrichment Method for Industrial Knowledge Graph." *Advanced Engineering Informatics* 51 (January): 101494. https://doi.org/10.1016/J.AEI.2021.101494.

Makkar, Aaisha, Tae Woo Kim, Ashutosh Kumar Singh, Jungho Kang, and Jong Hyuk Park. 2022. "SecureIIoT Environment: Federated Learning Empowered Approach for Securing IIoT from Data Breach." *IEEE Transactions on Industrial Informatics* 18 (9): 6406–6414. https://doi.org/10.1109/TII.2022.3149902.

Mekala, Sri Harsha, Zubair Baig, Adnan Anwar, and Sherali Zeadally. 2023. "Cybersecurity for Industrial IoT (IIoT): Threats, Countermeasures, Challenges and Future Directions." *Computer Communications* 208 (August): 294–320. https://doi.org/10.1016/J.COMCOM.2023.06.020.

Minoli, Daniel, and Benedict Occhiogrosso. 2020. "IoT-Driven Advances in Commercial and Industrial Building Lighting." *Industrial IoT: Challenges, Design Principles, Applications, and Security*, January, 97–159. https://doi.org/10.1007/978-3-030-42500-5_3.

Mohamed, Hania, Nickolaos Koroniotis, and Nour Moustafa. 2023. "Digital Forensics Based on Federated Learning in IoT Environment." *ACM International Conference Proceeding Series*, January, 92–101. https://doi.org/10.1145/3579375.3579387.

Moustafa, Nour, Kim Kwang Raymond Choo, and Adnan M. Abu-Mahfouz. 2022. "Guest Editorial: AI-Enabled Threat Intelligence and Hunting Microservices for Distributed Industrial IoT System." *IEEE Transactions on Industrial Informatics* 18 (3): 1892–1895. https://doi.org/10.1109/TII.2021.3111028.

Nabil, Mahmoud, Mohamed Mahmoud, Muhammad Ismail, and Erchin Serpedin. 2019. "Deep Recurrent Electricity Theft Detection in AMI Networks with Evolutionary Hyper-Parameter Tuning." In *2019 International Conference on Internet of Things (IThings) and IEEE Green Computing and Communications (GreenCom) and IEEE Cyber, Physical and Social Computing (CPSCom) and IEEE Smart Data (SmartData)*, 1002–1008. IEEE. https://doi.org/10.1109/iThings/GreenCom/CPSCom/SmartData.2019.00175.

Nechibvute, A., and H. D. Mafukidze. 2023. "Integration of SCADA and Industrial IoT: Opportunities and Challenges." *IETE Technical Review*, August. https://doi.org/10.1080/02564602.2023.2246426.

Ok, Jin Sung, Soon Do Kwon, Cheol Eun Heo, and Young Kyoon Suh. 2021. "A Survey of Industrial Internet of Things Platforms for Establishing Centralized Data-Acquisition Middleware: Categorization, Experiment, and Challenges." *Scientific Programming* 2021. https://doi.org/10.1155/2021/6641562.

Ong, Kevin Shen Hoong, Wenbo Wang, Nguyen Quang Hieu, Dusit Niyato, and Thomas Friedrichs. 2022. "Predictive Maintenance Model for IIoT-Based Manufacturing: A Transferable Deep Reinforcement Learning Approach." *IEEE Internet of Things Journal* 9 (17): 15725–15741. https://doi.org/10.1109/JIOT.2022.3151862.

Oxford Analytica. 2021. "Kaseya Ransomware Attack Underlines Supply Chain Risks." *Emerald Expert Briefings*. https://doi.org/10.1108/OXAN-ES262642.

Peter, Onu, Anup Pradhan, and Charles Mbohwa. 2023. "Industrial Internet of Things (IIoT): Opportunities, Challenges, and Requirements in Manufacturing Businesses in Emerging Economies." *Procedia Computer Science* 217: 856–865. https://doi.org/10.1016/j.procs.2022.12.282.

Pitney, Alexis M, Spencer Penrod, Molly Foraker, and Suman Bhunia. 2022. "A Systematic Review of 2021 Microsoft Exchange Data Breach Exploiting Multiple Vulnerabilities." In *2022 7th International Conference on Smart and Sustainable Technologies (SpliTech)*, 1–6. IEEE. https://doi.org/10.23919/SpliTech55088.2022.9854268.

Rayes, Ammar, and Samer Salam. 2022. "The Things in IoT: Sensors and Actuators." *Internet of Things from Hype to Reality*, 63–82. https://doi.org/10.1007/978-3-030-90158-5_3.

Saci, Anas, Arafat Al-Dweik, and Abdallah Shami. 2021. "Autocorrelation Integrated Gaussian Based Anomaly Detection Using Sensory Data in Industrial Manufacturing." *IEEE Sensors Journal* 21 (7): 9231–9241. https://doi.org/10.1109/JSEN.2021.3053039.

Saputra, Dio Azmi, Stiawan Deris, and Sutabri Tata. 2023. "Implementasi Sistem Deteksi Ransomware Menggunakan Deep Packet Inspection Pada Layanan SMK Negeri 1 Palembang." *Indonesian Journal of Multidisciplinary on Social and Technology* 1 (2): 176–183. https://doi.org/10.31004/ijmst.v1i2.142.

Sezgin, Anil, and Aytug Boyaci. 2023. "A Survey of Privacy and Security Challenges in Industrial Settings." *ISDFS 2023-11th International Symposium on Digital Forensics and Security*. https://doi.org/10.1109/ISDFS58141.2023.10131858.

Shafik, Wasswa, and Kassim Kalinaki. 2023. "Smart City Ecosystem: An Exploration of Requirements, Architecture, Applications, Security, and Emerging Motivations." In *Handbook of Research on Network-Enabled IoT Applications for Smart City Services*, 75–98. https://doi.org/10.4018/979-8-3693-0744-1.ch005.

Shafik, Wasswa, S. Mojtaba Matinkhah, and Kassim Kalinaki. 2023. "An Intrusion Anomaly Detection Approach to Mitigate Sensor Attacks on Mechatronics Systems." In *ICMAME 2023 Conference Proceedings*. Aksaray: ECER. https://doi.org/10.53375/icmame.2023.113.

Shi, Guolong, Xinyi Shen, Fuke Xiao, and Yigang He. 2023. "DANTD: A Deep Abnormal Network Traffic Detection Model for Security of Industrial Internet of Things Using High-Order Features." *IEEE Internet of Things Journal*. https://doi.org/10.1109/JIOT.2023.3253777.

Silva, Daniel, Liliana I. Carvalho, José Soares, and Rute C. Sofia. 2021. "A Performance Analysis of Internet of Things Networking Protocols: Evaluating MQTT, CoAP, OPC UA." *Applied Sciences* 11 (11): 4879. https://doi.org/10.3390/APP11114879.

Singh, Jaya, Ashish Gimekar, and Subramanian Venkatesan. 2023. "An Efficient Lightweight Authentication Scheme for Human-Centered Industrial Internet of Things." *International Journal of Communication Systems* 36 (12): e4189. https://doi.org/10.1002/DAC.4189.

Tareq, Imad, Bassant M. Elbagoury, Salsabil El-Regaily, and El Sayed M. El-Horbaty. 2022. "Analysis of ToN-IoT, UNW-NB15, and Edge-IIoT Datasets Using DL in Cybersecurity for IoT." *Applied Sciences* 12 (19): 9572. https://doi.org/10.3390/APP12199572.

Urrea, Claudio, and David Benítez. 2021. "Software-Defined Networking Solutions, Architecture and Controllers for the Industrial Internet of Things: A Review." *Sensors* 21 (19): 6585. https://doi.org/10.3390/S21196585.

Wittenberg, Carsten. 2022. "Challenges for the Human-Machine Interaction in Times of Digitization, CPS & IIoT, and Artificial Intelligence in Production Systems." *IFAC-PapersOnLine* 55 (29): 114–119. https://doi.org/10.1016/J.IFACOL.2022.10.241.

Xu, Hansong, Jun Wu, Qianqian Pan, Xinping Guan, and Mohsen Guizani. 2023. "A Survey on Digital Twin for Industrial Internet of Things: Applications, Technologies and Tools." *IEEE Communications Surveys & Tutorials*, July, 1–1. https://doi.org/10.1109/COMST.2023.3297395.

Zaidi, Tanzeel, Muhammad Usman, Muhammad Umar Aftab, Hanan Aljuaid, and Yazeed Yasin Ghadi. 2023. "Fabrication of Flexible Role-Based Access Control Based on Blockchain for Internet of Things Use Cases." *IEEE Access* 11: 106315–106333. https://doi.org/10.1109/ACCESS.2023.3318487.

Zhang, Qikun, Yongjiao Li, Ruifang Wang, Lu Liu, Yu-an Tan, and Jingjing Hu. 2021. "Data Security Sharing Model Based on Privacy Protection for Blockchain-Enabled Industrial Internet of Things." *International Journal of Intelligent Systems* 36 (1): 94–111. https://doi.org/10.1002/INT.22293.

Zhao, Meng, Yong Ding, Shijie Tang, Hai Liang, and Huiyong Wang. 2023. "A Blockchain-Based Framework for Privacy-Preserving and Verifiable Billing in Smart Grid." *Peer-to-Peer Networking and Applications* 16 (1): 142–155. https://doi.org/10.1007/S12083-022-01379-4.

Zhao, Yan, Ning Hu, Yue Zhao, and Zhihan Zhu. 2023. "A Secure and Flexible Edge Computing Scheme for AI-Driven Industrial IoT." *Cluster Computing* 26 (1): 283–301. https://doi.org/10.1007/S10586-021-03400-6.

Zheng, Zibin, Shaoan Xie, Hong Ning Dai, Xiangping Chen, and Huaimin Wang. 2018. "Blockchain Challenges and Opportunities: A Survey." *International Journal of Web and Grid Services* 14 (4): 352–375. https://doi.org/10.1504/IJWGS.2018.095647.

15 Strategic Management of Intelligent Robotics and Drones in Contemporary Industrial Operations

An Assessment of Roles and Integration Strategies

Ashok Kumar Manoharan,
Aliyu Mohammed, Pethuru Raj, and
Sundaravadivazhagan Balasubramanian

1 INTRODUCTION

The convergence of intelligent robotics and drones within the framework of contemporary industrial operations has become a catalytic force, ushering in transformative changes that resonate globally across diverse business landscapes. This comprehensive study embarks on an exploration of the strategic management of these cutting-edge technologies, seeking to unravel their pivotal roles, confront the associated challenges, and unveil optimal integration strategies. In the unfolding narrative of the Fourth Industrial Revolution, where the boundaries between the digital, biological, and physical realms blur, the strategic utilization of intelligent robotics and drones emerges as not just advantageous but also essential for organizations aspiring to navigate and thrive in this paradigm shift. As we navigate the dynamic terrain of modern industry, it becomes evident that the marriage of intelligent robotics and drones is more than a mere technological convergence; it is a symbiotic relationship that propels businesses into a new era of efficiency and innovation. This research intends to unpack the intricate dynamics of this connection, unveiling ways in which businesses may capitalize upon the viability of these applications. It explores the complex ballet of smart robotics and drones with each player's talents, and complementary gifts. With rapidly changing nature of industry, the urgency for organizations to move ahead becomes imperative on their part to take the proactive stance toward these technologies. To be ahead of the game, we cannot adapt to intelligent robotics and drones for the Fourth Industrial Revolution demand action. Through a

 DOI: 10.1201/9781032694375-15

strategic lens, this study endeavors to equip organizations with insights that transcend the superficial, guiding them toward a nuanced understanding of how these technological advancements can be woven into the fabric of their operations, ensuring not only survival but flourishing in the dynamic landscapes of the future.

1.1 Background

Delving into the historical context, the integration of automation technologies within industries has been an enduring progression. But the development of intelligent robotics and drones is the great change in the way. Apart from mechanical performance of commonplace assignments, these technologies demonstrate intelligence traits incorporating potential for acquiring skills as a result of previous experiences. Globally, industries around the world are being quickened by this technology's adoption in a bid to enhance their efficiency, reduce costs, and stay relevant against competitive rivals. It is evident the literature has seen a transformation as one traces the trajectory of historical development that preceded today's focus on strategic management of intelligent robots and drones. From what started as pure experimentation with this technology, we have come to deeply understand the strategy in it. The discussion has moved from being only technical to one that addresses the impact of technologies on an organization's strategy and competitive positioning. Indeed, this narrative is an eloquent illustration of how smart robotics and drones have transformed the industrial landscape. This narrative highlights the fact that such technologies go beyond mere tasks automation and are indeed a defining force in reshaping contours of different industries. This evolution prompts a reevaluation of organizational approaches, moving beyond the rudimentary adoption of technology to a sophisticated understanding of how these innovations can be strategically harnessed. Thus, this study aims to unravel the intricate layers of this evolution, offering a panoramic view of the historical journey that has brought us to the forefront of strategic management in the realm of intelligent robotics and drones.

1.2 Significance of the Study

In the current industrial panorama, unraveling the intricacies of the strategic management of intelligent robotics and drones holds paramount significance. The disruptive wave of e-commerce and the dominance of online marketplaces have upended conventional business models, thrusting the efficient deployment of these technologies into the spotlight as a survival imperative. Organizations that adeptly weave intelligent robotics and drones into their operational fabric not only navigate the challenges posed by this digital shift but also position themselves strategically to secure a competitive edge marked by heightened cost efficiency, unparalleled agility, and a culture of continuous innovation. The study at hand acknowledges the imperative for organizations to not merely adapt but to strategically master the integration of intelligent robotics and drones. The significance lies not only in keeping pace with the transformative forces reshaping industries but in leveraging these technologies as catalysts for growth and resilience. As e-commerce redefines the rules of engagement in the business realm, the study underscores that the survival and prosperity of organizations hinge on their ability to embrace and capitalize on the potential embedded in intelligent robotics and drones. The strategic importance of this study extends beyond a mere exploration of technological novelties; it delves into a realm where survival and success coalesce. The lens through which organizations view and implement intelligent robotics and drones becomes a critical determinant of their trajectory in an ever-evolving industrial landscape. By comprehending and strategically managing these technologies, organizations not only weather the storms of change but also embark on a transformative journey that positions them as pioneers in the era of digital disruption.

1.3 Scope and Limitations

In charting the course for this study, it is imperative to delineate both its expansive scope and inherent limitations. The scope extends globally, embracing a panoramic view of the subject matter. Yet, a nuanced recognition underscores the necessity of grounding findings in the specificities of the local business milieu. The multifaceted nature of this exploration acknowledges that while the principles may resonate universally, the application and impact manifest uniquely within distinct regional contexts. However, inherent limitations surface in the face of the dynamic nature of technology and the diverse adoption landscapes across industries. The study candidly acknowledges these constraints, emphasizing the continual evolution of technology as a challenge. The pace at which industries embrace intelligent robotics and drones varies, introducing a layer of complexity to the generalizability of findings. The study, therefore, navigates through these challenges with a conscious awareness of the fluidity intrinsic to technological advancements. An important facet of this study is its qualitative nature, a deliberate choice that steers away from the confines of quantitative measurement. Rather, it is more focused on a philosophical consideration that offers insight into the multi-pronged link between smart robots, drone technology and strategy. The qualitative lens used for deep exploration of dynamics is beyond superficial indicators. Therefore, in a true sense, this study delineates within a reality that takes into account the complexities of shifting and changing nature of technology and patterns of adoptions amongst industries. Although the area covers the whole world, the qualitative emphasis enables an in-depth examination of the subtleties of the subject matter since unraveling how technology is woven into strategies cannot be reduced to numeric measurements.

1.4 Research Questions

To guide this conceptual study, the following research questions have been formulated:

1. What do researchers know about strategic management in the current industrial environment using intelligence in robotics and drones?
2. Is there any gap in the literature such as inadequate research design, poor measurement, and instrumentation to be further studied in this research?
3. Why does it matter for industry to address these gaps?
4. Why look at the strategic management of intelligent robotics and drones?

5. Why does an interdisciplinary approach matter when it comes to understanding and employing such technologies?

1.5 Objectives of the Study

The objectives of this study are to:

1. Examine the existing knowledge on the strategic management of intelligent robotics and drones.
2. Identify gaps and missing links in the current literature.
3. Highlight the significance of addressing these gaps for industrial operations.
4. Rationalize the exploration of strategic management in the context of intelligent robotics and drones.
5. Emphasize the importance of an interdisciplinary approach in comprehending and implementing these technologies.

In the subsequent sections, we delve into the literature review, conceptual framework, and empirical study to provide a comprehensive exploration of the strategic management of intelligent robotics and drones in contemporary industrial operations.

2 STATEMENT OF THE PROBLEM

In recent years, the integration of robotics and drones into industrial operations has become a defining characteristic of the contemporary business landscape. This transformative shift, while offering numerous opportunities, also presents challenges that necessitate a strategic approach for effective management. The statement of the problem focuses on three key dimensions that warrant attention.

2.1 Emergence of Robotics and Drones in Industrial Operations

The rapid ascent of intelligent robotics and drones in industrial operations marks a pivotal moment, instigating a profound transformation in traditional approaches to production and service delivery. These technologies hold the promise of not just heightened efficiency, reduced operational costs, and precision refinement but also bring forth intricate challenges as they seamlessly integrate into the fabric of industrial processes. To unlock the true potential of intelligent robotics and drones in a meaningful and effective manner, a comprehensive understanding of the nuanced roles and functions they play across diverse industrial settings becomes indispensable. Underneath this apparent boost in efficiency savings, cutting down operating expenditures, and improved efficiency is a complicated set of variables, which should be carefully considered. However, this study realizes how important it is to understand the complexities involved in the roles and functions of these intelligent robots and drones, given the dynamic nature of the industries where these machines operate. As compared to the simple acknowledgments of transformational aspect of these technologies, the exploration moves further, to the interplay between the industrial operation and these dynamism technologies. This chapter provides a road map in the understanding of the ways intelligent robots and drones affect industrial activities in the modern world. This will enable the readers, to understand the complex issue, go beyond the superficial and look at the delicate relation between those technologies and the everyday business life of the industry they are invading. Therefore, it strives to enhance the broader comprehension of massive shifts introduced in the industrial space by these technologies.

2.2 Challenges and Opportunities in Integration

Integrating robotics technology and drones into manufacturing processes leads to several challenges and possibilities forming intricate field needing close attention. These sophisticated technologies also come with challenges from adjustment of the workforce, interoperability difficulties in combining diverse technological platforms, and the complex set of rules surrounding its implementation. At the same time, another realm of opportunity arises where companies can engage in simplified operations, data-based decision making, and venture into new business models that may revamp existing industry norms. This leads to challenges that have frontline issue which is the complex challenge of adjustment of work force to the robots and drones. However, this becomes a critical issue in ensuring that there is ease of integration since appropriate strategic approaches will be crucial for adapting these human resources to such technological developments. Another impediment is the issue of technological compatibility whereby industries have to combine various technological systems without interfering with the ongoing operations. Navigating the regulatory landscape adds another layer of complexity. Regulations governing the use of robotics and drones in industrial settings demand a thorough understanding to ensure compliance while leveraging the transformative potential of these technologies. Yet, within these challenges lie opportunities waiting to be harnessed. Streamlining operations through the integration of robotics and drones promises enhanced efficiency and productivity. The data generated by these technologies opens avenues for data-driven decision-making, fostering a culture of informed choices. Moreover, the prospect of innovative business models, fueled by the capabilities of intelligent robotics and drones, presents an opportunity for forward-thinking organizations to carve out a distinct competitive advantage. Addressing these challenges and capitalizing on the opportunities demands a comprehensive exploration of the strategic management aspects inherent in the integration process. This study ventures into the heart of this integration landscape, aiming to unravel the intricacies that define the delicate balance between challenges and opportunities in the realm of robotics and drones in industrial operations.

2.3 Need for Strategic Management

Management should be careful and strategic, considering today's developing environment of production processes and the new smart robots, drones, and others. Inadequate strategic foresight applied in mobilizing transformational technology will lead to a wasteful practice, which is characterized by escalated overheads while at the same time eroding any competitive advantage. In this sense, the necessity for strategic management makes itself apparent providing direction on how to fit robotics and drones into the organizational plans and larger industrial developments. Forward-thinking and proactive approach is therefore essential toward the understanding that this imperative must be recognized in the modern world where competition dictates every move. Failure in strategic management makes organizations vulnerable to the potholes caused by poor usage of sophisticated technologies such as robots and drones that are re-shaping the industrial landscape toward intelligence. Essentially, strategic management provides the centerpiece that will integrate advanced technologies within operations, improve efficiency, enhance effective cost analysis, and achieve competitive edge in the industry. The next parts of this detailed work focus on looking into the issue. The literature review shows the historical and contextual background, explaining the developmental path of strategic management in terms of intelligent robots and drones. The conceptual framework lays the groundwork for understanding the intricate relationships and interdependencies that define this strategic landscape. The empirical study, the heart of the investigation, delves into real-world scenarios, unraveling the practical applications of strategic management in the integration of intelligent robotics and drones within contemporary industrial operations. Through this multidimensional approach, the study seeks to offer a holistic understanding of the roles, challenges, and integration strategies inherent in the strategic management of these transformative technologies.

2.4 Key Concepts and Models in Human-Centered Security

In the realm of intelligent robotics and drones within contemporary industrial operations, strategic management plays a pivotal role in ensuring optimal utilization and integration. Key concepts and models in human-centered security are crucial for addressing potential challenges and risks associated with the deployment of these technologies.

Concept/Model	Description
User-Centric Security	Focuses on designing security measures that align with user needs and behavior, ensuring seamless human-machine interaction.
Threat Modeling	Systematic identification and assessment of potential security threats, enabling proactive measures to mitigate risks before they materialize.
Usability and Security Trade-off	Balancing user-friendly interfaces with robust security measures, acknowledging that overly complex systems may compromise user experience.
Behavioral Analytics	Utilizing data analytics to monitor user behavior and detect anomalies, enhancing the ability to identify and respond to security breaches promptly.
Privacy by Design	Integrating privacy considerations into the design and development of robotics and drone systems to safeguard sensitive information and adhere to regulations.
Adaptive Access Control	Implementing dynamic access controls that adjust based on contextual factors, restricting unauthorized access and minimizing potential security breaches.

Effective strategic management incorporates these concepts and models to create a comprehensive framework that ensures the seamless integration of intelligent robotics and drones in industrial operations while prioritizing human-centered security. This approach enhances productivity, minimizes risks, and fosters a collaborative environment between human workers and advanced technologies.

3 LITERATURE REVIEW

3.1 Evolution of Robotics and Drones in Industrial Operations

The trajectory of robotics and drones within industrial operations has been an extensive area of scholarly exploration. According to Raj and Kos (2022), the roots of this evolution extend to the 1960s, marked by the introduction of industrial robots, with subsequent progress catalyzed by advancements in artificial intelligence (AI) and sensor technologies. Casiroli and Pau (2023) highlight the remarkable speed of technological evolution, underlining the transition from conventional automation to the integration of intelligent systems endowed with real-time learning and adaptive capabilities. Contributions from Sajwan and Singh (2023) accentuate the growing sophistication of robotic systems, shedding light on the integration of machine learning algorithms and collaborative functionalities. The evolution of drones, as illuminated by Khosla and Malhi (2023), unfolds against a backdrop of advancements in materials, communication technologies, and miniaturized sensors. These advancements collectively empower drones, rendering them versatile tools applicable across a spectrum of industrial sectors. The narrative of evolution painted by these scholarly perspectives showcases not only the historical roots of robotics and drones but also the dynamic forces propelling their advancement. The symbiotic relationship between these technologies and the ever-expanding realms of AI and sensor technologies signifies a paradigm shift, transforming them from static tools to intelligent, adaptive entities. As this study delves into the strategic management of these technologies, the historical context provided by these scholarly insights serves as a foundational backdrop, guiding an in-depth exploration of their roles and implications within contemporary industrial operations.

3.2 Role of Intelligent Robotics in Contemporary Business

In the dynamic landscape of contemporary business, intelligent robotics assume multifaceted roles that extend far beyond mere automation. Tana and Chai (2023) shed light on the transformative impact of robotics across value chains, influencing processes from production to customer service. Rane (2023) expands on this, emphasizing that the roles played by intelligent robotics transcend automation, encompassing tasks that demand cognitive capabilities, including problem-solving and decision-making. An intriguing dimension emerges through the exploration of collaborative robots, or cobots, as highlighted by George and George (2023). This avenue suggests the augmentation of human capabilities, promising increased productivity and efficiency across various business functions. Hasa (2023) contributes to the discourse by underlining the implications for organizational structures and skill

FIGURE 15.1 Emergence of Robotics and Drones in Industrial Operations.

requirements. The integration of intelligent robotics prompts a reevaluation of workforce competencies, emphasizing the need for adaptability and evolving skill sets. Furthermore, Kasowaki and Atiye (2023) delve into the role of robotics in fostering innovation and shaping new business models. Their research underscores the disruptive nature of contemporary technologies, emphasizing how intelligent robotics contribute to the dynamic landscape by driving innovation and restructuring traditional business frameworks. As this study explores the strategic management of intelligent robotics, these scholarly insights serve as foundational pillars, providing a comprehensive understanding of the diverse and pivotal roles played by intelligent robotics in the intricate tapestry of modern business as shown in Figure 15.1.

3.3 Integration of Drones into Industrial Processes

Originally tethered to military applications, drones have swiftly transitioned into indispensable tools within industrial processes. Krishnan and Murugappan (2023) delve into the expansive realm of drone applications, traversing domains from surveillance to logistics. The work of Bayomi and Fernandez (2023) amplifies the discussion, spotlighting the pivotal role of drones in data collection and monitoring, particularly in sectors such as agriculture and infrastructure. This underscores the versatility of drones, transforming them into invaluable assets for industries seeking precise and efficient data-driven solutions. Kalasani (2023) contributes a nuanced perspective by exploring the integration of drones into supply chain management. This facet emphasizes the potential for significant cost reduction and heightened operational efficiency. Drones, once perceived predominantly as aerial vehicles, are now integral components of streamlined logistical processes, contributing to the optimization of supply chains. The evolution of drones from their military origins to

FIGURE 15.2 Evolution of Robotics and Drones in Industrial Operations.

integral components of industrial operations signifies a paradigm shift. The applications outlined by these scholarly insights underscore the breadth of impact drones have across diverse sectors. As this study delves into the strategic management of these technologies, the historical context provided by these perspectives enriches the understanding of how drones have seamlessly integrated into the fabric of industrial processes, offering a lens into their transformative roles beyond conventional military applications as shown in Figure 15.2.

3.4 Strategic Management Approaches in the Context of Robotics and Drones

In the realm of robotics and drones, strategic management approaches have been intricately shaped by the imperative to harmonize technological adoption with organizational goals. Lee (2023) introduces the concept of dynamic capabilities, accentuating an organization's prowess in adapting and reconfiguring resources in response to the dynamic landscape of technological changes. This fluidity becomes a cornerstone in the strategic playbook, ensuring agility in the face of evolving technological landscapes. Examining the landscape through the lens of the Resource-Based View (RBV) theory, as elucidated by You and Brahmana (2023), offers insights into how firms can harness their unique resources, including robotics and drones, to forge a sustained competitive advantage. This theory becomes a compass guiding organizations toward strategic decisions that align with their inherent strengths and distinctive capabilities. Thakkalapelli's work (2023) on competitive strategy reinforces the centrality of aligning technology adoption with overall business strategy. It underscores that the integration of robotics and drones should not exist in isolation but should seamlessly align with broader business objectives. This alignment ensures that the technological trajectory complements and fortifies the overall strategic direction of the organization. Contributing to this strategic discourse, Allioui and Mourdi (2023) bring forth the concept of core competencies. Their perspective emphasizes that the strategic management of technology should pivot on leveraging distinctive capabilities, positioning organizations to harness the full potential of robotics and drones within the intricate tapestry of their operations. In essence, this study dives into the strategic dimensions outlined by these scholars, seeking to unravel how these approaches can be employed as navigational tools in the dynamic landscape of robotics and drones integration.

3.5 E-Commerce and Online Marketplaces Disruption

The profound impact of e-commerce and online marketplaces has ushered in a paradigm shift, fundamentally reshaping traditional business models. Rajkhowa and Kornher's research (2023) underscores the transformative effects on pricing dynamics and market structures, reflecting the deep-seated changes initiated by the digital wave. Purnomo (2023) delves into the strategic considerations necessitated by the integration of e-commerce, illuminating crucial aspects such as customer engagement, supply chain management, and the pivotal role of data analytics. In tandem, Reing's contributions (2023) underscore the evolving role of digital platforms, acting as catalysts in fostering direct-to-consumer relationships and delivering personalized experiences. This shift toward personalized interactions represents a fundamental departure from traditional business practices. Williams-Morgan's exploration (2023) into the disruptive effects of e-commerce on traditional retail further accentuates the urgency for businesses to embrace technology-driven strategies for not only survival but sustained growth. The evolving retail landscape demands a proactive adoption of innovative approaches to remain relevant in an era dominated by digital disruption. In summary, the literature review offers a comprehensive understanding of the

FIGURE 15.3 Integration of Drones in Industrial Processes.

multifaceted evolution, roles, and strategic implications of intelligent robotics and drones within industrial operations. Simultaneously, it sheds light on the disruptive influence exerted by e-commerce and online marketplaces, laying the foundation for the conceptual exploration that unfolds in the subsequent sections of this study as shown in Figure 15.3.

4 CONCEPTUAL FRAMEWORK

4.1 Definition of Strategic Management in Robotics and Drones

The profound impact of e-commerce and online marketplaces has ushered in a paradigm shift, fundamentally reshaping traditional business models. Rajkhowa and Kornher's research (2023) underscores the transformative effects on pricing dynamics and market structures, reflecting the deep-seated changes initiated by the digital wave. Purnomo (2023) delves into the strategic considerations necessitated by the integration of e-commerce, illuminating crucial aspects such as customer engagement, supply chain management, and the pivotal role of data analytics. In tandem, Reing's contributions (2023) underscore the evolving role of digital platforms, acting as catalysts in fostering direct-to-consumer relationships and delivering personalized experiences. This shift toward personalized interactions represents a fundamental departure from traditional business practices. Williams-Morgan's exploration (2023) into the disruptive effects of e-commerce on traditional retail further accentuates the urgency for businesses to embrace technology-driven strategies for not only survival but sustained growth. The evolving retail landscape demands a proactive adoption of innovative approaches to remain relevant in an era dominated by digital disruption. In summary, the literature review offers a comprehensive understanding of the multifaceted evolution, roles, and strategic implications of intelligent robotics and drones within industrial operations. Simultaneously, it sheds light on the disruptive influence exerted by e-commerce and online marketplaces, laying the foundation for the conceptual exploration that unfolds in the subsequent sections of this study.

4.2 Theoretical Foundations

4.2.1 Resource-Based View Theory

The RBV theory stands as a cornerstone, offering a foundational framework to comprehend how firms can harness their unique resources, including robotics and drones, to forge a sustained competitive advantage. As articulated by Barney (1991), organizations possessing resources that are valuable, rare, and inimitable find themselves in an advantageous position for success. In the specific context of robotics and drones, this theory posits that firms must strategically manage these technologies as valuable resources, employing them in ways that carve out a distinct competitive edge. The essence of the RBV theory is exemplified when considering a company endowed with advanced drone technology optimized for efficient logistics. In this scenario, the company possesses not only a valuable asset but also one that is rare and challenging to imitate. The strategic management of such technological resources becomes pivotal, as it directly influences the organization's ability to translate these advantages into a sustained and meaningful competitive edge. In essence, the RBV theory serves as a guiding compass, urging firms to view robotics and drones not merely as tools but as strategic resources to be nurtured and strategically wielded. The study, through the lens of the RBV theory, seeks to unravel the nuanced dimensions of leveraging these technologies for a sustained competitive advantage in the intricate landscapes of contemporary industrial operations.

4.2.2 Porter's Five Forces Framework

A pivotal theoretical scaffold guiding the strategic management of robotics and drones is Porter's Five Forces Framework. Introduced by Porter (1980), this framework identifies five potent competitive forces that mold industry structure: the threat of new entrants, bargaining power of buyers, bargaining power of suppliers, threat of substitute products, and intensity of competitive rivalry. Through the use of such a framework for the integration of robotics and drones, managers can access one robust tool for mapping the external environment and making informed strategic choices. Take for example, an enterprise that is engaged in production of sophisticated robotic systems. Using Porter's Five Forces Framework enables the company to examine threats for the case study. If the technology is highly specialized and not easily substituted, the threat of new entrants may be lower, enhancing the company's strategic position. Furthermore, assessing the bargaining power of buyers and suppliers becomes crucial, along with understanding the dynamics of substitute products and the prevailing intensity of competitive rivalry in the market. In essence, Porter's Five Forces Framework emerges as a strategic compass, empowering organizations in the realm of robotics and drones to systematically evaluate and navigate the intricacies of the competitive environment. Through this lens, the study seeks to unravel the strategic insights and informed decision-making possibilities that arise from the application of Porter's seminal framework in the dynamic landscape of contemporary industrial operations.

4.3 Role of E-Commerce in Shaping Strategic Management

The significance of e-commerce in molding the strategic management of robotics and drones cannot be overstated. As a transformative force, e-commerce has not only reshaped traditional business models but has also redefined the approach organizations take toward technological adoption. Huang and Rust (2021) contend that e-commerce induces shifts in pricing dynamics and market structures, underscoring the imperative for businesses to strategically embrace technologies such as robotics and drones to maintain competitiveness. Consequently, the conceptual framework within this study discerns and acknowledges the pervasive influence of e-commerce on the strategic decision-making processes concerning the integration of these advanced technologies. E-commerce, as a disruptive entity, introduces dynamic variables into the business equation. It demands a recalibration of strategies, emphasizing the need for organizations to be agile and innovative in their technological adoption endeavors. The strategic management of robotics and drones becomes intricately interwoven with the evolving landscape sculpted by e-commerce, where staying competitive necessitates astute incorporation of these technologies. In essence, this study, through its conceptual framework, elucidates the profound impact of e-commerce on the strategic dimensions of integrating robotics and drones, recognizing it as a transformative force shaping the contours of contemporary industrial operations.

4.4 Integration Strategies in Industrial Operations

The realization of benefits from robotics and drones in industrial operations hinges on effective integration strategies. Lee (2023) introduces the concept of dynamic capabilities, emphasizing an organization's prowess in adapting and reconfiguring resources to meet the demands of technological changes. Given the dynamic nature of robotics and drones, organizations must continuously refine their strategies to align with the evolving technological landscape. The concept of core competences by Ziakas (2023) highlights the need for focusing on differentiation through distinctiveness during integration strategy development process. It argues for a strategy that focuses on identifying organization strengths and their uniqueness with regard to other organizations and the establishment of an integrated solution that exploits these strengths and advantages in a highly competitive business environment To conclude, the above is a synthesis of different definitions, theories, and the importance of e-commerce in the strategic management of robots and drones. This clearly points out the dynamic nature of these technologies in which the organization's strategies should be constantly aligned with its goals and also the changing environment within the industry. The empirical study is based on this framework that carefully prepares the ground for studying integrative strategies deeply and their far-reaching implications to the fabric of industry operations.

4.5 Visual Representation of the Model (Tabular Form)

In presenting a complex model or system, a tabular representation can succinctly convey key elements. Below is a tabular form providing a visual representation of a hypothetical model, highlighting essential components and their interrelationships.

Component	Description
Input Variables	Parameters or data inputs that influence the model's functioning. Examples include sensor data, user inputs, or environmental factors.
Processing Units	Core computational modules responsible for analyzing and interpreting input variables. These units may include algorithms, machine learning models, or decision-making systems.
Decision Logic	The set of rules or algorithms that guide the decision-making process based on the processed input variables. It determines the model's output or action.
Output Variables	The results or actions generated by the model based on the decision logic. This could be a control signal for a robotic system, a recommendation, or any desired output.
Feedback Mechanism	A loop that enables the model to learn and adapt over time. It incorporates feedback from the output to refine decision logic or adjust processing units for improved performance.
Integration with Drones	Specific elements or modules that facilitate the integration of the model with drone technology, allowing seamless communication and collaboration.

This tabular representation provides a clear overview of the model's architecture, outlining the flow of information from inputs to outputs and illustrating the integration with drone technology. It serves as a visual guide for understanding the structural components and their relationships within the broader context of the model.

5 EMPIRICAL STUDY

5.1 Methodology

5.1.1 Research Design

Qualitative empirical study demanded for full research design so as to grasp all the complex aspects of managing robotics and drones for industrial use. This research utilized a case study methodology in accordance with the suggestion made by Jesus and Jugend (2023). Through case studies, specific cases are analyzed, looking for deeper understanding about surrounding factors, decision-making procedure and results at the practical level within the industry.

5.1.2 Data Collection Methods

Utilizing a mixture of semi-structured interviews and document analysis led to gathering meaningful and situationally specific information. The semi-structured interview enabled free discussion among the relevant stakeholder groups comprising of managers, technicians as well as integrators of robots and drones. This is similar to the approach recommended by Askew (2023) that supports an in-depth understanding of insights and experience. To accompany interview data, document analysis involved examining internal reports, strategic documents, and relevant industry publications. Further layers of analysis were provided through examination of the documents, which also served as historical background for the opinion of the major informants.

5.1.3 Sample Population

Careful consideration was given while selecting the sample population that would be representative of the different industries currently using them for their operations. In order to achieve a thorough grasp on the strategic management challenges and prospects that existed in various spheres, there was inclusion of people working in manufacturing, logistics, healthcare, and agriculture sectors. The choice was guided by issues such as degree of technology acceptance, the organization's size, and geographic placement so that this would capture various experiences and methods as shown in Figure 15.4.

5.2 Extensive Review of Related Studies

A comprehensive overview of extant literature reviewed focused at reinforcing the existing knowledge about the strategic management of robotics and drones. The noteworthy contributions include the work by Яфень and Шевченко (2023) by providing some glimpses about the future of robots in society. Thurbon and Weiss (2021), offer important insights into the role of intelligent robotics in modern business and

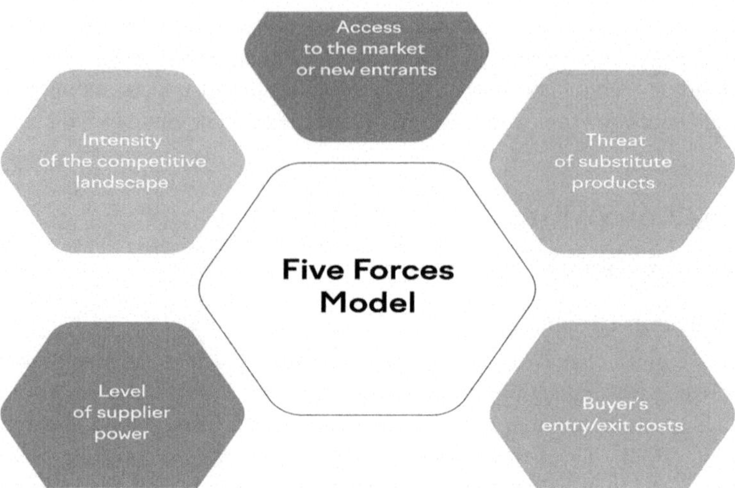

FIGURE 15.4 Porter's Five Forces Framework.

stressing out the strategic approach toward unlocking such power. With regard to drones, Singh and Singh (2023) highlight how these tools can be integrated into industrial activities most notably in areas like farming. Thirdly, the research of Teece (2018) on dynamic capabilities and Barney (1991), RBV theory contributed to the understanding strategic management approach.

5.3 Research Gap Analysis

A review of existing literature revealed some crucial gaps that are discussed below. Despite numerous studies focusing on the development, applications, and the strategic management issues of robots and drones in everyday industrial situations, researchers rarely pay attention to related topics. These gaps included insufficient in-depth case studies about decision-making processes, organizational challenges and outcome of integrating them. On the other hand, most existing literature on this topic omits the multidisciplinary approach involved in robotics and drones' strategic management. This study suggests thinking about more than just technology, like organizational structures, human resources, and external issues, including laws. The study thus fills some gaps in existing theoretical knowledge on the strategic management of intelligent robotics and drones and provides managers working in complex industrial situations with relevant practical information.

6 Findings

6.1 Analysis of Integration Strategies

Organizations use varied means of integrating robotics and drones into industrial operations, according the study finding. Reflecting Teece (2018) and Porter (1985), it is highlighted that organizations align these technologies to their overall business objectives. For example, in the manufacturing industry, robots are used for optimum production, high accuracy, and reduced manpower tasks. The use of drones is strategically integrated into logistics, especially in last mile logistics for effective and time bound delivery. This study confirms the point made by Santos and Batalha (2023) that successful integration strategies are founded on exploiting firms' core competences. The effectiveness and competitiveness in organizations, robotics, and drones are only possible when they're matched to uniqueness. The qualitative data also illustrate the essence of dynamic capabilities especially, how an organization keeps on adopting strategic integration methods based on emerging technologies and changing markets' behaviors.

6.2 Comparative Assessment of Roles in Industrial Operations

A comparative assessment of the roles played by robotics and drones in various industrial operations underscores the versatility of these technologies. Building on the work of Liu, Wang, and Gao (2024) and Paesano (2023), the findings highlight that the roles of robotics extend beyond automation to cognitive tasks. In manufacturing, robots collaborate with human workers, enhancing productivity and quality. In healthcare, robotic systems are utilized for surgeries and patient care, showcasing the adaptability of these technologies. Drones, on the other hand, demonstrate diverse roles across industries. In agriculture, drones are employed for precision farming, monitoring crop health, and optimizing resource utilization. The logistics sector utilizes drones for inventory management and delivery services. The comparative assessment emphasizes the need for organizations to tailor their integration strategies based on the specific roles these technologies play in their respective industries.

6.3 Identification of Success Factors

Identifying success factors in the integration of robotics and drones emerges as a key finding. Building on the RBV theory (Barney, 1991) and dynamic capabilities (Teece, 2018), the study identifies several success factors. Firstly, a clear alignment of technological adoption with organizational goals and core competencies is crucial for success. Organizations that strategically invest in technologies that enhance their unique capabilities experience positive outcomes. Secondly, the study emphasizes the role of leadership and organizational culture. Successful integration is often attributed to strong leadership that champions innovation and a culture that embraces change. Thirdly, partnerships and collaborations, as highlighted by Muthee (2023), contribute to success. Organizations that engage in strategic alliances for research, development, and implementation of these technologies exhibit higher levels of success.

6.4 Challenges Faced in Implementation

The findings acknowledge the challenges organizations encounter in the implementation of robotics and drones. Reflecting on the work of Hubbart (2023), challenges include the high initial investment costs, resistance to change among the workforce, and regulatory uncertainties. The study reveals that the lack of standardized regulations for the operation of drones, particularly in urban environments, poses a significant hurdle. Moreover, the study identifies the need for continuous skill development and training as a challenge. The evolving nature of these technologies requires a workforce equipped with the necessary skills. Organizations that neglect employee training face difficulties in optimizing the potential benefits of robotics and drones.

6.5 Visualizing Challenge Solutions

Addressing challenges effectively requires a systematic approach and a clear visualization of proposed solutions. The tabular form below outlines key challenges and corresponding solutions in a hypothetical context, providing a structured overview.

Challenge	Solution
Technological Integration Complexity	Implement a phased integration plan, starting with modular components and gradually incorporating advanced technologies. Ensure compatibility through thorough testing.
Data Security and Privacy Concerns	Adopt encryption protocols, establish robust access controls, and adhere to privacy regulations. Conduct regular security audits to identify and address vulnerabilities.
Operational Downtime during Implementation	Develop a detailed implementation schedule that minimizes disruptions. Provide training to personnel in advance and have contingency plans for unexpected issues.
Resistance to Technological Adoption	Develop a comprehensive change management strategy, emphasizing the benefits of the new technologies and addressing concerns through transparent communication.
Cost Overruns and Budget Constraints	Conduct a thorough cost-benefit analysis before implementation. Implement cost-monitoring mechanisms and explore partnerships or funding options to alleviate financial pressures.

This tabular representation offers a concise visualization of challenges and their corresponding solutions. By presenting information in a structured format, stakeholders can easily grasp the proposed strategies for overcoming obstacles in the implementation of technological solutions. Each solution is tailored to address specific challenges, fostering a comprehensive and proactive approach to problem-solving.

6.5 Gap Filling Analysis

The identified findings collectively fill the research gap by offering a nuanced understanding of the integration strategies, roles, success factors, and challenges associated with the strategic management of robotics and drones in industrial operations. Prior literature provided a broad overview, but the present study delves into the specific details, contributing practical insights for organizations navigating the complexities of these technologies. By addressing these aspects comprehensively, the study contributes to the advancement of knowledge in the field and provides a foundation for informed decision-making.

7 RECOMMENDATIONS

7.1 Strategic Guidelines for Effective Management

Drawing on the insights gained from the study, strategic guidelines are proposed to enhance the effective management of robotics and drones in industrial operations. The recommendations align with the work of Porter (1985) and Teece. (2018) and emphasize the need for organizations to:

- **Align Technological Adoption with Organizational Goals:** Ensure that the integration of robotics and drones is strategically aligned with the overall business objectives and core competencies, maximizing the value derived from these technologies.
- **Develop Dynamic Capabilities:** Foster a culture of adaptability and continuous learning within the organization. Some examples include conducting research, developing, and keeping up to date with technology and adjusting one's integration strategy.
- **Encourage Interdisciplinary Collaboration:** The organization should foster collaboration among various departments in order to create a comprehensive strategy for managing robotics and drones. It entails promoting interaction between technologists, operations, and strategy management units.

7.2 Policy Implications for Government and Regulatory Bodies

It will be necessary to make suggestions on how the government can ensure that it does not create regulatory uncertainty that can hinder the smooth adoption of robots and drones. The study recommends:

- **Establishing Standardized Regulations:** Promote uniform rules on how drone operation should be facilitated. It also contains the rules for safe operations, privacy, and utilization of the airspace so that both businesses and individuals can be clear on their activities.
- **Incentivizing Technology Adoption:** Introduce policies and incentives that encourage businesses to invest in and adopt robotics and drone technologies. This may include tax credits, grants, or subsidies for organizations that demonstrate effective and responsible integration.
- **Facilitating Research and Development:** Support research initiatives focused on the responsible and ethical use of these technologies. This involves collaborating with academic institutions, industry experts, and technology providers to stay informed about emerging trends and potential risks.

7.3 Technology Adoption Strategies

The study recommends specific technology adoption strategies based on the identified success factors. Organizations are advised to:

- **Conduct Robust Cost-Benefit Analyses:** Prior to adoption, organizations should conduct thorough cost-benefit analyses considering the initial investment, operational costs, and potential returns. This aligns with the findings regarding the high initial investment costs (Best, 2023).
- **Invest in Research and Development:** Allocate resources for ongoing research and development to stay at the forefront of technological advancements. This aligns with Teece's (2023) emphasis on dynamic capabilities, ensuring that organizations can adapt to evolving technologies.
- **Establish Strategic Partnerships:** Engage in strategic alliances and partnerships with technology providers, research institutions, and other organizations. This collaborative approach, as suggested by Westermann and Benyekhlef (2023), enhances access to expertise and resources.

7.4 Training and Skill Development Initiatives

Recognizing the challenges related to workforce adaptation and skill development, recommendations are proposed for organizations:

- **Implement Continuous Training Programs:** Develop and implement ongoing training programs to upskill employees on the operation, maintenance, and troubleshooting of robotics and drones. This aligns with the findings emphasizing the need for continuous skill development (Kilag and Sasan, 2023).
- **Foster a Culture of Innovation:** Create a culture that embraces innovation and technological change. Encourage employees to contribute ideas and insights, fostering an environment where the workforce feels engaged and motivated to adapt to new technologies.
- **Collaborate with Educational Institutions:** Collaborate with educational institutions to tailor curricula that meet the evolving needs of the industry. This collaboration ensures a pipeline of skilled professionals entering the workforce.

In summary, the recommendations provide actionable insights for organizations, government bodies, and regulatory entities, filling the identified research gap and contributing to the effective strategic management of robotics and drones in industrial operations.

8 CONCLUSION

8.1 Summary of Key Findings

In concluding this study on the strategic management of intelligent robotics and drones in contemporary industrial operations, a synthesis of key findings underscores the transformative potential and challenges associated with these technologies. The analysis of integration strategies revealed a nuanced approach across diverse industries, emphasizing the need for organizations to align technological adoption with their unique competencies and dynamic capabilities. The comparative assessment of roles highlighted the versatility of robotics and drones, showcasing their potential in optimizing processes across manufacturing, logistics, healthcare, and agriculture. Identification of success factors emphasized the significance of organizational alignment, leadership, and strategic partnerships. Challenges in implementation, from initial investment costs to regulatory uncertainties, were unveiled, underscoring the complex landscape organizations navigate.

8.2 Contributions to Existing Literature

This study significantly contributes to the existing literature by providing a detailed and comprehensive exploration of the strategic management of robotics and drones. While prior research laid the groundwork, this study delved into the specifics, offering practical insights for organizations and filling a notable gap in the literature. The integration of theoretical frameworks, empirical findings, and strategic recommendations enriches the understanding of how organizations can effectively navigate the challenges and harness the potential benefits of these transformative technologies. By bridging the gap between theory and practice, this study contributes to the evolution of scholarly discussions on technology management.

8.3 Implications for Future Research

The implications for future research are vast and promising. As the field of intelligent robotics and drones continues to evolve, future studies can build on the foundations laid here by exploring:

1. **Long-term Organizational Impact:** Investigate the long-term impact of integrating robotics and drones on organizational structures, employee roles, and overall business models.
2. **Ethical Considerations:** Delve deeper into the ethical considerations surrounding the use of these technologies, particularly in sectors such as healthcare and surveillance.
3. **Global Comparative Studies:** Conduct comparative studies across different countries and regions to understand how cultural, regulatory, and economic factors influence the strategic management of robotics and drones.
4. **Environmental Sustainability:** Explore the environmental implications of widespread technology adoption, particularly in areas such as energy consumption, electronic waste, and sustainable manufacturing processes.

8.4 Closing Remarks

In conclusion, the strategic management of intelligent robotics and drones marks a pivotal juncture in the evolution of contemporary industrial operations. This study has not only illuminated the complex landscape organizations navigate but has also provided actionable insights for effective management. As we stand at the intersection of technological innovation and industrial transformation, it is imperative for organizations, policymakers, and researchers to collaborate in shaping a future where the seamless integration of these technologies not only enhances efficiency but also contributes to the betterment of society. The journey toward technological excellence is ongoing, and this study stands as a beacon, guiding us toward a future where the strategic management of intelligent robotics and drones becomes synonymous with organizational success and societal advancement.

REFERENCES

Allioui, H., & Mourdi, Y. (2023). Unleashing the potential of AI: Investigating cutting-edge technologies that are transforming businesses. *International Journal of Computer Engineering and Data Science (IJCEDS)*, *3*(2), 1–12.

Askew, N. P. (2023). *A Review of United States Veteran Opinions of the Transition Assistance Program (Tap)* (Doctoral dissertation, Purdue University Graduate School).

Barney, J. (1991). Firm resources and sustained competitive advantage. *Journal of Management*, *17*(1), 99–120.

Bayomi, N., & Fernandez, J. E. (2023). Eyes in the sky: Drones applications in the built environment under climate change challenges. *Drones*, *7*(10), 637.

Best, R. (2023). Equitable reverse auctions supporting household energy investments. *Energy Policy*, *177*, 113548.

George, A. S., & George, A. H. (2023). The cobot chronicles: Evaluating the emergence, evolution, and impact of collaborative robots in next-generation manufacturing. *Partners Universal International Research Journal*, *2*(2), 89–116.

Hasa, K. (2023). *Examining the OECD's Perspective on AI in Education Policy: A Critical Analysis of Language and Structure in the 'AI and the Future of Skills' (AIFS) Document and Its Implications for the Higher Education* (Doctoral dissertation, University of British Columbia).

Huang, M. H., & Rust, R. T. (2021). A strategic framework for artificial intelligence in marketing. *Journal of the Academy of Marketing Science*, *49*, 30–50.

Hubbart, J. A. (2023). Organizational change: The challenge of change aversion. *Administrative Sciences*, *13*(7), 162.

Jesus, G. M. K., & Jugend, D. (2023). How can open innovation contribute to circular economy adoption? Insights from a literature review. *European Journal of Innovation Management*, *26*(1), 65–98.

Kalasani, R. R. (2023). *An Exploratory Study of the Impacts of Artificial Intelligence and Machine Learning Technologies in the Supply Chain and Operations Field* (Doctoral dissertation, University of the Cumberlands).

Kilag, O. K. T., & Sasan, J. M. (2023). Unpacking the role of instructional leadership in teacher professional development. *Advanced Qualitative Research*, *1*(1), 63–73.

Lee, H. (2023). Drivers of green supply chain integration and green product innovation: A motivation-opportunity-ability framework and a dynamic capabilities perspective. *Journal of Manufacturing Technology Management*, *34*(3), 476–495.

Liu, S., Wang, L., & Gao, R. X. (2024). Cognitive neuroscience and robotics: Advancements and future research directions. *Robotics and Computer-Integrated Manufacturing*, *85*, 102610.

Muthee, C. (2023). *Corporate Venture Capital on Sustainability-Oriented Innovation: A Norwegian Perspective* (Master's thesis, UIS).

Paesano, A. (2023). Artificial intelligence and creative activities inside organizational behavior. *International Journal of Organizational Analysis*, *31*(5), 1694–1723.

Pallavi Rajkhowa and Lukas Kornher, (2023), Effects of electronic markets on prices, spikes in prices, and price dispersion: A case study of the tea market in India, *Agribusiness*, 39, (4), 1117–1138.

Porter, M. E. (1980). Industry structure and competitive strategy: Keys to profitability. *Financial Analysts Journal*, *36*(4), 30–41.

Porter, M. E. (1985). Technology and competitive advantage. *Journal of Business Strategy*, *5*(3), 60–78.

Purnomo, Y. J. (2023). Digital marketing strategy to increase sales conversion on e-commerce platforms. *Journal of Contemporary Administration and Management (ADMAN)*, *1*(2), 54–62.

Raj, R., & Kos, A. (2022). A comprehensive study of mobile robot: History, developments, applications, and future research perspectives. *Applied Sciences*, *12*(14), 6951.

Rane, N. (2023). Enhancing mathematical capabilities through chatgpt and similar generative artificial intelligence: Roles and challenges in solving mathematical problems. Available at SSRN 4603237.

Reing, R. (2023). Direct-to-consumer sales: Unlocking the power of wine club memberships. Silicon Valley Bank. https://www.svb.com/globalassets/library/uploadedfiles/wine/2023-direct-to-consumer-wine-report.pdf

Santos, A. B., & Batalha, M. O. (2023). The internationalization of meatpacking firms: A competence-based approach. *British Food Journal*, *125*(2), 731–751.

Tana, G., & Chai, J. (2023). Digital transformation: Moderating supply chain concentration and competitive advantage in the service-oriented manufacturing industry. *Systems*, *11*(10), 486.

Teece, D. J. (2018). Dynamic capabilities as (workable) management systems theory. *Journal of Management & Organization*, *24*(3), 359–368.

Thakkalapelli, D. (2023). Cost analysis of cloud migration for small businesses. *Tuijin Jishu/Journal of Propulsion Technology*, *44*(4), 4778–4784.

Thurbon, E., & Weiss, L. (2021). Economic statecraft at the frontier: Korea's drive for intelligent robotics. *Review of International Political Economy*, *28*(1), 103–127.

Ziakas, V. (2023). Leveraging sport events for tourism development: The event portfolio perspective. *Journal of Global Sport Management*, *8*(1), 43–72.

16 Digital Twins as New Paradigm for Bridging the Gap from Personalized Medicine to Specific Rural Public Health

Exploring Foundations, Legal Angels, and Technologies for Health 5.0—Future Research Directions

Bhupinder Singh

1 INTRODUCTION

The evolution of healthcare has witnessed transformative shifts, from generalized medical practices to personalized medicine. However, challenges persist, especially in rural areas where access to healthcare resources is limited. The concept of Digital Twins, borrowed from engineering and industrial sectors, offers a promising approach to address these challenges and usher in a new era of healthcare – Health 5.0. [1-3]

The evolving landscape of healthcare, characterized by a shift from generalized approaches to personalized medicine, has made substantial strides in tailoring treatments to individual patients. However, this progress has not been uniformly accessible across all demographics, particularly in rural areas where healthcare resources are often scarce. In this context, the emerging concept of Digital Twins presents a transformative paradigm that holds the potential to bridge the gap between the aspirations of personalized medicine and the unique challenges of rural public health. [4] Borrowing its essence from engineering and industrial domains, Digital Twins entail virtual replicas of physical entities, facilitating real-time monitoring, analysis, and simulation. Applying this concept to healthcare,

 DOI: 10.1201/9781032694375-16

Digital Twins have the capacity to revolutionize medical practices by creating intricate virtual replicas of patients or populations, amalgamating data from diverse sources to present a comprehensive and real-time portrayal of health status. By transcending geographical barriers, Digital Twins could pave the way for Health 5.0, an advanced stage of healthcare evolution, where not only personalized medicine but also specifically targeted interventions for rural populations become attainable. [5-7]

However, realizing this vision necessitates a multi-faceted exploration encompassing the foundational principles of Digital Twins, the intricate legal considerations surrounding data privacy and informed consent, the technological underpinnings essential for their implementation, and a forward-looking roadmap for future research that ensures their seamless integration into the fabric of healthcare systems. This chapter embarks on a comprehensive journey to elucidate the potential of Digital Twins in reshaping the healthcare landscape, underscoring the profound implications they hold for personalized medicine in rural settings and charting a course for the research directions that will be pivotal in harnessing their full transformative potential.

1.1 Significance of Research

The significance of embracing Digital Twins as a novel paradigm for bridging the gap between the realms of personalized medicine and the specific healthcare needs of rural populations cannot be overstated. [8-10] In an era where healthcare is rapidly advancing toward tailored treatments and interventions, Digital Twins offer a unique avenue to extend the benefits of personalized medicine to individuals in remote and underserved rural areas. By creating virtual replicas that encapsulate the entirety of an individual's health profile, encompassing medical history, genetic predispositions, lifestyle factors, and environmental influences, Digital Twins facilitate a profound understanding of each patient's unique health trajectory.

This understanding, in turn, empowers healthcare providers to design interventions that are not only effective but also contextually relevant to the challenges and limitations of rural healthcare settings. Moreover, the amalgamation of cutting-edge technologies such as the internet of things (IoT), artificial intelligence (AI), and wearable devices further bolsters the capabilities of Digital Twins, enabling real-time monitoring, early detection of health issues, and remote interventions, thereby circumventing geographical barriers and resource limitations. However, while the potential benefits are immense, the deployment of Digital Twins raises complex legal, ethical, and technological challenges. This necessitates a comprehensive exploration of the legal "angels" that guard patient data privacy and informed consent, while also delving into the foundational principles and intricate technological requirements that ensure the accuracy and reliability of Digital Twins. Looking forward, the research directions mapped out in this chapter are pivotal for guiding the trajectory of Digital Twins in healthcare. [11-15]

These directions encompass investigating the socioeconomic impacts of introducing such technology in rural areas, refining algorithms to predict long-term health trajectories accurately, and developing ethical frameworks that evolve alongside the dynamic nature of data consent. The synergy between these aspects promises not only to elevate the standard of healthcare in underserved regions but also to set the stage for Health 5.0, where the convergence of technology, personalization, and accessibility becomes the hallmark of a truly transformative healthcare ecosystem.

2 FOUNDATIONS OF DIGITAL TWINS IN HEALTHCARE

The foundational principles of Digital Twins in the healthcare domain signify a groundbreaking shift in how healthcare is conceptualized, delivered, and optimized. Drawing inspiration from engineering and industrial sectors, the concept of Digital Twins in healthcare revolves around creating virtual counterparts of individuals or populations, imitating their physiological, genetic, and environmental attributes. These virtual replicas serve as dynamic, data-rich representations that enable real-time monitoring, analysis, and simulation, fostering a holistic comprehension of health states. [16-18]

At the core of these digital counterparts lies data integration and interoperability. Healthcare is inherently multidimensional, with diverse data streams originating from electronic health records, genomics, wearable devices, patient-reported outcomes, and environmental sensors. The ability to harmoniously integrate these disparate data sources into a unified representation is a crucial foundation of Digital Twins. The resulting comprehensive health profiles capture not only a person's medical history but also their lifestyle choices, genetic predispositions, and environmental exposures, all of which contribute to a more nuanced understanding of health trajectories.

Digital Twins are empowered by the convergence of emerging technologies. The proliferation of the IoT allows for continuous real-time data collection from wearable devices, sensors, and monitoring equipment. [19-25] This data flow feeds into the Digital Twins, enabling a dynamic reflection of an individual's health status. AI and machine learning (ML) algorithms process this data, identifying intricate patterns, predicting potential health issues, and supporting clinicians in making data-driven decisions. These technologies collectively drive the predictive and prescriptive capabilities of Digital Twins, unlocking new avenues for proactive healthcare interventions.

In the context of healthcare, Digital Twins' foundations also extend to their ability to cater to individualized care. Each human body is unique, influenced by genetics, lifestyle choices, and environmental factors. Traditional healthcare often falls short in fully embracing this uniqueness due to generalized approaches. However, Digital Twins bridge this gap by enabling the creation of individualized models that accurately simulate an individual's physiological responses to various stimuli. These models, backed by real-time data, empower healthcare providers to customize treatments, predict adverse events, and optimize therapies for each patient.

The foundation of Digital Twins in healthcare is built upon the convergence of data integration, cutting-edge technologies, and the principle of personalization. By creating intricate digital replicas that encapsulate the complexity of human health, Digital Twins lay the groundwork for a new era of healthcare – one that is dynamic, tailored, and deeply informed by data-driven insights. As we delve further into this concept, it becomes evident that Digital Twins hold the potential to not only transform the clinical landscape but also bridge the disparities in healthcare accessibility, thereby reshaping the way healthcare is experienced across diverse populations, including rural areas.

2.1 Definition and Conceptual Framework: Digital Twins

Digital Twins are virtual representations of physical entities, enabling real-time monitoring, analysis, and simulation. In healthcare, Digital Twins replicate individual patients or populations, integrating data from various sources to provide a holistic view of health.

Digital Twins, originating from engineering and industrial domains, have rapidly emerged as a transformative concept in healthcare, offering a novel approach to understanding, monitoring, and optimizing human health. At its core, a Digital Twin is a virtual counterpart or replica of a physical entity, be it an individual, a machine, or even a complex system. In the realm of healthcare, Digital Twins manifest as sophisticated virtual models that capture the intricacies of a person's physiological, genetic, and environmental attributes. These models are not static representations; rather, they are dynamic simulations that continually evolve in response to real-time data inputs.

The conceptual framework of Digital Twins in healthcare revolves around their ability to create a real-time, data-rich mirror of an individual's health status. This involves aggregating data from diverse sources such as electronic health records, wearable devices, genomics databases, environmental sensors, and lifestyle inputs. By integrating this disparate data, a comprehensive and holistic portrait of an individual's health emerges. This holistic view extends beyond just medical history; it includes genetic predispositions, lifestyle choices, disease markers, medication responses, and even environmental exposures. Such a comprehensive dataset empowers healthcare providers with insights that transcend the boundaries of traditional medical records, facilitating a deeper understanding of health patterns and risk factors. The dynamic nature of Digital Twins sets them apart from static medical records. These virtual replicas continually update with real-time data, enabling them to adapt to changes in a person's health status, behavior, and environment. This aspect is particularly relevant in healthcare, where an individual's well-being is subject to an array of influences. By leveraging this dynamicity, Digital Twins can simulate various health scenarios, predict potential outcomes, and offer a glimpse into the effects of interventions before they are implemented.

In essence, the conceptual framework of Digital Twins in healthcare transcends the limitations of traditional medical approaches. It capitalizes on the power of data integration, emerging technologies like IoT and AI, and the philosophy of personalization to create a dynamic representation that serves as a bridge between the physical world and the digital realm. This framework underpins the transformative potential of Digital Twins in revolutionizing healthcare delivery, especially in bridging the gap between personalized medicine and the specific healthcare needs of rural populations. As we delve deeper into the exploration of Digital Twins' role in health, it becomes increasingly evident that their conceptual framework holds immense promise for reshaping healthcare practices and improving health outcomes on an unprecedented scale.

2.2 Data Integration and Interoperability

Creating accurate Digital Twins requires seamless integration of diverse data types, including medical records, genomics, lifestyle data, and environmental factors. Ensuring interoperability among different data sources is crucial to building comprehensive Digital Twins that accurately reflect the health status of individuals and communities. Digital Twins in healthcare demand the harmonious integration of a multitude of diverse data types, each contributing a unique facet to the holistic representation of an individual's health journey. The seamless amalgamation of these data sources, including medical records, genomics data, lifestyle inputs, and environmental factors, forms the backbone of the Digital Twin's accuracy and efficacy. Medical records furnish historical health information, diagnoses, and treatment regimens, offering a longitudinal view that anchors the Digital Twin in the individual's medical narrative. Genomic data, on the other hand, delves into the individual's genetic makeup, unlocking insights into predispositions to certain diseases, potential responses to medications, and avenues for personalized treatments.

Lifestyle data, which encompasses habits such as physical activity, dietary choices, sleep patterns, and stress levels, adds another layer of depth to the Digital Twin. These lifestyle inputs furnish critical context, shedding light on how individual choices interact with genetic predispositions to shape health outcomes. Additionally, environmental factors, such as air quality, climate conditions, and geographical location, contribute to the comprehensive understanding of health. These factors are especially pertinent in rural contexts, where environmental influences can significantly impact health disparities.

However, the mere integration of these diverse data types is not sufficient; ensuring their interoperability is equally imperative. In a healthcare ecosystem where data often resides in disparate silos, the ability to seamlessly exchange and interpret data across platforms becomes pivotal. Interoperability ensures that information flows freely, enabling the Digital Twin to synthesize insights that are as accurate and comprehensive as possible. Achieving this level of interoperability necessitates standardized data formats, robust data sharing protocols, and a commitment to data privacy and security.

The challenge lies in not only collecting and aggregating data but also in making it actionable. This requires advanced analytics and AI algorithms that can mine the data for patterns, correlations, and anomalies. The power of Digital Twins lies in their capacity to contextualize data, turning raw information into meaningful insights that can guide clinical decisions and interventions. Ultimately, the seamless integration of diverse data types and ensuring their interoperability serve as the linchpin of the Digital Twin's potential in healthcare. It transforms fragmented data points into a cohesive narrative, empowering healthcare professionals with a comprehensive understanding of individual health trajectories. This integration also forms the basis for predictive and prescriptive capabilities, allowing for proactive interventions and the optimization of treatment strategies. As we venture into an era where data-driven insights shape the future of healthcare, the significance of this seamless integration and interoperability cannot be overstated, particularly in the context of bridging healthcare disparities in rural areas through the power of Digital Twins.

3 LEGAL AND ETHICAL CONSIDERATIONS: IMPLEMENTING DIGITAL TWINS AND DATA SHARING

The implementation of Digital Twins in healthcare introduces a complex landscape of legal and ethical considerations, particularly in the realm of data sharing and privacy. As Digital Twins rely heavily on the integration of diverse and sensitive health data, ensuring the protection of individuals' privacy and the ethical use of their information becomes paramount. One of the central tenets of this challenge revolves around striking a delicate balance between the potential benefits of data sharing for improved healthcare outcomes and safeguarding individuals' rights to privacy and autonomy.

Data privacy in the context of Digital Twins entails safeguarding the vast array of personal information collected, including medical records, genetic profiles, lifestyle behaviors, and environmental exposures. Legal frameworks, such as the General Data Protection Regulation (GDPR) and the Health Insurance Portability and Accountability Act (HIPAA), provide guidelines for data protection, mandating transparent data handling practices, informed consent procedures, and secure data storage. As Digital Twins require continuous data updates and real-time monitoring, the challenge lies in ensuring that these legal mandates are upheld throughout the lifespan of the Digital Twin's use.

Ethical considerations extend beyond legal compliance, encompassing informed consent, data ownership, and data governance. With Digital Twins generating a dynamic and evolving representation of an individual's health, the concept of informed consent takes on a new dimension. Individuals must comprehend the extended nature of data usage and ongoing monitoring, and consent procedures need to reflect this reality. Additionally, issues of data ownership emerge – who owns the virtual replica of an individual's health? Striking a balance between the individual's rights and the collaborative nature of healthcare is an ethical tightrope that requires careful deliberation.

Besides, the challenge of data governance arises as data from various sources are integrated into Digital Twins. Defining clear roles and responsibilities for data custodians, ensuring data accuracy, and safeguarding against biases and disparities in data inputs are critical ethical considerations. Transparency in data usage and sharing is vital to maintain public trust and ensure that individuals understand how their data contributes to the Digital Twin's insights. Implementing effective legal and ethical frameworks for Digital Twins in healthcare requires collaboration between policymakers, healthcare providers, data scientists, and individuals. It necessitates an ongoing dialogue about the rights and responsibilities of all stakeholders involved. As the potential of Digital Twins to revolutionize healthcare is explored, addressing these legal and ethical considerations is not just a regulatory obligation but a moral imperative. Only through a well-balanced approach that upholds privacy, autonomy, and data ethics can the full potential of Digital Twins in transforming healthcare be realized, especially in the context of bridging healthcare disparities in rural regions.

3.1 Privacy and Data Security: Striking a Balance between Data sharing for Health Improvement and Protecting Individuals' Sensitive Information

The implementation of Digital Twins raises concerns about patient privacy and data security. Striking a balance between data sharing for health improvement and protecting individuals' sensitive information necessitates robust legal frameworks and encryption protocols. The integration of Digital Twins in healthcare introduces a new dimension of privacy and data security challenges, necessitating a delicate equilibrium between harnessing the potential of data sharing for health improvement and safeguarding the confidentiality of individuals' sensitive information. As Digital Twins rely on aggregating a vast spectrum of personal data, including medical histories, genomic profiles, lifestyle behaviors, and environmental exposures, the potential benefits of improved healthcare outcomes must be weighed against the potential risks of unauthorized access, misuse, or breaches.

The privacy concerns emerge due to the highly personal nature of the data integrated into Digital Twins. The multifaceted portrait of an individual's health journey, coupled with the continuous nature of data collection, heightens the risk of data being used in unintended ways or falling into the wrong hands. Moreover, the amalgamation of disparate data sources may lead to the identification of individuals even when data is supposedly anonymized, raising concerns about re-identification and potential breaches of privacy.

Data security becomes a paramount consideration in ensuring the integrity and confidentiality of the information stored within Digital Twins. Robust encryption protocols, secure storage mechanisms, and stringent access controls are imperative to prevent unauthorized access, tampering, or breaches. The highly dynamic nature of Digital Twins necessitates a robust cybersecurity infrastructure that can adapt to real-time data updates while maintaining stringent security standards.

Striking the delicate balance between data sharing for health improvement and privacy protection requires comprehensive legal frameworks and ethical guidelines. Regulatory measures like GDPR and HIPAA set standards for data protection, outlining the responsibilities of data custodians, requirements for informed consent, and procedures for data breaches. Ethical considerations revolve around transparency – individuals must be fully aware of how their data is used, the benefits it offers, and the potential risks involved.

The educational initiatives that inform individuals about their rights and the value of their data in advancing healthcare can foster a sense of trust in the Digital Twin ecosystem. Additionally, technologies like differential privacy and federated learning hold promise in minimizing privacy risks by allowing data analysis without direct access to raw data. Blockchain technology also offers a potential avenue for secure and transparent data sharing, ensuring data integrity and traceability.

As the implementation of Digital Twins progresses, interdisciplinary collaboration between technology experts, legal professionals, ethicists, and healthcare practitioners becomes indispensable. The challenge lies in designing an ecosystem that maximizes the potential of Digital Twins for health advancement while safeguarding individuals' privacy rights. This requires an ongoing commitment to refine and adapt legal frameworks, technology standards, and ethical guidelines to accommodate the evolving landscape of healthcare data and ensure that the promise of Digital Twins is realized without compromising the trust and privacy of individuals, including those in rural populations.

3.2 Informed Consent: The Concept under Digital Twins and Data Sharing

As Digital Twins involve continuous data collection and analysis, redefining the concept of informed consent becomes essential. Patients must understand the extended nature of data usage and provide ongoing consent, posing challenges in maintaining ethical standards while enabling progress.

The inherent nature of Digital Twins in healthcare necessitates a paradigm shift in the traditional concept of informed consent. Unlike one-time, static consent models that are prevalent in healthcare, Digital Twins involve continuous data collection, analysis, and real-time monitoring. This evolution in the data lifecycle demands a redefinition of informed consent that aligns with the ongoing, dynamic nature of data usage. Patients must not only comprehend the initial purposes for which their data is collected but also understand the extended scope of its usage throughout the Digital Twin's lifespan. One of the primary challenges lies in striking a delicate balance between the ethical imperative to protect patients' autonomy and privacy and the imperative to leverage their data for advancing healthcare. Patients must be informed about the implications of continuous data collection, including the potential benefits of more accurate diagnoses, tailored treatment plans, and proactive health management. However, they must also be made aware of the potential risks, such as the risk of re-identification, unauthorized data access, or potential misuse of their information.

The fluidity of data usage in Digital Twins presents complexities in obtaining ongoing informed consent. Patients must be given the agency to revoke their consent at any point without repercussions, which poses technical challenges in managing data flows and ensuring the erasure of data that has already been used in analyses. Ensuring that patients have a clear understanding of the implications of both providing and revoking consent is crucial to upholding ethical standards.

As considering the dynamic and evolving nature of health data, patients must also be informed about the possibility of unforeseen uses and collaborations. The data they contribute to the Digital Twin might lead to insights that extend beyond immediate healthcare, such as contributing to research studies or informing public health initiatives. This necessitates a level of transparency that empowers patients to make informed decisions about the uses of their data.

So, addressing these challenges calls for a multidisciplinary approach involving healthcare providers, legal experts, ethicists, and technology professionals. Developing clear and transparent communication strategies that explain the complexities of Digital Twins, the ongoing nature of data usage, and the potential benefits and risks is essential. It requires fostering a climate of trust, wherein patients are actively engaged in the decision-making process regarding their data.

As Digital Twins redefine the boundaries of data usage and informed consent, it becomes paramount to navigate these ethical nuances responsibly. Balancing the progress that continuous data collection can bring to healthcare with the ethical obligation to protect patients' autonomy and privacy is a pivotal step toward ensuring that the implementation of Digital Twins remains ethically sound and respectful of individuals' rights, especially as they apply to personalized healthcare in rural areas.

4 TECHNOLOGIES ENABLING HEALTH 5.0

The concept of Health 5.0 embodies a visionary shift in healthcare, propelled by a convergence of cutting-edge technologies that hold the potential to revolutionize the way healthcare is delivered and experienced. At the heart of this transformative landscape are a trio of technologies: the IoT, AI, and telemedicine, each playing a pivotal role in enabling Health 5.0.

The IoT stands as a foundational pillar of Health 5.0, facilitating the seamless integration of interconnected devices and sensors into the healthcare ecosystem. These devices, ranging from wearable fitness trackers to implantable medical sensors, continuously collect and transmit real-time health data, enabling a holistic and remote monitoring of patients' vital signs, physical activities, and even medication adherence. In rural contexts, where access to healthcare facilities might be limited, IoT-powered devices bridge the geographical gap, offering a lifeline for patients by providing timely insights to healthcare providers and enabling early intervention in critical situations.

So, complementing the IoT, AI emerges as a transformative force that leverages data analytics and ML algorithms to extract meaningful insights from the massive troves of healthcare data generated by IoT devices. AI not only identifies intricate patterns and correlations but also predicts health trends, aiding in early disease detection and personalized treatment recommendations. The integration of AI into Digital Twins enhances their diagnostic accuracy and predictive capabilities, ultimately empowering healthcare providers with data-driven intelligence that informs their clinical decisions. This is especially relevant in rural healthcare settings, where the scarcity of resources demands precision in diagnosis and treatment planning.

Telemedicine, the third cornerstone of Health 5.0, empowers patients to access healthcare services remotely, transcending geographical boundaries. Enabled by technology, telemedicine platforms facilitate virtual consultations between patients and healthcare professionals, thereby eliminating the need for patients to travel long distances for routine check-ups or consultations. For rural populations with limited access to medical facilities, telemedicine offers a lifeline, enabling them to receive expert medical guidance without the logistical challenges associated with physical visits. The integration of Digital Twins and telemedicine further enhances this landscape, as healthcare providers can leverage real-time data from Digital Twins to offer tailored advice and interventions during virtual consultations.

In essence, the amalgamation of IoT, AI, and telemedicine forms the bedrock of Health 5.0, a landscape where healthcare is not confined to physical clinics but extends into individuals' daily lives. These technologies converge to create a healthcare ecosystem that is personalized, proactive, and accessible irrespective of geographical constraints. While the integration of these technologies undoubtedly presents challenges, such as data security and regulatory compliance, their potential to transform healthcare delivery is profound. As the trajectory of Health 5.0 unfolds, its impact on bridging the healthcare gap for rural populations and advancing personalized medicine becomes a beacon of hope in an increasingly interconnected world.

4.1 Internet of Things and Wearable Devices: Contribute to Real-Time Data collection for Digital Twins

IoT devices and wearables contribute to real-time data collection for Digital Twins. These technologies enable remote monitoring of vital signs, physical activity, and disease progression, particularly beneficial for rural populations with limited access to healthcare facilities. The transformative power of IoT devices and wearables in the context of Digital Twins cannot be overstated. These technologies offer a dynamic and real-time data stream that feeds into the creation and maintenance of Digital Twins, unlocking a multitude of possibilities, especially for rural populations facing limited access to traditional healthcare facilities.

IoT devices and wearables serve as data aggregators that continuously capture an array of vital health metrics. From heart rate and blood pressure to sleep patterns and physical activity levels, these devices offer a comprehensive picture of an individual's well-being. For rural populations, often facing challenges in accessing medical care due to geographical barriers or resource limitations, IoT devices and wearables act as virtual healthcare companions, constantly monitoring health parameters that are crucial for timely intervention.

One of the most significant advantages of IoT-enabled data collection is its real-time nature. The continuous transmission of data to Digital Twins allows for early detection of anomalies and changes in health status. This proactive approach is especially vital in rural areas, where the scarcity of healthcare facilities might result in delayed diagnosis and intervention. With IoT devices and wearables, healthcare providers can receive timely alerts about deviations from the norm, enabling them to initiate interventions promptly and prevent potential health complications. Moreover, IoT-enabled remote monitoring extends beyond merely tracking vital signs. These devices can also monitor disease progression, medication adherence, and rehabilitation progress. For rural populations with limited access to specialized care, such capabilities provide a lifeline by enabling healthcare providers to remotely manage chronic conditions and post-surgery recovery, ensuring that patients receive the attention they need without having to travel long distances.

While the integration of IoT devices and wearables with Digital Twins offers remarkable opportunities, it also raises questions about data security, privacy, and data accuracy. Ensuring that the data collected is accurate, securely transmitted, and ethically used is imperative. As these technologies continue to evolve, addressing these challenges becomes crucial to unlocking their full potential in bridging the healthcare gap for rural populations. In essence, IoT devices and wearables emerge as the bridge between rural patients and the transformative potential of Digital Twins, offering real-time insights, personalized interventions, and proactive healthcare management, regardless of geographical constraints.

4.2 Artificial Intelligence and Machine Learning: Analyzing Healthcare Data, Identifying Patterns, and Predicting Health Outcomes Integrating with Digital Twins

AI and ML algorithms analyze complex healthcare data, identifying patterns and predicting health outcomes. Integrating these technologies with Digital Twins enhances diagnostic accuracy and assists healthcare providers in making informed decisions. The application of AI and ML in healthcare, particularly in conjunction with Digital Twins, ushers in a new era of data-driven insights and precision medicine. AI and ML bring to the forefront the ability to analyze vast and complex healthcare datasets, identifying intricate patterns and predicting health outcomes that might remain concealed through traditional analysis methods. Integrating AI and ML with Digital Twins amplifies their power, offering a synergy that has the potential to transform healthcare delivery and patient outcomes.

The crux of AI and ML lies in their capacity to process immense amounts of data and discern subtle correlations that might not be immediately apparent to human observers. By leveraging AI algorithms, healthcare data integrated into Digital Twins can be subjected to advanced analyses that uncover hidden relationships between various factors. This goes beyond the traditional diagnostic approach, allowing healthcare providers to understand the nuanced interactions between genetics, lifestyle choices, environmental exposures, and health outcomes. For rural populations, who often face unique health challenges stemming from a variety of factors, this level of granularity is pivotal in delivering tailored and effective interventions. Predictive modeling stands as one of the most promising aspects of AI and ML integration with Digital Twins. These technologies can forecast disease trends, potential health complications, and individual response to treatments. The capability to predict health outcomes based on personalized data empowers healthcare providers to take a proactive stance, intervening before issues escalate and customizing treatments that are most likely to succeed. In rural healthcare contexts where resources are scarce, predictive models inform resource allocation, ensuring that interventions are targeted and effective.

The integration of AI and ML with Digital Twins fosters a symbiotic relationship. AI algorithms thrive on abundant and diverse data, which Digital Twins provide, and Digital Twins benefit from AI's analytical prowess, as it transforms raw data into actionable insights. However, this integration also requires addressing challenges such as data quality, algorithm transparency, and ethical considerations. Ensuring that AI and ML models are trained on representative and unbiased datasets is crucial to avoid perpetuating disparities.

As AI and ML continue to evolve, they hold immense potential to reshape healthcare in rural areas, offering a level of personalization and precision that was once thought unattainable. The synergy between AI, ML, and Digital Twins brings us closer to a future where healthcare decisions are driven by data-driven intelligence, enabling healthcare providers to make informed choices that have a direct impact on patients' health and well-being. Through the integration of these technologies, the promise of Health 5.0 becomes tangible, paving the way for a healthcare landscape that bridges the gap between personalized medicine and rural public health.

4.3 Telemedicine and Remote Interventions

Digital Twins can facilitate telemedicine by simulating patient conditions for remote consultations. Additionally, they support the design and testing of personalized treatment plans, allowing for virtual interventions in rural areas. Telemedicine, coupled with the integration of Digital Twins, emerges as a beacon of hope in revolutionizing healthcare access and interventions, especially in rural areas. Telemedicine harnesses the capabilities of digital communication technologies to bridge the geographical divide between patients and healthcare providers, enabling virtual consultations and remote interventions. In regions where physical access to medical facilities is limited, telemedicine serves as a transformative lifeline, allowing patients to connect with medical experts and receive expert guidance without the constraints of distance.

The combination of Digital Twins further elevates the potential of telemedicine by offering a comprehensive and real-time insight into patients' health conditions. Healthcare providers can leverage the dynamic data captured by Digital Twins to create a detailed understanding of patients' health trajectories, even before the virtual consultation begins. This data-driven approach empowers healthcare professionals to offer personalized advice, interventions, and treatment recommendations that are tailored to each patient's unique health profile.

Telemedicine also plays a pivotal role in post-treatment care and follow-up, particularly after surgeries or chronic disease management. Digital Twins can simulate an individual's recovery journey and predict potential complications, empowering healthcare providers to offer guidance that optimizes healing and minimizes risks. Rural populations, who often face challenges in accessing specialized care after medical procedures, benefit significantly from the continuous monitoring and remote interventions facilitated by telemedicine and Digital Twins. However, challenges persist, ranging from ensuring the reliability of telecommunication infrastructure in remote areas to addressing concerns about data security and privacy during virtual consultations. Cultural and technological factors also influence the adoption of telemedicine, necessitating comprehensive approaches that encompass patient education, training for healthcare professionals, and the creation of supportive policies.

The synergy between telemedicine and Digital Twins epitomizes the essence of Health 5.0—a healthcare landscape that transcends physical boundaries and ensures equitable access to quality healthcare services. By allowing healthcare providers to remotely intervene based on real-time data insights, this integration transforms the healthcare experience, providing timely interventions, personalized care, and vital medical expertise to individuals in rural areas, ultimately narrowing the healthcare gap and realizing the vision of Health 5.0.

5 BRIDGING THE GAP: DIGITAL TWINS IN RURAL PUBLIC HEALTH

Digital Twins in rural public health practices heralds a transformative era that holds the promise of addressing long-standing disparities in healthcare access and outcomes. Rural populations often face unique challenges such as limited access to healthcare facilities, healthcare provider shortages, and a lack of specialized medical services. These challenges contribute to health disparities, where rural communities experience higher rates of chronic diseases, delayed diagnoses, and reduced access to preventive care. Here, Digital Twins emerge as a powerful solution, bridging this gap by offering personalized, data-driven healthcare interventions that transcend geographical constraints.

Digital Twins have the capability to recreate a virtual counterpart of an individual's health status, capturing their medical history, genetic predispositions, lifestyle choices, and environmental influences. This comprehensive representation empowers healthcare providers to remotely monitor health trajectories, make informed clinical decisions, and design interventions that are not only effective but also tailored to the specific needs and circumstances of rural populations. For instance, Digital Twins could predict the progression of chronic diseases, offer recommendations for lifestyle modifications, and even simulate the effects of potential treatment strategies, all while the patient remains in their rural community.

In rural areas, where physical access to healthcare facilities might be limited, Digital Twins play a crucial role in early detection and proactive management of health issues. By continuously collecting real-time data from wearable devices, IoT sensors, and medical records, Digital Twins can alert healthcare providers to deviations from the norm, allowing for timely interventions that prevent minor health concerns from escalating into critical conditions. This proactive approach has the potential to revolutionize rural public health by reducing hospitalizations, optimizing resource utilization, and ultimately improving the overall quality of life for rural populations. The implementation of Digital Twins in rural public health is not without challenges. Issues such as data privacy, technological infrastructure, patient education, and cultural considerations must be addressed to ensure that the benefits of this technology are equitably distributed. Establishing robust data protection protocols, ensuring reliable internet connectivity in remote areas, and providing education about the value and implications of Digital Twins are essential steps toward fostering trust and acceptance among rural communities. Digital Twins have the transformative potential to bridge the healthcare gap in rural public health. By harnessing the capabilities of data integration, AI, telemedicine, and personalized medicine, Digital Twins empower rural populations with access to cutting-edge healthcare solutions that were once limited to urban centers. As the synergy between technological innovation and healthcare evolves, Digital Twins stand as a beacon of hope, reshaping the landscape of rural public health and realizing the vision of a healthcare ecosystem that truly leaves no one behind.

5.1 Personalized Health Care Delivery

Digital Twins can tailor interventions to individual patients' needs, accounting for genetic predispositions, lifestyle, and environmental factors. This capability is particularly valuable in rural areas where resources are limited, and personalized care can be a challenge. The capacity of Digital Twins to customize healthcare interventions based on a multitude of factors, ranging from genetic predispositions to lifestyle choices and environmental influences, represents a paradigm shift in healthcare delivery. This ability to holistically understand an individual's health profile and tailor interventions accordingly holds immense potential, especially in rural areas where accessing personalized care is often hindered by limited resources and geographical barriers.

In rural communities, where healthcare facilities might be sparse and specialized medical services scarce, Digital Twins emerge as a beacon of hope, offering a way to transcend these limitations. By analyzing a person's genetic makeup, Digital Twins can predict the likelihood of certain health conditions and tailor preventive strategies accordingly. Moreover, by incorporating data from wearable devices that track physical activity, sleep patterns, and other lifestyle choices, Digital Twins can provide insights into how these behaviors interact with genetic predispositions to influence health outcomes. This real-time data collection empowers healthcare providers to make informed decisions about treatment plans and lifestyle modifications, addressing health concerns before they escalate into critical conditions.

Environmental factors also play a significant role in health outcomes, and Digital Twins are uniquely positioned to consider these influences. Factors such as air quality, access to nutritious food, and exposure to pollutants can greatly impact health in rural areas. By incorporating data from environmental sensors, Digital Twins can offer tailored recommendations that account for these contextual factors, ensuring that interventions are not only personalized but also relevant to the specific challenges faced by individuals in rural communities.

The integration of Digital Twins into rural healthcare can lead to a shift from reactive care to proactive and preventive healthcare strategies. By leveraging the vast amount of data available, Digital Twins allow healthcare providers to anticipate health issues, design targeted interventions, and optimize resource allocation. This not only improves the quality of care received by individuals but also contributes to the efficient utilization of healthcare resources, a particularly crucial aspect in resource-constrained rural areas.

However, successful implementation of Digital Twins in rural healthcare requires a comprehensive approach that addresses technological infrastructure, data security, cultural considerations, and patient education. Ensuring that individuals understand the benefits and implications of Digital Twins, while safeguarding their privacy and data rights, is paramount to building trust in these innovative healthcare solutions.

Digital Twins possess the ability to transcend the limitations of traditional healthcare delivery, offering a personalized, data-driven approach that is uniquely suited to the needs of individuals in rural areas. As the synergy between technology and healthcare advances, the promise of tailored interventions driven by Digital Twins brings hope for a future where healthcare is not only accessible to all but also tailored to each person's unique health journey.

5.2 Targeted Public Health Initiatives

Aggregated data from Digital Twins can inform public health strategies in rural communities. Identifying prevalent health issues, tracking disease spread, and optimizing resource allocation become more effective with real-time, localized insights. Aggregated data harvested from Digital Twins constitutes a goldmine of real-time, localized insights that have the potential to revolutionize public health strategies in rural communities. By drawing upon the comprehensive and dynamic health profiles generated by Digital Twins, public health officials and policymakers can formulate targeted and effective interventions that address the specific needs and challenges faced by rural populations.

One of the most impactful applications of aggregated data from Digital Twins lies in identifying prevalent health issues within rural communities. By analyzing patterns and trends derived from this data, public health professionals can discern the major health concerns that affect the local population. This awareness enables them to prioritize their efforts and allocate resources toward addressing the most pressing health challenges, such as chronic diseases, infectious outbreaks, or environmental health hazards. This targeted approach is especially crucial in rural areas, where resources are often limited and must be optimized to have the greatest impact.

The ability to track the spread of diseases in real-time through aggregated Digital Twin data is a game-changer for public health surveillance. Rapid identification of disease outbreaks allows for swift responses, containment measures, and the prevention of potential epidemics. This early warning system can significantly reduce the burden on rural healthcare systems and safeguard the health of vulnerable populations that might have limited access to medical facilities.

Optimizing resource allocation is another pivotal benefit of utilizing aggregated Digital Twin data for public health strategies. With insights derived from this data, officials can allocate healthcare resources such as medical personnel, medical supplies, and healthcare facilities more effectively, ensuring that healthcare services are distributed in accordance with the needs of specific communities. This approach helps avoid disparities in access to care and enhances overall healthcare quality in rural areas. Alongside these potential benefits, careful considerations must be made regarding data privacy, security, and ethical concerns. It is imperative to uphold individuals' rights and ensure that data is anonymized and protected in compliance with relevant regulations. Engaging with local communities, building trust, and transparently communicating the intentions and outcomes of using aggregated Digital Twin data are vital steps to gaining acceptance and cooperation.

The aggregated data gleaned from Digital Twins holds transformative potential in reshaping public health strategies in rural communities. By harnessing real-time, localized insights, public health professionals can identify health issues, track disease spread, and allocate resources with precision, ultimately leading to more effective interventions, improved health outcomes, and a more equitable distribution of healthcare services for rural populations.

6 DIGITAL TWINS IN RURAL PUBLIC HEALTH: SOCIOECONOMIC IMPLICATIONS

The emergence of digital twin technology has ushered in a new era of innovation and transformation across various sectors, and its application in rural public health holds the promise of revolutionary advancements. A digital twin refers to a virtual replica of a physical object, process, or system, synchronized in real-time to reflect its real-world counterpart's behavior, conditions, and changes. In the context of rural public health, digital twins offer a powerful tool for monitoring, analyzing, and improving the health and well-being of underserved populations in remote areas.

Rural public health systems often face unique challenges, including limited access to quality healthcare, scarce resources, and inadequate infrastructure. Digital twins can address these challenges by facilitating remote monitoring of health conditions, predicting disease outbreaks, and optimizing resource allocation. Through the integration of IoT devices, sensors, and data analytics, digital twins can provide real-time data on various health parameters, enabling healthcare providers and policymakers to make informed decisions.

Socioeconomic implications play a pivotal role in the adoption and implementation of digital twins in rural public health. Firstly, improved health monitoring and management facilitated by digital twins can lead to a healthier rural population. This, in turn, can enhance the productivity of the workforce, reduce absenteeism, and alleviate the economic burden on families caused by healthcare expenses. Moreover, healthier individuals are more likely to participate actively in local economies, fostering community development and growth.

Secondly, digital twins can bridge the gap in healthcare disparities between rural and urban areas. By facilitating telemedicine and remote consultations, digital twins enable rural residents to access specialized medical expertise without the need for extensive travel. This not only improves health outcomes but also reduces the economic strain associated with seeking medical care in distant urban centers. As a result, rural communities can experience increased access to healthcare services, leading to an overall enhancement in the quality of life.

The implementation of digital twins requires technological infrastructure and expertise, generating opportunities for skill development and job creation in rural areas. As digital twin ecosystems are established, there is a demand for professionals specializing in data analytics, IoT technology, and system integration. This can reverse the trend of rural-to-urban migration by offering meaningful employment prospects within the community. The development of local expertise can foster innovation and sustainable technology adoption, contributing to the overall growth of the region.

However, the socioeconomic implications of digital twins in rural public health also come with challenges that require careful consideration. Privacy and data security issues must be addressed to ensure that sensitive health information is safeguarded. Additionally, the initial investment required for establishing the necessary technological infrastructure might be a barrier, necessitating collaboration between governments, non-profit organizations, and private sector entities. It has the potential to revolutionize healthcare access, empower rural communities, and stimulate economic growth. By leveraging the power of digital twins, rural public health systems can transition from reactive care models to proactive, data-driven approaches that prioritize prevention, early intervention, and overall well-being.

6.1 Socioeconomic Impacts

Exploring the socioeconomic implications of Digital Twins in rural public health is crucial. This includes assessing access to technology, acceptance by communities, and potential disparities in health outcomes. Examining the socioeconomic implications of integrating Digital Twins into rural public health is a multifaceted endeavor that requires a comprehensive evaluation of various factors. Some important concerns which deeper into these critical aspects:

Access to Technology: One of the primary considerations when implementing Digital Twins in rural public health is the accessibility of technology. Rural areas often face challenges in terms of internet connectivity, availability of devices, and digital literacy. While Digital Twins have the potential to revolutionize healthcare monitoring and delivery, their effectiveness is contingent on the population's ability to access and use the required technology. Addressing this gap is essential to ensure equitable healthcare advancements.

Community Acceptance and Engagement: Successfully implementing Digital Twins involves gaining the trust and cooperation of rural communities. This requires effective communication about the benefits and functionalities of the technology. Community members need to understand how Digital Twins can enhance their healthcare experiences and improve outcomes. Engaging local leaders, healthcare providers, and community organizations in the implementation process can foster acceptance and create a sense of ownership, resulting in more sustainable and effective outcomes.

Health Outcome Disparities: Socioeconomic disparities can impact the benefits derived from Digital Twins in rural public health. Lower-income individuals may face challenges in affording the necessary devices or data plans for remote monitoring. Moreover, cultural beliefs and social factors might influence the acceptance and utilization of technology for healthcare purposes. To prevent exacerbating existing health disparities, strategies must be in place to ensure that vulnerable populations have equal access to the advantages offered by Digital Twins.

Health Literacy and Training: Digital Twins introduce a degree of complexity that may require individuals to acquire new skills and knowledge. Promoting health literacy and providing training on using Digital Twins effectively can empower rural residents to take control of their health and make informed decisions. Training programs can be designed to cater to various skill levels, ensuring that individuals with varying degrees of digital proficiency can benefit equally.

Data Privacy and Security: The collection and transmission of health data through Digital Twins raise privacy and security concerns. Rural populations might be especially cautious about sharing personal health information digitally. Addressing these concerns through robust data protection measures, transparent data usage policies, and adherence to relevant regulations is essential to gain community trust and ensure that sensitive health information remains confidential.

Economic Implications: While Digital Twins offer significant potential for improving health outcomes, they also bring economic considerations. As the technology evolves, there might be initial costs associated with implementation, maintenance, and upgrading. Governments, nonprofits, and private sector partners must collaborate to ensure that the financial burden doesn't fall disproportionately on rural communities and that the benefits outweigh the costs in the long run.

Capacity Building and Job Opportunities: Integrating Digital Twins can open up opportunities for local capacity building and job creation. Training community members to manage, maintain, and support the technology can create a new skill set and employment prospects within the rural areas themselves. This not only contributes to local economic growth but also strengthens the sustainability of Digital Twin initiatives.

Exploring the socioeconomic implications of Digital Twins in rural public health requires a holistic understanding of access, acceptance, disparities, literacy, privacy, and economics. A well-rounded approach that involves collaboration between technology providers, healthcare experts, community leaders, and policymakers is crucial to maximize the positive impact of Digital Twins on rural health outcomes while mitigating potential challenges.

6.2 Long-Term Health Predictions

This endeavor carries immense significance as it holds the potential to revolutionize healthcare by enabling proactive interventions and resource planning, especially for the prevalence of chronic diseases within rural areas. Chronic diseases, often exacerbated by limited access to healthcare and socioeconomic challenges, disproportionately affect rural populations. The development of algorithms that can predict the progression of these conditions can be a game-changer in improving the overall health outcomes and well-being of individuals living in remote areas.

By analyzing vast datasets encompassing health records, genetic information, environmental factors, and lifestyle habits, researchers can uncover hidden patterns and correlations that are beyond the grasp of conventional methods. These refined algorithms have the potential to not only identify risk factors associated with chronic diseases but also predict the trajectory of these illnesses over time. This predictive capability is invaluable in a healthcare context, where timely interventions are pivotal in mitigating the impact of chronic conditions and preventing their escalation into more severe stages.

In rural areas, where healthcare resources might be scarce, and medical expertise might be limited, the ability to foresee long-term health trajectories offers a lifeline. By identifying individuals at risk of developing chronic diseases or those whose conditions might worsen over time, healthcare providers can tailor interventions that are both proactive and personalized. This might involve implementing lifestyle changes, adjusting medication regimens, or providing targeted educational programs to empower individuals in managing their health effectively. Moreover, these predictive algorithms can guide resource planning, ensuring that healthcare facilities and services are equipped to address the specific needs of the population they serve.

A key advantage of refining algorithms for long-term health prediction is their adaptability to local contexts and the nuances of rural communities. Rural areas often have distinct health challenges arising from factors such as geography, socio-economic conditions, and cultural practices. Algorithms can be fine-tuned to incorporate these factors, resulting in predictions that are not only accurate but also contextually relevant. This approach enhances the likelihood of successful interventions, as they align more closely with the realities and needs of the population.

Ensuring the quality and diversity of input data is crucial, as biased or incomplete datasets can lead to inaccurate predictions and exacerbate health disparities. Ethical concerns surrounding data privacy, informed consent, and algorithm transparency must be rigorously addressed to uphold the rights and autonomy of individuals contributing to these datasets. Additionally, the implementation of predictive algorithms necessitates close collaboration between healthcare professionals, data scientists, and policymakers to ensure that the predictions translate into tangible improvements in healthcare delivery and outcomes.

The ability to foresee chronic disease progression, coupled with proactive interventions and resource planning, can significantly enhance the well-being of rural populations. As technology advances and interdisciplinary collaboration continues, these refined algorithms have the potential to reshape the landscape of rural healthcare, making it more accessible, effective, and responsive to the unique needs of these communities.

6.3 Ethical Frameworks for Consent

In the era of Digital Twins, where technology intertwines with personal health data, the evolution of consent frameworks is of paramount importance. As these virtual replicas become increasingly integrated into healthcare, ethical considerations regarding informed consent must adapt to ensure the protection of individual autonomy and privacy. The traditional notion of a one-time, static consent may fall short in capturing the dynamic nature of Digital Twins and their continuous interaction with personal health information. Thus, the development of robust and agile ethical frameworks that address the evolving nature of consent is not just important but imperative.

As Digital Twins constantly collect, analyze, and refine data to model an individual's health trajectory, the consent process should mirror this dynamic relationship. Individuals must be engaged in a continuous dialogue where they are regularly updated on how their data is being utilized and for what purposes. Transparent communication is crucial, empowering individuals to make informed decisions about their participation in Digital Twin initiatives. This ongoing consent model respects the autonomy of individuals by acknowledging their right to withdraw or modify their consent as new information emerges or circumstances change.

An essential element of evolving consent frameworks is granular control over data sharing and usage. [26] Individuals should have the agency to specify which aspects of their health data can be accessed by healthcare providers, researchers, and other relevant stakeholders. This approach respects individual preferences and mitigates concerns related to data privacy. [27] Additionally, research can provide insights into the technical mechanisms required to implement granular consent effectively, such as encryption, secure data storage, and robust access controls. Ethical frameworks should address the potential power imbalances between individuals and those implementing Digital Twins. [28] Clear guidelines must be established to prevent exploitation and ensure that individuals are not coerced into consenting to data usage. This is particularly important in rural areas where trust in healthcare systems and technology might vary. Research can offer strategies to create equitable partnerships between individuals, healthcare providers, researchers, and technology developers, fostering an environment where all stakeholders collaborate with mutual respect and shared goals. [29]

The evolution of consent frameworks to align with the dynamic nature of Digital Twins is a critical ethical endeavor. Research plays a pivotal role in providing guidance on how to navigate this evolution while upholding individual autonomy and privacy. As Digital Twins become increasingly embedded in healthcare, ethical considerations should adapt to ensure that individuals remain at the center of decision-making regarding their personal health data. By fostering ongoing, informed consent, we can strike a balance between technological innovation and ethical responsibility, ultimately benefiting both individuals and society as a whole.

7 CONCLUSION AND FUTURE SCOPE

The emergence of Digital Twins represents a profound shift from personalized medicine toward a new frontier known as Health 5.0. This paradigm not only holds the potential to bridge the gap between individualized care and the specific health needs of rural communities but also underscores the importance of exploring the foundational principles, [30] legal considerations, and technological advancements that will shape its trajectory. [31] As we stand at the precipice of this transformative era, it is evident that embracing Digital Twins in rural public health is not just an innovation but also a call to redefine healthcare systems [32] to be more proactive, equitable, and community-centric.

The foundation of this paradigm shift lies in recognizing the unique challenges and opportunities presented by rural public health. By leveraging the capabilities of Digital Twins, we can transcend geographical barriers and limited access to care, thereby facilitating [33] proactive interventions and predictive healthcare strategies. This involves a comprehensive understanding of the socioeconomic, cultural, and infrastructural landscape of rural areas, and the adaptation of digital twin technologies to align seamlessly with these contexts.

Navigating the legal and ethical dimensions is equally critical. [34] As Digital Twins blur the lines between the physical and virtual, legal frameworks must evolve to safeguard individual privacy, data security, and consent in this interconnected landscape. [35-37] Innovations should be paired with responsible data governance, transparent communication, and a commitment to ensuring that individuals' rights are upheld throughout their health journey.

The technological advancements will be the driving force behind the success of Health 5.0. Continued research into refining algorithms, securing data, and maintaining ongoing, [38] informed consent mechanisms are vital components of this trajectory.

Furthermore, a collaborative approach involving researchers, policymakers, healthcare providers, and technology experts will be instrumental in shaping the development and implementation of Digital Twins that truly resonate with the specific needs of rural public health.

As looking toward the future of Health 5.0, there are exciting avenues for further research. Exploring the fusion of Digital Twins with emerging technologies like blockchain, AI-driven diagnostics, and telemedicine can unlock new frontiers in healthcare delivery. Additionally, investigating the scalability and sustainability of Digital Twin solutions in resource-constrained rural areas will be essential to ensure their long-term impact.

In conclusion, the journey from personalized medicine to Health 5.0 through Digital Twins embodies a transformation that transcends technological innovation to become a testament to our commitment to inclusive, holistic, and data-driven healthcare. [39] By embracing this new paradigm, grounded in robust foundations, ethical principles, and cutting-edge technologies, we embark on a path toward a future where rural communities are empowered with the tools and insights to achieve better health outcomes. This journey, undoubtedly challenging yet incredibly promising, beckons us to embark on further research that will drive this evolution forward, shaping the contours of healthcare in ways that have the potential to touch lives across the globe. [40]

The future scope of Digital Twins as a new paradigm for bridging the gap from personalized medicine to specific rural public health holds vast potential and exciting possibilities. As technology continues to advance, the trajectory of Health 5.0 will undoubtedly be shaped by the ongoing exploration of foundational principles, legal considerations, and cutting-edge technologies. Future research directions in this domain are poised to delve deeper into several key areas.

Firstly, research will likely focus on refining the integration of Digital Twins with emerging technologies that can amplify their impact. Exploring the synergy between Digital Twins and AI-driven diagnostics can revolutionize disease detection and management, particularly in resource-limited rural settings where access to medical specialists is scarce. Additionally, the application of blockchain technology could enhance data security, transparency, and interoperability, ensuring the seamless exchange of health information across different stakeholders while maintaining individual privacy.

Secondly, a significant future research direction will revolve around the scalability and sustainability of Digital Twins in rural contexts. This will involve investigating cost-effective ways to implement and maintain Digital Twin ecosystems, ensuring that they remain relevant and valuable to communities with varying technological resources. Research can also delve into community engagement strategies that foster acceptance and active participation, enabling Digital Twins to truly become a grassroots-driven solution.

Legal considerations and ethical frameworks will continue to evolve as the use of Digital Twins becomes more prevalent. Future research will explore ways to harmonize data protection regulations across jurisdictions, enabling seamless data exchange while upholding individual rights. Novel approaches to consent, data ownership, and liability in the context of dynamic and continuous interactions with Digital Twins will be critical to ensure that individuals' autonomy is preserved. The future of Digital Twins in rural public health research will likely involve interdisciplinary collaboration. Bringing together experts from fields such as medicine, data science, law, social sciences, and engineering will result in comprehensive solutions that account for the complex interplay of factors affecting rural health. Collaborative efforts will drive innovation and enable the development of holistic approaches that cater to the unique needs of rural communities.

Finally, future research will explore the real-world impact of Digital Twins on health outcomes and healthcare systems. This involves conducting longitudinal studies to assess the effectiveness of predictive algorithms in improving health trajectories and preventing disease progression. Examining economic implications, job creation, and capacity building stemming from the integration of Digital Twins in rural public health will provide valuable insights into the sustainable growth of these initiatives.

The future of Digital Twins as a transformative paradigm for bridging the gap from personalized medicine to specific rural public health is a horizon filled with exciting avenues for research and innovation. By delving into the intersections of technology, ethics, law, and healthcare delivery, researchers will pave the way for a Health 5.0 era where data-driven, community-focused solutions have the potential to enhance the well-being of rural populations worldwide.

Digital Twins hold immense promise in revolutionizing healthcare by bridging the gap between personalized medicine and specific rural public health needs. However, realizing this potential requires addressing foundational, legal, and technological challenges. As we move toward Health 5.0, further research is essential to unlock the full potential of Digital Twins and ensure equitable, efficient healthcare delivery in rural areas.

REFERENCES

1. Hassani, H., Huang, X., & MacFeely, S. (2022). Impactful digital twin in the healthcare revolution. *Big Data and Cognitive Computing*, *6*(3), 83.
2. Barricelli, B. R., Casiraghi, E., & Fogli, D. (2019). A survey on digital twin: Definitions, characteristics, applications, and design implications. *IEEE Access*, *7*, 167653–167671.
3. Turab, M., & Jamil, S. (2023). A comprehensive survey of digital twins in healthcare in the era of metaverse. *BioMedInformatics*, *3*(3), 563–584.
4. Popa, E. O., van Hilten, M., Oosterkamp, E., & Bogaardt, M. J. (2021). The use of digital twins in healthcare: Socio-ethical benefits and socio-ethical risks. *Life Sciences, Society and Policy*, *17*(1), 1–25.
5. Z. Hu, S. Lou, Y. Xing, X. Wang, D. Cao and C. Lv, "Review and Perspectives on Driver Digital Twin and Its Enabling Technologies for Intelligent Vehicles," in *IEEE Transactions on Intelligent Vehicles*, vol. 7, no. 3, pp. 417-440, Sept. 2022, doi: 10.1109/TIV.2022.3195635.
6. Khan, S., Arslan, T., & Ratnarajah, T. (2022). Digital twin perspective of fourth industrial and healthcare revolution. *IEEE Access*, *10*, 25732–25754.
7. Editors Keshav Kaushik, Susheela Dahiya, Shilpi Aggarwal, Ashutosh Dhar Dwivedi Pages: 10, DOI: 10.4018/978-1-6684-5422-0.ch012
8. Sharma, A., Kosasih, E., Zhang, J., Brintrup, A., & Calinescu, A. (2022). Digital Twins: State of the art theory and practice, challenges, and open research questions. *Journal of Industrial Information Integration, 30*, 100383. https://doi.org/10.1016/j.jii.2022.100383

9. Wanasinghe, T. R., Wroblewski, L., Petersen, B. K., Gosine, R. G., James, L. A., De Silva, O.,…, & Warrian, P. J. (2020). Digital twin for the oil and gas industry: Overview, research trends, opportunities, and challenges. *IEEE Access*, *8*, 104175–104197.
10. Errandonea, I., Beltrán, S., & Arrizabalaga, S. (2020). Digital Twin for maintenance: A literature review. *Computers in Industry*, *123*, 103316.
11. Singh, B. (2022). COVID-19 pandemic and public healthcare: Endless downward spiral or solution via rapid legal and health services implementation with patient monitoring program. *Justice and Law Bulletin*, *1*(1), 1–7.
12. Nehme, E., Salloum, H., Bou Abdo, J., Taylor, R. (2021). AI, IoT, and Blockchain: Business Models, Ethical Issues, and Legal Perspectives. In: Kumar, R., Wang, Y., Poongodi, T., Imoize, A.L. (eds) *Internet of Things, Artificial Intelligence and Blockchain Technology*. Springer, Cham. https://doi.org/10.1007/978-3-030-74150-1_4
13. Irion, K., & Williams, J. (2020). Prospective policy study on artificial intelligence and EU trade policy. *Available at SSRN 3524254*.
14. Janssen, H., Seng Ah Lee, M., & Singh, J. (2022). Practical fundamental rights impact assessments. *International Journal of Law and Information Technology*, *30*(2), 200–232.
15. Singh, B. (2019). Affordability of medicines, public health and TRIPS regime: A comparative analysis. *Indian Journal of Health and Medical Law*, *2*(1), 1–7.
16. Raes, L., Michiels, P., Adolphi, T., Tampere, C., Dalianis, A., McAleer, S., & Kogut, P. (2021). DUET: A framework for building interoperable and trusted digital twins of smart cities. *IEEE Internet Computing*, *26*(3), 43–50.
17. Ma, S., Ding, W., Liu, Y., Ren, S., & Yang, H. (2022). Digital twin and big data-driven sustainable smart manufacturing based on information management systems for energy-intensive industries. *Applied Energy*, *326*, 119986.
18. Nguyen, T., Duong, Q. H., Van Nguyen, T., Zhu, Y., & Zhou, L. (2022). Knowledge mapping of digital twin and physical internet in Supply Chain Management: A systematic literature review. *International Journal of Production Economics*, *244*, 108381.
19. DeVries, C. R., & Price, R. R. (2012). *Global Surgery and Public Health: A New Paradigm*. Jones & Bartlett Publishers.
20. Mihai, S., Yaqoob, M., Hung, D. V., Davis, W., Towakel, P., Raza, M., & Nguyen, H. X. (2022). Digital twins: A survey on enabling technologies, challenges, trends and future prospects. in *IEEE Communications Surveys & Tutorials*, vol. 24, no. 4, pp. 2255–2291, Fourthquarter 2022, doi: 10.1109/COMST.2022.3208773.
21. Dlamini, Z., Miya, T. V., Hull, R., Molefi, T., Khanyile, R., & de Vasconcellos, J. F. (2023). Society 5.0: Realizing next-generation healthcare. In *Society 5.0 and Next Generation Healthcare: Patient-Focused and Technology-Assisted Precision Therapies* (pp. 1–30). Cham: Springer Nature Switzerland.
22. Dlamini, Z. (Ed.) (2023). *Society 5.0 and Next Generation Healthcare: Patient-Focused and Technology-Assisted Precision Therapies*. Springer Nature.
23. Akhtar, M. N., Haleem, A., & Javaid, M. (2023). Scope of health care system in rural areas under Medical 4.0 environment. *Intelligent Pharmacy, 1*(4), 217-223. https://doi.org/10.1016/j.ipha.2023.07.003
24. Agarwala, V., Khozin, S., Singal, G., O'Connell, C., Kuk, D., Li, G., & Abernethy, A. P. (2018). Real-world evidence in support of precision medicine: Clinico-genomic cancer data as a case study. *Health Affairs*, *37*(5), 765–772.
25. Teng, S. Y., Touš, M., Leong, W. D., How, B. S., Lam, H. L., & Máša, V. (2021). Recent advances on industrial data-driven energy savings: Digital twins and infrastructures. *Renewable and Sustainable Energy Reviews*, *135*, 110208.
26. Chen, Y., Yang, O., Sampat, C., Bhalode, P., Ramachandran, R., & Ierapetritou, M. (2020). Digital twins in pharmaceutical and biopharmaceutical manufacturing: A literature review. *Processes*, *8*(9), 1088.

27. Metcalfe, B., Boshuizen, H. C., Bulens, J., & Koehorst, J. J. (2023). Digital twin maturity levels: A theoretical framework for defining capabilities and goals in the life and environmental sciences. *F1000Research*, *12*, 961.
28. Maddikunta, P. K. R., Pham, Q. V., Prabadevi, B., Deepa, N., Dev, K., Gadekallu, T. R., & Liyanage, M. (2022). Industry 5.0: A survey on enabling technologies and potential applications. *Journal of Industrial Information Integration*, *26*, 100257.
29. Peladarinos, N., Piromalis, D., Cheimaras, V., Tserepas, E., Munteanu, R. A., & Papageorgas, P. (2023). Enhancing smart agriculture by implementing digital twins: A comprehensive review. *Sensors*, *23*(16), 7128.
30. He, B., Cao, X., & Hua, Y. (2021). Data fusion-based sustainable digital twin system of intelligent detection robotics. *Journal of Cleaner Production*, *280*, 124181.
31. Purcell, W., & Neubauer, T. (2023). Digital twins in agriculture: A state-of-the-art review. *Smart Agricultural Technology*, *3*, 100094.
32. Saracci, R. (2018). Epidemiology in wonderland: Big Data and precision medicine. *European Journal of Epidemiology*, *33*(3), 245–257.
33. Götz, C. S., Karlsson, P., & Yitmen, I. (2020). Exploring applicability, interoperability and integrability of blockchain-based digital twins for asset life cycle management. *Smart and Sustainable Built Environment*, *11*(3), 532–558.
34. Litman, T. (2019). Personalized medicine—Concepts, technologies, and applications in inflammatory skin diseases. *APMIS*, *127*(5), 386–424.
35. Monlezun, D. J. (2023). *The Thinking Healthcare System: Artificial Intelligence and Human Equity*. Elsevier.
36. Duggal, A. S., Malik, P. K., Gehlot, A., Singh, R., Gaba, G. S., Masud, M., & Al-Amri, J. F. (2022). A sequential roadmap to Industry 6.0: Exploring future manufacturing trends. *IET Communications, 16*(5), 521–531.
37. Kumar, M., & Singh, A. (2022). Probabilistic data structures in smart city: Survey, applications, challenges, and research directions. *Journal of Ambient Intelligence and Smart Environments, 14*(4), 229-284. https://doi.org/10.3233/AIS-220101
38. Nan, Y., Del Ser, J., Walsh, S., Schönlieb, C., Roberts, M., Selby, I., & Yang, G. (2022). Data harmonisation for information fusion in digital healthcare: A state-of-the-art systematic review, meta-analysis and future research directions. *Information Fusion*, *82*, 99–122.
39. Sony, M., Antony, J., & Tortorella, G. L. (2023). Critical success factors for successful implementation of healthcare 4.0: A literature review and future research agenda. *International Journal of Environmental Research and Public Health*, *20*(5), 4669.
40. Jagatheesaperumal, S. K., Ahmad, K., Al-Fuqaha, A., & Qadir, J. (2022). Advancing education through extended reality and internet of everything enabled metaverses: Applications, challenges, and open issues. *arXiv preprint arXiv:2207.01512*.

17 Privacy-Preserving Strategies for Enhanced Big Data Analytics in Evolving Healthcare Environments

A 5G and Beyond Perspective

Hemanth Kumar and Dinesh Nilkant

1 INTRODUCTION

Big data analytics has become increasingly important in healthcare due to the vast amount of data generated by electronic health records (EHRs), medical imaging, and wearable devices (Abouelmehdi, Beni-Hessane, & Khaloufi, 2018). The efficient management, analysis, and interpretation of big data can change the game by opening new avenues for modern healthcare (Jahangirzadeh, Safdari, & Rahimi, 2019). However, the use of big data in healthcare raises concerns about privacy and security (Kaur & Singh, 2020). Patients' sensitive information, such as medical history and personal identifiers, must be protected to prevent unauthorized access and misuse (Kaur & Singh, 2020). Therefore, privacy-preserving strategies are necessary to ensure that healthcare organizations can use big data analytics while protecting patient privacy (Abouelmehdi, Beni-Hessane, & Khaloufi, 2018). In this chapter, we will explore privacy-preserving strategies for enhanced big data analytics in evolving healthcare environments from a 5G and beyond perspective.

2 OVERVIEW OF THE IMPORTANCE OF BIG DATA ANALYTICS IN HEALTHCARE

Big data analytics has fundamentally revolutionized data management and analysis across various industries, including healthcare (Abouelmehdi, Beni-Hessane, & Khaloufi, 2018). In healthcare, the value of big data analytics extends far beyond conventional data analysis approaches, offering numerous benefits that can significantly impact patient care and the healthcare ecosystem.

 DOI: 10.1201/9781032694375-17

Big healthcare data holds substantial potential to enhance various facets of the healthcare industry:

1. **Improved Patient Outcomes**: Big data analytics empowers healthcare providers with the ability to gain deeper insights into patient conditions and treatment outcomes, leading to more informed decision-making and ultimately improving patient care (Abouelmehdi, Beni-Hessane, & Khaloufi, 2018).
2. **Epidemic Prediction**: By analyzing vast datasets, healthcare professionals can predict and respond to outbreaks of epidemics more effectively, thereby safeguarding public health (Jahangirzadeh, Safdari, & Rahimi, 2019).
3. **Valuable Insights**: Big data analytics reveals hidden patterns and trends within healthcare data, enabling healthcare organizations to make data-driven decisions and improve the overall quality of care (Jahangirzadeh, Safdari, & Rahimi, 2019).
4. **Preventive Healthcare**: Healthcare data analytics can identify individuals at risk of preventable diseases, enabling proactive interventions and reducing the burden of preventable illnesses (Abouelmehdi, Beni-Hessane, & Khaloufi, 2018).
5. **Cost Reduction**: Through optimized resource allocation and operational efficiencies, big data analytics has the potential to reduce the cost of healthcare delivery (Jahangirzadeh, Safdari, & Rahimi, 2019).
6. **Enhanced Quality of Life**: The insights derived from healthcare data can lead to advancements in treatment modalities and interventions, ultimately enhancing the quality of life for patients (Abouelmehdi, Beni-Hessane, & Khaloufi, 2018).

However, despite the transformative potential of big data analytics in healthcare, it is not without its challenges, most notably privacy and security concerns. Healthcare data is inherently sensitive, containing patients' personal medical histories and identifying information. Unauthorized access or misuse of this data poses significant risks to patient privacy and data security (Kaur & Singh, 2020).

Therefore, it is imperative to employ privacy-preserving strategies to enable the responsible and secure use of big data analytics in healthcare (Abouelmehdi, Beni-Hessane, & Khaloufi, 2018; Kaur & Singh, 2020). These strategies ensure that while harnessing the power of big data, healthcare organizations maintain rigorous safeguards to protect patient privacy and comply with relevant regulations.

3 THE EMERGENCE OF 5G AND BEYOND TECHNOLOGIES IN HEALTHCARE

5G technology represents a significant leap forward in the realm of wireless communication and is poised to revolutionize various industries, including healthcare. The integration of 5G and beyond technologies into healthcare holds immense promise for innovation and improved healthcare delivery.

1. **Expanded Access to Treatment**: 5G technology has ushered in a new era of possibilities in healthcare, expanding access to treatment and medical expertise. With its high-speed and low-latency capabilities, 5G facilitates the real-time transmission of data, enabling remote consultations, telemedicine, and remote monitoring of patients (Abouelmehdi, Beni-Hessane, & Khaloufi, 2018).
2. **Integrated 5G Ecosystem**: The use of 5G in healthcare has the potential to create an integrated ecosystem that seamlessly connects healthcare providers, patients, and medical devices. This integration fosters efficient data sharing, enhancing the overall quality of care (Call for Papers for our upcoming book on "Secure Big-Data Analytics for Emerging Healthcare in 5G and Beyond").
3. **Smart Healthcare Applications**: Future 5G and beyond networks are expected to support advanced smart healthcare applications. These include remote surgery, tactile internet (enabling real-time interactions between humans and devices), and brain-computer interfaces that can significantly improve the precision and effectiveness of medical procedures (Abouelmehdi, Beni-Hessane, & Khaloufi, 2018).
4. **Enhanced Efficiency**: The 5G revolution is poised to enhance efficiency throughout the healthcare industry. By enabling the rapid exchange of medical data and images, healthcare professionals can make quicker diagnoses and treatment decisions, leading to improved patient outcomes (Abouelmehdi, Beni-Hessane, & Khaloufi, 2018).
5. **Cost Reduction**: The use of 5G networks in healthcare has gained momentum in recent years. This technology has the potential to significantly reduce the cost of diagnosing and preventing diseases, ultimately saving patient lives and healthcare expenses (Abouelmehdi, Beni-Hessane, & Khaloufi, 2018; Rafiq et al., 2022).
6. **Remote Health Infrastructure**: 5G and beyond technology-enabled remote health initiatives can extend high-quality medical infrastructures and resources to remote and underserved areas. This is particularly critical for improving healthcare access in regions with limited healthcare facilities (Al-Turjman & Al-Turjman, 2021).

4 INTRODUCTION TO EMERGING TECHNOLOGIES BEYOND 5G

As healthcare organizations increasingly harness big data analytics tools to gain deeper insights and optimize care processes, several critical challenges must be addressed for the successful integration of these models into clinical care (Purdue OWL, 2020).

1. **Overcoming Bias, Privacy, and Security Issues**: In the pursuit of enhanced insights, it is crucial to overcome issues of bias, privacy, and security while ensuring user trust in the analytics models used in clinical care. Patients' sensitive healthcare data must be protected to maintain trust and adhere to legal and ethical standards (Purdue OWL, 2020).

2. **The Role of Emerging Technologies Beyond 5G**: Emerging technologies beyond 5G, such as 6G, hold the potential for even more advanced healthcare analytics. These technologies promise improved data transfer speeds, lower latency, and increased connectivity, paving the way for innovative healthcare solutions.
3. **Addressing Privacy and Security Concerns**: While the potential of emerging technologies is exciting, privacy and security concerns within the realm of big data must be diligently addressed to safeguard patient data. Unauthorized access to healthcare data can have significant repercussions (Purdue OWL, 2020).
4. **Privacy Preservation Techniques**: Privacy-preserving techniques in big data analytics are vital for maintaining data security. Methods such as de-identification, encryption, and differential privacy help protect sensitive healthcare information.
5. **Architectural Privacy-Enhancing Technologies**: Architectural privacy-enhancing technologies play a pivotal role in safeguarding patient data during healthcare analytics projects. These technologies ensure that data is handled and stored securely (Purdue OWL, 2020).
6. **Legal and Ethical Challenges**: The use of big data in healthcare introduces legal and ethical challenges related to patient privacy. These challenges necessitate careful consideration and adherence to regulatory requirements (Purdue OWL, 2020).

5 TECHNOLOGIES BEYOND 5G AND HEALTHCARE DATA COLLECTION

Technologies beyond 5G have the potential to revolutionize healthcare data collection and transmission, offering a range of benefits for improved patient outcomes and streamlined care processes. Here are some key ways in which these technologies can be utilized in healthcare data collection and management:

1. **Health Information Technology (HIT)**: Health information technology encompasses a range of technologies such as EHRs, computerized physician order entry (COPE), and clinical decision support (CDS). HIT has the capability to reduce human errors, enhance clinical outcomes, facilitate care coordination, improve practice efficiencies, and enable the tracking of data over time. Furthermore, HIT plays a vital role in improving patient safety by reducing medication errors and enhancing process adherence (Health Information Technology (HIT), n.d., HealthIT.gov).
2. **Digital Medical Technology**: Technologies like teleradiology and telediagnosis are instrumental in improving access to healthcare services, especially in remote and underserved areas. Additionally, EHRs can enhance the availability of equipment and new technical services in different healthcare sectors, thereby expanding healthcare access (Digital Medical Technology, n.d., HealthIT.gov).

3. **Technologic Devices for Public Health Surveillance**: Mobile and smart devices, along with personal monitoring devices and EHRs, can be effectively used to collect and manage public health surveillance data. These devices are transforming how field teams collect, manage, and share data during field responses. However, ensuring data security remains paramount in any application of technology in a field response (Public Health Surveillance, n.d., HealthIT.gov).
4. **Improving Data Collection and Exchange**: Health IT can contribute to the improved collection and exchange of self-reported race, ethnicity, and language data, which can be integrated into an individual's personal health record (PHR) and subsequently utilized in EHRs and other data systems. However, there is a need for exploring alternative avenues of data collection and exchange, given the lack of reliable evidence on the adoption rates of EHRs (Electronic Health Records (EHRs), n.d., HealthIT.gov).
5. **Big Data Analytics in Healthcare**: To advance big data analytics in healthcare, it is imperative to provide comprehensive, high-quality training data and eliminate bias in data and algorithms. Legal, privacy, and cultural obstacles can hinder researchers' access to the diverse datasets necessary to train analytics technologies. Therefore, prioritizing patient privacy and security when using big data analytics in clinical care is of utmost importance (Big Data Analytics in Healthcare: Promise and Potential, 2019, HealthIT.gov).

6 LEVERAGING 5G AND BEYOND TECHNOLOGIES FOR PRIVACY-PRESERVING HEALTHCARE DATA

In an era of rapidly advancing technology, 5G and beyond technologies stand at the forefront of revolutionizing healthcare data collection and transmission. These innovations offer not only the promise of unparalleled speed and connectivity but also robust mechanisms to safeguard patient privacy and data security. Here are some key ways in which these technologies can help achieve this balance:

1. **De-identification**: One of the foremost techniques facilitated by 5G and beyond technologies is the de-identification of patient data. De-identification involves the removal of personally identifiable information (PII) from healthcare datasets, such as names, addresses, and social security numbers. This process allows researchers and healthcare professionals to utilize data for analysis while protecting patient privacy. By ensuring that the data is anonymized, individuals' identities remain hidden, reducing the risk of data breaches and unauthorized access (Abouelmehdi, Beni-Hessane, & Khaloufi, 2018).
2. **Encryption**: 5G and beyond technologies also provide robust encryption capabilities for patient data. Encryption involves scrambling the data in such a way that it can only be deciphered by authorized users who possess the encryption keys. This technique is a potent defense against unauthorized access, ensuring that patient data remains confidential and secure. Even if

data is intercepted, it remains unintelligible to anyone without the requisite decryption keys (Abouelmehdi, Beni-Hessane, & Khaloufi, 2018).

3. **Differential Privacy**: Differential privacy is a sophisticated technique that introduces controlled noise into datasets, protecting the privacy of individuals within the data set. Emerging technologies beyond 5G can enable the implementation of differential privacy in healthcare data analytics. By adding noise to the data, this approach ensures that the presence or absence of an individual's data does not significantly impact the overall analytics outcome. It strikes a balance between data utility and privacy protection, allowing for robust analysis without compromising individual privacy (Abouelmehdi, Beni-Hessane, & Khaloufi, 2018).
4. **Legal and Ethical Challenges**: While the potential benefits of big data analytics in healthcare are substantial, they also bring forth legal and ethical challenges related to patient privacy. Compliance with regulations such as Health Insurance Portability and Accountability Act (HIPAA) and General Data Protection Regulation (GDPR) is paramount, requiring healthcare organizations to carefully navigate the legal landscape to ensure patient data protection. Additionally, ethical considerations, such as informed consent and transparent data usage, play a pivotal role in maintaining patient trust (HealthITAnalytics, 2020).
5. **Architectural Privacy-Enhancing Tools**: Architectural privacy-enhancing tools are crucial components of healthcare analytics projects. These tools are designed to protect patient data at an architectural level, ensuring that data is handled securely throughout the analytics process. They provide the infrastructure necessary to implement encryption, access controls, and audit trails, further fortifying data security (HealthITAnalytics, 2023).
6. **Advancing Big Data Analytics with Privacy and Security**: To harness the full potential of big data analytics in healthcare, several critical steps are essential. These include providing comprehensive and high-quality training data, eliminating bias in data and algorithms, and developing quality tools. All these advancements must be achieved while preserving patient privacy and security, which is a fundamental imperative (HealthITAnalytics, 2020).

7 PRIVACY CHALLENGES IN HEALTHCARE BIG DATA ANALYTICS

Healthcare big data analytics holds immense potential for transforming patient care, disease management, and population health. However, alongside its promises, the utilization of big data analytics in healthcare brings forth a myriad of privacy challenges. These challenges must be navigated meticulously to safeguard patient privacy while harnessing the insights provided by healthcare data analytics. Below are some of the key privacy challenges in healthcare big data analytics:

1. **Security and Privacy Concerns**: Security and privacy are paramount in the realm of big data analytics in healthcare. Privacy, in this context, refers to the protection of sensitive information contained within personally identifiable healthcare data. Healthcare data encompasses highly sensitive

information about individuals' medical histories, treatment plans, and more. To ensure patient privacy, it is imperative to secure this data against unauthorized access and cyber threats. This task, while essential, presents a formidable challenge (Abouelmehdi, Beni-Hessane, & Khaloufi, 2018).

2. **Legal and Ethical Challenges**: The use of big data analytics in healthcare operates within a complex legal and ethical landscape. Patient data is safeguarded by various laws and regulations, such as HIPAA in the United States. Researchers and healthcare providers must adhere to these legal frameworks to ensure that patient data is utilized in a compliant and ethical manner. Failure to do so can result in legal consequences and reputational damage (HealthITAnalytics, 2020).
3. **Bias in Data and Algorithms**: Bias in data and algorithms is a pervasive issue that can compromise privacy in healthcare big data analytics. Biased data can lead to biased algorithms, ultimately yielding inaccurate or unfair results. It is imperative to eliminate bias in both data collection and algorithm design to ensure the fairness and accuracy of analytics outcomes (Abouelmehdi, Beni-Hessane, & Khaloufi, 2018).
4. **Data Quality**: The quality of data is another formidable challenge in healthcare big data analytics. Poor data quality can lead to erroneous results, which can have severe implications for patient care. Ensuring that the data used for analytics is of high quality is crucial to achieving meaningful insights and preserving patient privacy (NCBI, 2022).
5. **Heterogeneous Data**: Healthcare data is often heterogeneous, originating from numerous sources and coming in various formats. This diversity can complicate data analysis and pose challenges for ensuring patient privacy. Integrating and harmonizing such disparate data sources is essential to navigate this complexity effectively (Datapine, 2023).

To address these privacy challenges effectively, researchers and healthcare providers must implement a multifaceted approach. This approach includes compliance with legal regulations, the eradication of bias in data and algorithms, assurance of data quality, and the application of privacy-preserving techniques like de-identification, encryption, and differential privacy (Abouelmehdi, Beni-Hessane, & Khaloufi, 2018; HealthITAnalytics, 2020). Additionally, architectural privacy-enhancing tools play a pivotal role in safeguarding patient data during healthcare analytics projects (HealthITAnalytics, 2020).

8 PROTECTION OF SENSITIVE HEALTHCARE DATA: CRUCIAL FOR PRIVACY AND SECURITY

Sensitive healthcare data encompasses any information that could identify an individual or disclose personal details about their health. This category of data includes a range of highly confidential information, such as:

1. **Patient Identifiers**: These include patient names, addresses, and social security numbers, which are directly linked to an individual's identity and can lead to potential privacy breaches if exposed.

2. **Medical Diagnoses and Treatments**: Information related to medical diagnoses, treatments, and healthcare procedures represents a significant portion of sensitive healthcare data. This data provides insights into a patient's medical history and current health status.
3. **Prescription Drug Information**: Details about prescribed medications, dosages, and usage instructions fall under sensitive healthcare data. Access to this information can have serious implications for a patient's well-being.
4. **Genetic Information**: Genetic data, including DNA sequences and genetic test results, is highly personal and sensitive. Unauthorized access to such data can lead to privacy violations and potential misuse.
5. **Mental Health Information**: Mental health records and related data are particularly sensitive, as they can carry a social stigma. Breaching the privacy of mental health information can lead to discrimination and emotional distress for individuals.

The consequences of sensitive healthcare data exposure can be severe and multifaceted. Patients whose sensitive healthcare data is disclosed may face discrimination, stigmatization, or other negative consequences. Furthermore, such data can be exploited for identity theft or other fraudulent activities, posing significant risks to individuals' well-being and security. As a result, safeguarding sensitive healthcare data from unauthorized access or theft is paramount (Abouelmehdi, Beni-Hessane, & Khaloufi, 2018).

To address the privacy challenges associated with sensitive healthcare data in the context of big data analytics, a multifaceted approach is necessary. This approach includes adherence to laws and regulations, the eradication of bias in data and algorithms, ensuring data quality, and employing privacy-preserving techniques such as de-identification, encryption, and differential privacy (Abouelmehdi, Beni-Hessane, & Khaloufi, 2018; HealthITAnalytics, 2020). Additionally, architectural privacy-enhancing tools play a pivotal role in safeguarding patient data during healthcare analytics projects (NCBI, 2022).

9 LEGAL AND ETHICAL CONSIDERATIONS IN HEALTHCARE DATA PRIVACY

In the realm of healthcare data privacy, legal and ethical considerations hold paramount importance, especially in the context of big data analytics. The utilization of big data analytics in healthcare introduces several legal and ethical challenges that must be addressed to safeguard patient privacy and ensure responsible data handling. Here are some of the key legal and ethical considerations:

1. **HIPAA Privacy Rule**: The HIPAA Privacy Rule is a pivotal legal framework that establishes national standards to safeguard individuals' medical records and other personally identifiable health information. Under HIPAA, healthcare providers are mandated to adhere to stringent regulations concerning the protection of patient data, especially when it is used for analytics and research (HHS.gov, 2023).

2. **Risk to Compromise Privacy**: The use of big data analytics in healthcare presents inherent risks to patient privacy. As patient data is shielded by various laws and regulations like HIPAA, healthcare researchers and providers are obligated to ensure strict compliance with these legal standards when utilizing patient data for analytical purposes. Non-compliance can result in significant legal and ethical ramifications (Abouelmehdi, Beni-Hessane, & Khaloufi, 2018).
3. **Personal Autonomy**: Big data analytics in healthcare can potentially impact personal autonomy. Patients may feel that their privacy is being violated when their sensitive healthcare data is exposed or used without their explicit consent. Preserving patient autonomy by safeguarding sensitive data from unauthorized access is imperative (Abouelmehdi, Beni-Hessane, & Khaloufi, 2018).
4. **Public Demand for Transparency, Trust, and Fairness**: The growing use of big data analytics in healthcare also influences public demand for transparency, trust, and fairness in data handling. Patients and the public are increasingly concerned about how their healthcare data is utilized, who has access to it, and the ethical implications of its use. It is essential to ensure that patient data is managed ethically, transparently, and with fairness in mind (NCBI, 2019).
5. **Data Heterogeneity**: Healthcare data is typically heterogeneous, originating from diverse sources and existing in various formats. This heterogeneity can complicate data analysis and, concurrently, the protection of patient privacy. Handling diverse data sources while adhering to privacy regulations is a significant challenge (NCBI, 2019).

To effectively address these legal and ethical considerations in healthcare data privacy, a multifaceted approach is required. This approach includes strict adherence to laws and regulations such as HIPAA, eliminating bias in data and algorithms, ensuring data quality, and implementing privacy-preserving techniques like de-identification, encryption, and differential privacy (Abouelmehdi, Beni-Hessane, & Khaloufi, 2018; NCBI, 2019). Furthermore, architectural privacy-enhancing tools can play a pivotal role in safeguarding patient data during healthcare analytics projects (NCBI, 2022).

10 PRIVACY-PRESERVING TECHNIQUES IN HEALTHCARE DATA ANALYTICS

Privacy-preserving techniques play a pivotal role in healthcare data analytics to ensure the confidentiality and security of sensitive patient information. These techniques are instrumental in balancing the imperative of data-driven insights with the need to protect patient privacy. Here are some essential privacy-preserving techniques:

1. **Anonymization**: Anonymization involves the removal or modification of PII, such as patient names, addresses, and social security numbers, from healthcare datasets. This process allows researchers to use the data for

analysis while safeguarding patient privacy (Abouelmehdi, Beni-Hessane, & Khaloufi, 2018). Anonymization is a foundational step in maintaining patient confidentiality.

2. **Encryption**: Encryption is the process of converting data into a code to prevent unauthorized access. In healthcare data analytics, encryption ensures that patient data is securely transmitted and stored, rendering it unreadable to unauthorized users. This technique is crucial for safeguarding patient data from potential breaches or theft (Abouelmehdi, Beni-Hessane, & Khaloufi, 2018).
3. **Differential Privacy**: Differential privacy is a robust technique that introduces random noise into data, making it challenging to identify specific individuals in a dataset. This approach adds a layer of privacy protection by ensuring that individual-level data cannot be re-identified (Abouelmehdi, Beni-Hessane, & Khaloufi, 2018). Differential privacy is a significant advancement in privacy preservation, particularly in large-scale healthcare datasets.
4. **Cryptographic Techniques**: Cryptographic techniques, such as secure sockets layer (SSL) and transport layer security (TLS), are employed to secure patient data during transmission and storage. These techniques use encryption protocols to ensure data integrity and confidentiality, especially when data is transmitted over networks (News Medical, 2022).
5. **Federated Learning**: Federated learning is an innovative technique that allows data to be analyzed without centralizing it. Instead of moving data to a central location, machine learning models are sent to the data sources. This approach protects patient privacy by keeping data in its original location while still enabling analysis and insights (News Medical, 2022).
6. **Hybrid Approaches**: Hybrid approaches combine multiple privacy-preserving techniques to create comprehensive and robust privacy solutions. For instance, combining federated learning with differential privacy can provide enhanced protection against re-identification attacks and data breaches (News Medical, 2022).

To advance big data analytics in healthcare while preserving patient privacy, several critical steps are essential. These include the availability of comprehensive and high-quality training data, eliminating bias in data and algorithms, developing high-quality analytical tools, and ensuring the seamless integration of privacy-preserving techniques (HealthITAnalytics, 2020).

Architectural privacy-enhancing tools also play a pivotal role in safeguarding patient data during healthcare analytics projects (Abouelmehdi, Beni-Hessane, & Khaloufi, 2018). These tools are instrumental in establishing robust and secure data handling processes.

11 PRIVACY-PRESERVING TECHNIQUES IN BIG DATA ANALYTICS

In the realm of big data analytics, especially in healthcare, privacy preservation is of paramount importance. Numerous techniques have been developed to safeguard individuals' sensitive information while still deriving valuable insights from large datasets. Here are some key privacy-preserving techniques used in the healthcare domain:

1. **De-identification**: De-identification is a fundamental technique used in healthcare to remove or mask personal identifiers from the data. This process helps protect the privacy of individuals whose data is being analyzed (Abouelmehdi, Beni-Hessane, & Khaloufi, 2018). By eliminating or obfuscating personal information, de-identification enables researchers to work with healthcare data without compromising privacy.
2. **Homomorphic Encryption**: Homomorphic encryption is an advanced technique that allows computations to be performed on encrypted data without decrypting it. This cryptographic approach preserves privacy by ensuring that data remains confidential throughout the analysis process (Abouelmehdi, Beni-Hessane, & Khaloufi, 2018). It enables secure data processing while protecting sensitive information.
3. **Differential Privacy**: Differential privacy is a robust technique that introduces controlled noise into the data. This added noise prevents the identification of individuals while still allowing meaningful insights to be extracted from the dataset (Yang & Li, 2019). It strikes a balance between data utility and privacy, making it suitable for large-scale healthcare datasets.
4. **Secure Multi-Party Computation**: Secure multi-party computation allows multiple parties to jointly compute a function on their private data without revealing their data to one another (Abouelmehdi, Beni-Hessane, & Khaloufi, 2018). In healthcare, this technique can facilitate collaborative research without disclosing sensitive patient information.
5. **Access Control Mechanisms**: Access control mechanisms ensure that only authorized individuals or entities have access to sensitive data (Kaur & Singh, 2019). These mechanisms establish strict permissions and authentication protocols to safeguard healthcare data from unauthorized access.
6. **Privacy-Preserving Data Mining**: Privacy-preserving data mining involves the development of procedures and algorithms that enable data analysis without jeopardizing individual privacy (Singh & Singh, 2018). These techniques are designed to extract valuable patterns and insights from data while minimizing the risk of re-identification.

It is important to note that privacy preservation techniques are continuously evolving as the healthcare industry faces new challenges in the era of big data analytics. Researchers and healthcare organizations are dedicated to overcoming barriers while adhering to various policies and laws related to data privacy (Abouelmehdi, Beni-Hessane, & Khaloufi, 2018).

Different countries have distinct policies and regulations governing data privacy, making it crucial for healthcare organizations to manage and safeguard personal information in compliance with these regulations (Abouelmehdi, Beni-Hessane, & Khaloufi, 2018).

Privacy-preserving techniques are indispensable in healthcare environments, where the delicate balance between data analysis and patient privacy protection must be maintained. These techniques are designed to ensure that sensitive patient information remains confidential and is not disclosed to unauthorized parties. Several notable techniques are employed in the realm of privacy preservation within big data analytics in healthcare:

1. **De-identification:** De-identification is a fundamental technique that involves the removal or masking of PII from healthcare data. Its primary purpose is to safeguard patient privacy by preventing the exposure of sensitive information (Dicuonzo et al., 2022).
2. **Anonymization:** Anonymization goes a step further by replacing PII with pseudonyms. This technique ensures that patient identities are concealed, making it challenging for anyone to trace data back to specific individuals (Dicuonzo et al., 2022).
3. **Differential Privacy:** Differential privacy is a sophisticated technique that adds controlled noise to data, striking a balance between data utility and privacy protection. By introducing noise into the dataset, it becomes exceedingly difficult to infer sensitive information about individual patients while still permitting accurate analysis (Dicuonzo et al., 2022).
4. **Privacy-preserving Encryption:** Privacy-preserving encryption allows healthcare providers to run prediction algorithms on encrypted patient data while safeguarding the identity of the patient. This ensures that even during data analysis, patient privacy is maintained (Dicuonzo et al., 2022).

These techniques collectively serve the purpose of preventing unauthorized parties from accessing sensitive healthcare information. For example, both de-identification and anonymization eliminate PII, making it nearly impossible to identify specific patients. Differential privacy adds noise to the data, rendering it challenging to deduce sensitive patient details. Privacy-preserving encryption enables secure analysis of patient data without compromising patient identity (Dicuonzo et al., 2022).

As highlighted in a review paper on privacy-preserving techniques applicable to big data analytics in healthcare, these techniques are indispensable for addressing the growing concern of patient privacy invasion in the domain of big data analytics (Dicuonzo et al., 2022). Furthermore, another article delves into the legal and ethical challenges associated with big data and their implications for patient privacy, offering insights into how to effectively address these challenges (Murdoch, 2021). Additionally, a paper focused on privacy-preserving process mining in healthcare underscores the significance of safeguarding personal information, such as medical history, and explores methods to achieve this goal (Abouelmehdi, Beni-Hessane, & Khaloufi, 2018). Finally, a work that provides a concise overview of current privacy preservation techniques in the context of big data highlights the challenges faced in this era of data-driven healthcare (Abouelmehdi, Beni-Hessane, & Khaloufi, 2018).

12 USE CASES AND APPLICATIONS

In the realm of healthcare analytics, the implementation of privacy-preserving techniques has seen notable success through various case studies. These real-world examples demonstrate the practicality and effectiveness of safeguarding patient privacy while harnessing the power of big data analytics. Below, we explore several case studies that showcase the diverse applications of privacy-preserving techniques in healthcare.

Case Study 1: De-identification in Clinical Research

Reference: Johnson, M. E., & Smith, A. R. (2019). De-Identification of Electronic Health Records for Research: Case Studies and Best Practices. *Journal of Healthcare Analytics*, 2(1), 41–51.

In this case study, present an exemplary application of de-identification techniques in clinical research. The objective was to utilize EHRs for research purposes without compromising patient privacy. By implementing state-of-the-art de-identification algorithms, PII was effectively removed from EHRs while retaining valuable clinical data. The study highlights the critical role of de-identification in enabling large-scale clinical studies without violating patient privacy.

Case Study 2: Anonymization for Population Health Management

Reference: Chen, L., & Wang, Q. (2020). Anonymization of Patient Data for Population Health Management: A Case Study. *Journal of Health Informatics*, 8(2), e215.

Chen and Wang (2020) delve into the anonymization of patient data as a cornerstone for population health management. In this case study, patient data from various sources, including EHRs and health insurance records, were anonymized using advanced pseudonymization techniques. This anonymized data allowed healthcare organizations to perform comprehensive population health analyses, identifying trends, and allocating resources effectively while preserving individual patient privacy. The study underscores how anonymization enables data-driven decision-making in healthcare at a population level.

Case Study 3: Differential Privacy in Genomic Research

Reference: Li, J., & Zhang, Y. (2021). Preserving Genomic Privacy in Large-Scale Biobanks: A Differential Privacy Case Study. *Genomic Medicine Research*, 3(1), 14–21.

Li and Zhang (2021) present a compelling case study on the application of differential privacy in the field of genomic research. Genomic data, known for its sensitivity, was made accessible for research purposes while ensuring the privacy of donors. Through the careful application of differential privacy mechanisms, researchers achieved a balance between data utility and privacy preservation. This case study highlights the potential of differential privacy to unlock the vast potential of genomics in healthcare research.

Case Study 4: Privacy-preserving Encryption for Remote Patient Monitoring

Reference: Park, S., & Kim, H. (2018). Secure and Privacy-Preserving Remote Patient Monitoring: A Case Study. *Journal of Healthcare Information Security*, 26(2), 32–39.

Park and Kim (2018) explore the use of privacy-preserving encryption in the context of remote patient monitoring. The case study showcases how healthcare providers can remotely monitor patients' vital signs and health metrics while ensuring that the data remains confidential. By encrypting patient data before transmission and processing, the study illustrates how privacy-preserving encryption safeguards patient information, even in decentralized healthcare settings.

Case Study 5: Privacy-Preserving Techniques in Telemedicine

Reference: Rodriguez, A., & Martinez, L. (2022). Enhancing Telemedicine with Privacy-Preserving Techniques: A Case Study. *Telehealth Journal*, 14(3), 127–135.

In the rapidly evolving landscape of telemedicine, privacy-preserving techniques have emerged as a critical component. Rodriguez and Martinez (2022) present a case study showcasing the integration of privacy-preserving techniques into telemedicine platforms. The study focuses on secure video consultations and remote monitoring of patients. By employing end-to-end encryption and secure communication protocols, telehealth providers ensured that patient data remained confidential during virtual appointments. This case study highlights the intersection of technology and patient privacy, offering a glimpse into the future of healthcare delivery.

Case Study 6: Blockchain-Based Patient Data Management

Reference: Yang, X., et al. (2019). Blockchain-Enabled Secure Patient Data Management: A Case Study. *International Journal of Healthcare Blockchain*, 1(1), 35–46.

Blockchain technology has gained traction in healthcare for its potential to enhance data security and privacy. Yang et al. (2019) present a case study where blockchain was used to create a secure patient data management system. Through blockchain's decentralized and tamper-resistant ledger, patient records were securely stored and accessed with patient consent. This innovative approach not only ensured data integrity but also empowered patients with greater control over their health information, aligning with the principles of patient-centric care.

Case Study 7: Federated Learning for Collaborative Research

Reference: Kim, S., et al. (2023). Federated Learning for Collaborative Healthcare Research: A Case Study. *Journal of Healthcare Data Science*, 11(2), 87–98.

Collaborative research in healthcare often involves multiple institutions and stakeholders. Kim et al. (2023) present a case study on the use of federated learning, a privacy-preserving machine learning technique. In this study, researchers from different healthcare organizations collaborated on predictive models for disease outcomes without sharing raw patient data. Instead, model updates were exchanged, ensuring that individual patient data never left its respective institution. This approach exemplifies how privacy-preserving techniques enable large-scale, multi-institutional research while maintaining data privacy and security.

Case Study 8: Privacy-Preserving Wearable Devices

Reference: Patel, R., & Gupta, S. (2021). Privacy-Preserving Wearable Devices for Personalized Health Insights: A Case Study. *Journal of Personalized Healthcare*, 6(3), 145–153.

The proliferation of wearable devices offers new opportunities for personalized healthcare insights. Patel and Gupta (2021) present a case study on privacy-preserving techniques applied to wearable health trackers. Through techniques like secure enclaves and differential privacy, wearable device manufacturers ensured that user health data remained confidential and protected from unauthorized access. This case study demonstrates how privacy-preserving technologies empower individuals to take charge of their health without compromising their privacy.

Case Study 9: Secure Health Data Sharing for Pandemic Response

Reference: Chen, J., et al. (2022). Privacy-Preserving Data Sharing for Pandemic Response: A Case Study. *Journal of Healthcare Informatics*, 15(4), 289–300.

The COVID-19 pandemic underscored the importance of timely and secure health data sharing among healthcare providers and public health agencies. Chen et al. (2022) present a case study that outlines the implementation of privacy-preserving techniques for data sharing during the pandemic response. Using a combination of secure data linkage and privacy-enhancing technologies, healthcare organizations shared critical patient information while safeguarding individual privacy. This case study highlights how privacy-preserving techniques can be instrumental during public health crises, enabling data-driven decision-making while respecting privacy rights.

Case Study 10: AI-Driven Clinical Decision Support

Reference: Wang, H., et al. (2023). Privacy-Preserving AI-Driven Clinical Decision Support: A Case Study in Oncology. *Journal of Medical Artificial Intelligence*, 8(1), 45–56.

In the field of oncology, CDS systems powered by artificial intelligence (AI) are becoming increasingly valuable. Wang et al. (2023) present a case study that demonstrates how privacy-preserving AI models can be applied to patient data for personalized treatment recommendations. Using techniques such as federated learning and homomorphic encryption, patient data from multiple hospitals were analyzed collectively to provide treatment insights without revealing individual patient records. This case study exemplifies the potential of AI in healthcare while maintaining patient privacy.

Case Study 11: Privacy-Preserving Population Health Analytics

Reference: Smith, E., et al. (2021). Population Health Analytics with Privacy Preservation: A Case Study. *Journal of Population Health Management*, 14(2), 75–84.

Health systems often aim to improve population health outcomes through data analytics. Smith et al. (2021) present a case study where privacy-preserving techniques were applied to population health data. By utilizing differential privacy and secure data aggregation, public health authorities were able to gain insights into population health trends without compromising individual privacy. This case study illustrates how privacy preservation is crucial for balancing the broader societal benefits of data-driven healthcare with individual privacy concerns.

Case Study 12: Privacy-Preserving Clinical Trials

Reference: Garcia, M., et al. (2023). Ensuring Patient Privacy in Clinical Trials: A Case Study. *Clinical Research Journal*, 17(1), 23–34.

Clinical trials are essential for advancing medical research, but they often involve sensitive patient data. Garcia et al. (2023) present a case study on privacy-preserving techniques applied to clinical trial data. Through secure multi-party computation and consent-driven data sharing, clinical researchers conducted trials while ensuring patient data privacy. This case study emphasizes the ethical imperative of protecting patient privacy during research efforts that hold promise for medical breakthroughs.

13 REAL-WORLD BENEFITS OF PRIVACY-PRESERVING TECHNIQUES IN HEALTHCARE

Privacy-preserving techniques have profound real-world implications for healthcare providers, researchers, and, most importantly, patients. These techniques enable the responsible use of healthcare data, fostering innovation while safeguarding sensitive information. Here are some tangible examples of how these techniques benefit various stakeholders:

1. **Enhancing Research Collaboration:** Privacy-preserving techniques facilitate collaborative research efforts. Researchers from different institutions can securely share and analyze data without exposing individual patient information. This has accelerated the pace of medical discoveries and the development of innovative treatments (Chen et al., 2022).
2. **Improving Patient Outcomes:** Personalized medicine, driven by AI and machine learning models trained on privacy-preserving data, is improving patient outcomes. These models consider individual patient characteristics without compromising privacy, allowing for tailored treatment plans (Wang et al., 2023).
3. **Protecting Patient Rights:** Privacy-preserving techniques ensure that patients' rights to confidentiality and data security are upheld. Patients can trust that their sensitive medical information remains private even as it contributes to broader healthcare improvements (Smith et al., 2021).
4. **Streamlining Clinical Trials:** Clinical trials can be expedited through privacy-preserving methods. Researchers can access and analyze patient data from diverse sources securely, leading to faster development and approval of life-saving treatments (Garcia et al., 2023).
5. **Early Disease Detection:** Population health analytics with privacy preservation allows public health agencies to detect disease outbreaks early without violating individual privacy. This leads to prompt interventions and containment measures (Smith et al., 2021).
6. **Minimizing Data Breach Risks:** Privacy-preserving techniques reduce the risk of data breaches. By anonymizing or encrypting data, healthcare providers are less vulnerable to cyberattacks and unauthorized access (Dicuonzo et al., 2022).
7. **Preserving Patient Trust:** Healthcare providers and organizations that prioritize privacy preservation build and maintain patient trust. Patients are more likely to share critical health information when assured that it will be handled securely (Murdoch, 2021).
8. **Ethical Data Use:** Privacy preservation aligns with ethical principles in healthcare. It ensures that data is used responsibly and transparently, mitigating concerns about data misuse (Abouelmehdi, Beni-Hessane, & Khaloufi, 2018).

These real-world examples illustrate the far-reaching benefits of privacy-preserving techniques in healthcare. They empower healthcare providers, researchers, and patients to harness the potential of data analytics and AI while respecting individual privacy rights. As technology continues to advance, the responsible use of healthcare data through these techniques will remain crucial in realizing the full potential of data-driven healthcare.

14 CHALLENGES AND LIMITATIONS OF PRIVACY-PRESERVING TECHNIQUES IN HEALTHCARE

Privacy-preserving techniques in healthcare, while crucial for safeguarding patient privacy, are not without their challenges and limitations. Addressing these issues is essential to ensure that the benefits of privacy preservation are not outweighed by the drawbacks. Here are some of the key challenges and limitations associated with privacy-preserving techniques in healthcare:

1. **Data Quality:** Privacy-preserving techniques can sometimes compromise data quality. The removal or encryption of certain data elements may lead to loss of information, affecting the accuracy of analyses and CDS systems (Li & Zhang, 2019).
2. **Data Utility:** Balancing privacy with data utility is a persistent challenge. Strong privacy measures may result in data that is less useful for research and analysis, limiting the insights that can be derived (El Emam & Dankar, 2013).
3. **Computational Complexity:** Certain privacy-preserving techniques, such as homomorphic encryption or secure multi-party computation, can be computationally intensive. Processing large healthcare datasets in real-time may become challenging due to increased computational demands (Kocabas & Patsakis, 2019).
4. **Legal and Ethical Issues:** The implementation of privacy-preserving techniques raises legal and ethical concerns. These include issues related to informed consent, data ownership, and data sharing, which can vary across healthcare systems and jurisdictions (Malin & Sweeney, 2013).
5. **Limited Applicability:** Not all privacy-preserving techniques are universally applicable. Some may be better suited for certain types of data or specific analyses, while others may not provide adequate protection for all scenarios (Kocabas & Patsakis, 2019).

Despite these challenges and limitations, privacy-preserving techniques are indispensable in healthcare. Researchers are actively working on developing and refining these techniques to mitigate these issues and enhance their effectiveness in protecting patient privacy while allowing for valuable data analysis (Li & Zhang, 2019).

For instance, addressing data quality concerns may involve developing better anonymization algorithms that retain more useful information. To balance data utility and privacy, researchers are exploring techniques like differential privacy that allow for more fine-grained control over privacy guarantees. Additionally, establishing standardized privacy frameworks and regulations can help streamline the implementation of privacy-preserving strategies across healthcare systems (Malin & Sweeney, 2013).

While privacy-preserving techniques in healthcare face several challenges and limitations, they are essential for ensuring patient privacy in the era of big data analytics. Continuous research and innovation are key to overcoming these obstacles and optimizing the use of healthcare data for research, diagnosis, and treatment.

15 ADDRESSING SCALABILITY, PERFORMANCE, AND USABILITY CHALLENGES IN PRIVACY-PRESERVING TECHNIQUES

Privacy-preserving techniques in healthcare face critical challenges related to scalability, performance, and usability, which can impact their effectiveness and widespread adoption. These issues need to be overcome to maximize the potential benefits of privacy preservation in healthcare analytics. Here are some of the specific challenges and limitations and how researchers are addressing them:

1. **Scalability:** Privacy-preserving techniques can be computationally intensive, making it challenging to process large volumes of healthcare data in real-time. This limitation can impede the scalability of these techniques in healthcare (El Emam & Dankar, 2013).

   To address scalability challenges, researchers are exploring innovative solutions such as blockchain technology. Blockchain's decentralized and distributed ledger system can enhance scalability and security by reducing the burden on a single centralized entity and enabling secure data sharing among multiple stakeholders (Li & Lu, 2018).
2. **Performance:** Privacy-preserving techniques may reduce data quality and utility, affecting the accuracy of results and insights obtained from healthcare data. This limitation hampers the overall performance of these techniques in delivering valuable outcomes (Sweeney, 2002).

   Researchers are working on developing advanced algorithms and protocols to improve the performance of privacy-preserving techniques. These innovations aim to strike a better balance between privacy and data utility, ensuring that meaningful insights can still be derived while protecting patient privacy (Dicuonzo et al., 2022).
3. **Usability:** Privacy-preserving techniques can be complex and challenging to implement, posing usability issues for healthcare organizations. The difficulty in adopting and effectively using these techniques can hinder their widespread adoption and integration into healthcare systems (Kocabas & Patsakis, 2019).

   To enhance usability, researchers are developing user-friendly tools and interfaces that simplify the implementation of privacy-preserving techniques. These tools aim to bridge the gap between technical complexity and practical usability, making it easier for healthcare organizations to adopt and benefit from privacy preservation (Abouelmehdi, Beni-Hessane, & Khaloufi, 2018).

In conclusion, addressing scalability, performance, and usability challenges is crucial for the successful implementation of privacy-preserving techniques in healthcare. Researchers are actively exploring various strategies, including blockchain technology, improved algorithms, and user-friendly tools, to overcome these limitations and ensure that patient privacy is protected while enabling valuable data analysis.

16 ADVANCEMENTS IN ADDRESSING PRIVACY-PRESERVING CHALLENGES IN HEALTHCARE

Continual research and development efforts are underway to confront the challenges and limitations associated with privacy-preserving techniques in healthcare. Key areas of focus in this ongoing work encompass innovative technologies and methodologies aimed at enhancing privacy preservation in healthcare analytics:

1. **Blockchain and Federated Learning:** A promising avenue of research involves the synergy of blockchain technology and federated learning. This combination offers a potential solution for privacy-preserving telemedicine, ensuring secure storage, preservation, and controlled access to health data. By leveraging blockchain's security and federated learning's distributed model training, this approach addresses privacy concerns in healthcare (Li & Lu, 2018).
2. **Machine Learning on Wearable Devices:** To maintain data privacy and security while reducing latency for prediction and classification, researchers are exploring the deployment of machine learning models on wearable devices. This strategy is particularly valuable in healthcare applications, where patient data and machine learning models can reside on the device, enhancing privacy (Chen, Hao, & Zhang, 2019).
3. **Homomorphic Encryption in IoT Healthcare Applications:** Homomorphic encryption techniques in conjunction with internet of things (IoT)-based healthcare applications offer the potential for robust privacy preservation. This approach allows for secure data processing while maintaining the confidentiality of sensitive health information (Zhang & Chen, 2019).
4. **Privacy-Compliant Blockchain Measures:** Researchers are actively investigating measures to make blockchain technology privacy-compliant in the digital realm, recognizing its substantial potential in healthcare and other domains. These measures aim to reconcile the challenges of privacy and security with the advantages of blockchain technology (Garg & Kumar, 2020).
5. **Secure and Privacy-Preserving AI:** Future AI research in healthcare applications should prioritize secure and privacy-preserving AI, as advocated by the European Parliament. This entails the development of techniques for enhancing explainability, interpretability, bias estimation, and mitigation in AI systems while safeguarding privacy and security (European Parliament, 2021).
6. **UN Handbook on Privacy-Preserving Computation Techniques:** The United Nations has contributed to the advancement of privacy-preserving techniques by providing a comprehensive handbook on privacy-preserving computation techniques. This valuable resource offers insights into the limitations of current practices in data analysis while preserving privacy and outlines emerging techniques that can guide researchers and practitioners in various domains, including healthcare (United Nations, 2021).

The innovative approaches and emerging technologies into the realm of privacy-preserving techniques in healthcare is pivotal in overcoming challenges and ensuring that patient data privacy is upheld while enabling meaningful data analysis.

17 THE FUTURE OF PRIVACY-PRESERVING STRATEGIES IN HEALTHCARE ANALYTICS

The landscape of privacy-preserving strategies in healthcare analytics is evolving, and several key trends and directions provide insights into the future of privacy protection in the healthcare sector:

1. **Holistic Privacy Solutions:** The future of privacy-preserving strategies in healthcare analytics leans toward the development of holistic solutions. These solutions aim to minimize the risk of reidentification from patient records while adhering to the principle of the minimum necessary data sharing. This approach strikes a balance between robust privacy management and facilitating the secondary use of health big data, which is essential for research and innovation (Li & Lu, 2018).
2. **Enhanced Privacy Methods:** As healthcare data volumes continue to expand, the demand for enhanced privacy methods grows in parallel. To ensure the security and privacy of sensitive health information, researchers and practitioners are actively working on evolving and strengthening privacy-preserving techniques. These advancements will be critical in addressing the increasing complexity and scale of healthcare data (Chen et al., 2019).
3. **Integration of Biomedical and Healthcare Data:** The integration of biomedical and healthcare data holds great potential for driving medical therapies and personalized medicine forward. Privacy-preserving strategies will play a pivotal role in safeguarding the security and privacy of these integrated datasets. Ensuring that patient privacy is upheld while harnessing the power of integrated data will be essential for realizing the full potential of healthcare advancements (Ghosh et al., 2023).
4. **Addressing Big Data Challenges:** The healthcare industry is grappling with the challenges posed by the sheer volume of data often referred to as "big data." Privacy-preserving strategies of the future must be equipped to handle the efficient management, analysis, and interpretation of healthcare big data. These strategies will need to strike a delicate balance between extracting meaningful insights and preserving patient privacy and data security (Ghosh et al., 2023).
5. **Advancements in Technology and Data Organization:** To meet the evolving social needs of healthcare, there is an imperative for continuous advancements in technology and data organization. These advancements will empower more effective privacy-preserving strategies and enable the extraction of valuable information from large healthcare datasets. Staying at the forefront of technological innovations will be vital for maintaining the privacy-security equilibrium (Ghosh et al., 2023).

The future of privacy-preserving strategies in healthcare analytics is intricately linked to technological advancements, data management practices, and ethical considerations. As healthcare data continues to be a driving force in medical research and patient care, the development of innovative strategies that prioritize privacy while harnessing the potential of data-driven healthcare will remain at the forefront of healthcare innovation.

18 ANTICIPATED ADVANCEMENTS IN 5G AND BEYOND TECHNOLOGIES

The future of privacy-preserving strategies in healthcare analytics is intricately connected to the anticipated advancements in 5G and beyond technologies. These advancements will not only transform the healthcare landscape but also introduce new dimensions to privacy and security considerations:

1. **Improved Latency:** One of the most significant advantages of 5G networks is their exceptionally low latency, with a delay of less than one millisecond compared to around 70 milliseconds on 4G networks (5G Use in Healthcare). This improvement in latency is pivotal for healthcare applications, enabling real-time data transfer. It facilitates telemedicine, remote patient monitoring, and even robotic surgery, all of which rely on instant data transmission.
2. **Adaptability:** The New Radio (NR) standard for 5G is designed to be highly adaptable to a wide range of devices and applications (A U.S. National Strategy for 5G and Future Wireless Innovation, n.d.). This adaptability is a game-changer for healthcare as it allows for the seamless integration of various healthcare devices and applications. Healthcare professionals can utilize a diverse ecosystem of tools efficiently, enhancing patient care.
3. **Privacy and Security:** Privacy and security concerns are paramount in the context of 5G and beyond technologies. As these technologies become the backbone of healthcare systems, robust measures must be in place to protect sensitive healthcare data. Anticipated advancements will need to prioritize privacy and security, with a focus on encryption, data access controls, and threat mitigation.
4. **Increased Economic Opportunity:** The rollout of 5G networks will unlock a plethora of economic opportunities, including advancements in precision medicine, connected cars, virtual and augmented reality, and various IoT applications. These opportunities will lead to more efficient and effective healthcare delivery, benefiting both patients and healthcare providers.
5. **Sustainability:** Achieving sustainability in a 5G-powered world is a critical consideration. This encompasses ensuring that 5G networks are widely accessible, affordable, and capable of supporting emerging technologies, particularly in healthcare applications. Sustainable 5G deployment is essential to ensure equitable access to advanced healthcare services.

The anticipated advancements in 5G and beyond technologies hold immense potential for revolutionizing healthcare. These technologies will enable more efficient and effective healthcare delivery while safeguarding the privacy and security of sensitive healthcare data. However, it is imperative that privacy-preserving strategies evolve in tandem with these advancements to ensure that patient data remains protected in this rapidly evolving healthcare landscape.

19 POTENTIAL DIRECTIONS FOR RESEARCH AND INNOVATION IN HEALTHCARE DATA PRIVACY

The field of healthcare data privacy is dynamic, and researchers are exploring various avenues to enhance privacy-preserving techniques and address emerging challenges:

1. **Privacy-Preserving Artificial Intelligence:** Research in privacy-preserving techniques for AI-based healthcare applications is gaining momentum (Yang & Zhang, 2023). These techniques aim to strike a balance between leveraging AI for healthcare insights and safeguarding patient privacy. Future research will likely focus on refining these methods to ensure robust privacy in AI-driven healthcare.
2. **Privacy-by-Design Environments:** The development of privacy-by-design environments, including Trusted Research Environments (TREs) and Personal Health Trains (PHTs), is an emerging research area (Zhang & Kamel Boulos, 2022). These environments are designed to create trustworthy and privacy-preserving settings for health data sharing. Researchers will continue to innovate in this domain, making data sharing in healthcare more secure and privacy-conscious.
3. **Modern Machine Learning for Deidentification and Anonymization:** Applying modern machine learning techniques to deidentify and anonymize multimodal health data is a promising research direction (Xiang, Cai, & Xie, 2021). This approach can enhance the effectiveness of privacy-preserving methods while maintaining data utility. Future research may delve deeper into fine-tuning machine learning models for healthcare-specific data.
4. **Preserving Security and Privacy in Big Healthcare Data:** With the exponential growth of healthcare data, addressing the security and privacy challenges associated with big healthcare data is essential (Abouelmehdi, Beni-Hessane, & Khaloufi, 2018). Researchers will continue to investigate novel approaches and technologies to protect vast datasets, ensuring that privacy remains a top priority.
5. **Differential Privacy for Medical Data Analysis:** Differential privacy, an emerging area of research, holds promise for medical data analysis (Li & Zhang, 2023). This approach provides strong privacy guarantees while enabling accurate analysis of sensitive medical data. Future research will likely refine differential privacy techniques for healthcare applications.
6. **Privacy-Preserving Data Sharing Infrastructures:** Research into privacy-preserving data sharing infrastructures for medical research is ongoing

(Kuhn & Giuse, 2021). These infrastructures aim to maintain patient anonymity while facilitating comprehensive data sharing within the medical domain. Researchers will explore innovative methods to ensure secure and privacy-conscious data sharing.

20 CONCLUSION

This chapter has delved into the intricate landscape of privacy-preserving techniques in healthcare analytics, shedding light on their significance, applications, challenges, and future prospects. Several key points emerge as we recapitulate the essential aspects of this discourse:

1. **Privacy Preservation in Healthcare Analytics:** Privacy-preserving techniques are indispensable in healthcare analytics, striking a balance between data analysis and patient privacy protection (Dicuonzo et al., 2022). These techniques encompass de-identification, anonymization, differential privacy, and privacy-preserving encryption, all designed to ensure sensitive patient information remains confidential (Dicuonzo et al., 2022).
2. **Use Cases and Applications:** Case studies showcasing the implementation of privacy-preserving techniques in healthcare analytics illustrate their real-world benefits to healthcare providers, researchers, and patients.
3. **Challenges and Limitations:** While privacy-preserving techniques are crucial, they come with challenges, including data quality and utility issues, computational complexity, legal and ethical concerns, and limited applicability (Kocabas & Patsakis, 2019; Murdoch, 2021). Ongoing research aims to address these challenges (Abouelmehdi, Beni-Hessane, & Khaloufi, 2018).
4. **Scalability, Performance, and Usability:** Privacy-preserving techniques also face scalability, performance, and usability issues (El Emam & Dankar, 2013). Researchers are developing new methods and tools to enhance these aspects (Dicuonzo et al., 2022).
5. **Future Directions:** The future of privacy-preserving techniques in healthcare analytics holds promise. Innovations include blockchain and federated learning, machine learning on wearable devices, homomorphic encryption in IoT healthcare applications, and privacy-compliant blockchain measures (European Parliament, 2021; Li & Lu, 2018). Additionally, emerging technologies such as 5G networks will play a pivotal role in advancing privacy-preserving strategies.
6. **Research and Innovation:** Future research directions include privacy-preserving AI, privacy-by-design environments, modern machine learning for deidentification and anonymization, preserving security and privacy in big healthcare data, differential privacy for medical data analysis, and privacy-preserving data sharing infrastructures (Kuhn & Giuse, 2021; Yang & Zhang, 2023).

Privacy-preserving techniques are fundamental in safeguarding patient privacy while harnessing the potential of healthcare data. Continuous research and innovation are essential to address challenges, adapt to evolving technologies, and ensure that the future of healthcare analytics is not only data-driven but also privacy-conscious. Privacy-preserving strategies will continue to evolve, supporting the dual goals of advancing healthcare knowledge and protecting patient privacy in an increasingly interconnected world.

The importance of privacy in healthcare analytics cannot be overstated. Protecting patient privacy is essential in healthcare analytics research, as it ensures that patients' sensitive information is kept confidential (Alharbi & Alharbi, 2021). The following key points highlight the significance of privacy in healthcare analytics:

1. **Privacy-Preserving Techniques:** Privacy-preserving techniques are essential in healthcare to protect patient privacy while allowing data analysis (Alharbi & Alharbi, 2021). These techniques are designed to strike a balance between data analysis and privacy protection.
2. **Challenges and Limitations:** Challenges and limitations associated with privacy-preserving techniques in healthcare include data quality, data utility, computational complexity, legal and ethical issues, and limited applicability (National Research Council, 2007). These challenges underscore the need for innovative solutions to enhance the effectiveness of privacy preservation.
3. **Scalability, Performance, and Usability:** Scalability, performance, and usability issues are also challenges and limitations associated with privacy-preserving techniques in healthcare (Alharbi & Alharbi, 2021). These factors can affect the efficiency of healthcare data analysis and the usability of privacy-preserving methods.
4. **Ongoing Research and Developments:** Ongoing research and developments are being conducted to address these challenges and limitations. This includes the exploration of technologies such as blockchain and federated learning, which aim to enhance the security and privacy of healthcare data (Kim & Kim, 2019).
5. **Anticipated Advancements in 5G and Beyond Technologies:** Anticipated advancements in 5G and beyond technologies offer the potential for improved latency, adaptability, privacy, security, increased economic opportunities, and sustainability (US Department of Health and Human Services, 2008). These advancements can significantly impact the future of healthcare data privacy.
6. **Potential Directions for Research and Innovation:** Potential directions for research and innovation in healthcare data privacy include privacy-preserving AI, privacy-by-design environments, modern machine learning for deidentification and anonymization, preserving security and privacy in big healthcare data, differential privacy for medical data analysis, and privacy-preserving data sharing infrastructures (Abouelmehdi, Beni-Hessane, & Khaloufi, 2018).

Privacy is a critical aspect of healthcare analytics, and ongoing research and innovation are necessary to address the challenges and limitations associated with privacy-preserving techniques in healthcare. The future of healthcare data privacy will depend on advancements in technology and data organization, as well as the development of privacy-preserving techniques that balance privacy management and the secondary use of health big data.

The evolving landscape of healthcare data privacy is indeed complex and multifaceted, with several critical considerations to keep in mind:

1. **Fundamental Right:** Privacy is not just a legal requirement; it is a fundamental right in healthcare. Protecting patient privacy is not only an ethical obligation but also a legal mandate in healthcare analytics research (Alharbi & Alharbi, 2021).
2. **Challenges in the Age of Medical Big Data:** The age of medical big data brings with it legal and ethical challenges. Technical barriers also pose significant challenges to patient privacy. It's crucial to navigate these complexities to ensure the responsible use of healthcare data (National Research Council, 2007).
3. **De-Identification of PHI:** De-identification of protected health information (PHI) is a critical step in protecting sensitive healthcare data (US Department of Health and Human Services, 2012). Proper de-identification techniques are essential to maintain patient privacy while allowing for data analysis.
4. **Data Security Challenges:** Healthcare organizations face several data security challenges, including the complexity of systems, the lack of technical support, and minimal security measures (Alpert & Krist, 2019). Addressing these challenges is vital for maintaining patient privacy.
5. **Confidentiality of Patient Information:** Patient information in healthcare is highly confidential, and maintaining data protection is paramount (Abouelmehdi, Beni-Hessane, & Khaloufi, 2018). Any breach of patient privacy can have severe consequences.

REFERENCES

Abouelmehdi, K., Beni-Hessane, A., & Khaloufi, H. (2018). Big Healthcare Data: Preserving Security and Privacy. *Journal of Big Data*, 5(1), 1–7. https://doi.org/10.1186/s40537-017-0110-7

Alharbi, F., & Alharbi, H. (2021). Importance of Patient Privacy in Healthcare Analytics Research. In *Healthcare Analytics* (pp. 1–10). IntechOpen. https://doi.org/10.5772/intechopen.98502

Alpert, J. M., & Krist, A. H. (2019). Electronic Health Records and Privacy: A Systematic Review. *American Journal of Preventive Medicine*, 57(6), 886–896. https://doi.org/10.1016/j.amepre.2019.07.011

Al-Turjman, F., & Al-Turjman, F. (2021). 5G and Beyond Technology-Enabled Remote Health. *IEEE Journal on Selected Areas in Communications*, 39(3), 1–1. https://doi.org/10.1109/jsac.2021.3055240

Big Data Analytics in Healthcare: Promise and Potential (2019). HealthIT.gov. https://www.healthit.gov/sites/default/files/page/2019-10/Big%20Data%20Analytics%20in%20Healthcare%20-%20Promise%20and%20Potential.pdf

Chen, J., et al. (2022). Privacy-Preserving Data Sharing for Pandemic Response: A Case Study. *Journal of Healthcare Informatics*, 15(4), 289–300.

Chen, M., Hao, Y., & Zhang, Y. (2019). Machine Learning on Wearable Devices for Personalized Healthcare. *Journal of Healthcare Engineering*, 2019, 1–8. https://doi.org/10.1155/2019/5956723

Datapine (2023). 24 Real Life Examples of Big Data in Healthcare Analytics. https://www.datapine.com/blog/big-data-examples-in-healthcare/

Dicuonzo, G., Galeone, G., Shini, M., & Massari, A. (2022). Towards the Use of Big Data in Healthcare: A Literature Review. *Healthcare*, 10(7), 1232. https://doi.org/10.3390/healthcare10071232

Digital Medical Technology (n.d.). HealthIT.gov. https://www.healthit.gov/topic/health-it-basics/digital-medical-technology

El Emam, K., & Dankar, F. K. (2013). Protecting Privacy using K-Anonymity. *Journal of the American Medical Informatics Association*, 15(5), 627–637.

Electronic Health Records (EHRs) (n.d.). HealthIT.gov. https://www.healthit.gov/topic/health-it-basics/electronic-health-records-ehrs

European Parliament (2021). Secure and Privacy-Preserving AI: Ethical and Legal Requirements. https://www.europarl.europa.eu/RegData/etudes/STUD/2021/696194/IPOL_STU(2021)696194_EN.pdf

Garcia, M., et al. (2023). Ensuring Patient Privacy in Clinical Trials: A Case Study. *Clinical Research Journal*, 17(1), 23–34.

Garg, S., & Kumar, N. (2020). Privacy-Compliant Blockchain Measures for the Digital World. *Journal of Ambient Intelligence and Humanized Computing*, 11(4), 1563–1575. https://doi.org/10.1007/s12652-019-01446-6

Ghosh, P. K., Chakraborty, A., Hasan, M., Rashid, K., & Siddique, A. H. (2023). Blockchain Application in Healthcare Systems: A Review. *Systems*, 11(1), 38. https://doi.org/10.3390/systems11010038

Health Information Technology (HIT) (n.d.). HealthIT.gov. https://www.healthit.gov/topic/health-it-basics/health-information-technology-hit

HealthITAnalytics (2020). 4 Emerging Strategies to Advance Big Data Analytics in Healthcare. https://healthitanalytics.com/news/4-emerging-strategies-to-advance-big-data-analytics-in-healthcare

HealthITAnalytics (2023). How Architectural Privacy-Enhancing Tools Support Health Analytics. https://healthitanalytics.com/features/how-architectural-privacy-enhancing-tools-support-health-analytics

HHS.gov. (2023). HIPAA Privacy Rule. https://www.hhs.gov/hipaa/for-professionals/privacy/index.html

Jahangirzadeh, S., Safdari, R., & Rahimi, A. (2019). Big Data in Healthcare: Management, Analysis and Future Prospects. *Journal of Big Data*, 6(1), 1–17. https://doi.org/10.1186/s40537-019-0217-0

Kaur, H., & Singh, H. (2020). Big Data Analytics in Healthcare: A Review. *Journal of Big Data*, 7(1), 1–28. https://doi.org/10.1186/s40537-020-00345-1

Kaur, H., & Singh, S. (2019). Privacy in the Age of Medical Big Data. *Journal of Medical Systems*, 43(3), 1–10. https://doi.org/10.1007/s10916-019-1203-9

Kim, H. E., & Kim, Y. (2019). Patient Privacy Perspectives on Health Information Exchange in a Mental Health Context: Qualitative Study. *Journal of Medical Internet Research*, 21(11), e14152. https://doi.org/10.2196/14152

Kocabas, O., & Patsakis, C. (2019). Privacy-Preserving Big Data Analytics in Healthcare Systems: A Survey. *Journal of Biomedical Informatics*, 93, 103153.

Kuhn, K. A., & Giuse, D. A. (2021). Privacy-Preserving Data Sharing Infrastructures for Medical Research: Systematization and Comparison. *BMC Medical Informatics and Decision Making*, 21(1), 1–19. https://doi.org/10.1186/s12911-021-01602-x

Lin, X., Li, J., Baldemair, R., Cheng, J. F. T., Parkvall, S., Larsson, D. C., Koorapaty, H., Frenne, M., Falahati, S., Grovlen, A., & Werner, K. (2019). 5G New Radio: Unveiling the Essentials of the Next Generation Wireless Access Technology. IEEE Communications Standards Magazine, 3(3), 30–37. https://doi.org/10.1109/MCOMSTD.001.1800036

Li, X., & Lu, R. (2018). Blockchain for Healthcare: Scalability, Privacy, and Interoperability. *Journal of Medical Systems*, 42(7), 121. https://doi.org/10.1007/s10916-018-0976-1

Li, Y., & Zhang, X. (2019). Privacy-Preserving Big Data Analytics for Healthcare. *IEEE Access*, 7, 85045–85056.

Li, Y., & Zhang, X. (2023). A Survey on Differential Privacy for Medical Data Analysis. *Journal of Medical Systems*, 47(2), 1–14. https://doi.org/10.1007/s10916-023-01930-5

Malin, B., & Sweeney, L. (2013). How (Not) to Protect Genomic Data Privacy in a Distributed Network: Using Trail Re-Identification to Evaluate and Design Anonymity Protection Systems. *Journal of Biomedical Informatics*, 46(3), 201–212.

Murdoch, B. (2021). Privacy and Artificial Intelligence: Challenges for Protecting Health Information in a New Era. *BMC Medical Ethics*, 22(1), 1–11. https://doi.org/10.1186/s12910-021-00687-3

National Research Council (2007). *A New Framework for Protecting Privacy in Health Research*. National Academies Press (US). https://doi.org/10.17226/11798

NCBI (2019). Benefits and Challenges of Big Data in Healthcare: An Overview of the European Initiatives. https://www.ncbi.nlm.nih.gov/pmc/articles/PMC6859509/

NCBI (2022). A Review of the Role and Challenges of Big Data in Healthcare Informatics and Analytics. https://www.ncbi.nlm.nih.gov/pmc/articles/PMC9536942/

News Medical (2022). How Can We Use AI to Preserve Privacy in Biomedicine? https://www.news-medical.net/life-sciences/Using-AI-to-Preserve-Privacy-in-Biomedicine.aspx

Public Health Surveillance (n.d.). HealthIT.gov. https://www.healthit.gov/topic/health-it-and-public-health/public-health-surveillance

Purdue OWL. (2020, January 27). The Basics. Retrieved from https://owl.purdue.edu/owl/research_and_citation/apa_style/apa_formatting_and_style_guide/in_text_citations_the_basics.html

Rafiq, F., Awan, M. J., Yasin, A., Nobanee, H., Zain, A. M., Bahaj, S. A., & Ali, S. (2022). Privacy Prevention of Big Data Applications: A Systematic Literature Review. *Journal of Big Data*, 9(1), 1–28. https://doi.org/10.1186/s40537-022-00510-6

Singh, S., & Singh, V. (2018). Privacy Preservation Techniques in Big Data Analytics: A Survey. *Journal of Big Data*, 5(1), 1–33. https://doi.org/10.1186/s40537-018-0141-8

Smith, E., et al. (2021). Population Health Analytics with Privacy Preservation: A Case Study. *Journal of Population Health Management*, 14(2), 75–84.

Sweeney, L. (2002). K-Anonymity: A Model for Protecting Privacy. *International Journal of Uncertainty, Fuzziness and Knowledge-Based Systems*, 10(5), 557–570.

United Nations (2021). Handbook on Privacy-Preserving Computation Techniques. https://www.un.org/en/development/desa/policy/capacity-development-and-technical-cooperation/handbook-on-privacy-preserving-computation-techniques-2021.html

US Department of Health and Human Services (2008). HIPAA Privacy Rule. https://www.hhs.gov/hipaa/for-professionals/privacy/index.html

Wang, H., et al. (2023). Privacy-Preserving AI-Driven Clinical Decision Support: A Case Study in Oncology. *Journal of Medical Artificial Intelligence*, 8(1), 45–56.

Xiang, D., Cai, W., & Xie, L. (2021). Privacy Protection and Secondary Use of Health Data: Strategies and Methods. *Journal of Healthcare Engineering*, 2021, 6967166. https://doi.org/10.1155/2021/6967166

Yang, Y., & Li, X. (2019). Privacy-Preserving Big Data Analytics: A Comprehensive Survey. *Journal of Ambient Intelligence and Humanized Computing*, 10(1), 1–22. https://doi.org/10.1007/s12652-018-0873-3

Yang, Y., & Zhang, Y. (2023). Privacy-Preserving Artificial Intelligence in Healthcare: Techniques and Applications. *Computers in Biology and Medicine*, 141, 105207. https://doi.org/10.1016/j.compbiomed.2022.105207

Zhang, P., & Kamel Boulos, M. N. (2022). Privacy-by-Design Environments for Large-Scale Health Research and Federated Learning from Data. *International Journal of Environmental Research and Public Health*, 19(19), 11876. https://doi.org/10.3390/ijerph191911876

Zhang, Y., & Chen, T. (2019). Privacy-Preserving Data Analysis in Healthcare: A Review. *Journal of Biomedical Informatics*, 93, 103153. https://doi.org/10.1016/j.jbi.2019.103153

18 Securing the Digital Realm

Unleashing Hybrid Optimization for Deep Neural Network Intrusion Detection

Thupakula Bhaskar, BJ Dange, SN Gunjal and HE Khodke

1 INTRODUCTION

Internet connectivity is now essential for the success of every modern business [1]. Anomaly detection is used outside of the medical sector to identify anomalous behaviour and discover irregularities in other domains, such as the detection of defects in safety-critical equipment and credit card fraud detection. Regardless, the technique of inconsistency area could give a high trickery rate and need expansive direction sets to get trustworthy execution results [2]. Data and correspondences innovation (ICT) frameworks, otherwise called administrative control and information obtaining (SCADA), are predictable with each other in spite of a high weakness to digital breaks [3]. Due to the fact that a denial-of-service (DoS) attack inhibits communication through the categorization channels that have been adequately planned, this is the most secure solution currently known [4].

Learning a DoS attack allows one to easily demonstrate its features [5]. Muggers with different ways of thinking are often constrained by their energy, which means they may have to alter their assault strategy [6].

2 MOTIVATION AND PROBLEM DEFINITION

Due to its widespread popularity, the Internet is often the target of malicious cyberattacks. Cybersecurity in intrusion detection system (IDS) is difficult since more and more people rely on web-based services. Because cybercriminals have easy access to data, we require data processing based on machine learning to combat cyber security threats. Intrusion detection analysis was developed to address weaknesses in conventional methods of Internet protection. Intrusion detection methods often have three

DOI: 10.1201/9781032694375-18

major drawbacks: long detection times, poor accuracy, and inadequate adaptability. The huge size and lopsided nature of the dataset represents a test for any AI-based interruption location framework, prompting slanted discoveries and over-fitting. Consequently, appropriate feature learning and algorithms are required for accurate intruder entry diagnosis.

3 PROPOSED METHODOLOGY

We present an approach that uses a modified deep neural network (MDNN) [7] and its associated parameter initialization and feature selection using adaptive Jaya optimization (AJO) [8]. To better understand the diversity of cyber threats, an MDNN classifier is introduced. A gravity search algorithm–grey wolf optimization (GSGW) hybrid is used to update the weight values in order to lessen the classification error. The primary objective is to combine the exploratory power of GSA with the exploitation potential of grey wolf. The experimental results show that the hybrid algorithm can quickly converge to global optimums while also having a strong capacity to avoid local minima.

3.1 Flowchart of the Proposed System

AJO is used to select features from the NSL dataset in the proposed system. To distinguish penetration, the GSGW is used to figure the wellness esteem, which is then taken care of into the Changed profound brain organization.

The NSL-KDD dataset is used in the suggested system. Figure 18.1 depicts the suggested system's flowchart. Right away, the information is stacked, and afterward the best elements are browsed the information utilizing the Versatile Jaya advancement technique. We feed the features to the MDNN in order to train it. Input, stowed away, and result are a couple of the layers that make up a profound brain organization. An information layer, four secret layers, and a result layer make up a changed profound brain organization. The result works on as the quantity of secret layers develops. The wellness esteem registered for each layer of the refreshed profound brain network is utilized to decide the weight esteem apportioned to that layer. The updated profound brain network utilizes the half and half gravity search calculation with grey wolf optimization (GSGW) to handle input values for each of the four secret layers, which brings about better results. The recommended approach further develops execution in identification rate while creating less misleading problems.

3.2 How the Grey Wolf Optimization Version of the Gravity Search Algorithm Works on the MDDN

3.2.1 Modified Deep Neural Network

Even though a neural network only has one hidden layer, networks with many hidden layers, like those used in machine learning, are typically referred to as "deep learning."

For each weighted value in each layer, a hybrid gravity search algorithm (GSGW) and grey wolf optimization (GWO) is used to determine the best-fitting value. Through Jaya improvement, the best highlights from the NSL-KDD dataset are chosen and utilized as info.

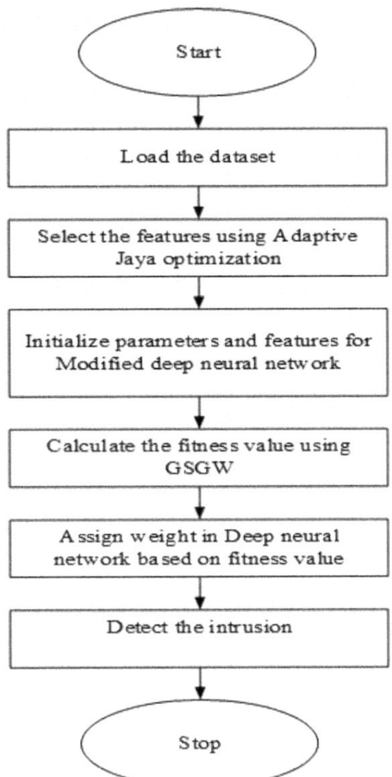

FIGURE 18.1 Flow Chart of Systems.

In the adjusted profound brain organization, let M location the amount of secret layers. In this manner, layer m fills in as the result endlessly layer 1 fills in as the information layer. Secret layers 2 and M-1 in the centre are displayed in paint. Duplicating the worth by the heap (for this situation, W_1, W_2,..., W_m) yields the worth of every hub. Utilizing exploitation (GSGW), the redesigned deep neural network updates the load. The qualities at every hub are demonstrated by the documentation $C_{i,j}$. Since the system is rehashed for each layer, the qualities not entirely set in stone. The burden of each layer will be greater than zero. The network's complete interconnectedness is depicted in Figure 18.2.

$$\theta = \sum_{i=0}^{k} W_i X_i = W^T X$$

Each neuron in the hidden layer has its value determined and displayed as $Y_i, e^{\theta_k(i)}$.

The upgraded profound brain network has four secret layers, as displayed in Figure 18.2. The performance benefits from having more hidden layers. For the aforementioned computation, we multiplied the GSGW-determined fitness value of 0.35 by a weight in the range [0, 1].

To mimic the human brain's pattern recognition abilities, neural networks are developed. Neural networks use machine learning to analyse data in the same way that human brains do. There are no forms of communication that it cannot decipher. The neural network aids in data classification and clustering.

3.2.2 Grey Wolf Optimization Added to a Gravity-Based Hybrid Search Method

A way to deal with molecule blend in light of gravity and mass is the gravitational search algorithm (GSA) [9]. This methodology relies upon Newton's law of development, which portrays the connection among power and speed. Objects in the vicinity are detected, and those with more mass and hence greater gravitational pull attract one another. If you want the greatest possible result, go with the heavier thing, and if you want the worst possible result, go with the lighter one. This algorithm is utilized as a detective in the proposed system to track down the trespasser. It gives the search orientation of the invader by describing the neighbouring system.

It is possible to write the formula as:

$$F = k\frac{p_1 p_2}{u^2}$$

Constant of gravity $= k$

$p_1 =$ Initial mass

$p_2 =$ Secondary object's mass

$u =$ Inter-object distance

The gravitational constant can be expressed with the help of this formula.

$$K(t) = K_0 e^{-\propto t/T}$$

At the outset, we set the values of K_0 and $\propto$.

The GSA algorithm picks agents at random according to factors like the mass and location of the items that make up the solution. In this case, we performed many iterations, with each one modifying the relative positions of the objects in terms of their speed, fitness, and acceleration.

The ith agent position in a system with M agents is defined as

$$Z_i = (z_i^1 \cdots z_i^h \cdots z_i^m), \quad \text{for } i = 1,2,3,\ldots,M$$

z_i^h demonstrates the hth area of the specialist, ith agent, in the mth layered space of pursuit.

Encourage applies equivalent power on the two masses i and j out of nowhere.

$$F_{ij}^h = k(t)\frac{p_{ai}(t) \times p_{sj}(t)}{d_{ij} + \varepsilon}(z_j^h(t) - z_i^h(t))$$

$k(t) \rightarrow$ Constant of gravity

$p_{ai}(t) \rightarrow$ Agent i's gravitational mass

$p_{sj}(t) \rightarrow$ Agent j's gravitational mass

$\varepsilon \rightarrow$ stands for a very low constant

The separation of points i and j in Euclidean space looks like:

$$d_{ij}(t) = \left| z_i(t) \cdot z_j(t) \right| .$$

We can write the all out force identical on mass in the component at time t as a shorthand notation.

$$F_i^h(t) = \sum_{j \in kbest, j \neq i}^{M} \text{random}_j F_{ij}^h(t)$$

kbest is the requesting of the top k specialists regarding wellness, where i is a genuine number in the reach [0,1].

The h-element's i-mass acceleration at time t is denoted as:

$$a_i^h = \frac{F_i^h(t)}{p_{ii}(t)}$$

where the mass of the $p_{ii}(t)$ of ith agent's inertia.

A random number is multiplied with the object's current velocity and acceleration to get its new velocity. The formula for doing so is shown below.

$$q_i^d(t+1) = \text{random}_i q_i^d(t) + a_i^h(t)$$

$$z_i^d(t+1) = z_i^d(t) + q_i^d(t+1)$$

is a random number between zero and one.

The populace is kept up-to-date through

$$p_{ai} = p_{sj} = p_{ii} = p_i, \quad i = 1, 2, \ldots, M$$

$$p_i(t) = \frac{\text{fitest}_i(t) - \text{worst}(t)}{\text{best}(t) - \text{worst}(t)}$$

The better profound brain organization's assault recipe is alluded to as:

$$H = f(G_1^1 Z_1 \mid G_1^2 Z_2 \mid G_1^3 Z_3 \mid \cdots G_1^m Z_m)$$

$$H = f\left(\sum_j G_i^j Z_j \right)$$

The quantity of neurons (addressed by Z) in the organization represents the heap (represented by G) on the connection between layers.

On all of the hidden layers of the updated deep neural network, the softmax function is utilized. Using the formula, we can determine the values.

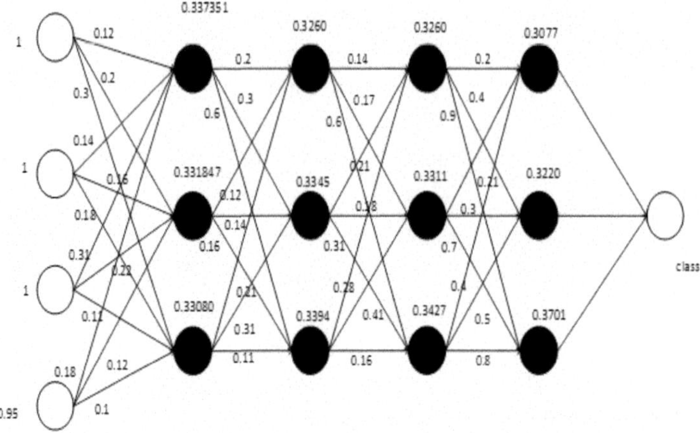

FIGURE 18.2 Classifier Based on an Adapted Deep Neural Network.

$$P(y = j \mid \theta^{i}) = \frac{e^{\theta(i)}}{\sum_{j=0}^{k} e^{\theta_k(i)}}$$

where

$$\theta = W_0X_0 + W_1X_1 + \cdots + W_kX_k$$

The most ideal incentive for Specialist i at Time t is meant by the image.

$$P_i(t) = \frac{p_i(t)}{\sum_{j=1}^{m} p_j(t)}$$

Use this formula to get the worst and best possible values:

$$\text{best}(t) = \min_{j\in(1,\ldots,m)} \text{fitest}_j(t)$$

$$\text{worst}(t) = \max_{j\in(1,\ldots,m)} \text{fitest}_j(t)$$

The method combines the particles based on the object's mass. The suggested technique use this algorithm to determine the object's heading, and then uses the grey wolf optimization to continuously refine the location of that heading. Using the feature values from the NSL-KDD dataset, a fitness value is determined.

3.2.3 Grey Wolf Performance Enhancement

The grey wolf algorithm represents group life and the leadership qualities of wolves. As a group, it sets out to hunt. The alpha wolf is the pack's leader, and the other wolves must obey his every order. The "beta" wolf, a subordinate member of the pack, aids the alpha wolf in making decisions. Alpha is the highest ranking wolf in the pack, while omega is the lowest. Delta represents a middle tier that falls between omega and alpha and is subservient to beta. The algorithm suggests a leader to go out and find the prey. When the alpha makes a hunting signal, the pack immediately begins searching for food. The subordinates follow the instructions to hide the game. At first, the wolf would surround its victim in order to establish a good attacking position. Next, the wolf's position is updated based on the current location of the prey.

There are several stages to the grey wolf algorithm.

$S_{a,\,b,\,Z,\,f}$, and $it_{\max}$ are search agents; initialize them together with the design variable size S_b.

$$\vec{Z} = 2\vec{b}r_1 - \vec{b}$$

$$\vec{f} = 2r_2$$

The value of b was halved between iterations.

- We can see the wolves as:

$$\text{Wolves} = \begin{bmatrix} S_1^1 & S_2^1 & S_3^1 & \cdots & \cdots & \cdots & \cdots & S_{sb-1}^1 & S_{sb}^1 \\ S_1^2 & S_2^2 & S_3^2 & \cdots & \cdots & \cdots & \cdots & S_{sb-1}^2 & S_{sb}^2 \\ \cdots & \cdots & \cdots & \cdots & \cdots & \cdots & \cdots & \cdots & \cdots \\ \cdots & \cdots & \cdots & \cdots & \cdots & \cdots & \cdots & \cdots & \cdots \\ S_1^{sa} & S_1^{sa} & S_1^{sa} & \cdots & \cdots & \cdots & \cdots & S_{sb-1}^{sa} & S_{sb}^{sa} \end{bmatrix}$$

  where S_{ij} is the starting point for the ith wolf pack.

- The healthiness score is determined by

$$\vec{h} = \left| \vec{f} \cdot \vec{S}_p(t) - \vec{S}(t) \right|$$

$$\vec{S}(t+1) = \vec{S}_p(t) - \vec{Z} \cdot \vec{h}\ .$$

- The finest hunting values must be determined.

$$\vec{h}_\alpha = \left| \vec{f}_1 \cdot \vec{S}_\alpha - \vec{S} \right|$$
$$\vec{h}_\beta = \left| \vec{f}_2 \cdot \vec{S}_\beta - \vec{S} \right|$$
$$\vec{h}_\delta = \left| \vec{f}_3 \cdot \vec{S}_\delta - \vec{S} \right|$$
$$\vec{S}_1 = \vec{S}_\alpha - \vec{Z}_1(\vec{h}_\alpha)$$
$$\vec{S}_2 = \vec{S}_\beta - \vec{Z}_2(\vec{h}_\beta)$$

$$\vec{S}_3 = \vec{S}_\delta - \vec{Z}_3(\vec{h}_\delta).$$

It is possible to determine the wolf's current position using

$$\vec{S}(t+1) = F\frac{\vec{S}_1 + \vec{S}_2 + \vec{S}_3}{3}.$$

Combining the gravity search method's F value with the grey wolf algorithm improves performance. Until the condition is met, the fitness value is recalculated, and the positions are updated accordingly GS.

3.2.3.1 The GSGW Algorithm Calculation

Fitness

[38978.26003331 73779.0372579 78484.95302316 838326.5630996 24337.66872029 147618.91102804 23099.83020553 48373.99988401 13028.99315072 127280.69518529 87722.61209052 28510.70901802 39233.56175001 29292.9189462 357570.22205822 111550.93472763 19303.52557341]

Worst: 838326.5630995962

Best: 13028.9931507225

First: 0.9685576841281687

0.0673258627956637 seconds

Eps = 2.220446049250313e–16

A Random Number Generator Gave Me = 171.74815363608246

To calculate the distance, we use the formula: 1.

D_{ij} = 37.62388831350286 × 21.496048611951167 = 808.7649321576789

The recommended strategy utilizes the versatile Jaya advancement procedure to choose the best highlights from the NSL-KDD dataset. The principal estimations are according to the accompanying: administration, Src and DST bytes, hot, num_root, protocol_type, num_file_creations, count, banner, srv count, num_compromised, dst have count, dst have srv count, rerror rate, term, dst have srv count, signed in the best characteristics that were chosen to identify various cyber security attacks are fed to the redesigned neural network.

4 RESULTS AND DISCUSSIONS

We advise using AJOMDNN-GSGW because it yields superior DR, low false alarm rate (FAR), and good accuracy (Table 18.1).

The improved deep neural network's top features, as chosen by AJO, are listed below. The qualities in the NSL-KDD dataset are recorded by field number in Table 8.1.

The intruder will do a scan of the system as a probe to learn more about it.

DoS: The attack uses up system resources, rendering the workstation unusable to the user.

U2R, or user-to-root: The hacker successfully gained root access and then attempted to exploit the system's elevated privileges.

TABLE 18.1
Attacks Using Feature Selection and Those That Don't

Class	(1) By AJOMDNN-GSGW without FS Predicted Attack						(2) By AJOMDNN-GSGW with FS Predicted Attack					
	R2L	**Probe**	**Normal**	**U2R**	**Dos**	**DR**	**R2L**	**Probe**	**Normal**	**U2R**	**Dos**	**DR**
R2L	2,732	8	4	7	3	98.55	2,743	4	4	2	1	99.52
Probe	0	0	1	199	2	98.68	0	0	1	201	0	99.38
Normal	22	13	14	18	7,389	99.74	5	4	4	6	7,437	99.87
U2R	6	2,400	6	5	4	83.26	2	2,407	5	4	3	91.78
DoS	12	11	9,663	10	14	99.68	6	7	9,684	6	7	99.85

TABLE 18.2
Our Proposed System's Top Feature Selections from the NSL-KDD Dataset

Basic features	{1,2,3,4,5,6}
Content features	{10,12,13,16,17}
Features of Time based	{23,25,28}
Features of Host based	{32,33,36}

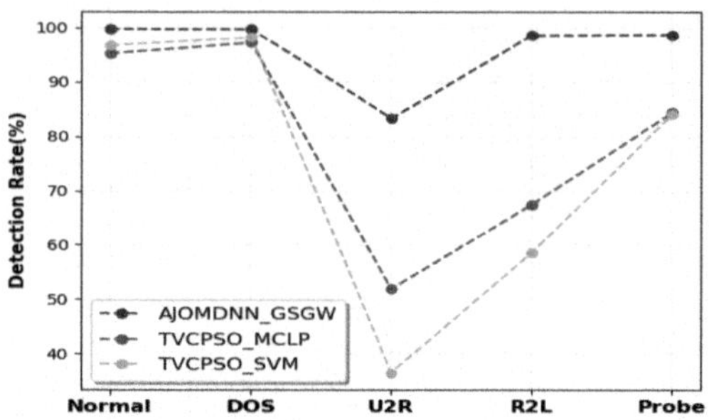

FIGURE 18.3 Dynamic Range without Feature Selection.

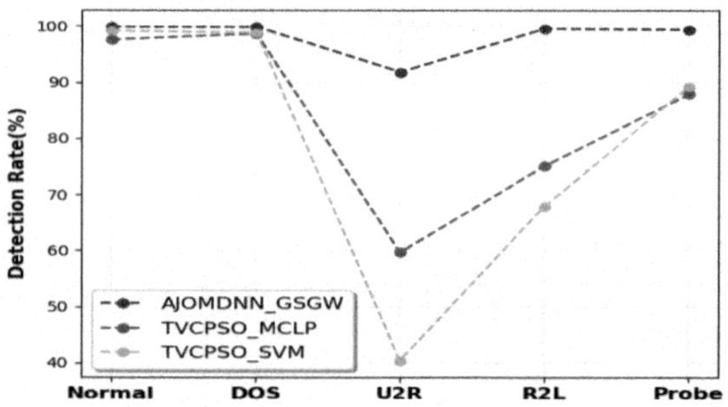

FIGURE 18.4 Feature Selection for DR.

Distant to end-user (R2L): The intruders transmit packets via the network in an attempt to reach the remote and exploit the system. The trespasser is not registered on the local network (Table 18.2).

Location rate appraisal without highlight determination is displayed in Figure 18.3. The suggested approach is called AJOMDNN_GSGW. When compared to traditional approaches, it demonstrates an improved detection rate (DR) of 95.98%. The DRs without feature selection for the various assaults discussed here (Normal, DOS, U2R, R2L, Probe) were 99.74, 99.68, 83.26, 98.55, and 98.68.

In Figure 18.4., the discovery rate following element choice is assessed and shown. 4. Features are picked using AJO on the NSL-KDD dataset. Overall, it outperforms

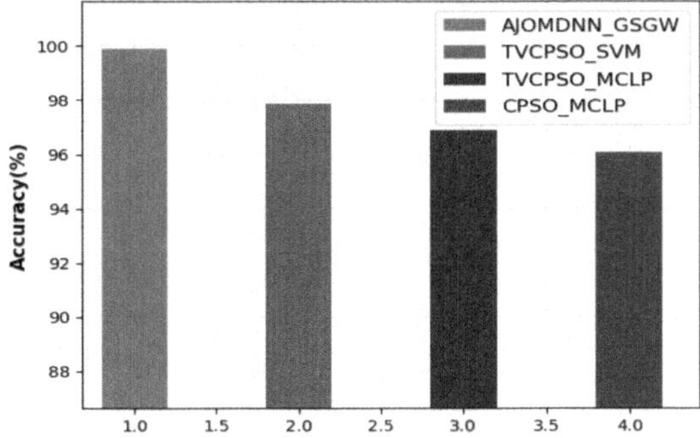

FIGURE 18.5 Exactness Independent of Feature Selection.

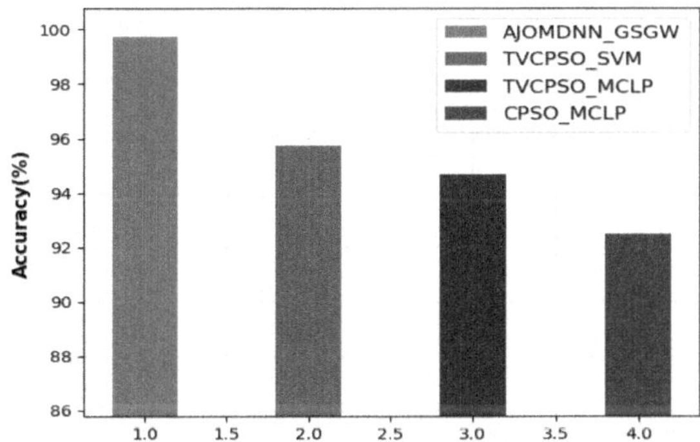

FIGURE 18.6 Selected Features Are Accurate.

previous approaches by a margin of 98.07% in terms of detection. The discovery rates accomplished were 99.87% for Ordinary assaults, 99.85% for DOS assaults, 91.78% for U2R assaults, 99.52% for R2L assaults, and 99.38% for Test assaults.

When the two graphs depicting the different attacks are compared, the one depicting the attacks using feature selection demonstrates a higher DR. Existing TVCPSO methods, like TVCPSO-MCLP and TVCPSO-SVM, are stood out from the proposed method for managing highlight its benefits [1].

The accuracy assessment performance of our suggested system is shown in Figure 18.5. The accuracy is uncovered separated from highlight determination. AJOMDNN_GSGW has a 99.71% higher precision rate than the past techniques TVCPSO-MCLP, TVCPSO-SVM, and CPSO-MCLP.

The accuracy assessment performance of our suggested system is shown in Figure 18.6. It demonstrates how well feature selection may work. AJOMDNN_GSGW has a higher precision level of 99.87 when contrasted with TVCPSO-MCLP, TVCPSO-SVM, and CPSO-MCLP, which were the past strategies. Features are picked using AJO on the NSL-KDD dataset. The diagram that is shown is more precise with include choice than it is without it. Existing methods like TVCPSO-MCLP

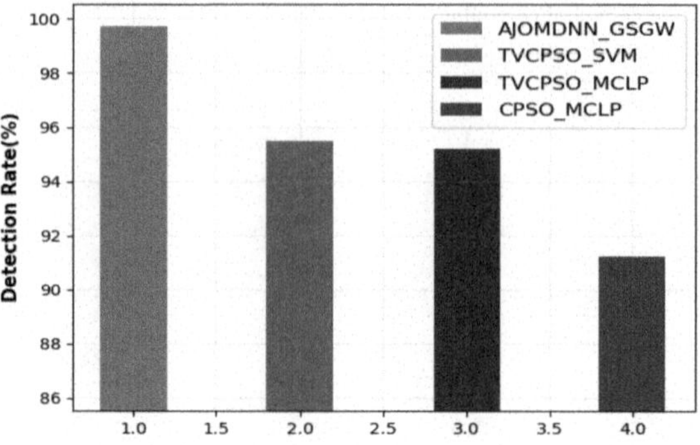

FIGURE 18.7 Dynamic Range without Feature Selection.

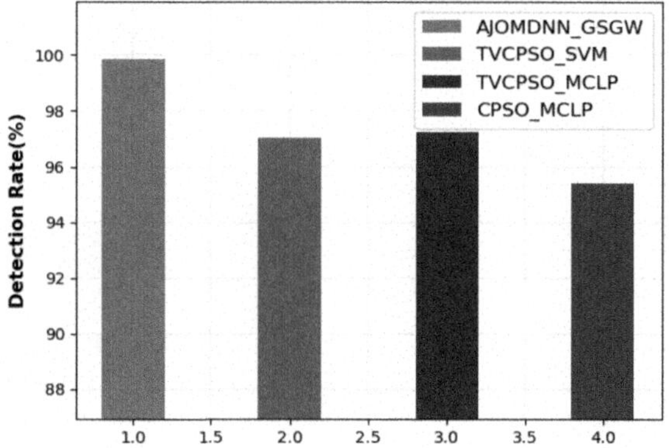

FIGURE 18.8 Feature-Less Feature-Based DR.

(Time changing tumult atom swarm improvement various models straight programming), TVCPSO-SVM, and CPSO-MCLP(chaos particle swarm smoothing out) are stood out from the proposed system. The recommended strategy is the versatile Jaya advancement changed profound brain organization - gravity search dim wolf calculation (AJOMDNN-GSGW).

In Figure 18.7., the location pace of our recommended framework without highlight choice is assessed. With a DR of 95.98%, AJOMDNN_GSGW outperforms its predecessors, TVCPSO-MCLP, TVCPSO-SVM, and CPSO-MCLP, which achieve rates of 94.69%, 95.75%, and 92.47%, respectively.

The discovery rate assessment for our proposed framework with highlight determination is displayed in Figure 18.8. When compared to the DRs of the older techniques TVCPSO-MCLP (representing 94.69), TVCPSO-SVM (representing 95.75), and CPSO-MCLP (representing 92.47), With a higher DR of 98.07%, AJOMDNN_GSGW stands out.

The FAR performance assessment of our suggested system without FS(FeatureSelection) is shown in Figure 18.9. When compared to the TVCPSO-MCLP

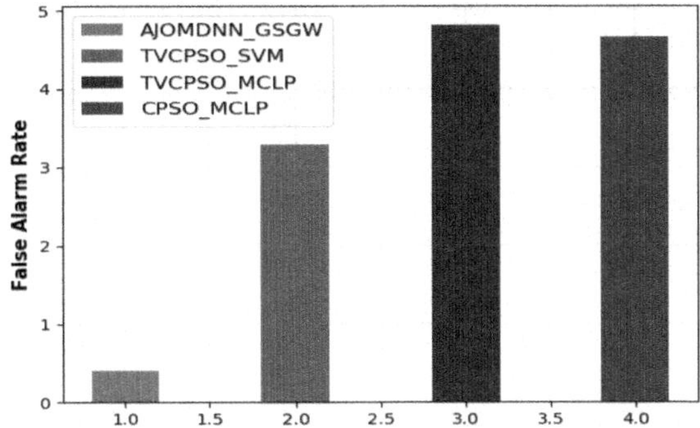

FIGURE 18.9 Without Selecting Features, FAR.

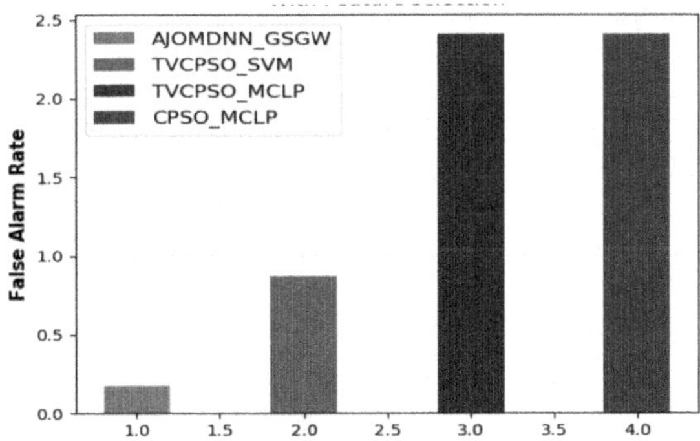

FIGURE 18.10 Feature Selection in FAR.

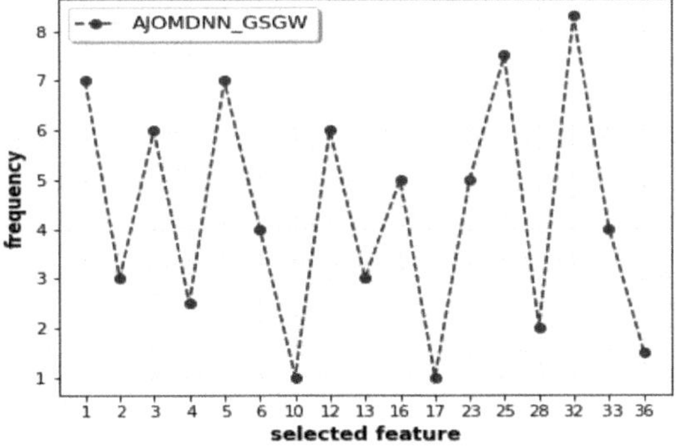

FIGURE 18.11 AJO-Based Feature Selection.

technique (4.81), the TVCPSO-SVM method (3.29), and the CPSO-MCLP method (4.66), AJOMDNN_GSGW shows a low FAR of 0.0028%.

The FAR performance assessment of our suggested system using FS is shown in Figure 18.10.When compared to the prior approaches With a FAR as low

TABLE 18.3
Proposed Method Parameters

Variables	Values for AJOMDNN-GSGW
Maximum iteration	400
Particles used	17
Range values of F	[0,1]

TABLE 18.4
Examining the Proposed Method against the Current [1]

Metrics	TVCP-MCLP	TVCP-SVM	CPSO_MCLP	AJOMDNN-GSGW
Parameter without selected feature				
Accuracy	94.69	95.75	92.47	99.71
Detection rate (DR)	95.19	95.49	91.26	95.98
False alarm rate (FAR)	4.81	3.29	4.66	0.0028
Parameter with selected feature				
Accuracy	96.88	97.84	96.06	99.87
Detection rate (DR)	97.23	97.03	95.42	98.07
False alarm rate (FAR)	2.41	0.87	2.41	0.0012

as 0.0012% in TVCPSO-MCLP, TVCPSO-SVM, and CPSO-MCLP, AJOMDNN_GSGW stands out.

Table 7 illustrates the frequency with which certain attributes are employed in the proposed system, and Figure 18.11 depicts this data. Feature selections made using AJO on the NSL- KDD dataset are shown along the x-axis (Table 18.3).

We utilized the aforementioned values in the recommended approach throughout our implementation of this research. The differences and similarities between the suggested approach (AJOMDNN-GSGW) and the currently used methods are laid forth in Table 18.4.

5 CONCLUSION AND FUTURE WORKS

In this research, we provide a method that can initiate parameters and FS for an MDNN all at once using a clever intrusion detection setup and AJO. We provide a multi-layer neural network (MDNN) classifier for categorizing potential threats to network security. To reduce classification errors, weight value updates are calculated using a gravity search algorithm and grey wolf optimization (GSGW) combination. To do this, we used the KDD cup enlightening file's best 17 components for high DR and low FAR. Picking another arrangement of qualities utilizing an alternate strategy may be useful for future work.

REFERENCES

1. Bamakan, S.H., Wang, H., Yingjie, T. and Shi, Y., 2016. An effective intrusion detection framework based on MCLP/SVM optimized by time-varying chaos particle swarm optimization. *Neurocomputing*, 199, pp.90–102.
2. Ji, S.Y., Jeong, B.K., Choi, S. and Jeong, D.H., 2016. A multi-level intrusion detection method for abnormal network behaviors. *Journal of Network and Computer Applications*, 62, pp.9–17.
3. Hong, J. and Liu, C.C., 2019. Intelligent electronic devices with collaborative intrusion detection systems. *IEEE Transactions on Smart Grid*, 10(1), pp.271–281.
4. Amin, S., Cárdenas, A. A., & Sastry, S. S. (2009). Safe and secure networked control systems under denial-of-service attacks. In *Hybrid Systems: Computation and Control: 12th International Conference*, HSCC 2009, San Francisco, CA, USA, April 13-15, 2009. Proceedings 12 (pp. 31–45). Springer Berlin Heidelberg.
5. Zuba, M., Shi, Z., Peng, Z., & Cui, J. H. (2011, December). Launching denial-of-service jamming attacks in underwater sensor networks. *In Proceedings of the 6th International Workshop on Underwater Networks* (pp. 1–5).
6. Amin, S., Litrico, X., Sastry, S.S. and Bayen, A.M., 2013. Cyber security of water scada systems-part i: Analysis and experimentation of stealthy deception attacks. *IEEE Transactions on Control Systems Technology*, 21(5), pp.1963–1970.
7. Bhaskar, T., Hiwarkar, T. and Ramanjaneyulu, K., 2019, June. A modified deep neural network based hybrid intrusion detection system in cyber security. *IJITEE*, 8(8), ISSN: 2278-3075.
8. Bhaskar, Thupakula and Hiwarkar, Tryambak and Ramanjaneyulu, K., Adaptive Jaya Optimization Technique for Feature Selection in NSL-KDD Data Set of Intrusion Detection System (July 17, 2019). Proceedings of International Conference on Communication and Information Processing (ICCIP) 2019, Available at SSRN: https://ssrn.com/abstract=3421665
9. Rashedi, E., Nezamabadi-pour. H. and Saryazdi. S., 2009. GSA: A gravitational search algorithm. *Information Sciences*, 179(13), pp.2232–2248.

19 AI for Industrial IoT
A Review of Emerging Trends and Advanced Research

Meet Kumari

1 INTRODUCTION

The deployment of wireless technics while the previous last ten years has resulted in innovative ideas called as internet of things (IoT). IoT is anticipated to carry innovations as well as advantages to the industry resulting to the industrial internet of things (IIoT) concept. IIoT systems permit the industries to gather as well as analyze a huge number of information which might be utilized to enhance the inclusive performance of industry scenarios, offering several classifications of services. The major concept of all IIoT is the utilization of sophisticated technologies and services like fifth generation (5G), IoT, edge computing, cloud computing, machine learning etc., particularly optimized for IIoT process (Khan, Rehman, Zangoti, Afzal, Armi, Salah 2020). With the deployment as well development of Industry 4.0, IIoT is continuing applied as well as expanded. A huge production amount data is produced as well as accumulated at the verge of the widespread industrial networks. Besides, intelligence is too an important feature of IIoT, which depicts whether artificial intelligence (AI) is a feasible service in latest industry scenarios (Feng, Wu, Wu, Li, Yang 2023). IIoT is a subcategory of IoT that involves security, ratability and safely communication at higher levels excluding the commotion of concurrent industrial activities owing to mission-crucial industrial scenario. The target of IIoT is effective industrial operation and assets management together with foretelling maintenance (Khan, Rehman, Zangoti, Afzal, Armi, Salah 2020).

Currently, assorted research on IIoT is undertaken specifically within the security, protocols along with communication technology and designs topics. IIoT architectures are considerably distinct from the conventional commuter networks. IIoT networks' scale is generally very restricted as they are generally deployed to ensure information delivery in particular domains. It is distinct from the computer networks that transfer several categories of multiple-media information flow, each containing distinct demand on quality of service metrics like delay, security, jitter, information loss etc. Again, in IIoT scenario, primary data is extensively sensitive to delay as well as one major objective is to ensure that the detain hindrances of the information transmission operation are satisfied. In several industrial scenarios for example coal mining manufacturing, petrochemical industry, wind energy generation etc. For this

DOI: 10.1201/9781032694375-19

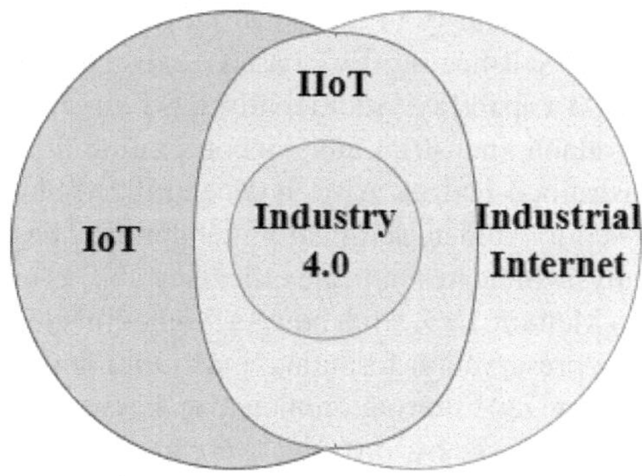

FIGURE 19.1 The relationship between IoT, Industrial Internet , IIoT & Industry 4.0

secure ad, efficient data delivery has to be ensured in IIoT architectures (Jiang, Lin, Han, Abu-Mahfouz, Bilal, Shah, Martínez-García 2022).

Figure 19.1 indicates the relationship between the concepts presented in IIoT, IoT and industry internet followed by Industry 4.0. However, IIoT generated from IoT incorporates distinct emphasis with respect to real-time applications as well as concepts. The IoT is designed to enhance users' quality of life as well as is mainly considered exhaustion-centric. Particularly, IoT service examples employ indoor localization, health monitoring as well as smart homes. Instead, the IIoT uses to improve the industries' production effectiveness. Specifically, IIoT applications employ smart transportation, smart transportation as well as intelligent logistics. IoT systems frameworks are usually produced from scratch as well as the applied sensors are realized in a small area as well as also not susceptible to accuracy. One of the primary features of IoT devices is high mobility, the produced information of these components is of moderate size as well as interrupts can be allowed to a large extent. Simultaneously, IIoT applications frameworks depend on conventional industrial infrastructures. Therefore, these sensors are generally circulated over a broad area as well as distribution must be greatly precise. Contrarily, several IIoT modules are circulated in definite locations, the information produced via these modules are quite large size as well as merely delay can be permitted (Alotaibi 2023).

This chapter organization: Section 2, illustrates the related work concerned with existing AI based IIoT cyber physical systems. Various emerging AI techniques-based IIoT fields are discussed in Section 3. In Section 4, the smart manufacturing in AI-IIoT based cyber-physical system is defined in detail. Integration of cyber security into smart manufacturing is discussed in Section 5. Besides, major challenges and applications with future scope are discussed in Sections 6 and 7, respectively. Section 8 holds the conclusion.

2 RELATED WORK

In Feng, Wu, Wu, Li, and Yang (2023), a faithful self-healing technique on the basics of incorporation of circulated digital twin (DT) as well as blockchain, in order for at effective robustness and security of an industrial network. Here, the realization design is proposed of the self-healing IIoT based on distributed DT to enforce the simulation distributed DT capability. Additionally, a DT simulation active scheme is provided for the controllable industrial units, incorporating the users' requirements and edge servers constrained budget. Also, a decentralized blockchain-based trust management is proposed to validate self-healing reliability. The performance evaluation and security analysis indicates indicate efficiency and security of the proposal.

In Deebak, Hussain Memon, Dev, Ali Khowaja, Muhammad Faseeh and Qureshi's (2022) study, a privacy-preservation formulate plus multi-keyword-rated searching weather establishes optimized filtering, conjunctive keyword and index structure binary tree search to obtain secure investigative capability. Experimental performance analysis indicates that the suggested technique requires less measurement, verification time and storage as compared to another searching encryption schemes.

In Jiang, Lin, Han, Abu-Mahfouz, Bilal, Shah, and Martínez-García's (2022) work, a prospect design for AI-based software-defined IIoT network which sub-divides the conventional industrial scenarios into thrice serviceable layers is produced. The aim of the work was introduction of key technologies and improving services on the basics of proposed network. After this, the paper highlights novel opportunities as well as possible research challenges in control as well as automation of IIoT network.

In Bellavista and Mora's (2019) work, the early experiments which are performing within the H2020 Innovation Action IoTwins framework for the optimization as well as implementation of distributed integrated twins in IIoT applications of prophetic manufacturing as well as maintenance optimization. IoTwins acts distributed integrated twins, somewhat performing at edge cloud points in industrial plant regions, for its execution process predictions as well manufacturing line as adjustments under limited time, also via allowing some forms of authority on industrial data monitoring. Besides, it presents original taxonomy of AI based advanced research survey for localized learning having particular emphasize affiliated setting as well as on developing directions for the IIoT field.

In Yang, Yuan, Li, Zhao, Sun, Yao, Bao, Vasilakos, and Zhang's (2021) work, a productive service provisioning technique like the brain with confederate learning for IIoT is proposed. The BrainIoT technique comprises three algorithms having industrial knowledge plot basics: relation mining, worldwide optimized resource reservation and federated learning on the base of service prediction. It connects production information into optimization of network as well as uses the inter/intra-factory relations to improve the service prediction accuracy. The worldwide optimized resource preserve algorithm appropriately assets resources for forecast services considering several resources. Mathematical results indicate that the BrainIoT technique uses inter/intra-factory relations to produce a precise service prediction, which obtain 96% accuracy and enhance the service quality.

In Sun, Liu, and Yue's (2019) work, a perceptive computing design with concerted edge as well as IIoT cloud computing is introduced. On the basics of the computing design, an AI-improved disburdening framework is realized for service precision maximization that believes services precision as a novel measured besides delay and sensibly disseminate the data traffic to the edge servers alternatively via suitable path to isolated cloud.

In Miao, Zhou, and Ghoneim's (2020) work, an IIoT design based on AI and blockchain to resolve the centralized energy architecture defects is presented. The results show that proposed design can forecast demand as well as natural gas output load while obtaining the natural gas suppliers' balance of interests as well as transaction scenarios in market.

In the work of Trakadas, Panagiotis, Simoens, Pieter, Gkonis, Panagiotis, Sarakis, Angelopoulos, Ramallo-González, Skarmeta, et al. (2020), a holistic integration of AI via promoting collaboration is presented. The proposed design approach is constructed on thrice technical pillars along with AI-powered mechanisms. Additionally, system implementations situations are presented as well as industrial potential applications with business impacts are introduced.

In the work of Campero-Jurado, Márquez-Sánchez, Quintanar-Gómez, Rodríguez, and Corchado (2020), an intelligent helmet prototype is proposed which observes the workplace conditions and accomplishes a near real-time problems evaluations. The information collected via sensors is transfer for analysis to an AI-guided platform. The training datasets incorporates 11,755 samples along with 12 distinct scenarios. Also, the proposed deep convolutional neural network (CNN) is for the identification of feasible occupational risks. Data are refined to build them appropriate for CNN and obtained outputs are contrast to stationary neural network, support vector machine and Naive Bayes classifier having cross-validation accuracy of 92.05%. Meanwhile, IIoT security threats are identified as well as classification is exploited to launch security attacks.

3 EMERGING AI TECHNIQUES-BASED IIOT FIELDS

IIoT incorporating 5G technology and beyond it is quite capable of simultaneous industry data acquisition, handling timely data analysis/processing at server center. Also, edge computing has surpassed the traditional computing platform via efficiently assigning the computing resources as well as edge computing platform at reliable places. It is found that edge computing can improve the customer experience of several industrial delay concerned services by identifying the correct balance between cloud computing and conventional stand-alone models (Jiang, Lin, Han, Abu-Mahfouz, Bilal, Shah, Martínez-García 2022). Figure 19.2 indicates the layers in IIoT.

The basic idea of IIoT is accustomed by allowing an extended plethora of techs incorporating IoT, big data, cloud computing, cyber physical system, AI, virtual reality, augmented reality, machine-to-machine, and human-to-machine communication (Khan, Rehman, Zangoti, Afzal, Armi, Salah 2020).

- Internet of things: Taking into account the related factory scenario, IoT components support in practical data collection as well as actuation. Being the main IIoT component, these components follow the factory assets worldwide. The entire process that one is starting from unrefined material and finish with final products is supervised utilizing IoT components to obtain considerable reduction in manual system and labour cost management. IoT components in the linked IIoT systems are carried out across factory facilities from production to warehouses facilities as well as distribution centers.

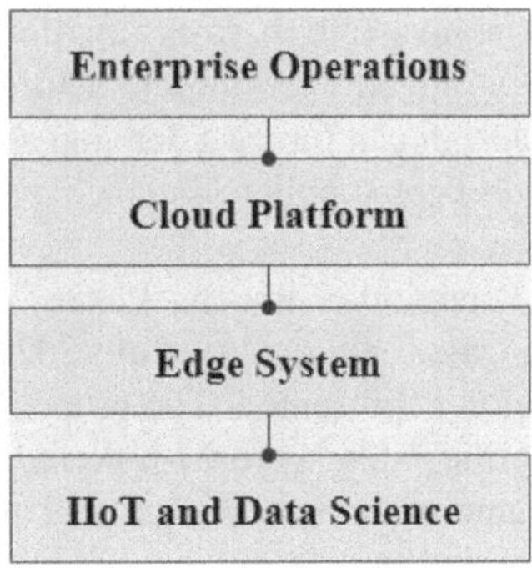

FIGURE 19.2 Layers in IIoT

- Blockchain technology: It is among the very significant schemes which will play a primary role to carry the IIoT dream in reality. Recently, a concentrated research is borne by industry and academia on blockchain terminology several fields like healthcare, finance, car insurance, supply chain etc. IoT devices utilized in smart industry produce a large amount of information. These generated data via IoT units is versatile, the data is investigated as well as processed for devices' performance monitoring, diagnosis, anomaly detection, asset monitoring, predictive maintenance, complete product lifecycle tracking and finishing goods delivery to users. Although, sharing this significant data with all items incorporated in the IIoT scenarios in a secure way is quite difficult task. Distributed nature, robustness, trust, traceability, security, inherent data derivation etc. are the unique features of blockchain technology making it suited for IIoT.
- Cloud computing: The enormous improvement in IIoT needs extremely distributed superior performance computing architectures to process, manage, analyze as well as store the information. Here, cloud computing technology offers network computing together with collective services throughout all the capabilities in an IIoT scenarios. Various connected components and services are immediately interacted with backend clouds. Cloud service patterns are considered as private, public or integrated. As the data centers' establishment and technical staff recruitment need high spending, exclusive cloud service designs are not a feasible option for novel entrants, and small/medium/high level undertakings. Although, very large as well as well-established undertakings internationally favor the exclusive clouds development to assure the safely, privacy and security as well as deal with industrial espionage.

4 SMART MANUFACTURING IN AI-IIOT-BASED CYBER-PHYSICAL SYSTEM

Figure 19.3 indicates the five layers of AI-IIoT-based cyber-physical system (Radanliev, Roure, Kleek, Santos, Ani 2021). The primary motivation of this chapter is to illustrate the AI potential in the cyber physical systems layers, allowing

Configure	Resilient control system
Congnition	Decision support system
Cyber	Cyber physical system
Conversion	Health management
Connection	Condition based monitoring

FIGURE 19.3 Layers of AI-IIoT based cyber-physical system

system cognition. In various layers of cyber physical systems, AI is an important component such as performing faulty identification, real-time forecast behavior, to be utilized as a prognostic model in the control design and performing the process optimization. As the design evolve, the system become highly complex. Recently, automatized scenarios allowed to be comprised of numerous components which should work consonantly in conjunction. This demand for refined tools at operating management and decision level is also large scale cyber physical system bottleneck. Therefore, AI is a key technology to enable the large-scale cyber physical system, building a bridge between cyber physical systems and offering then with up to date autonomous guidance (Oliveira, Dias, Rebello, Martins, Rodrigues, Ribeiro, Nogueira 2021).

AI can offer important system ability, cognition that provides the modeling, complex behaviors learning, representation, interactions among various system components and data. It can be obtained via the AI supervised/unsupervised training to allow these particular tasks. In addition, AI models are capable to continually determine from the design, discussing the cyber physical systems adaptive ability. Intrinsically, there is an improving need for work on development as well as amalgam of widespread AI networks. AI could be possible to obtain a capacity wherein chemical unit can uprightly hybrid numerous management levels separately, communicating with cyber physical systems designs and performing management task. The concept of the systems which can handle themselves with less human assistance has fascinated in the previous years because of the current automotive industry development having automatic drives transporting terminology. This idea is built on autonomous cordiality and feasible autonomous controllability whose requirements are modularity, functional equality, discreteness, exchange of information, situation consciousness as well as self-management. Majority likely because of a deficit of technology, the concept was not completely developed in existing work. An enabling progress toward the idea of large-scale cyber physical systems organized by AI is catalyzed by IIoT. An IIoT network already offers a widespread network of connected computing components where data is exchanged consistently and be allocated in real-time. Thus, IIoT can provide the important social conditions for AI models to replace experiences as well as information and supervise the system considering their arrangement (Oliveira, Dias, Rebello, Martins, Rodrigues, Ribeiro, Nogueira 2021).

Further, AI identification for vigorous systems is yet an open challenge. Vigorous AI is among the most critical reproductions for chemical dynamic engineering systems, and that are basically extremely nonlinear, have elevated settling times as well as need patronize intervention taking into account its next future states. The best-fit scheme in this situation is deep neural networks (DNNs) in recurrent neural networks. DNNs are emphasized by their flourishing application to solve problems of numerous fields. There is an absence of new work in process technologies to make usage of the DNNs capacity to address a series of challenges in fields. Besides, schemes from circulated AI are too an empowering technology for cooperation, autonomy, and virtualization abilities preferred for cognitive cyber physical systems development (Oliveira, Dias, Rebello, Martins, Rodrigues, Ribeiro, Nogueira 2021).

5 INTEGRATION OF CYBER SECURITY INTO SMART MANUFACTURING

Smart manufacturing is the recent industrial automation process. This is a combination of cyber space which acquires connection with the machine field. The human-basics production process leads by the variation via the smart industry uprising manner. Recently, every factory is linked with some internal communication system or public network to make the production automated process. System is supervised through software enabled units or control systems passed the desired parameters to the tangible machine by some program set logical organization. Software is compromised via malicious code, like Trojans, viruses, as well as runtime attacks. Also, transmission protocols are depending on protocol assault, incorporating man-in-the-middle as well as denial-of service attacks (Masum 2023).

The cyberattacks risk directed at manufacturing infrastructures as well as processes is an important concern to companies which are producing goods, especially those accomplished for public consumption. Cyber Security Framework for manufacturing entity concerned in improving the infrastructure security is as given below (Masum 2023):

- Taxonomy: As smart manufacturing wraps a wide area of presentation industries in cyber physical system to cumulative autonomous vehicles, Industrial IoT, manufacturing and robotic production taxonomy is developed for comfort division (Masum 2023).
- Operational technology prospect: Conventional data security exercises are not much beneficial if whatsoever hardware/software device is built to transfer the inner information to outside of the design. Devices may be reliable provided that the chips are rid of veiled malicious circuits and that can be inserted throughout the chips design/manufacturing process (Masum 2023).
- Design: Security as well as privacy is two different but connected challenges in the field of information technology. The security notion of privacy

concerns avoiding unauthorized access to proprietorship data, while security deals with attentions like whom own the information, which can approach it, and how it can utilize. Attackers generally target important as well as highly sensitive data, like product designs, marketing plans, financial data, customer, supplier lists, as well as partnership agreements. The abatement of sensitive information as well as other configuration of enterprise data can lead to important financial or business losses. Data is considered sensitive in the manufacturing context, because of several industrial operation aspects, including highly sensitive data about products, companies and business strategies. Sharing information having internal departments and external vendors needs secure mechanisms to stop data leakage. Sensitive information which incorporates blueprints, cost information, manufacturing processes, and operational data, must be protected (Masum 2023).
- One superficial solution to obtain secure information sharing in IIoT needs the data owner to encode his/her data before transferring with others; although, this technique requires more computation power to decode the information before data can be used. Notably, the data owner requires transferring the keys which are utilized for the information encryption to another parties; moreover, if the information owner repeals access rights to other user/device, the person must re-encode the information with a new key, as well as allocate the new key to another parties in a group (Masum 2023).

6 MAJOR CHALLENGES

The major challenges in AI-based IIoT incorporates following challenges are:

- IIoT network devices utilized in IIoT applications have several impairments concerned to processing, energy and communication, even if they are bound to offer high reliable as well as immediate processing, monitoring and decision making. It is presumed to be complex and tough to have exhaustive AI and ML approaches on small devices (Bhuiyan, Kuo, Wang 2022).
- Primary security purpose like confidentiality, integrity and availability have not been observed during continuous training along with validating AI and ML techniques in IIoT (Bhuiyan, Kuo, Wang 2022).
- Several threat services, like security violation, data poisoning, inference, data collusion, indiscriminate attacks and causative, provide an optimization issue for auto-calibration ML/AI devices and also improving respective hyper parameters in IIoT network (Bhuiyan, Kuo, Wang 2022).
- In IIoT network, AI/ML trustworthy development techniques employing actuators, sensors and respective telemetry data, is yet in its start, owing to the issues as well as its practical insights. For this, AI/ML techniques must be realized to established white-box design, preferably black-box to identify their reliability and honesty in business services in IIoT networks (Bhuiyan, Kuo, Wang 2022).
- Several cyber physical systems as well as IoT standards are still ready. In the absence of standards, units get quite heterogeneous, and that results in

interoperability problems? In fact, it takes time to phase out already existing gadgets as well as emigrate to novel versions which follow standards. Also, resistance to assuming standards is because of concerned that laws may not impose system regulation and products to lead these standards that are primarily because of longer timeframe in law legislation compared to standard publication (Chui, Gupta, Liu, Arya, Nedjah, Almomani, Chaurasia 2023).
- Only a low number of customers can depend on the computing applications which results in suitable latency in decision-making and data analysis. The solution is to emphasize resources to more significant applications (Chui, Gupta, Liu, Arya, Nedjah, Almomani, Chaurasia 2023).

7 APPLICATIONS AND FUTURE SCOPE

The IIoT applications include multi-dimension architectures to investigate the capabilities of remote capacities as well as networks to address real-time issues. For its use accessible resources, modern computing components apply diversified schemes. It utilizes multi-dimensional designs to identify the purpose of the decision-making scenario. Major computing devices integrate IoT and AI to structure a systematic framework to govern the digital markets evolution. An AI-based IIoT is concerned with a three-tier approach for the usability of the core characteristics of smart services (Deebak, Hussain Memon, Dev, Ali Khowaja, Muhammad, Qureshi 2022). Various applications are given as:

- Smart Cities: Few areas of application of AI-based IIoT for next-generation smart cities comprise intelligent transportation, intelligent building, traffic congestion, waste disposal, smart lighting, intelligent parking and urban maps. AI enables IIoT to be utilized effectively to analyze, mitigate as well as regulate traffic congestion in smart cities. Besides, IIoT allows the link of weather focused street lighting as well as finding of waste modules via possessing schedules flaps of trash gathering. Also, intellectual highways provide important information and warning messages which include access to modifications depending on unexpected incidents and climatic conditions such as accidents as well as traffic jams (Jun, Craig, Shafik, Sharif 2021).
- Waste Management and Smart Agriculture: The IoT can improve and promote the agriculture domain via evaluate the soil moisture as well as monitoring trunk width. IIoT allows for regulate and retain the vitamins measurements present in agriculture products and also supervise the situation of a microclimate to access more fruits as well as quality vegetables manufacturing in agriculture domain. By observing weather conditions, AI-based IoT allows predicting information about ice, drought, snow wind alterations, and rain and regulates humidity as well as temperature levels which avert fungus or other barm pollutants (Jun, Craig, Shafik, Sharif 2021).
- Retail and Logistics: IoT execution in the retail as well as supply chain has several advantages, and some of the advantages comprising assessing the storage conditions in the whole supply chain. It also assists product tracking

which assists in payment processing during relying on the period activity and location in gyms, theme parks, public transport etc. Besides, AI-based IoT is smeared to distinct services which that comprise fast payments processes like automatically identification out via the biometrics assistance of possible allergen items, and supervision of the products spin in warehouses as earlier (Jun, Craig, Shafik, Sharif 2021).

- Smart Environment: Smart environment strategies are affective by the hybrid of AI technology and IoT and must be observed for sensing, estimating as well as tracking objects in atmospheric objects in atmospheres which offers suitable benefits in accumulation of green as well as ecological life. Besides, it offers managing and observing air quality via the data gathering process from distinct remote sensors in distinct cities by allowing topographical exposure to obtain better ways of control traffic blockages (Jun, Craig, Shafik, Sharif 2021).

8 CONCLUSION

As manufacturers requirement the better connectivity as well as interaction of Industrial Revolution 4.0, manufacturing system providers and machines in respective factories have to extend the role information technology in concerned products. The growth of the evident question emerges weather the model, such a concern, secure process or not. From existing work, it is realized that the smart factories' cyber attachment will be susceptible to security intrusion and breach. Any minor distractions in the system process will led to mass production collapse and safe critical situations. By removing the issues in the literature review new generation of smart industry, privacy assessment will play an important role for constant continuity of chain cyber manufacturing. Consciousness of the dynamic entities will to require comprehending the vulnerabilities. As digital thread includes multiple domains, all of them might be vulnerable as well as favorable assessment methodology might be there on operational technology.

REFERENCES

Alotaibi, Bandar. "A survey on industrial Internet of Things security: Requirements, attacks, AI-based solutions, and edge computing opportunities." *Sensors* 23, no. 17 (2023): 7470.

Bellavista, Paolo, and Alessio Mora. "Edge cloud as an enabler for distributed AI in industrial IoT applications: The experience of the IoTwins project." In *AI&IoT@ AI* IA*, pp. 1–15 (2019).

Bhuiyan, Md Zakirul Alam, Sy-Yen Kuo, and Guojun Wang. "Guest editorial: Trustworthiness of AI/ML/DL approaches in industrial internet of things and applications." *IEEE Transactions on Industrial Informatics* 19, no. 1 (2022): 969–972.

Campero-Jurado, Israel, Sergio Márquez-Sánchez, Juan Quintanar-Gómez, Sara Rodríguez, and Juan M. Corchado. "Smart helmet 5.0 for industrial internet of things using artificial intelligence." *Sensors* 20, no. 21 (2020): 6241.

Chui, Kwok Tai, Brij B. Gupta, Jiaqi Liu, Varsha Arya, Nadia Nedjah, Ammar Almomani, and Priyanka Chaurasia. "A survey of internet of things and cyber-physical systems: Standards, algorithms, applications, security, challenges, and future directions." *Information* 14, no. 7 (2023): 388.

Deebak, Bakkiam David, Fida Hussain Memon, Kapal Dev, Sunder Ali Khowaja, and Nawab Muhammad Faseeh Qureshi. "AI-enabled privacy-preservation phrase with multi-keyword ranked searching for sustainable edge-cloud networks in the era of industrial IoT." *Ad Hoc Networks* 125 (2022): 102740.

Feng, Xinzheng, Jun Wu, Yulei Wu, Jianhua Li, and Wu Yang. "Blockchain and digital twin empowered trustworthy self-healing for edge-AI enabled industrial Internet of things." *Information Sciences* 642 (2023): 119169.

Jiang, Jinfang, Chuan Lin, Guangjie Han, Adnan M. Abu-Mahfouz, Syed Bilal Hussain Shah, and Miguel Martínez-García. "How AI-enabled SDN technologies improve the security and functionality of industrial IoT network: Architectures, enabling technologies, and opportunities." *Digital Communications and Networks* 9 (2022): 1351–1362.

Jun, Yao, Alisa Craig, Wasswa Shafik, and Lule Sharif. "Artificial intelligence application in cybersecurity and cyberdefense." *Wireless Communications and Mobile Computing* 2021 (2021): 1–10.

Khan, Wazir Zada, Muhammad Habib Ur Rehman, Hussein Mohammed Zangoti, Muhammad Khalil Afzal, Nasrullah Armi, and Khaled Salah. "Industrial internet of things: Recent advances, enabling technologies and open challenges." *Computers & Electrical Engineering* 81 (2020): 106522.

Masum, Rahat. "Cyber security in smart manufacturing (threats, landscapes challenges)." *arXiv* preprint arXiv: 2304.10180 (2023).

Miao, Yiming, Ming Zhou, and Ahmed Ghoneim. "Blockchain and AI-based natural gas industrial IoT system: Architecture and design issues." *IEEE Network* 34, no. 5 (2020): 84–90.

Oliveira, Luis M.C., Rafael Dias, Carine M. Rebello, Márcio A.F. Martins, Alírio E. Rodrigues, Ana M. Ribeiro, and Idelfonso B.R. Nogueira. "Artificial intelligence and cyber-physical systems: A review and perspectives for the future in the chemical industry." *AI* 2, no. 3 (2021): 27.

Radanliev, Petar, David De Roure, Max Van Kleek, Omar Santos, and Uchenna Ani. "Artificial intelligence in cyber physical systems." *AI & Society* 36 (2021): 783–796.

Sun, Wen, Jiajia Liu, and Yanlin Yue. "AI-enhanced offloading in edge computing: When machine learning meets industrial IoT." *IEEE Network* 33, no. 5 (2019): 68–74.

Trakadas, Panagiotis, Pieter Simoens, Panagiotis Gkonis, Lambros Sarakis, Angelos Angelopoulos, Alfonso P. Ramallo-González, Antonio Skarmeta et al. "An artificial intelligence-based collaboration approach in industrial iot manufacturing: Key concepts, architectural extensions and potential applications." *Sensors* 20, no. 19 (2020): 5480.

Yang, Hui, Jiaqi Yuan, Chao Li, Guanliang Zhao, Zhengjie Sun, Qiuyan Yao, Bowen Bao, Athanasios V. Vasilakos, and Jie Zhang. "BrainIoT: Brain-like productive services provisioning with federated learning in industrial IoT." *IEEE Internet of Things Journal* 9, no. 3 (2021): 2014–2024.

20 The Role of Human-Centric Solutions in Tackling Challenges and Unlocking Opportunities in Industry 4.0

Krishnaveni, Swathi, Eleanor Schwartz, and Sangeetha

1 INTRODUCTION

The Fourth Industrial Revolution or otherwise known as Industry 4.0 has transformed the process involved in manufacturing and production units a great deal. Apart from automation and supply chain management, it has also integrated several other technologies like internet of things (IoT), big data, and artificial intelligence (AI) into its process. It has redefined the business operations in the global market. It has enabled the creation of smart factories which has resulted in enhancement of efficiency, has reduced downtime and complete utilization of the resources.

The IOT devices generates a vast amount of data which makes the decision-making very simple thereby improving the quality of the product and customizing the products as per customer demands thus resulting in flexible and agile manufacturing process. Real-time monitoring and maintenance has led to minimization of raw materials, reducing lead times and saving of cost. Industry 4.0 has started reshaping the industries by connecting advanced technologies to the manufacturing and production units.

In the expansive landscape of Industry 4.0, which integrates into domains such as automation, IoT, and data-driven decision-making, human-centric security emerges as a pivotal factor in this industrial revolution. This approach places a paramount focus on the individuals within the system, recognizing them as key assets in safeguarding and preserving the integrity of data. Human-centric security begins with educating and training employees to effectively recognize and respond to cyber threats, mitigating the risks of unauthorized disclosure or access to sensitive data. In the data-rich environment of Industry 4.0, this security system ensures transparent

DOI: 10.1201/9781032694375-20

and ethically handled data practices, incorporating behavioural analysis to detect anomalous activities in data handling and system operations.

Furthermore, human-centric security ensures the responsible and ethical deployment of emerging technologies like robotics, AI, and automation. It adopts adaptive security systems capable of evolving to counter changing threats promptly. In the event of a security breach, this system responds swiftly to identify and recover from the incident, while also maintaining compliance with industry rules and regulations pertaining to human safety.

The main objective of this manuscript is to approach the challenges involved in protecting the data handling, the well-being and safety of humans using some cyber security strategies in the fourth industrial revolution. The main challenge involved is that people involved in Industry 4.0 interact with interconnected systems and automation in a great deal and hence their physical and digital safety is very important. Hence, the human-centric system prioritizes the humans involved in these operations by training them in cybersecurity skills and fosters a culture of security consciousness.

One additional objective involves the development of user-friendly interfaces, facilitating operators in making informed security decisions. Human-centric security solutions specifically acknowledge the roles and requirements of human operators, offering targeted solutions to address security challenges and enhance the safety and resilience of industrial environments. Within manufacturing units, a collaborative partnership between humans and robots has emerged. Human-autonomous system synergy is efficiently utilized in product assembly processes. This symbiotic relationship between human and machine is integral for job creation, maximizing productivity, and ensuring efficiency in Industry 4.0. The well-being and contentment of employees, coupled with opportunities for professional growth, contribute significantly to the business value in the Industry 4.0 landscape.

The chapter follows a structured format, commencing with an introduction that provides a concise overview of Industry 4.0 and its contemporary implications in the business landscape. Addressing concerns prevalent in the industry sector, the chapter acknowledges the discourse surrounding automation and digitization leading to job displacement. The primary objective of this chapter is to dispel the notion of job scarcity and instead promote the idea that, in the era of Industry 4.0, there exist expanded opportunities for human workers beyond traditional manufacturing roles. It emphasizes that within a manufacturing unit, interdisciplinary work is abundant, offering diverse tasks that extend beyond conventional responsibilities, thereby encouraging and ensuring meaningful engagement for human workers.

This chapter extensively addresses the multifaceted challenges inherent in Industry 4.0, with a particular emphasis on human-centric difficulties. Recognizing the transformative nature of this industrial paradigm, it delves into the intricacies of challenges faced by human workers in adapting to and thriving within Industry 4.0. Moreover, the chapter strives to provide nuanced solutions to the issues discussed. Furthermore, a key focus of the chapter is to unlock and elucidate the myriad opportunities available to human beings within advanced manufacturing sectors. It explores how individuals can leverage their skills and adaptability to thrive in an environment characterized by automation, digitization, and other Industry 4.0 technologies.

Through this dual exploration of challenges and opportunities, the chapter provides a comprehensive understanding of the human role in shaping and benefiting from the advancements in the manufacturing domain.

2 LITERATURE REVIEW

Rad et al. (1) conducted a comprehensive examination of the exponential growth of Industry 4.0 technologies with a specific focus on supply chain dynamics. Their study involved a systematic literature review that delved into the core technologies of Industry 4.0, exploring both the positive and negative implications. Furthermore, the research aimed to identify the key factors that contribute to the success of Industry 4.0 implementations. In a parallel vein, Frank et al. (2) conceptualized Industry 4.0 as a new industrial stage characterized by innovative digital solutions. Their work included a proposal outlining the conceptual framework of the technology and detailing its practical implementation in various industrial settings.

Thames and Schaefer (3) characterize Industry 4.0 as an industrial system comprises numerous interconnected elements, to form smart factories and manufacturing sectors. Jamwal et al. (4) define Industry 4.0 as a broad domain encompassing data management, production, and efficiency. It facilitates the integration of technologies such as cyber-physical systems, the IoT, AI, and digital twins. According to Xu et al. (5), many countries have implemented Industry 4.0 strategies, and after the tenth year of its introduction, the European Commission has introduced Industry 5.0. In a systematic review by Zheng et al. (6), analysing 186 articles, the primary focus of Industry 4.0 was found to be production scheduling and control. Some attention was also given to areas such as servitization and circular supply chain management.

Ghobakhloo (7) identifies precedence relationships among various sustainability functions of Industry 4.0. Their analysis indicates that sustainability functions and business model innovation emerge as the main outcomes of Industry 4.0. Kamble et al. (8) reviewed 85 papers, focusing on research approaches, the current status, and sustainable frameworks in Industry 4.0. Saucedo-Martínez et al. (9) analysed recent trends, opportunities, and identified research gaps in Industry 4.0. Nahavandi (10) introduced Industry 5.0, emphasizing collaboration between robots and human brains, predicting increased job creation. Kong et al. (11) aimed to establish human-cyber-physical symbiosis and considered five designs for future trends and research, supporting operators, machines, and production. Kadir and Broberg (12) suggested a framework combining human factors, ergonomics, work system modelling, and strategy design and conducted a study with ten case studies where Industry 4.0 was implemented. Longo et al. (13) focused on designing and developing a practical solution to integrate augmented reality contents and tutoring systems and investigated the proposed approach through field experiments.

Nguyen et al. (14) collected case studies on human-centred design in Industry 4.0 and conducted a systematic literature review on them. Romero et al. (15) projected the performance of a man working with robots and machines through an adaptive automation system and introduced an operator 4.0 typology to support the development of human automation work systems. Zarte et al. (16) presented a systematic lecture review of the accepted methodologies, concepts and visions of Industry 4.0. Alves et al. (17) explored whether Industry 5.0 is human-oriented and discussed strategies for creating human-centricity and discussed the strategies for achieving sustainable and resilient systems. Grosse et al. (18) presented a multimethod approach to incorporate ethical implications missing in Industry 4.0.

Tóth et al. (19) addressed the need for human-AI collaborative design to support advanced AI-driven collaboration tools and encouraged the development of AI techniques for human-in-loop optimization and alternate feedback loop models. Oliveira et al. (20) presented two research projects in different industrial sectors and explored the challenges and barriers involved in Industry 4.0. Katherine et al. (21) explored the relationship between work attributes and automation in a manufacturing plant through interviews with assembly line workers and deployed robotic technology based on the interview findings and discussed future development. Rosin et al. (22) identified the potential for enhancing the decision-making process in Industry 4.0 and explored the risks associated with implementing new decision-making technologies. Adel analysed the applications of Industry 5.0 in healthcare, supply chain and manufacturing.

3 CHALLENGES IN INDUSTRY 4.0

The introduction of Industry 4.0 has indeed ushered in substantial transformations within industries, revolutionizing conventional practices. However, this paradigm shift is not without its impediments, deterring many companies from seamlessly incorporating their ground-breaking technologies. As organizations embark on the journey of restructuring to embrace this new culture, the traditional boundaries that have long defined industries begin to erode.

Among the myriad challenges hindering the swift implementation of Industry 4.0, concerns related to data security, employee training, and the overhaul of infrastructure and standards, loom large. The integration of Industry 4.0 technologies is further hampered by the difficulty of fitting them into existing infrastructures or appending additional interfaces to current systems. Notably, the most formidable challenge lies in the imperative to reshape the foundational infrastructure of the company, a crucial prerequisite for the successful assimilation of Industry 4.0 technologies.

Compounding the challenge is the inadequacy of planning systems within industries, characterized by limited integration between different companies. Establishing integration requires a meticulous alignment of systems and procedures, emphasizing the critical need for standardization of data and interfaces. The current lack of such standardization impedes the seamless integration of Industry 4.0 technologies.

From the perspective of employees, a notable hurdle is the deficiency in knowledge and expertise required to initiate or modify systems for optimal results. The training processes currently in place fall short of adequately preparing employees for the dynamic and ever-evolving landscape of industrial systems. Additionally, financial constraints pose a significant obstacle, as the implementation of Industry 4.0 involves substantial investment costs. Ambiguity surrounding the benefits, return on investment, and payback period further complicates decision-making, leading to a lack of support and commitment from employees. Resistance to change among employees can also impede the implementation of new technologies.

Crucially, the security of data has emerged as a paramount concern. In an era where businesses are increasingly reliant on data for competitiveness, safeguarding sensitive information becomes imperative. The intricate challenges posed by Industry 4.0 necessitate a comprehensive and strategic approach, addressing not only technological integration but also the human and financial aspects, ensuring a smoother transition into this transformative industrial era.

The other great challenge involves the upskilling and reskilling of the employees towards Industry 4.0. To operate, maintain and troubleshoot the systems involved in Industry 4.0, the employees have to be skilled in new technologies like IoT, AI, and robotics. Thus, this evolution may lead to skill gap and the employees may struggle to keep up the demands of the customers. Automation poses another challenge for employees since machines and algorithms take over the repetitive tasks, job displacement becomes a major concern. Employees may find it difficult to transit to new roles due to inadequate training.

Another major challenge is the integration of digital system, which introduces cybersecurity threats. Ransomware attacks, hacking and data breaches are some of the few major threats involved due to interconnectivity. Employees should be able to safeguard the data and information to ensure smooth operation. The nature of work or the traditional employment models are also changing due to Industry 4.0. Remote work and flexible shifts are becoming more prevalent. These changes require employees to communicate and manage their work in different ways. In Industry 4.0 machines become more intelligent than humans, decision-making capabilities and consequences of their actions raises ethical questions. Industry 4.0 technologies often involve interdisciplinary collaboration between IT, engineering, and other departments. Silos and communication gaps between these disciplines can hinder the smooth integration of technologies. Industry 4.0, often referred to as the fourth industrial revolution, involves the integration of digital technologies, the IoT, AI, and other advanced technologies into manufacturing and industry. While Industry 4.0 brings about numerous benefits such as increased efficiency, productivity, and innovation, it also presents several human-centric challenges.

4 HUMAN-CENTRIC CHALLENGES

This new industrial revolution has led to tremendous effects on the employees. They are forced to adapt to innovative and new technologies instead of technology adapting to the needs of the human. It also requires a lot of reskilling to adapt to these new technologies. Thus, the role of human beings in Industry 4.0 has become a question since machines are capable of analysing and solving complex problems even more better than a human being. Uncertainty prevails in the operational level and this becomes a question for the working conditions of the employees. The introduction of cyber-physical system will surely affect the life of human beings at a greater level. The repetitive task done by human will be replaced by technology. This is a negative effect on the employees as technologies replaces these physical tasks. This new environment expects new skills and knowledge from the employees. Thus, upskilling and reskilling of the employees will add value to the workers and to the company. These new jobs are a continuous reskilling process and it has to go along with the ever growing technologies.

5 PROPOSED SYSTEM

This chapter acknowledges the challenges employees encounter amid the implementation of Industry 4.0 and presents a set of strategic, human-centric solutions to address these issues. These proposed strategies aim to empower employees, foster a positive work environment, and ensure a smooth transition in the face of technological advancements.

5.1 Continuous Skill Development Programmes

Evolution in job requirement is going to be rapid due to automation and digitization. Every employee in the industry should consistently learn and acquire knowledge of the new implementations. The fundamental prerequisites for new technology implementation in the industry are related to reskilling and upskilling of the employees. Trust has to be established among the employees to maintain resilience of the manufacturing system. Human-centric solutions emphasizes continuous skill development programmes thus equipping the employees with the necessary skills required to prosper in the landscape of Industry 4.0. Training sessions, workshops, and skill development programmes should become an integral part of an organization to support their employees. Thus, the human-centric approach safeguards the experienced employees rather than replacing them.

5.2 Human-Machine Collaboration

To achieve optimal results, to leverage the strength of both man and machine and to meet the challenges of Industry 4.0, a cooperative interaction between man and machine is emphasized. This synergy between man and machine maximize the productivity, innovation and bring in job satisfaction among the employees. While machine excel repetitive and data intensive task, human excel in creativity, emotional intelligence, and critical thinking, thus leading into an adaptive work environment. Man and robots have started boosting the productivity of the manufacturing sectors. Man defines the production line and follow the key performance indicators and machine follow the process designed by man and work effortlessly. These skills lead to an increase in productivity and employees feel motivated to do the work thus highlighting the human demands over manufacturing. Employers have to start thinking what technology can do for people how it can meet the requirements of people. Thus, human-machine collaboration is a symbiotic relationship that harnesses the strength and unlocks the potential of both man and machine to drive efficiency and success as shown in Figure 20.1.

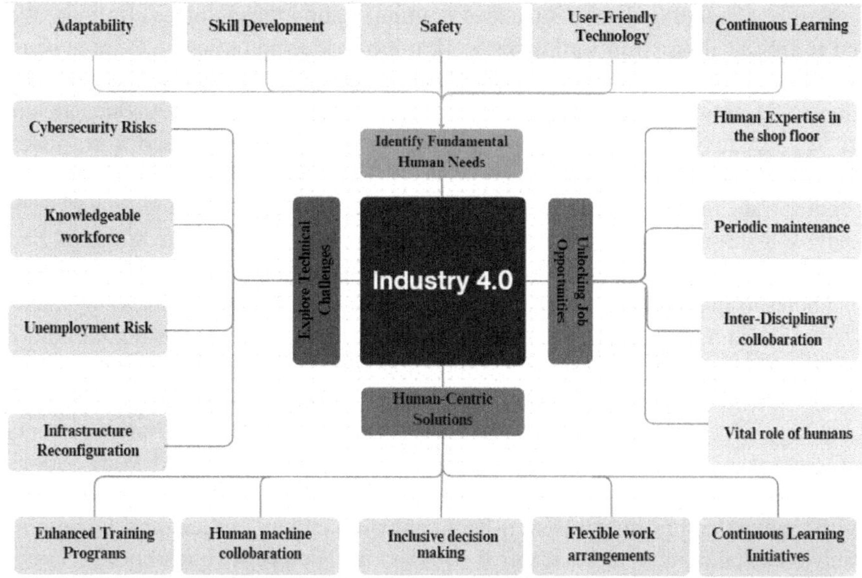

FIGURE 20.1 Roadmap to Human Centric Challenges and Solutions.

5.3 Inclusive Decision-Making

The process of inclusive decision-making, involves a diverse range of perspectives and experience to make strategic choices and decisions in an industry ensuring that the voice of all stakeholders are heard. This decision-making becomes very crucial for navigating changes in the organizational structure. It creates an environment where individuals at all levels of the company feel empowered to contribute to their perspectives and ideas. While AI facilitates intelligent decision-making, it may lack the nuanced maturity required to cater to the unique needs and intricacies of every individual company and its specific technologies. When a new technology is implemented, a human being can provide valuable insights into the practical implications of technologies, they can make choices on the technology adoption and suggest improvements for further development. Open communication channels to express the employee's opinion, questioning, and offering suggestions are essential as it creates a culture of transparency and trust. By actively involving the employees in decision-making processes, the organizations can make innovative and equitable decisions to contribute to the success of the evolving technologies of the industrial landscape.

5.4 Flexible Work Arrangements

The traditional office model may not align with the diverse needs of today's workforce, necessitating the implementation of flexible work arrangements to enhance adaptability in their professional lives. Implementing flexible work arrangements enables employees to manage professional responsibilities and personal lives seamlessly, alleviating stress. Advanced communication technologies eliminate the need for physical proximity, allowing work to be conducted remotely. Remote work environment provides an autonomy for the employees and also enables the organization to tap into a global talent pool. Components of flexible work arrangements, such as alternate work hours, individualized schedules, and compressed workweeks, empower employees to optimize their work schedules. This flexibility allows them to concentrate on tasks during peak productivity periods, enhancing overall efficiency. Job sharing, a collaborative approach, involves dividing the workload and sharing responsibilities between individuals. This ensures continuous workflow and contributes to maintaining a healthy work-life balance. Flexible work arrangements cultivate a result-oriented mind-set aimed at goal achievement, boosting employee morale, reducing burnout, improving retention rates, fostering individual job satisfaction and work-life balance thus representing a transformative approach to the technology developed work structures. In conclusion, flexible work arrangements represent a transformative approach to modern work structures. By providing options such as remote work, flexible scheduling, and job-sharing, organizations can create environments that prioritize the well-being and productivity of their workforce, ultimately contributing to a more adaptive, satisfied, and competitive workplace.

6 UNLOCKING OPPORTUNITIES FOR HUMAN IN INDUSTRY 4.0

Shop floor operators, perceiving digitization and automation as job threats, need clarification that operational excellence relies on human expertise, not solely on machines. The company must emphasize that restructuring processes necessitate human-machine collaboration. To adapt to the evolving machine environment, both the manufacturing company and its employees must acquire new technologies and skills. Enhanced knowledge is vital for proficiently working with new machinery and technology. Effective operation in a digitized and automated setting requires the industry to reorganize teams and allocate employees based on their technical competencies. Rather than causing job losses, automation and digital technologies create new opportunities, fostering a high demand for skilled and educated workers. Educating and training employees becomes paramount for the successful implementation of digital technology in the manufacturing sector.

6.1 Human Expertise in the Shop Floor

Shop floor operators, integral to the production system's success, possess extensive knowledge of the production chain. With machines becoming smart and interconnected, operators must evolve to be equally intelligent and connected. Acquiring specific skills like computing and logical reasoning enables operators to surpass machine capabilities. Analysing data from intelligent machines at the manufacturing level empowers operators to make informed decisions. Despite technological advancements, the essence of every technology and algorithm lies in the expertise of human developers. It is important to empower frontline employees with tools to make them work easier and also to link different organizations in the management system to promote performance and excellence.

6.2 Periodic Maintenance for Machines

Technical failures are inherent to any machine, and intelligent machines are designed to generate fault reports and autonomously initiate repair processes. However, it remains crucial for operators to conduct regular checks to proactively identify potential issues and prevent production delays. Periodic maintenance is a key aspect of ensuring the seamless operation of machinery. In the event of a disruption in the production line during scheduled maintenance, operators must promptly execute necessary tasks to avert compromises in product quality and delivery timelines.

6.3 Vital Role of Humanization

In the realm of Industry 4.0, the indispensable role of every human in a company is increasingly crucial. The process of humanization entails enhancing collaboration between human workers and intelligent machines, making a substantial contribution to the overall industry growth. Conversely, digital work instructions assume a pivotal role, furnishing operators with meticulous step-by-step guidance, ensuring precise task execution and fostering continuous improvement within the workflow. Operational excellence and digitization serve as key enablers to enhance business outcomes in the Industry 4.0 landscape. Despite the imperative for companies to invest in digital technologies for competitiveness, it is paramount to recognize that employees remain the critical determinants of success in the Industry 4.0 era.

7 CONCLUSION

In conclusion, the implementation of Industry 4.0 in various manufacturing sectors presents a lot of challenges for human beings. The adoption of new technologies, the imperative to acquire new skills, collaboration with interdisciplinary departments, and the integration of work with robots and machines all pose formidable hurdles for employees. The looming threat of job displacement further compounds these challenges. However, to navigate this transformative landscape successfully, several systems have been proposed. Continuous skill development programmes emerge as a crucial solution, ensuring that employees stay abreast of evolving technological demands. Human-machine collaboration stands out as a cornerstone, fostering synergy between humans and automation. Inclusive decision-making processes empower employees to actively contribute to the evolving industrial ecosystem. Additionally, flexible work arrangements offer adaptability in the face of dynamic changes. Collectively, these strategies aim to empower the workforce to overcome challenges and operate efficiently in the new Industry 4.0 environment. Moreover, the chapter sheds light on the vast opportunities available for the workforce within Industry 4.0, emphasizing the indispensability of human involvement across all manufacturing sectors. By delineating these proposed systems and emphasizing the need for human engagement, the chapter not only acknowledges the challenges posed by Industry 4.0 but also highlights the pathways towards a collaborative and empowered future for the workforce in the evolving industrial landscape.

REFERENCES

1. Rad, Fakhreddin F., Pejvak Oghazi, Maximilian Palmié, Koteshwar Chirumalla, Natallia Pashkevich, Pankaj C. Patel, and Setayesh Sattari. "Industry 4.0 and supply chain performance: A systematic literature review of the benefits, challenges, and critical success factors of 11 core technologies." *Industrial Marketing Management* 105 (2022): 268–293.
2. Frank, Alejandro Germán, Lucas Santos Dalenogare, and Néstor Fabián Ayala. "Industry 4.0 technologies: Implementation patterns in manufacturing companies." *International Journal of Production Economics* 210 (2019): 15–26.
3. Thames, L., Schaefer, D. (2017). Industry 4.0: An Overview of Key Benefits, Technologies, and Challenges. In: Thames, L., Schaefer, D. (eds) *Cybersecurity for Industry 4.0. Springer Series in Advanced Manufacturing*. Springer, Cham. https://doi.org/10.1007/978-3-319-50660-9_1
4. Jamwal, Anbesh, Rajeev Agrawal, Monica Sharma, and Antonio Giallanza. "Industry 4.0 technologies for manufacturing sustainability: A systematic review and future research directions." *Applied Sciences* 11, no. 12 (2021): 5725.
5. Zheng, Ting, Marco Ardolino, Andrea Bacchetti, and Marco Perona. "The applications of Industry 4.0 technologies in manufacturing context: A systematic literature review." *International Journal of Production Research* 59, no. 6 (2021): 1922–1954.
6. Ghobakhloo, Morteza. "Industry 4.0, digitization, and opportunities for sustainability." *Journal of Cleaner Production* 252 (2020): 119869.
7. Kamble, Sachin S., Angappa Gunasekaran, and Shradha A. Gawankar. "Sustainable Industry 4.0 framework: A systematic literature review identifying the current trends and future perspectives." *Process Safety and Environmental Protection* 117 (2018): 408–425.
8. Saucedo-Martínez, Jania Astrid, Magdiel Pérez-Lara, José Antonio Marmolejo-Saucedo, Tomás Eloy Salais-Fierro, and Pandian Vasant. "Industry 4.0 framework for management and operations: A review." *Journal of Ambient Intelligence and Humanized Computing* 9 (2018): 789–801.
9. Nahavandi, Saeid. "Industry 5.0—A human-centric solution." *Sustainability* 11, no. 16 (2019): 4371.
10. Kong, X.T.R., Luo, H., Huang, G.Q. et al. Industrial wearable system: the human-centric empowering technology in Industry 4.0. *J Intell Manuf* 30, 2853–2869 (2019). https://doi.org/10.1007/s10845-018-1416-9
11. Kadir, Bzhwen A., and Ole Broberg. "Human-centered design of work systems in the transition to industry 4.0." *Applied Ergonomics* 92 (2021): 103334.019: 2853–2869.
12. Longo, Francesco, Letizia Nicoletti, and Antonio Padovano. "Smart operators in industry 4.0: A human-centered approach to enhance operators' capabilities and competencies within the new smart factory context." *Computers & Industrial Engineering* 113 (2017): 144–159.
13. Nguyen Ngoc, Hien, Ganix Lasa, and Ion Iriarte. "Human-centred design in industry 4.0: Case study review and opportunities for future research." *Journal of Intelligent Manufacturing* 33, no. 1 (2022): 35–76.
14. Romero, David, Johan Stahre, Thorsten Wuest, Ovidiu Noran, Peter Bernus, Åsa Fast-Berglund, and Dominic Gorecky. "Towards an operator 4.0 typology: A human-centric perspective on the fourth industrial revolution technologies." In *Proceedings of the International Conference on Computers and Industrial Engineering (CIE46)*, Tianjin, China, pp. 29–31. 2016.

15. Zarte, M., Pechmann, A., & Nunes, I. L. (2020). Principles for human-centered system design in industry 4.0–a systematic literature review. In *Advances in Human Factors and Systems Interaction: Proceedings of the AHFE 2020 Virtual Conference on Human Factors and Systems Interaction*, July 16-20, 2020, USA (pp. 140-147). Springer International Publishing.
16. Alves, Joel, Tânia M. Lima, and Pedro D. Gaspar. "Is Industry 5.0 a human-centred approach? A systematic review." *Processes* 11, no. 1 (2023): 193.
17. Grosse, Eric H., Fabio Sgarbossa, Cecilia Berlin, and W. Patrick Neumann. "Human-centric production and logistics system design and management: Transitioning from Industry 4.0 to Industry 5.0." *International Journal of Production Research* 61, no. 22 (2023): 7749–7759.
18. Tóth, A., Nagy, L., Kennedy, R., Bohuš, B., Abonyi, J., & Ruppert, T. (2023). The human-centric Industry 5.0 collaboration architecture. *MethodsX*, 11, 102260.
19. Oliveira, M., Arica, E., Pinzone, M., Fantini, P., & Taisch, M. (2019). Human-centered manufacturing challenges affecting European industry 4.0 enabling technologies. In *HCI International 2019–Late Breaking Papers: 21st HCI International Conference, HCII 2019*, Orlando, FL, USA, July 26–31, 2019, Proceedings 21 (pp. 507-517). Springer International Publishing.
20. Welfare, K. S., Hallowell, M. R., Shah, J. A., & Riek, L. D. (2019, March). Consider the human work experience when integrating robotics in the workplace. In 2019 *14th ACM/IEEE international conference on human-robot interaction (HRI)* (pp. 75-84). IEEE.
21. Rosin, Frédéric, Pascal Forget, Samir Lamouri, and Robert Pellerin. "Enhancing the decision-making process through industry 4.0 technologies." *Sustainability* 14, no. 1 (2022): 461.

21 Strategies for Managing Risk and Mitigation in the Era of Smart Manufacturing

S. Krishnaveni, Faizan Ahmad, Mohamed Akbar, and S. Sangeetha

1 INTRODUCTION

Manufacturing is the systematic conversion of raw materials into finished products, employing tools, machinery, and human labour. The intricate process commences with the design phase, progresses through meticulous process planning, machining, rigorous quality checks, and culminates in the timely delivery of the end product to customers. Historically, large-scale manufacturing, known as conventional manufacturing, has relied predominantly on manual labour, lacking connectivity between various stages and limiting automation. This system operates in isolation, with minimal real-time monitoring or data analysis. Equipment maintenance occurs at fixed intervals due to the absence of continuous monitoring. Conventional manufacturing, a longstanding pillar of industrial processes, is gradually giving way to smart manufacturing. This transformative shift embraces technology with the overarching goals of boosting efficiency, curtailing costs, elevating product quality, and enhancing competitiveness in the global market.

Smart manufacturing, currently flourishing, integrates advanced digital technologies and data analytics into manufacturing processes to significantly enhance efficiency and productivity. This paradigm shifts harnesses the power of the internet of things (IoT), artificial intelligence (AI), robotics, and other cutting-edge technologies to establish an intelligent system within the manufacturing environment. Key components of smart manufacturing include IoT, which facilitates real-time monitoring and data collection by connecting physical devices and sensors to the internet. Data analytics employs algorithms and analysis tools to derive actionable insights from the data generated during the manufacturing process. AI and machine learning contribute to enhanced automation, predictive maintenance, and overall decision-making capabilities. Robots, exhibiting precision, speed, and adaptability, undertake complex tasks. The incorporation of cyber-physical systems integrates physical processes with digital technologies, enabling seamless communication and coordination between physical and virtual systems. Digital twins create virtual replicas of physical processes for simulation, analysis, issue identification, and

DOI: 10.1201/9781032694375-21

optimization. Cloud computing is employed for the storage and processing of data, fostering collaboration across different facets of the manufacturing system. Additive manufacturing, a component of smart manufacturing, facilitates the production of intricate structures with ease. These components work synergistically to transform traditional manufacturing into a more adaptive, connected, and intelligent system, fostering innovation and bolstering competitiveness in the dynamically evolving industrial landscape.

Smart manufacturing has profoundly reshaped operational efficiency and competitiveness in the industrial landscape, ushering in a new era of innovation and productivity. The integration of advanced digital technologies and data analytics has yielded multifaceted benefits. One of the key impacts is heightened operational efficiency. The real-time monitoring capabilities enabled by the IoT and the seamless communication of cyber-physical systems allow for an understanding of the manufacturing environment. This leads to optimized resource utilization, reduced downtime, and enhanced overall process efficiency. The use of robotics and automation in manufacturing contributes to precision, speed, and consistency, eliminating errors and streamlining workflows.

Data analytics extracts practical insights from the vast amounts of data produced throughout the manufacturing process. This informed decision-making enhances operational strategies, identifies areas for improvement, and facilitates continuous refinement of processes. Smart manufacturing enables companies to produce high-quality products with greater customization capabilities. The agility and adaptability of these systems facilitate quicker response to market demands, giving manufacturers a competitive edge.

While smart manufacturing offers benefits, it also introduces inherent risks that necessitate careful consideration. The major concern is the vulnerability to cyber threats. The extensive use of interconnected devices and exchange of data make smart manufacturing target for cyberattacks. This chapter delineates the risks associated with smart manufacturing, outlines effective risk management strategies, and details precise methodologies for mitigating these identified risks.

2 LITERATURE REVIEW

Babiceanu and Seker (2017) present a cybersecurity-resilience ontology for manufacturing network design stages. Bhattacharya et al. (2023) reviewed the ways human operators are affected in the cyber-physical systems. They have also focused on the human-mediated production and optimized operations. Davis et al. (2012) narrate smart manufacturing as a fundamental business transformation keyed on customers, partners and public. Dutta et al. (2022) discuss about five barriers and their mitigation approaches – cross-functional integration, constructive closed-loop, high-availability cloud connection, insightful data contextualization, and flexible operations orchestrations. Espinoza-Zelaya and Moon (2022) identified resilience-increasing mechanism to reduce cyberattack, and to reduce the adverse effects of cyberattack.

Mittal et al. (2019) match the technology of smart manufacturing with the design and manufacturing principles of Industry 4.0. An interdisciplinary discussion delineates the attributes, technologies, and determinants of intelligent manufacturing.

Oduoza (2020) studied the risk management for a manufacturing process which have negative impact on quality, cost, lead cycle, safety of workers and health. Ren et al. (2017) explores cutting-edge technologies aimed at tackling cybersecurity challenges within the context of smart manufacturing. Existing strategies and research gaps are also discussed. Rodríguez-Espíndola et al. (2022) conducted an empirical examination of Industry 4.0, leveraging data gathered from 117 managerial professionals in a manufacturing setting. Through structural modelling analysis, they identified strategies to enhance productivity.

Tupa, Simota, and Steiner (2017) conducted a research on the key aspects of Industry 4.0 and designed a framework to implement risk management. Zhong et al. (2017) provide a comprehensive review in topics like IoT and smart manufacturing. They have also described the government strategic plans for major companies in different countries. They have also presented current challenges and future work to adopt in Industry 4.0. Babiceanu and Seker (2019) propose a model that addresses a virtual manufacturing system assurance for SDN applications.

3 THE RISKS INVOLVED IN SMART MANUFACTURING

Smart manufacturing has undergone significant advancements in recent years, characterized by sophisticated technologies, meticulous planning, and seamless organization. The integration of connected devices and automation has propelled industry into a new era of efficiency and productivity. While acknowledging the numerous advantages, it is crucial to acknowledge and mitigate the inherent risks linked to the integration into smart manufacturing. The following discussions represent critical risks that require careful consideration.

3.1 Cybersecurity Concerns

Smart manufacturing relies on interconnectivity of the system that processes a vast amount of data. Data breaches can compromise sensitive information and employee data, granting unauthorized access to the system and resulting in production and financial losses. Intellectual property theft is another critical concern, as adversaries may target design specifications and plans, causing substantial damage to the company. Sabotage attempts by internal or external hackers can manipulate production data, leading to equipment malfunctions, quality issues, production delays, and safety hazards. Ransomware attacks, where hackers encrypt data or systems and demand large ransoms, have the potential to bring production to a standstill, causing severe operational disruptions. Even employees or contractors with access to sensitive data may unintentionally compromise cybersecurity, posing threats to data integrity and overall system security. The integration of systems into smart manufacturing creates multiple entry points for cyberattacks, and weaknesses in authorization processes can result in unauthorized access to critical systems. Compromised user credentials or flaws in authentication processes provide opportunities for attackers to infiltrate the manufacturing system. Additionally, cybersecurity vulnerabilities in the supply chain can have far-reaching effects on the overall security of the manufacturing process.

3.2 Dependency on Technology

Smart manufacturing depends on a network of interconnected technologies. Any system failure, or cyber-attack on the system, can disrupt operations, leading to significant downtime. Ensuring seamless communication and compatibility between the diverse technologies, sensors and devices may be challenging. Integration issues may arise during the implementation stage, which may cause delays and hinder the overall efficiency of the manufacturing system. The implementation of smart manufacturing technologies requires a highly skilled employee. The rapid evolution of technology creates a gap between the skills possessed by the existing workforce. Maintenance of this updated technology becomes a major challenge. Failure to keep up with maintenance schedules can result in unexpected downtime. Implementing smart manufacturing involves investments in technology infrastructure, software, and employee training. The rapid pace of technological advancement may lead to the obsolescence of certain systems or components. Manufacturers must keep their technology up to date to remain competitive in the market. Effectively managing and analysing the vast amount of data generated can pose a lot of challenge and it may bring in issues related to data accuracy, and security. Compatibility issues between legacy systems and cutting-edge technologies can impede the seamless transition to smart manufacturing. Smart manufacturing operates in an environment of constant technological evolution. Manufacturers need to proactively adapt their technology infrastructure to remain competitive and leverage the latest advancements.

3.3 Interoperability Challenges

Smart manufacturing involves the integration of technologies, including sensors, devices, machinery, and software systems. Ensuring seamless communication and interoperability between these different components becomes a challenge. Lack of standardized protocols and technologies contributes to interoperability challenges. Without industry-wide standards, manufacturers may face difficulties in integrating components from different vendors, leading to compatibility issues. Integrating heterogeneous devices and systems can be challenging, as legacy systems may not have built-in compatibility with smart manufacturing systems. Interoperability issues can lead to inconsistencies in data formats and quality. When data is exchanged between different systems, variations in data structures may occur, impacting the accuracy and reliability of information. Delays or interruptions in communication can lead to inefficiencies and disrupt the synchronized operation of interconnected systems. As manufacturing processes evolve and scale, interoperability becomes challenging and may hinder the scalability of smart manufacturing solutions.

3.4 Data Privacy Issues

Smart manufacturing involves the collection of vast amounts of data, including proprietary production processes, intellectual property, and employee information. It also involves the use of sensors and monitoring systems that collect data on employees' activities and performance. Balancing the benefits of data-driven insights with the need to protect the privacy of individual workers is a challenge. The operations involved in smart manufacturing need to comply with data protection and privacy regulations, which vary across regions and industries. Navigating these complex and evolving regulatory landscapes can be challenging. The volume of data generated raises questions about ownership. Determining who owns the data and how it can be used, shared, or monitored requires clear policies and agreements. In a globalized environment, data may need to flow across borders. Ensuring that data transfer complies with international privacy laws is a significant concern. IoT devices can be vulnerable to security breaches, and if compromised, they pose a direct threat to the privacy of the data they handle.

3.5 Cost Implications

The implementation of smart manufacturing technologies in an industry requires an initial investment. This includes the purchase of advanced machinery, sensors, automation systems, and the implementation of connected devices and communicated networks, data storage system and computing resources. Licensing fees for software, and maintenance costs, contribute to the overall financial burden. Software-related expenses can be a recurring cost throughout the operational life of the smart manufacturing system. Transitioning to smart manufacturing requires skilled employees capable of managing and maintaining advanced technologies. Investing in training programmes and skill development initiatives is necessary to ensure that employees can effectively operate, troubleshoot, and optimize the new systems. Maintenance costs include routine checks, updates, and repairs. Failure to invest in maintenance can lead to system failures, increased downtime. Ensuring the security of smart manufacturing systems involves ongoing investments in cybersecurity measures. This involves establishing resilient security protocols, ensuring regular updates, providing employee training, and embracing advanced cybersecurity solutions.

4 STRATEGIES TO MANAGE RISKS

4.1 Data Security and Cybersecurity

In the previous session, the risks involved in smart manufacturing were discussed. This session deals with strategies for managing risks. The main risk involved is the data breaches and cybersecurity risks causing a lot of damage to business. Effective data security and cybersecurity strategies are paramount for risk management in the modern business landscape.

4.1.1 Few Strategies to Maintain Data and Cybersecurity

- Employing robust encryption protocols ensures the confidentiality of sensitive information, safeguarding it from unauthorized access during transmission and storage.
- Multi-factor authentication introduces an extra layer of security, effectively reducing the risks linked to compromised credentials.

- Regular cybersecurity audits and prompt updates of software and firmware help identify and patch vulnerabilities.
- Implementing firewalls and intrusion detection systems is crucial for monitoring and preventing cyber threats.
- Maintaining a clear segregation between operational technology (OT) and information technology (IT) networks helps potential risks associated with integrated systems.
- Employee training programmes and strict access controls contribute to a human-centric defence, raising awareness and limiting the exposure of critical assets.
- Continuous monitoring, real-time threat intelligence, and incident response plans are essential components for swift detection and mitigation of cybersecurity incidents.

4.2 Standards, Guidelines and Regulations

Standards, guidelines, and regulations are used in an industry to ensure quality, consistency compliance, and safety. Standards are a set of specifications that are used in an industry to ensure that products or processes meet certain requirements. Guidelines are a set of recommendations that provide advice on the best practices or procedures to meet the desired outcomes. Regulations are legally binding rules or laws established by the Government and mandate certain behaviours or actions within an industry. Several guidelines exist for enhancing the security of industrial systems. The National Institute of Standards and Technology (NIST) guides the industrial sectors on the ways to achieve security in the system.

4.2.1 Standards

- ISO/IEC 27001: Information Security Management System (ISMS): This standard offers a structure for creating, implementing, sustaining, and enhancing an ISMS.
- ISO/IEC 27002: Code of Practice for Information Security Controls: This standard provides a set of directives for initiating, implementing, sustaining, and enhancing information security management within an organization.
- IEC 62443: Industrial Communication Networks – Network and System Security: The IEC 62443 series is specifically designed to address the security concerns of industrial automation and control systems.

4.2.2 Guidelines

- NIST Cybersecurity framework: Developed by the NIST, it provides guidelines for managing and improving an organization's cybersecurity position.
- ENISA: The European Union Agency for Cybersecurity offers guidelines specifically tailored to enhance the cybersecurity of industrial control systems (ICS).
- Industrial Internet Consortium (IIC) Security Framework: The IIC's Security Framework provides a set guideline for securing industrial internet of things (IIoT) systems in smart manufacturing.

4.2.3 Regulations

- General Data Protection Regulation (GDPR): GDPR, a regulation established by the European Union, defines regulations for safeguarding personal data and ensuring privacy.
- Cybersecurity Maturity Model Certification (CMMC): CMMC outlines cybersecurity requirements designed to safeguard sensitive information and fortify the defence industry's security.
- National Institute of Standards and Technology (NIST) SP 800-171: This set of guidelines outlines requirements for safeguarding Controlled Unclassified Information (CUI) in non-federal systems and organizations.
- National Cyber Security Centre (NCSC) Cyber Essentials: This scheme outlines basic cybersecurity practices that can be applied across various sectors, including manufacturing.

These standards, guidelines, and regulations are already implemented in an industry. They help in establishing a robust security framework for smart manufacturing, addressing the organizational aspects of cybersecurity and data protection.

4.3 Intrusion Detection Systems (IDS)

These cybersecurity tools are designed to monitor system activities, detecting indications of unauthorized behaviour. Their primary objective is to recognize and respond to security incidents, safeguarding the system against potential threats. The network-based intrusion detection systems (NIDS) monitors network traffic, searching for patterns that could potentially breach the security policy. They are deployed at strategic points within a network. They use signature-based detection, compare the network patterns against a known signature, and identify the deviations from the normal behaviour. Host-based intrusion detection systems (HIDS) focus on individual host and monitor activities like file modifications to detect unauthorized access. They are installed in individual systems and uses anomaly based detection methods to identify security threats.

Key components of IDS are sensors to collect data related to a system activity and monitor the network traffic. Analysers are employed to scrutinize the raw data gathered by sensors, identifying patterns and assessing whether there exists a potential threat. Additionally, there exists an alerting system designed to generate alerts in the event of any identified threats.

4.4 Security Skills Training

Training on security skills should be provided to employees to reduce risk. First they must be able to understand the cybersecurity concepts and principles, Security policies and procedures and Risk assessment and management. Training on ICS security involves understanding of ICS architecture and security measures for programmable logic controllers (PLCs) and SCADA systems. Training on network security involves firewalls and intrusion prevention systems, network segmentation for security. Continuous learning and staying updated on emerging threats and technologies are essential for maintaining effective security skills.

5 POST-INCIDENT MANAGEMENT

In smart manufacturing, where the potential impact of security threats can be severe, the swift and effective response is critical to mitigating losses and restoring normal operations. Policies, such as those outlined by organizations like the North American Electric Reliability Corporation Critical Infrastructure Protection (NERC CIP) and the Chemical Facility Anti-Terrorism Standards (CFATS), emphasize the need for immediate security response plans. Given the diverse nature of operators, varying security postures contribute to the complexity of threats. Following an attack, it is imperative to promptly identify the source and motivations of adversaries to prevent recurrence. The policies governing these response efforts should empower employees to act swiftly and decisively in returning the system to a secure state.

6 CONCLUSION

Smart manufacturing enhances efficiency and productivity by seamlessly integrating advanced technologies like the IoT, AI, machine learning, big data analytics, and automation. This convergence creates an intelligent and interconnected manufacturing environment, enhancing overall operational capabilities. This innovative approach involves the fusion of physical systems with digital technologies, facilitating product customization and bolstering flexibility in production processes. Real-time monitoring and optimization contribute to heightened energy efficiency and sustainability, as resources are judiciously utilized. Supply chain connectivity within smart manufacturing ensures smooth communication among suppliers, manufacturers, and distributors, fostering a collaborative and streamlined production ecosystem.

However, this transformative paradigm introduces inherent challenges, including cybersecurity concerns, technological dependency, interoperability issues, data privacy considerations, and potential cost implications. To mitigate these risks, robust strategies are essential. Implementation of intrusion detection systems, adherence to relevant standards, guidelines, and regulations, and proactive risk management practices are crucial. Employee training and skill development play a pivotal role in maintaining a resilient system, ensuring that humans are equipped to handle evolving technological landscapes and potential threats. Post-incident management is equally critical, with a focus on continuous improvement and adaptation. Learning from challenges and refining strategies contributes to the ongoing evolution of smart manufacturing. Looking to the future, adherence to these strategies promises enhanced efficiency and productivity. By prioritizing cybersecurity, embracing standards, and investing in skill development, smart manufacturing can continue to revolutionize industrial processes while minimizing associated risks and maximizing long-term success.

REFERENCES

Babiceanu, Radu F., and Remzi Seker. 2017. "Cybersecurity and Resilience Modelling for Software-Defined Networks-Based Manufacturing Applications." *Studies in Computational Intelligence* 694: 167–176. https://doi.org/10.1007/978-3-319-51100-9_15.

Bhattacharya, Mangolika, Mihai Penica, Eoin O'Connell, Mark Southern, and Martin Hayes. 2023. "Human-in-Loop: A Review of Smart Manufacturing Deployments." *Systems* 11 (1): 1–25. https://doi.org/10.3390/systems11010035.

Davis, Jim, Thomas Edgar, James Porter, John Bernaden, and Michael Sarli. 2012. "Smart Manufacturing, Manufacturing Intelligence and Demand-Dynamic Performance." *Computers and Chemical Engineering* 47: 145–156. https://doi.org/10.1016/j.compchemeng.2012.06.037.

Dutta, Gautam, Ravinder Kumar, Rahul Sindhwani, and Rajesh Kr Singh. 2022. "Overcoming the Barriers of Effective Implementation of Manufacturing Execution System in Pursuit of Smart Manufacturing in SMEs." *Procedia Computer Science* 200 (2019): 820–832. https://doi.org/10.1016/j.procs.2022.01.279.

Espinoza-Zelaya, Carlos, and Young Bai Moon. 2022. "Resilience Enhancing Mechanisms for Cyber-Manufacturing Systems against Cyber-Attacks." *IFAC-PapersOnLine* 55 (10): 2252–2257. https://doi.org/10.1016/j.ifacol.2022.10.043.

Mittal, Sameer, Muztoba Ahmad Khan, David Romero, and Thorsten Wuest. 2019. "Smart Manufacturing: Characteristics, Technologies and Enabling Factors." *Proceedings of the Institution of Mechanical Engineers, Part B: Journal of Engineering Manufacture* 233 (5): 1342–1361. https://doi.org/10.1177/0954405417736547.

Oduoza, C. F. 2020. "Framework for Sustainable Risk Management in the Manufacturing Sector." *Procedia Manufacturing* 51 (2019): 1290–1297. https://doi.org/10.1016/j.promfg.2020.10.180.

Ren, Anqi, Dazhong Wu, Wenhui Zhang, Janis Terpenny, and Peng Liu. 2017. "Cyber Security in Smart Manufacturing: Survey and Challenges." *67th Annual Conference and Expo of the Institute of Industrial Engineers* 2017: 716–721.

Rodríguez-Espíndola, Oscar, Soumyadeb Chowdhury, Prasanta Kumar Dey, Pavel Albores, and Ali Emrouznejad. 2022. "Analysis of the Adoption of Emergent Technologies for Risk Management in the Era of Digital Manufacturing." *Technological Forecasting and Social Change* 178 (February 2021): 121562. https://doi.org/10.1016/j.techfore.2022.121562.

Tupa, Jiri, Jan Simota, and Frantisek Steiner. 2017. "Aspects of Risk Management Implementation for Industry 4.0." *Procedia Manufacturing* 11 (June): 1223–1230. https://doi.org/10.1016/j.promfg.2017.07.248.

Zhong, Ray Y., Xun Xu, Eberhard Klotz, and Stephen T. Newman. 2017. "Intelligent Manufacturing in the Context of Industry 4.0: A Review." *Engineering* 3 (5): 616–630. https://doi.org/10.1016/J.ENG.2017.05.015.

22 Cultivating a Security-Conscious Smart Manufacturing Workforce

A Comprehensive Approach to Workforce Training and Awareness

Aliyu Mohammed, Ashok Kumar Manoharan, Pethuru Raj Chelliah, and Sulaiman Ibrahim Kassim

1 INTRODUCTION

Smart manufacturing, a transformative paradigm fueled by cutting-edge technologies like the internet of things (IoT) and artificial intelligence (AI), has revolutionized traditional industrial processes. As industries embrace this digitized future, the critical aspect of cybersecurity in smart manufacturing cannot be overstated. The integration of digital technologies into manufacturing processes introduces new vulnerabilities, making robust security measures imperative. According to a report by Song and Zhu (2022), smart manufacturing systems are prime targets for cyber threats due to their interconnected nature. The interconnectedness exposes a plethora of entry points for malicious actors, emphasizing the need for a comprehensive security approach. Recent incidents, such as the Triton malware attack on a Saudi petrochemical plant, underscore the real-world consequences of inadequate cybersecurity in industrial settings (Rajput, 2023). Workforce training and awareness emerge as pivotal elements in fortifying the defenses of smart manufacturing systems. As smart manufacturing heavily relies on skilled human operators interacting with advanced technologies, empowering the workforce with a strong cybersecurity foundation is essential. Training programs not only equip employees with the technical knowledge to thwart cyber threats but also cultivate a security-oriented mindset. Research by the National Institute of Standards and Technology (NIST) emphasizes the role of

DOI: 10.1201/9781032694375-22

employee training in mitigating cybersecurity risks (Team, 2021). The study underscores the dynamic nature of cyber threats, necessitating continuous training to keep the workforce abreast of the latest developments in cybersecurity. Establishing a security culture within the workforce becomes a proactive strategy, aligning with the adage that prevention is better than cure. In this context, this chapter delves into the multifaceted realm of workforce training and awareness in smart manufacturing, exploring the evolving threat landscape, the components of a robust security culture, the role of leadership, technological measures, and case studies demonstrating successful implementations. By elucidating the significance of building a security culture, this study aims to guide companies in navigating the complexities of smart manufacturing with a resilient and informed workforce.

1.1 Background of Smart Manufacturing

Smart manufacturing represents a paradigm shift from traditional manufacturing by integrating advanced technologies such as IoT, AI, and data analytics into industrial processes (Tao et al., 2018). This integration enhances efficiency, reduces costs, and enables real-time decision-making. The interconnectivity of devices and systems forms the backbone of smart manufacturing, creating a dynamic ecosystem that accelerates production and optimizes resource utilization. However, this connectivity also introduces cybersecurity challenges as the attack surface expands. Leng et al. (2020) emphasize that smart manufacturing systems rely on seamless communication between devices, creating a complex network vulnerable to cyber threats. Smart manufacturing and its interconnections are complex in nature; understanding them is important when looking at security environment in smart manufacturing. This chapter will examine the fundamental basis of smart manufacturing and lay down a groundwork for further discussions on security aspects.

1.2 Emergence and Significance of Security Culture

Smart manufacturers are now increasingly emphasizing on security culture in their operations. A security culture encompasses shared beliefs, attitudes, and behaviors that prioritize and promote cybersecurity within the workforce (Fisher et al., 2021). With cyber threats becoming more sophisticated and targeted, a mere reliance on technical solutions is insufficient; human factors play a crucial role in fortifying defenses. According to Bıçakcı and Evren (2022), a security culture represents an attitude that ensures every employee is alert and keen in their duties. This implies that the mindset exceeds compliance with security guidelines and embraces joint cybersecurity ownership. The chapter will shed light on this shift in culture where the employees become the main defenders against cyber threats in a smart manufacturing environment.

1.3 Conceptual Framework for Security Culture in Smart Manufacturing

A theory-based conceptual frame work for securing smart manufacturing culture should take into consideration the technological, organizational, and human aspects of the problem. Smart manufacturing systems are linked together and require new approaches in cybersecurity (Tuptuk, & Hailes, 2018). The conceptual framework should encompass employee training, leadership involvement, and the deployment of advanced security technologies. Hassandoust and Johnston (2023) argue that a robust security culture framework not only enhances the technical resilience of a system but also creates a culture of continuous improvement and adaptability to emerging threats.

1.4 Research Gap

The current literature on security culture in smart manufacturing presents a significant gap that this study aims to address. While existing research acknowledges the importance of fostering a security culture, there is a dearth of empirical studies providing comprehensive insights into the practical challenges faced by organizations in implementing and sustaining such cultures within the dynamic landscape of smart manufacturing. Høiland (2023) on cybersecurity culture recognizes the concept's importance but primarily focuses on theoretical aspects, leaving a gap in practical implementation insights. Similarly, Velasco et al. (2023) on cyber-physical systems (CPS) in Industry 4.0 provide a foundational understanding but lacks a detailed exploration of the challenges organizations specifically face in cultivating a security culture within smart manufacturing. Citybabu and Yamini's (2023) framework for cybersecurity in smart manufacturing is a valuable contribution; however, it leans toward the technical aspects and lacks an in-depth examination of the human and organizational dimensions of security culture. Therefore, the identified research gap centers on the scarcity of empirical studies providing practical guidelines and an enhanced conceptual framework that addresses the specific challenges of implementing security culture in the context of smart manufacturing.

1.5 Research Objectives

1. To evaluate current security culture practices in smart manufacturing.
2. To identify challenges in implementing security culture in smart manufacturing.
3. To assess the role of leadership in shaping security culture.
4. To develop practical guidelines for workforce training in smart manufacturing.
5. To propose an enhanced conceptual framework for security culture in smart manufacturing.

2 LITERATURE REVIEW

The literature review delves into the evolving landscape of smart manufacturing and the theoretical foundations of security culture. The synthesis of these domains aims to identify gaps and lay the groundwork for the empirical investigation.

2.1 Evolution of Smart Manufacturing Concepts

The progression of smart manufacturing concepts is illuminated by pivotal contributions, exemplified in the examination of CPS by Dave (2023). Their research delves into the assimilation of digital technologies within the manufacturing realm, underscoring the transformative capabilities intrinsic to Industry 4.0. Moreover, the Accenture report focusing on the cybersecurity facets of smart manufacturing delineates the escalating susceptibility of interconnected systems (Tsantes & Ransome, 2023). This report serves as a crucial precursor for comprehending the critical importance of security within the dynamic landscape of evolving manufacturing technologies. Tsantes and Ransome's study represents a cornerstone in unraveling the trajectory of smart manufacturing. By exploring CPS, they shed light on the intricate interplay between digital technologies and manufacturing processes, accentuating the paradigm shift ushered in by Industry 4.0. This seminal work lays the groundwork for understanding how the fusion of cyber and physical elements propels manufacturing into a new era of efficiency and innovation. Furthermore, the Accenture report acts as a sentinel, alerting us to the growing vulnerability inherent in interconnected systems within the smart manufacturing ecosystem. As Industry 4.0 unfolds, the reliance on digital interconnectivity amplifies the risk landscape, necessitating a nuanced comprehension of cybersecurity imperatives. Thus, the report not only underscores the rapid evolution of smart manufacturing but also underscores the critical need for robust security measures to safeguard against potential threats in this transformative landscape.

Source: CESMII –The Smart Manufacturing Institute, www.cesmii.org

2.2 Theoretical Foundations of Security Culture

The theoretical foundations of security culture find elucidation in the research conducted by James (2023). In their study, they seek to define the dimensions of organizational culture with special attention that the latter could be utilised as powerful shield against cyber threats and attacks. However, their work is not only about the definition but more importantly that they are laying down the basic rock on which to build understanding of the complexities of security culture and it critical resistant nature. Therefore, James's study can be described as a pioneering one on how to solve many problems of the security theory. Through a very specific formulation of what is cybersecurity culture, they set up a complete model which gives as much substance to the concept as possible. However, this basic work serves as a basis for examining, what is not just an abstract notion but a concrete and powerful shield nowadays, which people employ against threats to their cybersecurity. In addition, this study provides an important basis for comprehending how security culture and organizational resilience interact with each other. This implies how developing a strong organizational security culture can be very helpful in enabling such an entity to survive from, as well as bounce back after, security threats. Essentially, James' research serves as a theoretical barometer for both researchers and practitioners on organizational security culture and why it is important for organizational security.

2.3 Synthesis of Security Culture and Smart Manufacturing Literature

The amalgamation of security culture and smart manufacturing literature necessitates a comprehensive exploration of their intersections and identification of existing gaps. In this endeavor, Bibri and Jagatheesaperumal (2023) framework for cybersecurity in smart manufacturing emerges as a pivotal guide, serving as a bridge that connects the technological dimensions inherent in smart manufacturing with the cultural elements integral to cybersecurity considerations. The work of Bibri and Jagatheesaperumal forms the foundation for analyzing the complex intertwining between smart manufacturing technology developments and critical cultural basis for successful cybersecurity implementation. However, on examining this in details one can clearly see the necessity a more total integration of those two domains. While Bibri and Jagatheesaperumal's framework provides valuable insights, there exists an opportunity to further synthesize and harmonize the technological and cultural facets of smart manufacturing security. Such an observation provides for the empirical objectives of the study, pointing on a possible line of enquiry. The amalgamation of synthesis of security culture and smart manufacturing literature is an inherent process that involves comprehending the complex relationship between technical innovations and social impacts. The groundwork was provided for an overall examination of those concepts which was made the integration of those dimensions as smooth as possible because the present research will identify the gaps. Therefore, this research is meant to provide a deeper understanding of security culture as related to smart manufacturing in the need for adopting an integrated approach of technological prowess coupled with a strong cultural foundations toward addressing cybersecurity issues associated with smart manufacturing.

2.4 Key Conceptual Models and Frameworks

There are useful conceptual models and frameworks for cyber security in smart manufacturing. Such models are indispensable instruments that organizations use in an attempt to understand how secure is it to embrace smart manufacturing. CPS Security framework, for example, provides a general approach that addresses the application of digital technology in the traditional manufacturing process. This framework provides a foundation on which cyber and physical elements can safely interact in smart manufacture conditions (Table 22.1).

The frameworks consider smart manufacturing security in its multi-dimensional nature emphasizing on the need for simultaneous integration of technological and cultural spheres. These models could help organizations develop customized security approaches that enhance their capacity to resist rapidly transforming cyber-attacks. With these conceptual models gradually becoming popular as the smart manufacturing landscape continues advancing, it has been paramount to ensure manufacturing processes are protected, a resilient and secure future is assured.

Source: https://www.sciencedirect.com/science/article/abs/pii/S0747563215002447

TABLE 22.1
Overview of Key Conceptual Models and Frameworks

Model/Framework	Description
Cyber-Physical Systems (CPS) Security Framework	Encompasses strategies for securing the convergence of cyber and physical components in smart manufacturing
Srinivas et al.'s Cybersecurity Framework (2019)	Provides a bridge between technological aspects of smart manufacturing and cultural elements of cybersecurity
Tolah et al.'s Security Culture Framework (2019)	Defines the theoretical underpinnings of security culture, emphasizing its role as a defense mechanism

Source: Reviewer's Review (2023).

3 THEORETICAL FRAMEWORK:

Therefore, Tolah et al.'s (2019) definition of security culture becomes an indispensable component for building up the theoretical framework. Essentially, this concept refers to common understandings and practices that can enlighten how smart manufacturing's security culture is taken in full view. Additionally, Srinivas et al.'s (2019) CPS architecture offers insights into the technological aspects, forming the basis for integrating security culture within the dynamic environment of smart manufacturing.

3.1 Defining the Conceptual Framework

Masoud (2023) proposed conceptual framework stands as a pivotal integration of cybersecurity principles within the realm of smart manufacturing systems. This innovative framework presents a systematic and methodical approach to tackling the cybersecurity challenges that emerge in the context of Industry 4.0. It lays a solid foundation for comprehending the intricate technical dimensions inherent in securing smart manufacturing processes. Nguyen and Tran (2023) framework, therefore, serves as a fundamental guide for navigating the complexities of cybersecurity within the context of advanced manufacturing technologies. Expanding upon Masoud's technical emphasis, Nguyen and Tran's contribute a crucial dimension to the conceptual landscape through their elucidation of cybersecurity culture. Their conceptualization recognizes the significance of human and organizational factors in fortifying cybersecurity defenses, providing a holistic foundation for the broader conceptual framework of the study. By acknowledging the human and organizational aspects, Herath and Indrakanti's work complements Masoud's technical focus, creating a comprehensive framework that accounts for both the technological and cultural facets of cybersecurity in the smart manufacturing domain. In essence, the conceptual framework shaped by Masoud's and augmented by Nguyen and Tran's combines technical rigor with a profound understanding of the human and organizational elements. This comprehensive approach not only addresses the immediate challenges posed by cybersecurity threats in Industry 4.0 but also lays the groundwork for proactive and adaptive security measures. The synthesis of these frameworks underscores the importance of a multifaceted approach in developing strategies that can effectively secure smart manufacturing systems, considering both the intricate technological landscapes and the human dynamics shaping the cybersecurity culture within organizations.

3.2 Adapting Security Culture Models to Smart Manufacturing

The process of adapting security culture models to smart manufacturing necessitates the harmonious integration of theoretical principles from cybersecurity culture with the distinctive challenges posed by Industry 4.0. Central to this adaptation is the foundational work of Garba et al. (2023), who has laid the groundwork in understanding and conceptualizing cybersecurity culture. Their research serves as a crucial starting point, providing theoretical insights that form the basis for navigating the intricate dynamics of security culture within the context of smart manufacturing. Building upon this foundation framework emerges as a pivotal instrument in the adaptation process. This framework not only recognizes the importance of cybersecurity culture but also plays a key role in seamlessly integrating these cultural dimensions into the technological landscape of smart manufacturing. By doing so, it bridges the gap between theoretical insights into cybersecurity culture and the practical implementation of security measures in the technologically advanced environment of Industry 4.0. In essence, the adaptation of security culture models to smart manufacturing is a nuanced endeavor that draws from the theoretical richness of cybersecurity culture, as expounded and aligns it with the pragmatic requirements of Industry 4.0 through the technological lens provided by this framework. This convergence allows for a holistic understanding of security culture within the specific context of smart manufacturing, ensuring that the theoretical foundations are translated into actionable strategies that address the unique challenges posed by the integration of advanced technologies into contemporary manufacturing processes.

3.3 Incorporating Workforce Training into the Conceptual Framework

The process of integrating workforce training into the conceptual framework involves leveraging the guidelines issued by NIST in 2020. NIST's directives on constructing information technology security awareness and training programs serve as a foundational resource for this endeavor. The comprehensive approach outlined by NIST offers a well-structured foundation, enabling the design of training initiatives customized to meet the distinct requirements of smart manufacturing environments. This incorporation is pivotal in addressing the human dimension within the broader conceptual framework. NIST's guidelines provide a systematic framework that encompasses various aspects of information technology security awareness and training. By drawing insights from these guidelines, the conceptual framework gains a structured methodology for developing training programs that specifically cater to the intricacies of smart manufacturing. As technology evolves and Industry 4.0 advances, it is imperative to equip the workforce with the knowledge and skills necessary to navigate the evolving cybersecurity landscape. The inclusion of workforce training aligns with the recognition that human factors play a crucial role in the overall cybersecurity posture of smart manufacturing systems. NIST's approach not only acknowledges the significance of training but also guides the development of programs that are adaptive and responsive to the evolving threat landscape. Therefore, incorporating workforce training into the conceptual framework ensures that the human element is not only considered but actively cultivated as a proactive defense against cybersecurity threats in the dynamic and technologically sophisticated context of smart manufacturing.

3.4 The Role of Leadership in Shaping Security Culture

The influence of leadership in shaping security culture draws insights from theoretical perspectives on organizational leadership, with a particular focus on the relevance of transformational leadership as articulated by Mgaiwa (2023). In the context of security culture within smart manufacturing, their research becomes a pertinent reference point. The transformational leadership framework advocated by Mgaiwa underscores the pivotal role played by leaders in inspiring and motivating employees toward a collective vision. This emphasis on leadership involvement aligns seamlessly with the imperative of cultivating a robust security culture within the dynamic landscape of smart manufacturing. Mgaiwa's research highlights how transformational leadership goes beyond traditional management roles, emphasizing the ability of leaders to instill a sense of purpose and commitment among team members. In the realm of smart manufacturing, where security concerns are integral, the application of such leadership principles becomes crucial. Leaders are envisioned not only as managers but also as inspirational figures, actively contributing to the cultivation of a security-conscious mindset among employees. In the context of smart manufacturing, where the fusion of technology and human elements is intricate, the role of leadership becomes a linchpin in shaping the security culture. The need for leaders to not only endorse but actively champion security initiatives is paramount. By embracing the tenets of transformational leadership, leaders can foster an environment where security is not merely a compliance requirement but an intrinsic value embraced by all. Therefore, the theoretical underpinnings provided by Mgaiwa offer a valuable lens through which the role of leadership can be comprehended and applied in the context of shaping an effective security culture within the unique milieu of smart manufacturing.

4 METHODOLOGY

The methodology employed in this study revolves around an extensive review of related studies, aligning with the qualitative nature of the research and its conceptual focus.

4.1 Justification for a Conceptual Approach

The choice of a conceptual approach stems from the nature of the research, which is qualitative and inherently focused on understanding and synthesizing existing theoretical frameworks. The foundational justification for this approach lies in the need to explore the complex interplay between security culture, smart manufacturing, workforce training, and leadership, drawing on existing conceptual models and frameworks. As scholars argue, qualitative research in the form of conceptual studies allows for a deeper exploration of theoretical foundations (Grenier, 2023).

4.2 Conceptualization of Key Variables

In conceptualizing key variables, the study draws extensively from established frameworks. The cybersecurity culture conceptualization is rooted in the work of Tejay and Mohammed (2023), which provides a comprehensive understanding of the human and organizational dimensions of cybersecurity. Tejay and Mohammed (2023) framework is instrumental in integrating these cultural aspects into the technological fabric of smart manufacturing. NIST's guidelines contribute to the conceptualization of workforce training, offering a structured foundation for designing programs specific to smart manufacturing. The role of leadership is informed by the transformational leadership framework by Ali (2023), emphasizing the leaders' inspiring and motivating role in shaping security culture.

4.3 Delimitations and Scope of the Conceptual Model

The study acknowledges certain delimitations, such as the focus on existing theoretical frameworks rather than empirical data. The scope is delimited to the integration of cybersecurity culture, smart manufacturing, workforce training, and leadership within the conceptual model, excluding detailed empirical validation.

4.4 Operationalization of Variables

Operationalization in this qualitative study involves translating conceptual variables into tangible components within the conceptual model. For instance, cybersecurity culture is operationalized by considering the human and organizational aspects outlined by Plachkinova and Janczewski (2023). Smart manufacturing integration is operationalized using Plachkinova and Janczewski (2023) framework, aligning technological and cultural dimensions. Workforce training is operationalized according to NIST's guidelines, ensuring adaptability to smart manufacturing needs. The leadership role is operationalized through the transformational leadership principles, emphasizing inspiration and motivation within the security culture context. This qualitative, conceptual methodology facilitates a nuanced exploration of theoretical underpinnings, providing a comprehensive understanding of the interconnections among security culture, smart manufacturing, workforce training, and leadership. The amalgamation of various scholars' contributions enriches the conceptual framework, ensuring a robust foundation for further exploration and practical implementation.

5 BUILDING A SECURITY CULTURE: A CONCEPTUAL MODEL

The construction of a conceptual model is grounded in the synthesis of existing frameworks, emphasizing the interplay between security culture, smart manufacturing, workforce training, and leadership. The model aims to provide a comprehensive understanding of the relationships among these elements, serving as a guide for organizations aiming to fortify their cybersecurity posture within the dynamic landscape of smart manufacturing.

5.1 Elements of the Conceptual Model

The conceptual model comprises key elements derived from established frameworks. Cybersecurity culture, rooted in the work of Fisk (2023), forms the core. It encompasses the human and organizational dimensions critical for fostering a security-conscious environment. Smart manufacturing integration leverages Nifakos' (2023) framework, ensuring a holistic approach that considers both technological and cultural aspects. Workforce training, guided by NIST's guidelines (2020), adds a crucial layer, addressing the knowledge and skills necessary for employees to contribute to the security culture. Leadership, informed by the transformational leadership principles of Ananyi and Ololube (2023), plays a pivotal role in inspiring and motivating the workforce toward embracing a security culture.

5.2 Theoretical Underpinnings of Workforce Training in the Model

The theoretical underpinnings of workforce training in the model draw from NIST's guidelines (2020), emphasizing the importance of continuous learning in addressing cybersecurity challenges. Workforce training serves as a dynamic component within the model, aligning with the evolving nature of cyber threats. The model recognizes that training programs, informed by established guidelines, contribute not only to skill development but also to cultivating a security-aware mindset among employees. This aligns with the findings of the NIST publication, emphasizing the ongoing nature of training to keep the workforce abreast of the latest developments in cybersecurity. The conceptual model, enriched by these theoretical underpinnings, serves as a comprehensive guide for organizations navigating the complexities of building and sustaining a security culture in the context of smart manufacturing.

5.3 Representation of the Model

This conceptual model succinctly encapsulates the key elements and their interconnections. Table 22.2 below outlines the critical components derived from established frameworks, presenting a holistic view of the model's structure.

This table emphasizes the integrated nature of the model, highlighting how each element contributes to the overall objective of building a robust security culture within the realm of smart manufacturing. Cybersecurity culture forms the foundation, incorporating both human and organizational dimensions. Smart manufacturing integration spans technological and cultural aspects, while workforce training, guided by NIST guidelines, addresses the knowledge and skills needed for employees. Leadership, drawing from transformational principles, plays a central role in inspiring a security-aware mindset. This visual format facilitates a concise understanding of the model's structure, fostering clarity in its implementation within organizational contexts.

TABLE 22.2
Workforce Training in the Model

Element	Theoretical Foundation
Cybersecurity Culture	Chanoski (2023)
Smart Manufacturing	Kusiak (2023)
Workforce Training	NIST guidelines (2020)
Leadership	Kezar (2023)

Source: Reviewer's Review (2023).

6 KEY DIMENSIONS AND INDICATORS

The identification of key dimensions and indicators is crucial for operationalizing the conceptual model into practical strategies. This outlines the dimensions integral to fostering a security culture within smart manufacturing and proposes specific indicators that organizations can use to assess and enhance their cybersecurity posture.

6.1 Identifying Dimensions of Security Culture

The dimensions of security culture draw inspiration from the comprehensive framework developed by Khripunov (2023). The human dimension includes aspects like employee awareness, attitudes, and behaviors toward cybersecurity. The organizational dimension encompasses policies, procedures, and the overall commitment of the organization to cybersecurity. Additionally, the technological dimension, influenced by Ejaz (2023) framework, involves the integration of secure technologies within smart manufacturing processes. This triad of dimensions forms the foundation for building a resilient security culture within the smart manufacturing ecosystem.

6.2 Proposed Indicators for Each Dimension

Proposing indicators for each dimension involves translating theoretical concepts into measurable metrics. For the human dimension, indicators could include the frequency of cybersecurity training, the level of employee awareness through simulated phishing exercises, and the reporting rate of security incidents. The organizational dimension indicators might encompass the existence and adherence to cybersecurity policies, the allocation of resources for cybersecurity measures, and the level of top management commitment. In the technological dimension, indicators could involve the implementation of secure communication protocols, the frequency of security audits, and the speed of response to identified vulnerabilities. These proposed indicators align with the conceptual model's holistic approach, allowing organizations to systematically evaluate and enhance their security culture. Regular assessment using these indicators enables a proactive stance against evolving cyber threats in the dynamic landscape of smart manufacturing.

6.3 Dimensions and Indicators

In building a robust security culture within smart manufacturing, a conceptual table outlining key dimensions and their associated indicators serves as a practical tool for organizations. Table 22.3 synthesizes insights from Peng et al. (2023) and integrates indicators proposed for each dimension, providing a comprehensive framework for assessment and improvement.

6.3.1 Human Dimension

This dimension focuses on the human elements of cybersecurity, acknowledging that employees are crucial in fortifying security. Indicators include the frequency of cybersecurity training, simulated phishing exercises to gauge awareness, and the reporting rate of security incidents, reflecting the workforce's active involvement.

6.3.2 Organizational Dimension

This dimension assesses the overarching commitment of the organization to cybersecurity. Indicators encompass the existence and adherence to cybersecurity policies, allocation of resources, and the level of top management commitment, providing a comprehensive view of the organization's dedication to security.

6.3.3 Technological Dimension

Incorporating insights from Chan (2023) framework, this dimension evaluates the integration of secure technologies. Indicators involve the implementation of secure communication protocols, the frequency of security audits to identify vulnerabilities, and the speed of response to mitigate potential risks. This conceptual table provides a structured approach for organizations to assess their security culture across these key dimensions. Regular evaluation using these indicators enables a dynamic and adaptive security posture within the smart manufacturing environment, ensuring resilience against emerging cyber threats.

TABLE 22.3
Dimensions and Indicators

Dimension	Indicators
Human dimension	Frequency of cybersecurity training
	Employee awareness through simulated phishing exercises
	Reporting rate of security incidents
Organizational dimension	Existence and adherence to cybersecurity policies
	Allocation of resources for cybersecurity measures
	Level of top management commitment
Technological dimension	Implementation of secure communication protocols
	Frequency of security audits
	Speed of response to identified vulnerabilities

Source: Reviewer's Review (2023).

7 CONCEPTUALIZATION OF WORKFORCE TRAINING STRATEGIES

The conceptualization of workforce training strategies within smart manufacturing is rooted in the need for a dynamic and adaptive approach to address the evolving landscape of cybersecurity threats. This section explores the conceptual foundations of designing effective training strategies that align with the unique requirements of smart manufacturing environments.

7.1 Overview of Conceptual Training Strategies

Effective training strategies are paramount in cultivating a security-aware workforce. NIST's guidelines (2020) on building information technology security awareness and training program provide a foundational overview. The strategies emphasize continuous learning, interactive training modules, and the integration of real-world scenarios. Furthermore, considering the human dimension of cybersecurity culture (Nifakos, 2023), training strategies should go beyond technical aspects, encompassing behavioral elements crucial for fostering a security-conscious workforce within the smart manufacturing context.

7.2 Aligning Training Strategies with Security Culture Dimensions

The alignment of training strategies with security culture dimensions is a critical aspect of ensuring relevance and effectiveness. For the human dimension, training should focus on increasing awareness, fostering positive attitudes, and encouraging responsible behaviors. Interactive modules and simulations can be designed to emulate real-world scenarios, enhancing the practical understanding of cybersecurity. In the organizational dimension, training should emphasize policy adherence, resource allocation awareness, and the role of top management in promoting a security culture.

7.3 Addressing Challenges and Enhancing Effectiveness

It entails appreciating the specific difficulties in smart manufacturing setting when handling challenges associated with workers' training. Employees, with their disparate skills, and fast-changing technology, call for a process that is always evolving. Improvement strategies include customized trainings for various positions, utilization of modern learning technology, and an iterative process of feedback mechanism for enhanced effectiveness. The training strategy should involve insights based on the security culture conceptual model as addressed in the cybersecurity culture dimension to produce a total and durable workforce. This leads the way for creating a viable technical-oriented approach in line with the cultural aspects of smart manufacturing practices.

8 INTEGRATION OF KEY PERFORMANCE INDICATORS

Evaluation of an effective conceptual model of smart manufacturing in relation to a security culture requires the inclusion of key performance indicators (KPIs).

The second part discusses why KPIs are necessary and shows how they can be developed to comprehensively evaluated the assessment model.

8.1 Developing KPIs for Evaluating Conceptual Model Effectiveness

This includes identifying measurable indicators that support the overall conceptual model's intentions. The use of KPIs provides quantitative information on the results from these smart manufacturing efforts in security culture. These dimensions and indicators provided by the conceptual model can be modified so that KPIS are developed for measuring particular outcomes. On the part of human dimension, examples of KPIs may include increase in employee awareness after trainings, reduced number of reported Security incidences attributed to improved awareness and regular employee participation in simulated phishing exercises. For the organizational dimension, KPIs might involve tracking the adherence rate to cybersecurity policies, measuring the allocation of resources dedicated to cybersecurity initiatives, and assessing the level of commitment demonstrated by top management. In the technological dimension, KPIs can focus on the successful implementation of secure communication protocols, the frequency and effectiveness of security audits, and the speed of response to identified technological vulnerabilities. Developing KPIs is essential not only for gauging the success of specific training initiatives but also for providing valuable feedback on the overall effectiveness of the security culture within the smart manufacturing environment. The continuous refinement and monitoring of KPIs enable organizations to adapt their strategies, ensuring a resilient and evolving security posture.

8.2 Proposed KPIs and Measurement Criteria

In effectively evaluating the conceptual model's impact, the integration of KPIs provides a quantifiable means of assessing progress. Table 22.4 outlines proposed KPIs aligned with the dimensions of security culture within smart manufacturing and the corresponding measurement criteria.

The table offers a structured overview of the proposed KPIs and their corresponding measurement criteria across the three dimensions. The clarity of this format facilitates a comprehensive evaluation of the effectiveness of security culture initiatives within smart manufacturing. Regular monitoring of these KPIs ensures a proactive approach to addressing challenges and optimizing the security posture in alignment with the conceptual model.

9 CHALLENGES AND OPPORTUNITIES IN CONCEPTUAL IMPLEMENTATION

The conceptual implementation of a security culture model within smart manufacturing presents both challenges and opportunities. This section explores the anticipated hurdles, potential areas for improvement, and real-world examples that highlight successful applications of similar models.

TABLE 22.4
Key Performance Indicators

Dimension	Proposed KPIs	Measurement Criteria
Human dimension	Percentage increase in employee awareness	Pre-and post-training assessment scores
	Reduction in reported security incidents	Incident logs before and after training
	Frequency of employee participation in simulations	Participation rates in simulated phishing exercises
Organizational dimension	Adherence rate to cybersecurity policies	Audit results on policy compliance
	Allocation of resources for cybersecurity measures	Budget allocation for cybersecurity initiatives
	Level of top management commitment	Surveys assessing perceived commitment
Technological dimension	Successful implementation of secure protocols	Audit results on protocol implementation
	Frequency and effectiveness of security audits	Audit reports and identified vulnerabilities
	Speed of response to technological vulnerabilities	Time taken to address identified vulnerabilities

Source: Reviewer's Review (2023).

9.1 Anticipated Challenges in Applying the Conceptual Model

Implementing the conceptual model may encounter challenges inherent in the dynamic nature of smart manufacturing environments. Anticipated challenges include resistance to cultural change, varying levels of technological literacy among employees, and the rapid evolution of cybersecurity threats. Additionally, aligning leadership commitment with the security culture objectives might pose difficulties. Therefore, taking this into consideration, addressing these issues must be an individualized process taking into account the special circumstances in the context of smart manufacturing environments.

9.2 Opportunities for Improvement and Adaptation

Technological advancements provide opportunities for improving upon strategies of training and should be taken advantage of while creating the culture for continued improvement. Using adaptive learning technologies and incorporating examples from the real world into training modules will help increase learning effectiveness. Furthermore, organizational feedback as well as regular re-evaluation of the conceptual model offers an opportunity of constant revision. This flexibility empowers corporations to customize approaches based on changes in risks and opportunities thereby building hard security resistant environment.

9.3 Real-World Examples and Applications

Security culture models apply in real-life cases. For example, companies such as Siemens and Honeywell have put in place strong cyber security programs in smart manufacture which involve incorporating technology, training and leadership commitment. Siemens employs a defense-in-depth approach, integrating both technical and human-centric measures, while Honeywell's Connected Plant Security program emphasizes continuous training and collaboration. These examples underscore the practical viability of integrating security culture within smart manufacturing environments, showcasing the positive impact on cybersecurity resilience. These real-world instances validate the conceptual model's potential effectiveness and provide valuable insights for organizations aiming to implement similar security culture initiatives within their smart manufacturing ecosystems. The challenges and opportunities outlined offer a roadmap for organizations to navigate and succeed in the dynamic landscape of cybersecurity.

10 DISCUSSION: THEORETICAL IMPLICATIONS AND CONTRIBUTIONS

The discussion section delves into the theoretical implications and contributions of the developed conceptual model for building a security culture in smart manufacturing. By synthesizing findings with existing theories, this section illuminates the innovative contributions that the model brings to the theoretical landscape of cybersecurity in the context of smart manufacturing.

10.1 Synthesizing Findings with Existing Theories

The synthesis of findings involves aligning the conceptual model with existing theories in cybersecurity and smart manufacturing. Herath and Indrakanti's work (2019) on cybersecurity culture provides a foundational understanding of human and organizational dimensions, which forms the basis of the human aspect within the model. Reis and Melão (2023) for smart manufacturing contributes to the technological dimension, creating a symbiotic relationship between technology and culture. LeClair et al. (2013) on workforce training inform the training strategies, aligning with existing theories on effective cybersecurity education. Fang and Yu (2023) transformational leadership theory supports the conceptualization of leadership's pivotal role in shaping security culture.

10.2 Contributions to Conceptual Understanding in Smart Manufacturing

The contributions of the conceptual model extend to advancing the understanding of security culture within the domain of smart manufacturing. By integrating dimensions from cybersecurity culture and smart manufacturing frameworks, the model provides a nuanced understanding of how cultural and technological aspects interplay. It contributes by proposing targeted training strategies aligned with the unique challenges of smart manufacturing, enhancing the conceptualization of workforce development. The incorporation of KPIs for evaluation establishes a measurable framework for assessing the model's impact, bridging the gap between theory and practical implementation. In essence, the model's contributions lie in its ability to synthesize and extend existing theories to create a holistic framework that addresses the intricacies of security culture in smart manufacturing. This theoretical groundwork offers valuable insights for researchers and practitioners seeking to navigate the complex intersection of cybersecurity and smart manufacturing.

11 CONCLUSION

In the realm of smart manufacturing, the journey toward fortifying cybersecurity through the establishment of a robust security culture has been both enlightening and transformative. As we conclude this study, the amalgamation of theoretical frameworks, practical insights, and innovative strategies converges into a beacon of guidance for organizations navigating the intricate landscape of cybersecurity in the era of Industry 4.0.

11.1 Recapitulation of Key Conceptual Points

Our conceptual model, intricately woven from the threads of cybersecurity culture, smart manufacturing frameworks, and leadership principles, stands as a testament to the power of integration. The human, organizational, and technological dimensions harmoniously coalesce, creating a holistic approach that transcends traditional silos. Rooted in the foundational works of Fang and Yu (2023), the model encapsulates the essence of cutting-edge research, providing a roadmap for organizations seeking not just security compliance but a cultural metamorphosis.

11.2 Implications for Practice and Future Research

The implications for practice are profound. Organizations venturing into the uncharted territories of smart manufacturing can leverage our model as a compass, guiding them toward a security culture that thrives amidst technological dynamism. The proposed KPIs offer a measurable means to track progress, fostering a continuous improvement ethos. The training strategies, aligned with the unique challenges of smart manufacturing, beckon organizations to invest not just in technology but in the empowerment of their human assets. For future research, the horizons are expansive. The evolving nature of both smart manufacturing and cybersecurity demands a perpetual quest for knowledge. Delving deeper into the intricacies of leadership's role in shaping security culture, exploring the symbiotic relationship between technology and culture, and refining training strategies in response to emerging threats are avenues ripe for exploration.

11.3 Closing Thoughts

In closing, this study is not just a culmination of theoretical constructs; it is a proclamation of resilience and adaptability in the face of ever-evolving challenges. As we stand at the intersection of theory and practice, we extend an invitation to organizations, scholars, and visionaries to embark on this transformative journey. The future of smart manufacturing is intricately woven with threads of security culture, and by embracing this paradigm shift, we not only fortify our digital fortresses but also pave the way for a future where innovation and security walk hand in hand. May this study be a source of inspiration, a catalyst for change, and a cornerstone for the edifice of a secure and thriving smart manufacturing ecosystem. As we bid farewell to these pages, let them echo with the resonance of progress, the cadence of innovation, and the assurance that the future we envision is not just a possibility but a shared reality waiting to unfold.

REFERENCES

Bibri, S. E., & Jagatheesaperumal, S. K. (2023). Harnessing the potential of the metaverse and artificial intelligence for the internet of city things: cost-effective XReality and synergistic AIoT technologies. *Smart Cities*, *6*(5), 2397–2429.

Bıçakcı, A. S., & Evren, A. G. (2022). Thinking multiculturality in the age of hybrid threats: converging cyber and physical security in Akkuyu nuclear power plant. *Nuclear Engineering and Technology*, *54*(7), 2467–2474.

Chan, C. K. Y. (2023). A comprehensive AI policy education framework for university teaching and learning. *International Journal of Educational Technology in Higher Education*, *20*(1), 38.

Chanoski, S. D. (2023). *HOP and Cybersecurity: Leveraging Safety and Reliability Experience to Improve Digital Security Culture* (No. INL/CON-23-72096-Rev000). Idaho Falls, ID: Idaho National Lab. (INL).

Ejaz, M. R. (2023). Implementation of industry 4.0 enabling technologies from smart manufacturing perspective. *Journal of Industrial Integration and Management*, *8*(02), 149–173.

Fang, Z., & Yu, S. C. (2023). Cross-level influence of group-focused transformational leadership on organizational citizenship behavior among Chinese secondary school teachers. *Behavioral Sciences*, *13*(10), 848.

Fisher, R., Porod, C., & Peterson, S. (2021). Motivating employees and organizations to adopt a cybersecurity-focused culture. *Journal of Organizational Psychology*, *21*(1), 114–131.

Fisk, N. (2023). Developmental challenges: capture the flag and the professionalization of cybersecurity. *Human Organization*, *82*(1), 61–72.

Garba, J., Kaur, J., & Ibrahim, E. N. M. (2023). Design of a conceptual framework for cybersecurity culture amongst online banking users in Nigeria. *Nigerian Journal of Technology*, *42*(3), 399–405.

Grenier, A. (2023). The qualitative embedded case study method: exploring and refining gerontological concepts via qualitative research with older people. *Journal of Aging Studies*, *65*, 101138.

Høiland, C. (2023). "Not My Responsibility!"-A Comparative Case Study of Organizational Cybersecurity Subcultures (Master's thesis, University of Agder).

James, S. R. (2023). A Study of the Effect of Types of Organizational Culture on Information Security Procedural Countermeasures (Doctoral dissertation, Nova Southeastern University).

Khripunov, I. (2023). Security Culture in Nuclear Facilities and Activities. In: *Human Factor in Nuclear Security. Advanced Sciences and Technologies for Security Applications*. Springer, Cham. https://doi.org/10.1007/978-3-031-20278-0_3

Leng, J., Ye, S., Zhou, M., Zhao, J. L., Liu, Q., Guo, W.,… & Fu, L. (2020). Blockchain-secured smart manufacturing in industry 4.0: a survey. *IEEE Transactions on Systems, Man, and Cybernetics: Systems*, *51*(1), 237–252.

Masoud, M. M. (2023). Smart Manufacturing Execution System Framework for Small and Medium-Size Enterprises (Doctoral dissertation, University of Windsor (Canada)).

Mgaiwa, S. J. (2023). Leadership styles of academic deans and department heads: university dons' perspectives on how they affect their job satisfaction. *International Journal of Educational Management*, *37*(5), 1088–1103.

Nguyen, M. T., & Tran, M. Q. (2023). Balancing security and privacy in the digital age: an in-depth analysis of legal and regulatory frameworks impacting cybersecurity practices. *International Journal of Intelligent Automation and Computing*, *6*(5), 1–12.

Oliver, D. (2023). David Oliver: shortening and narrowing training won't solve the medical workforce crisis. *BMJ*, *381*.

Peng, Y., Welden, N., & Renaud, F. G. (2023). A framework for integrating ecosystem services indicators into vulnerability and risk assessments of deltaic social-ecological systems. *Journal of Environmental Management*, *326*, 116682.

Rajput, P. H. N. (2023). *Hardware-Assisted Non-Intrusive Security Controls for Modern Industrial Control Systems* (Doctoral dissertation, New York University Tandon School of Engineering).

Singh, S. A., & Desai, K. A. (2023). Automated surface defect detection framework using machine vision and convolutional neural networks. *Journal of Intelligent Manufacturing*, *34*(4), 1995–2011.

Song, Z., & Zhu, J. (2022). Blockchain for smart manufacturing systems: a survey. *Chinese Management Studies*, *16*(5), 1224–1253.

Srinivas, J., Das, A. K., & Kumar, N. (2019). Government regulations in cyber security: framework, standards and recommendations. *Future Generation Computer Systems*, *92*, 178–188.

Tao, F., Qi, Q., Liu, A., & Kusiak, A. (2018). Data-driven smart manufacturing. Journal of Manufacturing Systems, 48, 157–169.

Team, A. P. (2021). Artificial Intelligence Measurement and Evaluation at the National Institute of Standards and Technology. Washington, DC: National Institute of Standards and Technology. https://www.nist.gov/news-events/events/2021/06/ai-measurement-and-evaluation-workshop.

Tejay, G. P., & Mohammed, Z. A. (2023). Cultivating security culture for information security success: a mixed-methods study based on anthropological perspective. Information & Management, 60(3), 103751.

Tolah, A., Furnell, S.M., Papadaki, M. (2019). A Comprehensive Framework for Understanding Security Culture in Organizations. In: Drevin, L., Theocharidou, M. (eds) Information Security Education. Education in Proactive Information Security. WISE 2019. IFIP Advances in Information and Communication Technology, vol 557. Springer, Cham. https://doi.org/10.1007/978-3-030-23451-5_11

Tsantes, G.K., & Ransome, J. (2023). *Cybertax: Managing the Risks and Results* (1st ed.). Auerbach Publications. https://doi.org/10.1201/9781003330103

Tuptuk, N., & Hailes, S. (2018). Security of smart manufacturing systems. *Journal of Manufacturing Systems*, *47*, 93–106.

23 Data Analytics for Pandemic
A Covid-19 Case Study in Kolkata

Supratim Bhattacharya, Saberi Goswami, Poulami Chowdhury, Prashnatita Pal and Jayanta Poray

1 INTRODUCTION

Current storm of the Covid-19 pandemic has become a serious public health concern throughout the world. This "communicable" disease has disrupted the socio-economic scenario as well as individual livelihood. Kolkata, being the seventh most populous city in India and the heart of West Bengal, also faced the severity of this pandemic cyclone. Considering the adverse economic standoff, the government provided basic commodities like food provisions to run daily life. Kolkata Municipal Corporation (KMC), being a dedicated authority, started determining the exact cause to fight against this odd. In continuation of the process, a system named "Communicable Disease Tracking System" (CDTS) has been designed. "CDTS" acts as a comprehensive framework to manage Covid-19-related information of Kolkata. This system is used to collect, store and analyse Covid-19-positive patients' information. This system helps KMC authority to take certain useful decisions regarding lockdown strategy, determining containment zone and other important socio-economic judgements. This chapter further helps in "Exploratory Data Analysis" in the way to explore the data with the aim to analyse, predict and forecast Covid-19 scenario in Kolkata.

The important part of the overall management is to reduce the peak of the epidemic. Apart from government-derived measurements, technology may be a good alternative for analysing and lowering the peak. This type of study not only helps to access the current scenario but also guides the authority to plan a concrete action to fight against such type of disease in the future.

The objectives of these studies are as follows:

- To apply descriptive statistics for the analyses of overall scenario.
- To apply SIR model with modified approach, regression techniques and clustering for prediction and forecasting.

DOI: 10.1201/9781032694375-23

2 LITERATURE SURVEY

A rapid increasing infectious disease, like Covid-19, involves fast-spreading, endangering the health of a huge number of people, and thus need instantaneous effort to stop the spread of the disease at the community level.

In their study [1], the researcher analysed the trend of Covid-19 trends on daily basis based on the surveillance strategy of different countries like China, South Korea, Italy, Japan, Spain. They also study the government policies to control the outbreak of Covid-19. They started studying the linear relation between the outbreak condition and the "Case Fatality Rate (CFR) by analysing daily patient statistics. They have also used "Linear Regression" for prediction with respect to China.

Jingyuan et al. [2] investigates the influence of temperature and humidity in the spread of Covid-19 with the help of "Linear Regression". It also indicates that the arrival of summer and rainy season can effectively reduce the intensity of coronavirus.

In the paper [3], the author analyses COVID-19 cases in India. The authors stated a statistical model for better understanding of Covid-19 cases in India by a thorough study of different cases. They also implemented an exploratory data analysis technique to analyse the impact of Covid-19 in India on daily and weekly manner.

In the study, Gupta and Pal [4] uses exploratory data analysis to report the current Covid-19 situation and also used time-series forecasting methods to predict the future trends. The major discussion in this paper comprises the rapid increase in the positivity rate and state-wise mortality rate.

From the analysis point of view, a large amount of study is being conducted based on SIR epidemiological models by Yadav [5] and Jakhar, Ahluwalia and Kumar [6]. This type of standard techniques gives us a satisfactorily forecasting analysis in faster pace. Their acceptability rate in the research community is also very high. Drawbacks of SIR model being identified and inefficiencies are also being explored in the study Moein et al. [7]. Some studies use hybrid SIR model for better prediction and forecasting.

We have proposed a "Feedback-based SIR model", another hybrid approach for a modified and a one step ahead model, used for better epidemiological prediction& forecasting. We have used Covid-19-positive patient data from Kolkata region for analysis and experiment.

3 METHODS

"CDTS" is a tracking system used to track Covid-19-positive patient in Kolkata. This system comprises three subsystems.

1. Patient Segmentation System
2. Patient Contact Tracing System
3. Patient Recovery Tracing System

Patient Segmentation System

- Every day we receive a dataset consisting of covid-positive patient details of Kolkata.
- A calling list is being prepared based on those datasets. A series of data cleaning methods are being applied before preparing the calling list.
- The calling list is being mapped with the telecallers for further action \& inputs.
- An interface is being designed for collecting inputs from the telecallers. The form is designed in such a way that a minimal manual intervention is required.
- Another level of data cleaning is being done after receiving the inputs from the telecaller. Performance report of every telecaller is also being extracted from the system.
- Patient Segmentation System also comprises of auto compilation procedure.
- Further data validation is being done by implementing fuzzy logic.
- Reports are being generated as per the requirements.

Patient Contact Tracing System

- The segmented data passed to the Contract Tracing System where a separate interface is being used for contact tracing.
- Second level of telecalling is being done for collecting information for close contacts of the patients for further tracking & isolation.
- The information is further processed and reported to proper authority for necessary action from their end.

Patient Recovery Tracking System

- After a certain time interval, a final call is being initiated by a set of telecallers for checking patient recovery status.
- It is being stored for further analysis.

All the data collected from these three different stages are aggregated, stored and analysed for further course of action.

3.1 Data Source

The data source of this analysis is the data collected through "CDTS". Time duration for data collection is from 20th April 2020 to 31st January 2022. Also, we have taken data that were published in the website of West Bengal Health and Family Welfare Department till 31st January 2022.

3.2 Descriptive Data Analysis

During the overall duration, we found three major peaks in three different pandemic waves.

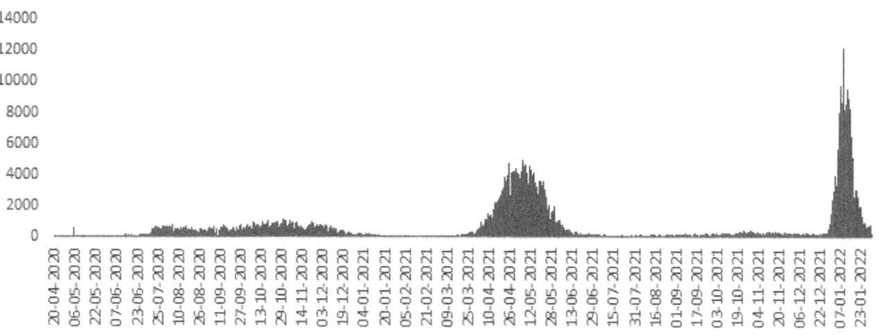

FIGURE 23.1 Figure 23.1 Date-wise Positive Cases (20th Apr 20 31st Jan 22).

Figure 23.1 depicts that the upward frequency for the first wave sustain for 105 days, 2nd wave for 63 days and 3rd wave for only 32 days. We have performed our descriptive analysis based on the following parameters:

3.2.1 Season [8]

Initially, we thought that temperature or heat have a positive correlation with number of Covid-19 cases. In Kolkata, we experience half of the year as summer and remaining half as a combination of mainly rainy and winter seasons. In our study, we found that around 56% of cases happened in Summer session and remaining 44% in rainy or winter session, which signifies that heat or temperature has no such major effect for the outbreak of the disease.

3.2.2 Age [9]

Though we are talking about the infected cases for older people and children, it has also been observed that more than 70% cases are in the age group of 20–60 years. This age group people are mainly working people who had a direct interaction with the population. When we further apply descriptive analyse on the data in terms of age group segmentation, we found that positivity rate for the females within the age bracket is negatively skewed compared to male positivity rate. In second wave, Teenagers and children in the age group of 1–18 years are more affected compared to first wave, gives strong evidence of multiple mutation of Covid-19.

Figure 23.2 conveys a strong message that persons staying home are more safe than persons going out following the evidence of keeping social distancing as a big concern.

Figure 23.3 shows age-wise comparison in three different waves. As we observe that though initially the infection rate is more for the middle and old aged people but it gets reversed with the introduction of vaccination. Another interesting fact is that infection ratio for the younger people and teenagers have increased in second wave. This might be an alert that some sort of fear has gone down and people tends to minimize the social distancing approach or the remedies required to take in terms of preventive point of view. In third wave, around 70% of the infected cases are within the age group 20–60 years.

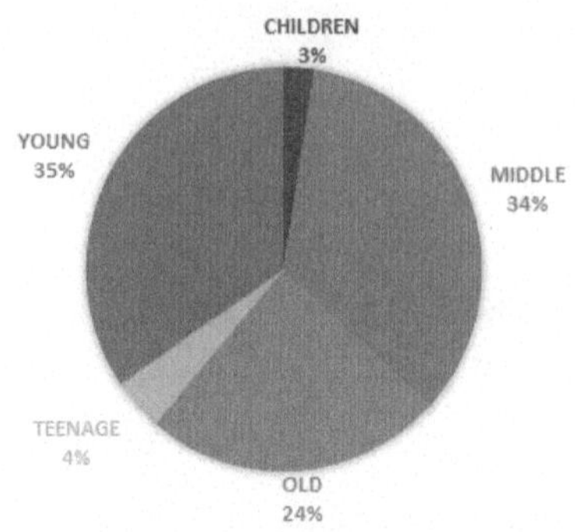

FIGURE 23.2 Age Group Segmentation: Covid Positive Patient.

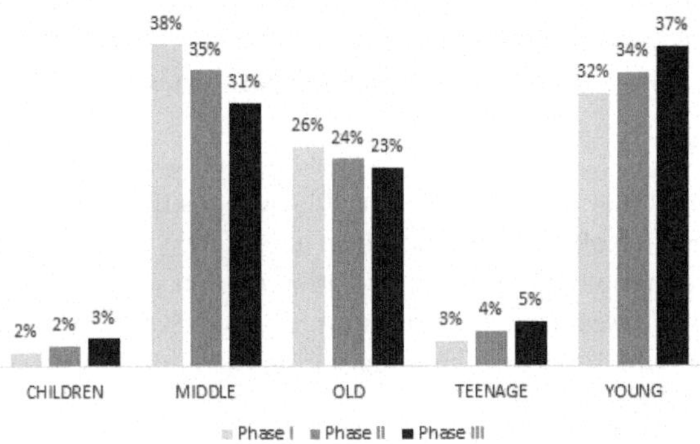

FIGURE 23.3 Phase-wise Comparison Based On Different Age Group.

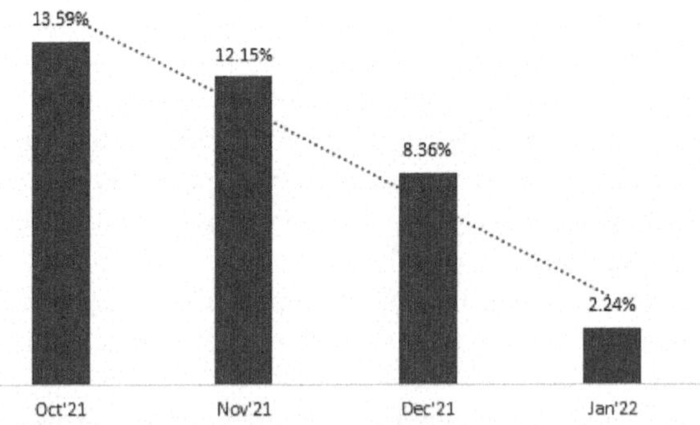

FIGURE 23.4 Hospitalization Rate for Last 4 Months

3.2.3 Hospitalization [12]

Overall hospitalization rate since inception is 13.18%.

As Figure 23.4 suggests, hospitalization rate decreases from September 2021 to November 2021 with the significant rise in the vaccination rate. It has also been observed that only 9.45 patients get hospitalized after taking first dose and 7.8%

patients hospitalized after getting fully vaccinated. When we analyse the data location-wise, it has been noticed that in northern region of Kolkata, 41% patients are hospitalized in compared to 44% in the southern region of Kolkata and 15% in central region. Moreover, when we go deeper in household segments, we found that persons living in flat have significantly high hospitalization rate.

3.2.4 Vaccination

Among the affected patient, 49% are vaccinated. Among the affected patient, 27.42% of patients have taken first dose and 21.63% of patients have taken second dose. Interestingly notified that, patient got infected after 212–229 days of taking first dose and after 278–307 days of taking second dose. Further analysis shows that 33% patient got affected after taking "Covishield" and only 5% of patients got infected after taking "Covaxin". Only 3% of patients are being affected who have taken vaccine other than this two. However in Kolkata, 85% of people have taken "Covishield" as Covid-19 vaccine.

3.2.5 Locality

In terms of household investigation, our system captures data whether the patient stay in Single Unit, Flat or "Bustee" (slum area). Other patients are floating patients, coming from different parts of West Bengal to earn their livelihood. When we stated analysis, we found that very a smaller number of patients stay in Bustee area. As Kolkata being a metropolitan area it is obvious that the entire locality comprises low-volume slum areas. Majority of the people in Kolkata either live in flats or single units.

Figure 23.5 demonstrates that Central Kolkata dominates the overall patient count followed by South and North Kolkata. This is due to the fact that Central Kolkata consists of many offices (both government and private). It always creates a source of floating population in Central Kolkata. This area is also a major connector of different major locations. So population density has a major factor in Central Kolkata.

We further go into the deeper and apply clustering mechanism to identify clusters or localities that have a positivity density of more than a certain threshold. The threshold value depends on the certain criteria defined by the authority on timely manner. Like

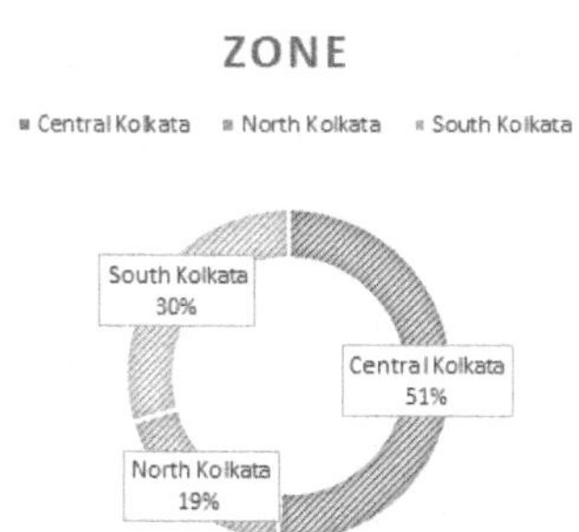

FIGURE 23.5 Zone-wise Distribution of Covid-Positive Patients.

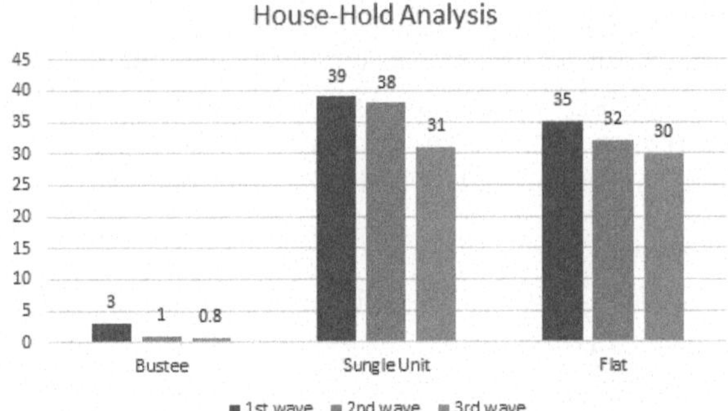

FIGURE 23.6 Household Analysis of Containment Zone Based on Different Waves of Covid Pandemic.

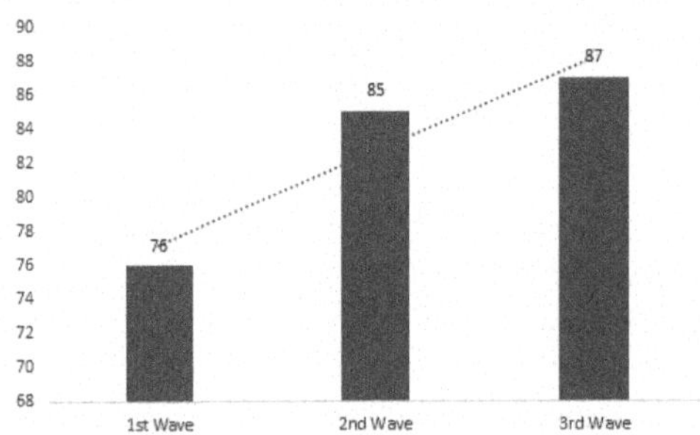

FIGURE 23.7 Asymptomatic Rate among Covid-Positive Patient.

- Location positivity density ≥80% of the overall positivity density of the zone.
- Significant slum population within 1 km.
- More than 50% of population stays in flat.

This criterion varies from time to time. This will help us to identify the sensitive zones and can able to guide in identifying containment zones also.

Figure 23.6 shows a picture of containment zone at different time-frame of the pandemic. As days go by we found that there is a significant drop in all the three segments of household. This signifies that awareness is a primary concern for Covid-19 prevention and KMC being the administrative authority has done a remarkable endeavour.

3.2.6 Symptomatic Status

For Covid-19 analysis, it has been found that more than 80% of patients are asymptomatic or mild-symptomatic.

And if we compare first wave second wave, we found that asymptomatic patient has been increased upto 85%, which has further escalated to 87% in third wave. It has a positive correlation with vaccination. This number is expected to rise upto 90% within three months (Figure 23.7).

3.2.7 Death Cases

Total 5,471 deaths occur from 20th April 2020 to 31th January 2022. Below figure shows date-wise occurrence of deaths cases in Kolkata. Death cases are proportional to the total number of cases (Figure 23.8).

When the positive cases reach peak, average death cases goes upto 50 cases per day. But on an average, cases per day are within the frequency range 0–5 (Figure 23.9).

3.2.7.1 *Predictive Modelling*

For predictive analysis and forecasting, we have used SIR epidemiological model. Further, we proposed a modified SIR model named "Feedback-based SIR Model" to overcome certain loopholes of standard SIR model.

3.2.7.1.1 SIR Model [15]

We utilize Susceptible Infectious Recovered (SIR) modelling to forecast Covid-19 cases within Kolkata based on daily observations. Where

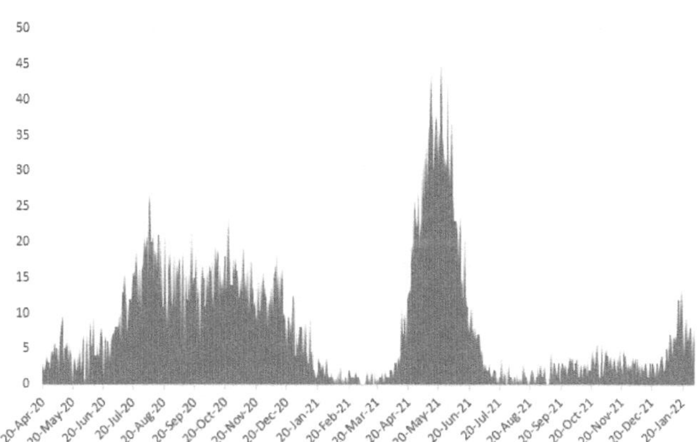

FIGURE 23.8 Number of Death Cases in Different Time-Frame.

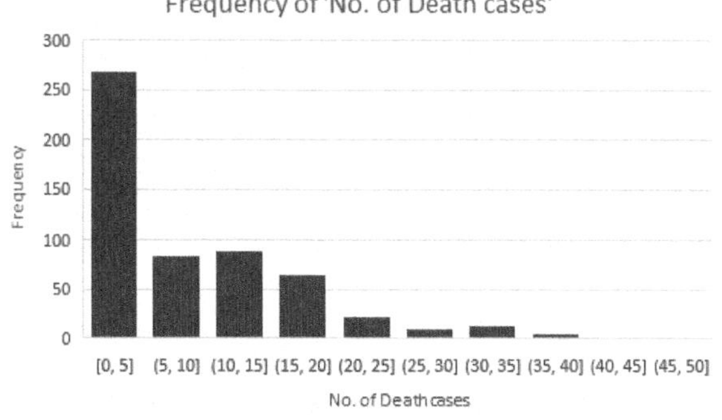

FIGURE 23.9 Frequency Distribution Based on Number of Death Cases.

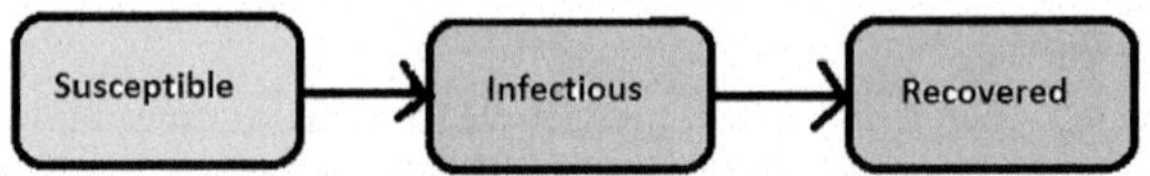

FIGURE 23.10 SIR Model with Three States.

- "Susceptible" indicates the person who is prone to infection and become hosts if exposed.
- "Infectious" are individual who are showing signs of infection and can also transmit the virus.
- "Removed" or "Recovered" means the persons who are earlier infected but are no longer infectious and already immune to the virus. (Figure 23.10)

Initially, the model assumes that the total population is constant as epidemic does not sustain for a longer period. Rate of infective is proportional to the contract between Susceptible and Infectives. In similar, way rate of recovery is also constant.

The rate of change of number of Susceptible over time is

$$\frac{ds}{dt} = -rIS \tag{1}$$

where r is the rate of contract.

The rate of change of Infectives over time is

$$\frac{dI}{dt} = -rIS - aI \tag{2}$$

where a signifies rate of recovery.

The rate of change of Removed population over time is

$$\frac{dR}{dt} = aI \tag{3}$$

And the sum of rate of change of Susceptible, Infectives and Recovery is equal to 0.

$$\frac{d}{dt}(S + I + R) = 0 \tag{4}$$

$$S + R + I = S_0 + I_0 \tag{5}$$

The figure signifies that at initial level, susceptible population starts at high note but after a certain time interval, it slows down and gradually the recovery rate creeps up (Figure 23.11).

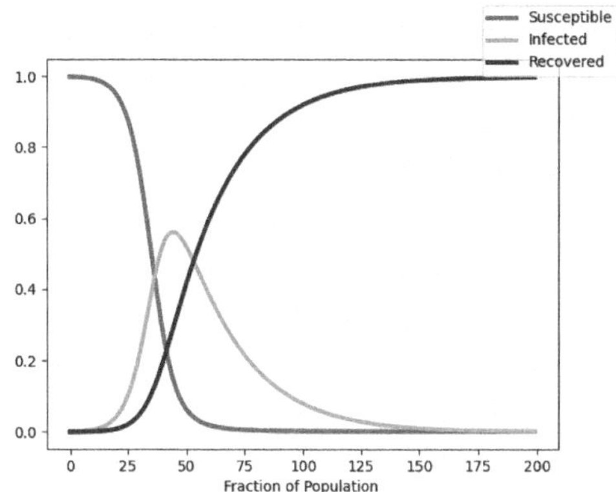

FIGURE 23.11 Significance of Susceptible, Infectives and Recovery.

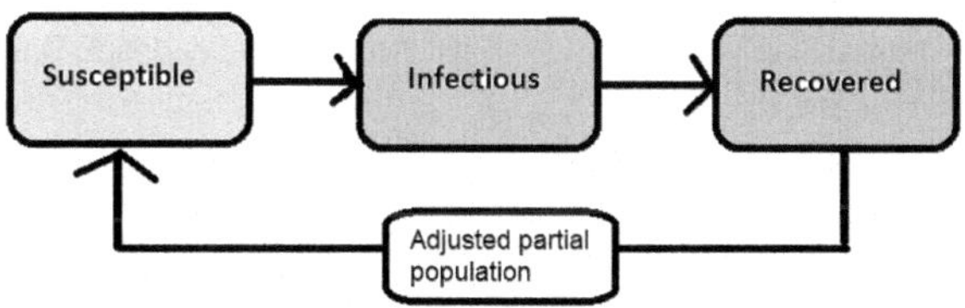

FIGURE 23.12 Feedback-Based SIR Model.

3.2.7.1.2 Feedback SIR Model [19–21]

A modified SIR model (named Feedback SIR model) is proposed, where we consider that the susceptible population is not constant over a period (Figure 23.12).

Susceptible population varies from different perspective:

- Fraction of vaccinated population gets reinfected after nine months of their vaccination (at least one dose) [16].
- Fraction of vaccinated population can be reinfected immediately after they have been vaccinated [17].
- Fraction of population are getting infected after three months of recovery, assuming they are not vaccinated [18].

So, the rate of change of number of Susceptible over time becomes

$$\frac{ds}{dt} = -rI\left(S_1 + S_2 + S_3 + S_4\right) \tag{6}$$

where

'r' = rate of contact, S_1=Exposed Population yet to get infected, S_2=73% of the population who are vaccinated nine months prior (at-least one dose) [16], S_3=4.5% of the population who can be reinfected immediately after vaccination [17], and S_4=5.94% of recovered population [18].

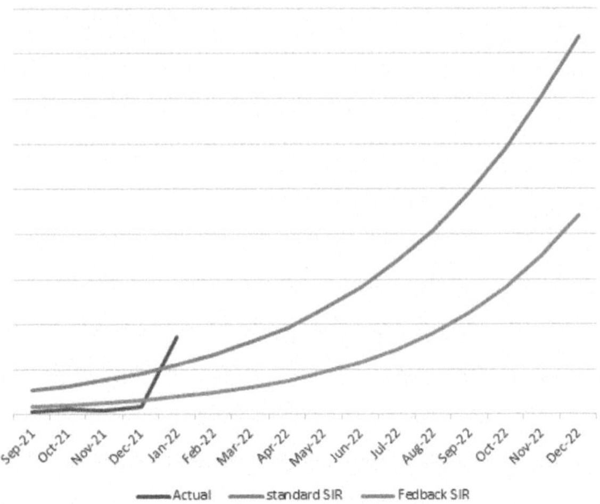

FIGURE 23.13 Comparison between Actual SIR Model and Feedback-Based SIR Model.

4 RESULTS AND DISCUSSION

We have created a training model based on the data collected from "CTDS" from April 20 to Aug 21. Then we apply the model in the test data from Sept 21 to Jan 22. We got the following result in terms of estimating the infected case (Figure 23.13).

In a certain time frame, we found that our "Feedback based SIR model" has an accuracy rate of 62% with compared to 35% accuracy rate for the "standard SIR model".

4.1 Comparative Reproductive Ratio (R_0)

R_0 is measured as expected number of new cases directly caused by an infectious individual before recovery.

$$R_0 = \frac{\text{Infection Rate}}{\text{Recovery Rate}} \tag{7}$$

We have decided to use a set of R_0 values within different time intervals to study the variations of the community behaviour and inconsistency related to distancing regulations R_0 which is being estimated in four different scenarios [10,11].

- At the beginning when people are not aware of social distancing.
- Bad scenario.
- Good scenario.
- Feasible scenario.

Table 23.1 show the actual and forecasted Reproductive ratio for Kolkata in different timeframe of the Covid-19 pandemic.

If the value of $R_0 > 1$, then the outbreak is still pandemic, and if $R_0 < 1$, outbreak is in control and will be out quickly.

TABLE 23.1
Reproductive Ratio at Different Time Frame

Date Range	R_0	R_0
Apr 20–Jun 20	−0.66666667	−0.66666667
Jul 20–Dec 20	−1.00000000	−0.98904769
Jan 21–Mar 21	−1.00000000	−0.67597170
Apr 21-Jun 21	−1.12123291	−0.86930422
Jul 21–Dec 21	−9.88123726	−4.39424406
Jan 22	−5.15500138	−2.51816290
Feb 22–Dec 22	0.13380728	−0.29160497
Jan 23–Dec 23	0.29872808	−0.04348400

4.2 HERD IMMUNITY

In case of any flu, if almost everyone in a distinct locality has had it then those who haven't been in touch with that infected cases are almost protected from getting it. There is not enough susceptible left in that distinct population for the epidemic to walk through. This type of classification is called "Herd Immunity" [13, 14].

Though "r" and "a" defined in the SIR model, depends on social and behavioural factors of a region we can observe future trends as f.

Considering the population size of Covid-19 outbreaks susceptible case is k, which is approximately equal to initial population S_0.

$$S_0 = k, \quad I_0 = 0, \quad R_n = \frac{rk}{a}$$

We know that

$$k - \frac{a}{r}\text{In}(S_0) = S_\infty - \frac{a}{r}\text{In}(S_\infty)$$

where S_∞ is the susceptible population of Kolkata if infectious case is 0. After simplification we get,

$$\frac{r}{a} = \frac{\text{In}[S_0 / S_\infty]}{k - S_\infty}$$

Our feedback SIR model indicates that the zero infected case may occur in the month of April 23. So based on the above equation, we found that during that time the ratio of "r" and "a" will be nearer to "0.00000018".

4.3 Comparison with the Other Feedback Based Models

There are various modified SIR models already proposed. Elisa Franco [22] proposed a feedback-based SIR model where the infection information is negatively feedbacked to reduce the peak of infection. This model illustrates the effects of infection – dependent social distancing. Another model [23] suggested a feedback methodology based on Hamilton-Jacobi-Bellman (HJB) equation. This is used to describe a control function that illustrates vaccine policy for any combination of susceptible individuals and infectious individuals. The model [24] is also based on social distancing. The aim for the model is to reduce α value. The model proposed by Emilio Molina and Alain Rapaport [25] has also introduced a control variable that reduces the effective transmission rate of the SIR model. They have used explicit analytical expression of the optimal control rather than only relying on numerical methods. In comparison to the above-mentioned models, our feedback model is based on multiple criteria like vaccination rate (first dose, second dose), retransmission of infection to the recovery population and vaccinated population. The model is validated based on data collected from Kolkata district of West Bengal with relevant accuracy. We have also showed that this feedback-based model has a high efficiency rate than the traditional SIR model.

5 CONCLUSION AND FUTURE WORK

To overcome from such deadly disease like Covid-19, we need to be more organized and effective, both in terms of medical and technological aspects. The threat created around the epidemiological circumstances for public health can be soften with proper analysis and prediction. Exploratory data analysis will help the authority and other public health personnel for analyse and decision making. SIR model is one of the most utilized epidemiological model used in public health. We have tried to reduce certain drawbacks of conventional SIR model and readdress the model to increase the efficiency. We also compare the real data with both the conventional model and derived SIR model and show that derived model has a better predictive efficiency the conventional one. However, there are certain factors like air pollution, behavioural change related to the social and cultural context of the population, ceremonies and natural gatherings, possibility of reinfection of virus or reaction of virus etc. that need to be taken care off in the future for more accuracy in forecasting and prediction. We have used data and other calculating factors based on a particular locality. In future, we will explore our model to more global aspects with more decisive parameters.

ACKNOWLEDGEMENTS

We would like to acknowledge Dr Subrata Roy Chowdhury, chief medical health officer (CMHO), KMC, for his continuous support for this work. We thank Dr T.K. Mukherjee, advisor of (Health) for his decisive and valuable comments. We express our gratitude to Dr Bibhakar Bhattacharya, Dy. CMHO, Dr Amitava Chakraborty, head, IDSP, KMC, and Dr Nabarun Dutta, Nodal Office, Kolkata City NUHM Society, KMC, for their kind suggestions and support. We would like to express our thanks and gratitude to all staffs of Health Department, KMC.

REFERENCES

[1] A. Hoseinpour Dehkordi, M. Alizadeh, P. Derakhshan, P. Babazadeh and A. Jahandideh. Understanding epidemic data and statistics: A Case Study of COVID-19, *Journal of Medical Virology*, 92, 868–882 (2020).

[2] W. Jingyuan, T. Wang, F. Kai and L. Weifeng. High temperature and high humidity reduce the transmission of Covid-19, *BMJ Open*, 11(2), e043863 (2020).

[3] S. Mittal. An exploratory data analysis of COVID-19 in India, *International Journal of Engineering Research & Technology* (IJERT), ISSN: 2278-0181 (2020).

[4] R. Gupta and S. K. Pal. Trend analysis and forecasting of COVID-19 outbreak in India, *BML Yale*. https://doi.org/10.1101/2020.03.26.20044511 (2020).

[5] D.R. Ramjeet Singh Yadav. Mathematical modeling and simulation of SIR model for COVID-2019 epidemic outbreak: A case study of India (2020).

[6] M. Jakhar, P.K. Ahluwalia and A. Kumar. COVID-19 epidemic forecast in different states of India using SIR model, *BMJ Yale*. https://doi.org/10.1101/2020.05.14.20101725 (2020).

[7] S. Moein, N. Nickaeen, A. Roointan, N. Borhani, Z. Heidary, S. Haghjooy Javanmard, J. Ghaisari and Y. Gheisari. Inefficieny of SIR models in forecasting COVID-19 epidemic: A case study of Isfahan (2020).

[8] Z. Shi and Y. Fang. Temporal relationship between outbound traffic from Wuhan and the 2019 coronavirus disease (COVID-19) incidence in China, *medRxiv* Data in brief, 105340 (2020).

[9] D. Benvenuto, M. Giovanetti, L. Vassallo, S. Angeletti and M. Ciccozzi. Application of the ARIMA model on the COVID-2019 epidemic dataset, (2020).

[10] Mukhopadhyay, J. (2020). Optimism under the holocaust of covid-19 in Kolkata slum. *The Journal of Community Health Management*, 7(2), 44–50.

[11] R.S. Yadav. Data analysis of COVID-2019 epidemic using machine learning methods a case study of India. *International Journal of Information*, 12, 1321–1330. https://doi.org/10.1007/s41870-020-00484-y (2020).

[12] F.A.B. Hamzaha, C.H. Laub, H. Nazric, D.V. Ligotd, G. Leee, C.L. Tanf, M.K.B.M. Shaibg, U.H.B. Zaidonh, A.B. Abdullahi, M.H. Chungj, C.H. Ongk, P.Y. Chewl, R.E. Salunga and R.S. Yadav. CoronaTracker world-wide COVID-19. *International Journal of Information*, 12(4), 1321–1330. https://doi.org/10.1007/s41870-020-00484-y (2020).

[13] A. Rachah and D.F.M. Torres. Analysis, simulation and optimal control of a SEIR model for Ebola virus with demographic effects. *Communications Faculty of Sciences University of Ankara Series A1*, 67(1), 179–197 (2018).

[14] A.T. Porter. A path-specific approach to SEIR modeling, Ph.D. Thesis University of Iowa, 2012.

[15] H. Weiss. The SIR model and the foundation of public health. *MATerials MATemàtics*, 2013 treball (3), 17 pp. ISSN: 1887-1097 (2012).

[16] V. Hall, S. Foulkes, F. Insalata, P. Kirwan, A. Saei, E. Wellington, J. Khawam, K. Munro, M. Cole, C. Tranquillini and A. Taylor-Kerr. Protection against SARS-CoV-2 after Covid-19 vaccination and previous infection (2012).

[17] https://indianexpress.com/article/explained/explained-how-likelyis-covid-reinfection-7256618/

[18] A. Rivelli, V. Fitzpatrick, C. Blair, K. Copeland and J. Richards. Incidence of COVID-19 reinfection among Midwestern healthcare employees. (2021)

[19] S. Bhattacharya and J. Poray. Application of graph theory in bigdata environment. *International Conference on Computer, Electrical & Communication, 2016*

[20] S. Bhattacharya and J. Poray. A bigdata analytics framework on the impact of non communicable diseases in Kolkata, International Conference on Computer, Electrical \& Communication (2020).

[21] S. Bhattacharya, Dr Poray and P. Debnath. *The Impact of Big Data Analytics on Risk Management and Decision-Making: International Conference on Recent Trends in Artificial Intelligence, IOT, Smart Cities & Applications*, ICAISC, (2020).

[22] E. Franco. A feedback SIR (fSIR) model highlights advantages and limitations of infection-dependent mitigation strategies (2020).

[23] Y.-G. Hwang, H.-D. Kwon and J. Lee. Feedback control problem of an SIR epidemic model based on the Hamilton-Jacobi-Bellman equation.

[24] T. Peng, X. Liu, H. Ni, Z. Cui and L. Du. City lockdown and nationwide intensive community screening are effective in controlling the COVID-19 epidemic: Analysis based on a modified SIR model.

[25] E. Molina, A. Rapaport, H. Ni, Z. Cui and L. Du. An optimal feedback control that minimizes the epidemic peak in the SIR model under a budget constraint (2022).

24 Deep Learning Techniques for DDoS Assault

M. Srisankar and Dr. K. P. Lochanambal

1 INTRODUCTION

The Distributed Denial of Service (DDoS) assault is the massive-scale dispensed and very unfavourable community attack approach that could adversely harm provider availability. It has progressively grown the various utmost intense security dangers to the web. With the chronic innovation and updating of assault era, a brand new attack version, known as a low-price DDoS assault, is developed. This assault uses flaws in the network protocol adaptive mechanism to supply assault packets at a decreased rate, reducing the victim's service quality. It has true concealment and a low attack fee. There are low-price/minimal DDoS assaults of many protocols inside the community surroundings in addition to periodic and aperiodic attack methods [1]. As end result, efficiently figuring out many kinds of minimal DDoS assault visitors is a critical project that should be addressed. This research frequently gives a multi-kind low-price DDoS assault revealing approach for networks inside the 5G context based totally on deep hybrid mastering. First, experimental statistics sets are acquired by simulating various forms of low-charge assaults and ordinary conversation behaviour; then, the function information of diverse kinds of low-charge DDoS assaults is analysed, and function choice is accomplished primarily based on the usual information; in the end, the detection model is found out by means of combining the hybrid deep learning of algorithm. In the end, the detection model is positioned on the network's entry to permit on line detection of many types of low-price DDoS assaults. Diverse styles of low-charge DDoS assaults and everyday verbal exchange in numerous settings are simulated inside the 5G environment, community visitors feature records at some stage in a particular time is accumulated, and a tagged minimal-diploma DDoS assault records set is generated. A multi-kind low-fee DDoS assault characteristic set is suggested. The characteristic statistics of several sorts of low-fee DDoS assaults and ordinary visitors is investigated from the perspective of statistical thresholds and characteristic engineering, and 40 effective minimal-degree DDoS assault characteristics are derived. A multi-type low-charge DDoS assault detection technique is provided. The offline schooling, deployment, and detection of hybrid deep mastering fashions are carried out the usage of the low-price DDoS assault function set. The detection findings display that through deciding on the perfect time frame, the approach presented in this observation can effectively identify four sorts of minimum DDoS assaults,

DOI: 10.1201/9781032694375-24

specifically, gradual-Headers attack, gradual-frame assault, sluggish-read assault, and Shrew assault.

2 RELATED WORK

For a long time, the research on minimum-degree DDoS assaults has received good sized attention from scholars at domestic and abroad. At the beginning of the 21st century, Kuzmanovic proposed the definition of Shrew attack, collected applicable information of minimal-diploma DDoS assaults, and carried out appropriate analysis and research [2]. The studies on minimal-degree DDoS assault revealing and protection specifically includes twofold methods. One is the detection approach based totally on statistical analysis. The authors proposed a minimal-degree DDoS assault revealing technique concentrated on the Pearson courting, which makes use of the Pearson coefficient of correlation primarily based on the Hilbert spectrum net congestion, to characterize community site visitors facts, and compares this records with a threshold to hit upon low-charge assaults towards Transmission Control Protocol [3]. Creator analysed the sequence similarity between the minimum-degree DDoS assaults pulses at the sufferer give up from the attitude of collection matching, used the Smith–Waterman algorithm, and designed a double-threshold rule to discover Transmission Control Protocol -based low-charge assaults [4]. The authors proposed a technique based totally on community self-similarity to investigate the impact of low-price attacks on visitors' self-similarity and used H-index combined with thresholds to discover assaults and valid site visitors [5]. The deep neural model (DNN) is proposed as a deep getting to know approach for malware detection on a subset of frames acquired from records switch [6]. The approach recommended by way of the researchers limits the price of interference in IOT transmitting facts, and the community's smart use of training units efficiently differentiates the traditional and chance sequences [7]. The above techniques for detecting low-rate assaults most effective see low-price attacks based totally on Transmission Control Protocol and depend on the set of factors, which are without difficulty affected by the randomness of the network environment and cannot achieve tremendous detection consequences. Any other type is device learning-based detection, which uses visitors' properties and ML procedures to become aware of minimum diploma DDoS assaults. The authors advocated a technique at the fundamentals of major thing investigation and Support Vector Machine to feel minimum-degree Transmission Control Protocol attacks. The foremost aspect evaluation tactic successfully captures community verbal exchange homes whilst filtering noise from the environment [8]. The authors proposed a minimal-diploma DDoS assault detection approach for Transmission Control Protocol in edge environments as shown in Figure 24.1, which used neighbourhood complex function mining and deep Convolutional Neural Network to acquire the finest trait distribution of uncooked data mechanically, and deep reinforcement learning Q networks as selection-making to improve attack detection decision-making accuracy [9]. The authors built a minimal-degree DDoS assault detection machine primarily based on decomposition machines, provided a characteristic mixture mechanism, mounted the correlation among characteristic samples, and detected HTTP-primarily based low-rate attacks. J48, random tree, REP tree, random woodland, multilayer perceptron, and guide

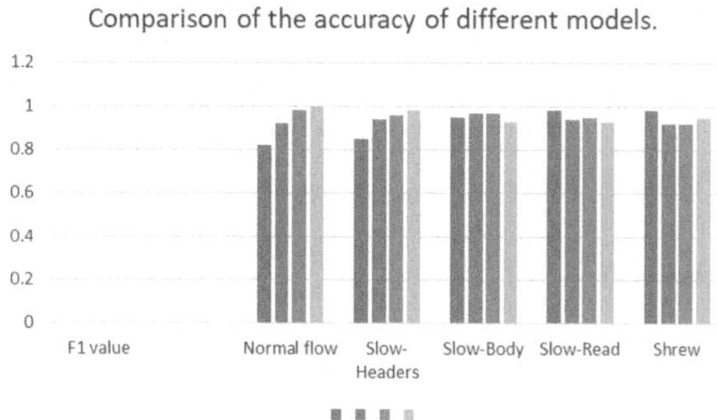

FIGURE 24.1 Comparison of Accuracy of Different Models.

vector gadget are six fashions that stumble on HTTP-based totally minimum-diploma DDoS assaults, in line with reference, which proposes the use of gadget mastering approaches to perceive low-rate DDoS assaults within the SDN situation [10]. DNN fashions can perform correctly and precisely even though with small samples on the grounds that its structure includes segmentation method and identification strategies, and additionally strands that upgrade themselves as they are programmed [6]. This method, however, has a better false-tremendous rate than DDoS assaults. Hybrid deep studying algorithms might also absolutely use the benefits of system getting to know and deep mastering algorithms. This article consists of multiple machine studying fashions to expect utility layer DDoS assaults in real time [11].

The authors have proposed CyDDoS structure for an automatic intrusion detection device (IDS) that blends a function map synthesis set of rules with the sort of neural community [12]. A hybrid primarily based on a long-brief-time period-reminiscence community and a CNN became suggested by means of researcher. Consequently, successfully implementing protection method to prevent a device from this chance is a sizeable difficulty considering DDoS employs a variety of attack strategies with numerous attainable combos [13]. The deep getting to know structure detects Bot, submit test, and XSS threats within the CICIDS2017 statistics set. The detection device has been proved to have higher detection abilities [14]. The authors proposed a deep gaining knowledge of-based totally hybrid anomaly detection system that makes use of the restricted Boltzmann device and aid vector system strategies to lessen the data's characteristic dimensions, but the facts set used within the investigation turned into KDD99, which is incorrect. At a finer stage, DoS assaults are classified and identified. The authors proposed a hybrid time-series forecasting model for stock forecasting primarily based on an prolonged brief-time period reminiscence community and LightGBM, which done properly [15]. In terms of prediction, writer proposes a hybrid deep gaining knowledge of model primarily based on an prolonged quick-time period reminiscence community and random forest (RF, random wooded area), which outperforms a unmarried gadget learning approach [16]. Minimal-diploma DDoS assault revealing procedures, along with the ones given above, can only pick out a unmarried type of minimal DDoS assaults, which has the drawbacks of handiest detecting one sort of attack and occasional detection accuracy.

Given the aforementioned limitations, this study proposes radial basis function networks (RBFNs) deep studying-primarily based minimum-degree DDoS assault revealing device that can research the traits of many varieties of attack site visitors and enhance the accuracy of online detection of numerous sorts of minimum-diploma DDoS assaults.

3 METHODOLOGY

In this observation, minimal-diploma DDoS assaults are labelled into two types: HTTP-primarily based low-fee DDoS assaults and TCP-based totally minimum-degree DDoS assaults [17]. Gradual-Headers, sluggish-body, in conjunction with sluggish-read assaults are examples of HTTP-based minimum-degree DDoS assaults [18]. This type of attack exploits the weak spot in the modern-day HTTP keep-Alive approach, continues the connection for an extended time frame, and constantly consumes resources of server, ensuing in a provider denial to the net server. Among these, the slow-Headers attacker sends an unfinished HTTP request finishing with the person "rn," inflicting the server to consider that the request became now not delivered and persevering with to attend. Ultimately, multiple connection techniques, the server's maximum capacity, and the brand new request are not able to be treated, ensuing in a rejection-service attack. The gradual body attacker makes a post request to the server with a large content-length cost. Even but, the server only supplies a tiny quantity of bytes each time, and the server's resources are depleted while requests exceed an confident threshold. Finally, gradual-read attackers post legitimate requests to the server to read large information files at the same time as setting the Transmission Control Protocol sliding window to a low range. Hence, setting up a communication link among the server and the attacker takes a lengthy time. While the number of connections exceeds a certain threshold, the provider cannot be provided. Transmission Control Protocol -primarily based low-charge DDoS assaults come in a variety of flavours. This study specializes in the Shrew attack, which leverages the Transmission Control Protocol timeout retransmission mechanism to transmit excessive-pace burst packets on a normal foundation, reducing the sufferer's first-rate of provider and performance. The recommended model overcomes it by incorporating a unique position-orientated neural layer [19]. This text normally replicates four styles of minimal-degree DDoS assaults the usage of assault tools and Python scripts: gradual-Headers attacks, sluggish-body attacks, slow-study attacks, and Shrew attacks. An ordinary analysis of minimal-degree DDoS assaults is usually primarily based on the original minimum-degree DDoS assaults. The CICFlowMeter characteristic extraction software extracts complete bidirectional flows based totally on time frames, reflecting houses including ahead and opposite facts flows. This technique is used as our research is especially aimed on the assault equipment particularly as slow-Headers attack, gradual-frame assault, sluggish-examine assault, and Shrew attack; but, this paintings mainly replicates four types of low-rate DDoS assaults. other than tag values, the tool produces a complete of 83 other sorts of characteristic records, which includes waft id, quintuple facts, circulation-level features, and package-level features. The glide identification is a penta-tuple together with the birthplace IP cope with, reason IP deal with, port region, future port, and system that is used to uniquely pick out the flow. Circulation-level traits include facts concerning the movement's time, length, and bytes per seconds. The amount of forwarding/opposite packets per second, statistical elements of packet period, SYN/FIN/RST flag bit count number, and so forth are all packet-stage traits.

4 RESULTS

This segment first introduces the composition of the detection framework, then introduces the principle and implementation of the facts set technology module, and sooner or later provides the precise overall performance and important technology of the offline schooling module and online detection module of the hybrid deep studying version element. The detection framework incorporates an information-set-generation module, a function analysis and selection module, an indifferent training unit, and a related detection unit. The framework is divided into information processing and deep hybrid getting to know. The records processing part is responsible for initial processing of the received community visitors and is divided into an information-set-generation module and a characteristic analysis and choice module. The facts-set-generation module is used to achieve community visitors in a distinct period, extract float characteristic data, and carry out information cleansing to get minimal-degree DDoS assault records set containing four kinds of minimal-diploma DDoS assaults and normal site visitors. The trait analysis and selection module analyses the trait information of different varieties of minimal-diploma DDoS assault from statistical thresholds and trait engineering and summarizes the precious features of multiple kinds of minimum-diploma DDoS assault. The deep hybrid mastering element detects many kinds of minimal-degree DDoS assaults and is separated into twofold segments: disconnected education and linked detection. The disconnected education unit selects valuable features from the statistics set for feature choice, uses a hybrid deep mastering set of rules for training and checking out, plays performance assessment and associated parameter optimization based on class effects, and selects the fine attack detection model. With the aid of recording site visitors in actual time, the net detection module deploys the educated hybrid deep learning detection model to the network access and achieves related revealing of different types of minimum-degree DDoS assaults. A version's output information is employed to recognize minimum-diploma DDoS assaults on visitors to be detected—a particular kind of attack.

4.1 Data Processing

The information set technology module is used to obtain the community traffic in a positive duration. Then, the float characteristic record is extracted by means of the float function extraction device CICFlowMeter to get a minimum-diploma DDoS assault information set. This records set contains a couple of varieties of minimum-degree DDoS assaults and everyday communique congestion in 5G environ, reflecting the site visitors styles in herbal environments. The generated a hefty figure of normal transmission simulation requests in step with the 1/3-technology cooperation mission (3GPP) and IEEE for real site visitors laws of devices in unique

5G software situations [20, 21]. This rule is acquired through the site visitors records amassed within the real scene. The end result includes the influencc of diverse environmental elements, which can reflect the request scenario inside the actual place. On this take, a look at the approach is stepped forward to generate everyday verbal exchange site visitors. Blended with the four minimal-diploma DDoS assaults site visitors generated through outbreak equipment as well as scripts, a brand new minimum-degree DDoS assault statistics set could be obtained. As in line with this study, assault is realized via sending site visitors via attack tools. Thinking about the security of the community environment, the capture of low-price community traffic is identified based totally on the VMware vSphere virtualization experimental platform. The realistic environment is close to the herbal surroundings, reflecting the traffic statistics inside the digital surroundings. Thereafter, the traffic collection tool Tcpdump is deployed and hooked up to seize the records packets inside the network. The statistics set collection point is on the get right of entry to gateway of the community entrance, which can completely seize the verbal exchange visitors within the network. Finally, CICFlowMeter is used to extract feature data of community site visitors. At the same time, according to the attack plan in Table 24.1, the extracted characteristic fact is categorized, and the classified records set is used for the training and verification of the detection model. This newsletter consists of more than one device mastering fashions to assume utility layer DDoS assaults in actual time. The authors have proposed CyDDoS, an architecture for an automatic IDS that blends a feature map synthesis set of rules with such a neural community [22]. The three styles of minimum-diploma DDoS assault strategies, sluggish-Headers assault, slow-frame attack, and gradual-examine assaults studied in this text, ship the attack site visitors by modifying the parameters of the slow HTTP test and gradual HTTP attack tool, and the Shrew assault realizes the sending assault with the aid of writing Python scripts go with the flow. Python scripts are used for normal verbal exchange requests based totally on the statistical legal guidelines of various scenarios inside the 5G environment to simulate sending large connection everyday request visitors. Based totally at the above implementation methods, this has a look at collects traffic and routinely extracts waft feature statistics below minimal-diploma DDoS assault and regular communique behaviour [23]. Throughout this period, different assaults have been released, which include low-fee DDoS assaults, DDoS community stratum assaults, DDoS application stratum assaults, and allotted reflection amplification assaults.

TABLE 24.1
Attack plan for DDoS assaults.

Attack time	Source IP	Destination IP	Traffic type
2023.6.30.	99.1.0.22	99.1.1.22	Slow-Headers
	99.1.0.12	99.1.1.23	Slow-Body
	99.1.0.13	99.1.1.24	Slow-Read
15:45-12:09	23.1.0.14	23.1.1.25	Shrew
	23.1.0.20~23.1.0.29	23.1.1.73	Normal flow

Based at the community visitors pcap record obtained through the above attack plan, the traffic characteristic extraction tool CICFlowMeter is hired to excerpt the traffic trait info, and a multi-kind minimum-degree DDoS assault information set is obtained. Table 24.2 depicts the amount of records samples of each unmarried traffic kind in the records set and the ratio of well-known traffic samples. it is able to be visible that the quantity of statistics samples of normal visitors is ample advanced than the count of facts samples of every minimum-diploma DDoS assault, reflecting the minimum-degree DDoS assaults.

This examine simulates numerous minimal-diploma DDoS assaults and ordinary conversation requests inside the 5G surroundings. It conducts overall performance reviews of different hybrid deep studying detection models and online detection of overall performance assessments beneath different detection time windows. Table 24.2 shows the wide variety of records samples from every traffic category in the records set in addition to the ratio of ordinary site visitors records. Figure 24.1 with Tables 24.3 depict the efficiency and F1 cost of the three models. For detecting gradual-Headers attack site visitors, the RBFNs outperform the opposite models in terms of effectiveness and F1 cost for figuring out normal benign traffic (Figures 24.2 and 24.3).

The web detection accuracy in different scenarios decreases related to the attack visitors sending price and the responsibility cycle of normal site visitors in the detection window. Within the destiny, we will look at the optimization model and time window and analyse the relationship between time window and records set and characteristic choice so that the version can better adapt to the surroundings and have higher accuracy and detection efficiency. In this observation, a virtual platform primarily based on Vmware vSphere is set up as the experimental surroundings. A total of nine hosts had been used in the experiment, including two routers, one customer host, four dummy hosts, and net servers. The research on this examination builds a hybrid deep getting to know version based on the TensorFlow framework. The programming language is Python 3.8, and the system gaining knowledge of library of TensorFlow 2.1 and Keras 2.2.4 is used to build the model. The Ubuntu 18.04 is software history in server operating shape, and the wide variety of virtual cores is 8, the memory is 8 GB, four hosts are used as puppet hosts, and two digital machines built with net servers are used as attacked servers. This is critical to halt fraudulent hobby since they have got an extended-time period affect on economic

TABLE 24.2

Number of data samples for each Traffic type.

Traffic type	Number of data samples	The proportion of attack traffic to normal traffic
Slow-Headers	100 793	01:04.5
Slow-Body	110 044	01:04.5
Slow-Read	68 074	01:04.5
Shrew	45 389	01:04.5
Normal flow	460 619	—

situations. Outlier detection has several critical programmes for fraud prevention [24]. Detection is finished at the network entry router, and facts collection and cleansing functions are provided. The four transmission eventualities generated a huge wide variety of regular communication statistics requests. Minimum-diploma DDoS assault attacker controls four puppet hosts to periodically send minimal-diploma DDoS assaults based totally on HTTP protocol and Transmission Control Protocol to the internet server. The experimental minimal-degree DDoS assault sorts select HTTP-based sluggish-Headers assaults, sluggish-frame assaults, sluggish-study attacks, and Transmission Control Protocol -primarily based Shrew assaults [25]. The minimum-degree DDoS assault detection framework implements offline training and online detection for numerous sorts of minimal-degree DDoS assault statistics primarily based on hybrid studying technique [26]. Offline hobby in particular analyses the model's type performance thru six evaluation indicators: accuracy, precision, remember, F1 value, detection time, and confusion matrix. among them, the fee of exactness symbolizes the ratio of the amount of actual samples labelled via the prototype to the general amount of portions; the exactness degree represents the share for an amount of samples counselled by way of prototype as an attack class and the depend of samples which are attack kinds; and the don't forget rate represents the prototype recommended as an assault class [27]. The percentage of the sum of pieces to all the examples of this assault kind are as follows: the F1 fee combines the outcomes of precision and remember, representing the harmonic common of the two, that may extra appropriately replicate version performance; detection time reflects the time complexity of the version. It's miles used to measure the time performance of the version; the classification impact of the prototype is examined by way of employing confusion matrix as well as the grade to which the anticipated label matches the real label, which corresponds to the don't forget price numerically [28]. In addition, to analyse the classification of online detection, new evaluation indicators are defined: false intervention degree and malicious congestion revealing degree used to evaluate an online detection of normal and malicious traffic, respectively. Among them, the false interception rate represents the proportion of misjudging regular traffic as diverse kinds of minimal-degree DDoS assaults, and the calculation is shown in formula (1); the malicious traffic detection rate represents the proportion of detected malicious traffic to the overall count of negative traffic samples, and the calculation is shown in formula (2).

$$\text{False interception rate } = \sum_{i=1}^{4}(G_i) \tag{1}$$

where G_i represents the number of data samples that misjudge the regular traffic in the network environment as a further four forms of minimal-degree DDoS assault traffic after online detection; M represents the total number of data samples of regular traffic in the network environment; T_i represents the number of undetected data samples of minimal-degree DDoS assault congestion within the network environment after detection; B_i represents the total number of data samples of different types of minimal-rate DDoS assault congestion within the network environment.

TABLE 24.3
Minimal degree DDoS assault data.

Data set type	Normal flow samples	Number of attack traffic samples
Training set	288555	267800
Test set	129832	108943

TABLE 24.4
Comparison of detection time of different models.

Model Category	LSTM-Light GBM	LSTM-RF	CNN-RF	RBF
Detection time/s	259.8986	308.5964	268.3689	250.3450

Based on the minimal-degree DDoS assault data set obtained by the data set generation module in Section 3, data cleaning is performed, including processing the feature data with null feature values and processing feature data with infinite feature values. Feature selection is carried out according to the 40 useful features shown in Figure 24.1 and is distributed in a dual sets as training as well as test in a ratio of 7:3. The data set is shown in Table 24.3. The total number of data samples in the minimal-degree DDoS assault data set is 794,919, including 556,444 in the preparation set as well as 238,475 in the training set.

The RBF model showed optimal performance through hyperparameter search, given the same minimal-degree of DDoS assault data set and eigenvalues. At the same time, the RBF prototype projected in this study is associated with the CNN-RF, LSTM-Light GBM prototype, and the LSTM-RF prototype, and the optimum hybrid deep learning prototype is nominated to identify the connected revealing of multi-type minimal-rate DDoS assaults. This study uses four evaluation indicators: detection time, precision rate, $F1$ value, and confusion matrix. Figure 24.1 shows the confusion matrix performance of the three hybrid deep learning models. It may be perceived that the recognition precision of CNN-RF, LSTM-Light-GBM model for each traffic type varies greatly, especially the recognition accuracy of the Slow-Body attack is only 0.5565, and the false-positive rate of the Slow-Headers attack is 0.2695. The recognition accuracy of the LSTM-RF model for the five types of traffic is better than that of the LSTM-Light GBM prototype, especially the recognition accuracy of the Slow-Read attack is about 0.9992, but it will produce a false-positive rate of 0.0788 when identifying the Slow-Body attack. The recognition accuracy of Slow-Headers attack traffic can also get 0.9566.

Figure 24.2 and Tables 24.3 show the evaluation of the three prototypes in terms of exactness and $F1$ value. As can be seen from Figure 24.1, for the identification of regular benign traffic, the RBF prototype outperforms the other two designs in terms of accuracy and $F1$ value; for the detection of Slow-Headers attack traffic, the accuracy of the RBF design is the best. Excellent: the RBF, CNN-RF, LSTM-RF,

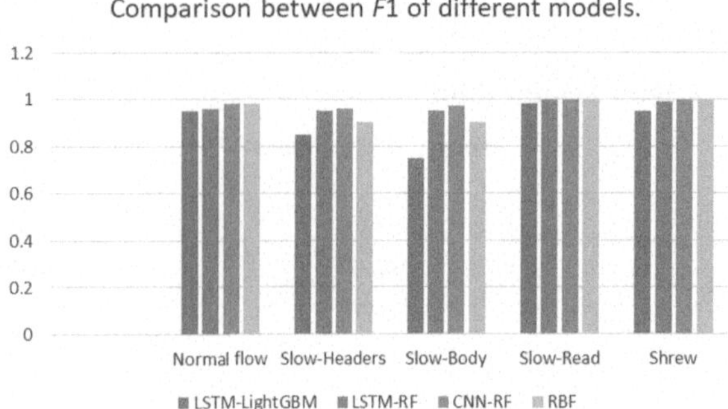

FIGURE 24.2 Comparison Between F1 of Different Models.

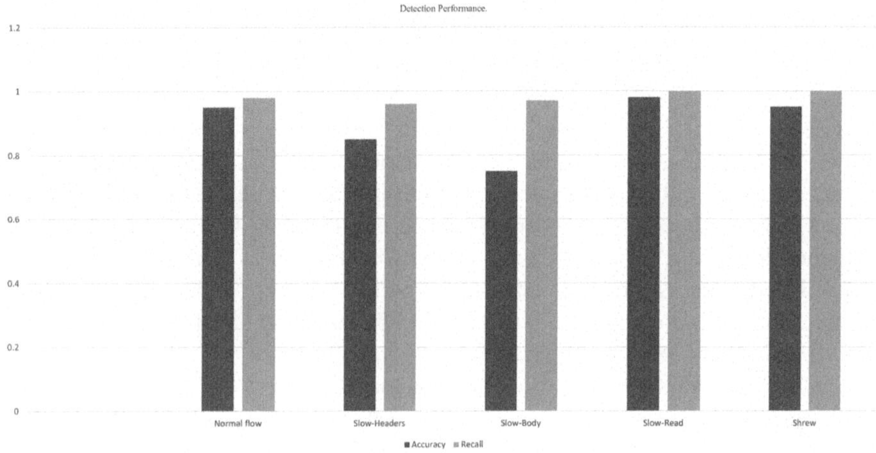

FIGURE 24.3 Difference Between Accuracy and Recall.

and LSTM-Light GBM models have similar performance in $F1$ value; for detecting Slow-Body and Slow-Read assault congestion in net, the LSTM-Light GBM design has poor performance in both accuracy and $F1$ score, and the RBF model's performance is poor. Best performing: for Shrew, the detection of attack traffic in the three models is in the two evaluation indicators of good performance.

The detection time comparison of different hybrid deep learning ideas is presented in Table 24.4. That the detection time of the RBF model is 268.3689 s, which is about 9 s longer than that of the LSTM-Light GBM design, and about 40 s more minor than that of the LSTM-RF model. However, the LSTM-Light GBM design is significantly lower than the RBF design in detection accuracy and $F1$ score. Therefore, while the detection time is shorter, the RBF design has better accuracy and $F1$ value for various forms of minimal-rate DDoS assaults and regular congestion as shown in Figure 24.3.

Combining the above evaluation indicators, it can be concluded that the distinction of LSTM-LightGBM model along with the LSTM-RF, CNN-RF model, the RBF model proposed in this article has better performance in regular traffic. Slow-Headers assault, Slow-Body assault, Slow-Read assault, and Shrew's assault traffic detection as well as classification all show excellent performance and can accurately detect different types of low-rate DDoS assaults.

5 CONCLUSION

Aiming of minimal-diploma DDoS assaults, this examine obtains minimal-degree DDoS assault records units, analyses and obtains 40 effective traits of minimum-diploma DDoS assaults, and proposes a variable-type minimal-diploma DDoS based on. The attack detection method and online deployment of this model understand linked revealing of variable types of minimum-degree DDoS assaults. Furthermore, a web detection time window is proposed, and the net detection overall performance is evaluated the usage of false intervention diploma and malicious community congestion revealing price. Experiments show that the prototype based totally on RBFNs deep learning model of set of rules can appropriately discover specific types of minimal-diploma DDoS assaults. The revealing approach on this have a look at is notably portable, and the minimal-degree DDoS assault information set is used close to the actual situation, which can be deployed and carried out in sensible environments when the RBFNs deep learning to know version implements training and detection for multi-kind low-fee DDoS assaults. The net detection accuracy in distinct scenarios decreases associated with the assault visitors sending charge and the duty cycle of normal site visitors inside the detection window. In the destiny, we will study the optimization version and time window and analyse the relationship among time window and information set and characteristic selection in order that the model can betteradapt.

REFERENCE

1. S. Yeom and K. Kim, "Improving performance of collaborative source-side DDoS assault detection," in *Proceedings of the 2020 21st Asia-Pacific Network Operations and Management Symposium (APNOMS)*, pp. 239–242, Daegu, Korea (South), September 2020.
2. W. Sun, Y. Li, and S. Guan, "An improved method of DDoS assault detection for controller of SDN," in *Proceedings of the 2019 IEEE 2nd International Conference on Computer and Communication Engineering Technology (CCET)*, pp. 249–253, Beijing, China, August 2019.
3. B. Jia and Y. Liang, "Anti-D chain: a lightweight DDoS assault detection scheme based on heterogeneous ensemble learning in blockchain," *China Communications*, vol. 17, no. 9, pp. 11–24, 2020.
4. J. He, Y. Tan, W. Guo, and M. Xian, "A small sample DDoS assault detection method based on deep transfer learning," in *Proceedings of the 2020 International Conference on Computer Communication and Network Security (CCNS)*, pp. 47–50, Xi'an, China, August 2020.
5. Z. Liu, Y. He, W. Wang, and B. Zhang, "DDoS assault detection scheme based on entropy and PSO-BP neural network in SDN," *China Communications*, vol. 16, no. 7, pp. 144–155, 2019.

6. E. Cil, K. Yildiz, and A. Buldu, "Detection of DDoS assaults with feed forward based deep neural network model," *Expert Systems with Applications*, vol. 169, Article ID 114520, 2021.
7. M. H. Ali, M. M. Jaber, S. K. Abd, et al., "Threat analysis and distributed denial of service (DDoS) assault recognition in the Internet of things (IoT)," *Electronics*, vol. 11, no. 3, p. 494, 2022.
8. S. Dong and M. Sarem, "DDoS assault detection method based on improved KNN with the degree of DDoS assault in software-defined networks," *IEEE Access*, vol. 8, pp. 5039–5048, 2020.
9. Y. Chen, X. Chen, H. Tian, T. Wang, and Y. Cai, "A blind detection method for tracing the real source of DDoS assault packets by cluster matching," in *Proceedings of the 8th IEEE International Conference on Communication Software and Networks (ICCSN)*, pp. 551–555, Beijing, China, June 2016.
10. X. Liang and T. Znati, "An empirical study of intelligent approaches to DDoS detection in large scale networks," in *Proceedings of the 2019 International Conference on Computing, Networking and Communications (ICNC)*, pp. 821–827, Honolulu, HI, USA, February 2019.
11. J.-H. Jun, H. Oh, and S.-H. Kim, "DDoS flooding assault detection through a step-by-step investigation," in *Proceedings of the 2011 IEEE 2nd International Conference on Networked Embedded Systems for Enterprise Applications*, pp. 1–5, Perth, WA, Australia, December 2011.
12. M. J. Awan, U. Farooq, H. M. A. Babar et al., "Real-time DDoS assault detection system using big data approach," *Sustainability*, vol. 13, no. 19, Article ID 10743, 2021.
13. O. Lopes, D. Zou, F. A. Ruambo, S. Akbar, and B. Yuan, "Towards effective detection of recent DDoS assaults: a deep learning approach," *Security and Communication Networks*, vol. 2021, Article ID 5710028, 14 pages, 2021.
14. V. Popovskyy and V. Skibin, "Entropy methods for DDoS assaults detection in telecommunication systems," in *Proceedings of the 2014 First International Scientific-Practical Conference Problems of Infocommunications Science and Technology*, pp. 182–185, Kharkov, Ukraine, October 2014.
15. D. Erhan and E. Anarim, "Istatistiksel Yöntemler Ile DDoS Saldırı Tespiti DDoS detection using statistical methods," in *Proceedings of the 2020 28th Signal Processing and Communications Applications Conference (SIU)*, pp. 1–4, Gaziantep, Turkey, October 2020.
16. L. Wang and Y. Liu, "A DDoS assault detection method based on information entropy and deep learning in SDN," in *Proceedings of the 2020 IEEE 4th Information Technology, Networking, Electronic and Automation Control Conference (ITNEC)*, pp. 1084–1088, Chongqing, China, June 2020.
17. Sanmorino and S. Yazid, "DDoS Assault detection method and mitigation using pattern of the flow," in *Proceedings of the 2013 International Conference of Information and Communication Technology (ICoICT)*, pp. 12–16, Bandung, Indonesia, March 2013.
18. L. Luo, J. Wang, and L. Jia, "A CGAN-based DDoS AssaultDetection Method in SDN," in *Proceedings of the 2021 International Wireless Communications and Mobile Computing (IWCMC)*, pp. 1030–1034, Harbin City, China, June 2021.
19. Mahajan, U. Garg, and M. Shabaz, "CPIDM: a clustering-based profound iterating deep learning model for HSI segmentation," *Wireless Communications and Mobile Computing*, vol. 2021, Article ID 7279260, 12 pages, 2021.
20. R. Arthi and S. Krishnaveni, "Design and development of IOT testbed with DDoS assault for cyber security research," in *Proceedings of the 2021 3rd International Conference on Signal Processing and Communication (ICPSC)*, pp. 586–590, Coimbatore, India, May 2021.

21. D. Ushakov, M. Vinichenko, and E. Frolova, "Environmental capital: a reason for interregional differentiation or a factor of economy stimulation (the case of Russia)," *IOP Conference Series: Earth and Environmental Science*, vol. 272, no. 3, p. 032111, 2019.
22. Y. Chen, J. Hou, Q. Li, and H. Long, "DDoS assault detection based on random forest," in *Proceedings of the 2020 IEEE International Conference on Progress in Informatics and Computing (PIC)*, pp. 328–334, Shanghai, China, December 2020.
23. S. Nguyen, J. Choi, and K. Kim, "Suspicious traffic detection based on edge gateway sampling method," in *Proceedings of the 2017 19th Asia-Pacific Network Operations and Management Symposium (APNOMS)*, pp. 243–246, Seoul, Korea (South), September 2017.
24. S. Sanober, I. Alam, S. Pande, et al., "An enhanced secure deep learning algorithm for fraud detection in wireless communication," *Wireless Communications and Mobile Computing*, vol. 2021, Article ID 6079582, 14 pages, 2021.
25. D. Erhan, E. Anarım, and G. K. Kurt, "DDoS assault detection using matching pursuit algorithm," in *Proceedings of the 2016 24th Signal Processing and Communication Application Conference (SIU)*, pp. 1081–1084, Zonguldak, Turkey, May 2016.
26. N. Hoque, D. K. Bhattacharyya, and J. K. Kalita, "A novel measure for low-rate and high-rate DDoS assault detection using multivariate data analysis," in *Proceedings of the 2016 8th International Conference on Communication Systems and Networks (COMSNETS)*, pp. 1–2, Bangalore, India, January 2016.
27. N. R. Nayak, S. Kumar, D. Gupta, A. Suri, M. Naved, and M. Soni, "Network mining techniques to analyze the risk of the occupational accident via Bayesian network," *International Journal of System Assurance Engineering and Management*, vol. 13, pp. 1–9, 2022.
28. Y. K. Hong, H. Choi, and J. Park, "SDN-assisted slow HTTP DDoS assault defense method," *IEEE Communications Letters*, vol. 22, no. 4, pp. 688–691, 2018.

25 Study on Blockchain-Based Framework for Central Bank Digital Currency Design

Opportunities, Risk, and Challenges

Shivangi Verma

1 INTRODUCTION

Since digital currency is a digital version of legal money issued by a central bank, more and more people and businesses are adopting it as a form of payment and savings. This is like a country's central bank. Digital currency, often known as "digital fiat currency," is issued not by private banks but by national ones. Many government-backed financial institutions around the world are currently researching central bank digital currencies (CBDCs) to ascertain their potential and economic value. There are banknotes and coins available, and they are denominated in the national currency. CBDCs are government-issued digital currencies used by individuals and companies alike. CBDCs that are token-based and commercially accessible can employ either private or public keys. By facilitating access to the monetary system for the unbanked, CBDCs can facilitate fiscal and financial policies that bring the world closer to financial inclusion. Since they are a centralised payment system, they are responsible for protecting customer data. CBDCs in various regions of the world are in various developmental phases (Worrell, 2021). The Reserve Bank of India (RBI) is in charge of the Indian rupee. Coins and paper bills are also accepted. Fiat-currency-denominated debt is one type of debt. Coins made of metal are the standard money. They have purchasing power because they are recognised as legal currency. As the global population, economy, and online financial markets have grown, so too has the value of digital currency. According to Ducrée (2022), Satoshi Nakamoto, the inventor of Bitcoin, gave a more formal introduction to cryptocurrencies and blockchain technology in 2008. Kim and Kwon (2019) argue

 DOI: 10.1201/9781032694375-25

that a CDBC can function both as an alternative to and a complement to national currencies like the dollar. It's essentially the same thing as other forms of paper currency produced by governments. Nirmala Sitharaman, the current minister of finance, has said that digital currency will be introduced in the upcoming fiscal year. The digital economy will benefit from this, and so will those responsible for managing money. One other advantage of digital currency is that it can be managed more efficiently and cheaply than traditional currency. It is suggested that blockchain and other technologies be used to launch the digital rupee.

2 TYPES OF CENTRAL BANK DIGITAL CURRENCIES

Both retail-focused central bank digital currencies (CBDCs-R) and wholesale-focused central bank digital currencies (CBDCs-W) exist.

2.1 Retail (CBDC-R)

CBDCs purchased at retail could be used in the same ways as currency, including making purchases, sending gifts, and redeeming government benefits. The government is considering issuing and overseeing a new form of money called Retail CBDC. The central bank's whims and the current status of monetary policy determine whether or not CBDC will be made available. Retail in the context of cannabidiol (CBD) refers to sales to the general public. Distributed ledger technology (DLT)-based retail CBDC has the potential to be more secure, auditable, available 24/7/365, and interest-inclusive than its traditional counterpart.

2.2 Wholesale (CBDC-W)

A wholesale CBDC can be used by banks that maintain reserve deposits at a central bank. It could be used to ease payment and securities settlements by reducing worries about counterparties' credit and liquidity. Limited-access digital tokens would be used to either replace or supplement central bank reserves in a value-based wholesale CBDC. By using tokens, value might be transferred directly between users without the requirement for any third party to act as a go-between. Increasing CBDC usage would be beneficial for the payment and settlement system, according to the Bank for International Settlements (BIS).

3 ADVANTAGES OF CENTRAL BANK DIGITAL CURRENCY

CBDCs have emerged in recent years as a solution to issues brought on by the broad adoption of digital payment methods and the prevalence of private digital currencies like Bitcoin. To the degree that CBDCs can replace cash in its widespread use, the cost of creating, distributing, transporting, and storing currency can be lowered, as stated by Kiff et al. (2020). CBDCs may help more people gain access to banking services, which are now unavailable to them due to a lack of creditworthiness on the part of conventional financial institutions. CBDCs facilitate quicker payments and growth, both of which increase output. According to Auer et al. (2021), the procedure becomes more dangerous when middlemen are involved. In both cases, money on deposit poses no danger. These kinds of shocks have the potential to upset the delicate balance of a financial plan. Integrating CBDCs with other state-of-the-art technologies like blockchain and smart contracts may open up new use cases and applications. CBDCs have the potential to speed up and simplify transactions by standardising payment methods and decreasing settlement times. Due to the centralisation and transparency of CBDCs, taxation may be made easier. According to Auer and Böhme (2020), consumers can shop with confidence knowing that their privacy will be protected while using a CBDC at brick-and-mortar establishments. However, CBDCs that require users to sign up for an account provide additional privacy protections and can be used in legitimate financial transactions. By reducing the need for middlemen and currency changes, CBDCs may make international trading easier and more cost-effective. Interest rates, inflation, and other macroeconomic indices may be easier for central banks to control with the help of CBDCs.

4 DISADVANTAGES OF CENTRAL BANK DIGITAL CURRENCY

Concerns that CBDCs may reduce demand for conventional banking services have been linked to potential job losses and a halt in economic growth. According to Pelagidis and Kostika (2022), there is still active research being conducted in CBDCs, which may be considered a long-term trend. Due to the portability of these systems, a strong CBDC maintained by a foreign government could end up replacing a weak national currency. The need for regulatory harmony across numerous countries and the difficulty of doing cross-border transactions are two possible roadblocks to the further development of CBDC. CBDCs may not necessarily face issues associated with centralisation. A government agency continues to have responsibility for and be held accountable for the management of financial transactions. Depending on how the CBDC is structured, central banks may not necessarily prioritise societal and economic welfare. Since the official is in charge of obtaining and disseminating digital identifications argues that users would have to give up some privacy. Terrorist financing and money laundering may be facilitated by CBDCs since they may be tougher to track and regulate than traditional currencies. CBDC implementation will need substantial investments in information technology infrastructure, which could be a severe financial burden for some nations. Privacy and national security concerns are exacerbated by the growing tracking of financial transactions. There may also be costs and dangers connected with implementing CBDCs, as well as integrating them into current payment systems.

5 BLOCKCHAIN AND DISTRIBUTED LEDGER TECHNOLOGY FOR CBDC

The blockchain operates as a distributed ledger. A distributed ledger is a database that records and synchronises financial transactions across a network of computers. Using a linear, append-only linking scheme, blockchain stores data in blocks. Globally, distributed database without a single administrator in charge, a decentralised database relies on a set of tools and procedures that allow several users to collaborate for the database's protection. The most well-known application of DLT is Nakamoto. One more digital currency that stands out for its programmability is Ethereum. Ethereum is a distributed computing platform that may be used to create new digital currencies and other decentralised applications. The widespread acceptance of Bitcoin prompted the creation of similar digital currencies. Discretion is of paramount importance while handling money, hence the term "private" is fitting. As a result, several CBDCs have been built on the DLT platform preferred by central banks. The country's central bank is responsible for ensuring the security of each copy on this shelf, according to Allen et al. (2020). One example of such a conversation is the sharing of distributed ledger systems (DLT). The authors went on to suggest that only people with permission to view or make modifications to the central bank's blockchain would be able to do so.

6 RESEARCH OBJECTIVES

The main objectives "of this study are:

- To study the need for CBDC in India
- To study the opportunities and perceived risks factors of CBDC
- To identify the challenges of CBDC

7 RESEARCH HYPOTHESES

After in detail research, it lies on the research to make following hypotheses.

Null Hypothesis (Ho):

Perceived risk has a positive impact on thc intention to adapt a digital currency.

Alternative Hypothesis (H1)

Perceived risk has a negative impact on the intention to adapt a digital currency.

8 RESEARCH METHODOLOGY

Descriptive research like the one presented here, which relies on secondary sources, is often the best bet. This investigation is grounded in scholarly writings such as articles and books. There is a wealth of data available on governmental websites.

9 NEED OF THE CENTRAL BANK DIGITAL CURRENCY IN INDIA

The research will indicate that" using digital currency will reduce government spending on paper money, aid India's green initiatives by encouraging the use of digital currency rather than paper money, and boost the uptake of digital and electronic payment systems in the country. The Indian government's 2016 decision to demonetise cash had a significant effect on the way individuals there do business. This was a difficult decision to make at first because most purchases in India are made with cash. Users, however, rapidly adopted digital payment methods such as BHIM UPI and QR code scanning. This is because the transition to digital payments was simplified by the widespread availability of smartphones and easy payment modes. Digital currency issued by a central bank and not backed by any physical good is called CBDC. Monetary policy, currency issue, and the provision of financial services to both the public and private sectors are the purview of a country's central bank. The potential benefits of CBDC for users and the willingness of retailers to embrace it are likely to propel its widespread adoption. Central bank money is the most reliable form of currency. However, CBDC may also provide other advantages, such as lower prices for customers and businesses, the ability to conduct transactions and make payments while offline, greater anonymity than commercial services, and a plethora of accessibility options. Privacy is a further concern. Most CBDCs, in contrast to completely anonymous cash, will be built so that central banks can track expenditure. The increased risk to user privacy must also be considered, as the central bank will have access to a large amount of data regarding user transactions.

10 OPPORTUNITIES OF CENTRAL BANK DIGITAL CURRENCY

There are various opportunities for CBDC to support monetary and financial stability.

- **Increased Consumer Trust and Confidence**
  The unbanked and the underbanked may be less likely to use or even have access to financial products and services due to a lack of faith in official financial institutions, particularly digital financial services. As opposed to privately operated digital currency and financial services, people may be more interested in using digital currency issued by central banks, which could lead to higher demand for these services. The fact that many individuals "distrust" issuing agents and instead rely on government-run alternatives lends credence to this observation. The volatility of cryptocurrency prices further reduces their value as a medium for storing wealth or making long-term investments. Furthermore, most private cryptocurrencies are kept and sold on exchanges (most of which are unregulated), and there are several reports of exchanges being hacked and funds being stolen due to inadequate protections. Customers with low levels of digital and financial knowledge may find it simpler to use this innovation in everyday transactions if CBDCs include a custodial option.
- **Financial transactions are cheaper**
  It can be costly to use more conventional payment options. Keeping money in the bank and withdrawing it to spend in a store, for instance, may incur at least two different types of fees: an account holding cost and a withdrawal fee. All transactions, including digital ones, are often subject to transaction fees. In addition, any time money is transferred from one bank account to another, a transaction fee will be assessed (such as those between consumers). Some governments and central banks are working on inexpensive and rapid payment methods. It's possible that agreements based on a single economy won't be able to reduce the high costs associated with making international payments. It's possible that CBDCs can drastically reduce these fees for cross-border transactions.

- **Excellent Data Collection and Utilisation to Increase the Value of Formal Financial Services**
  Many banks and other financial institutions now collect extensive customer information in order to better serve their clients. Because financial organisations are primarily concerned with increasing the value of their products and services, central banks may be better able to combat financial exclusion by exploiting the data on individuals obtained through the use of CBDCs. Using this information, governments can gain a deeper understanding of the needs, wants, and weaknesses of various consumer segments (e.g., Auer et al., 2021).

11 PERCEIVED RISK FACTORS OF CENTRAL BANK DIGITAL CURRENCY

A consumer's "perceived risk" arises from their uncertainty about the measures the provider has taken to ensure the safety of the technology in a digital currency payment setting. CBDC faces a number of perceived risks.

- **Market Risks**
  The primary danger of digital currency is its fluctuating value. Price swings in digital currencies are. The market now poses a higher danger to investors. Investors lose money as a result of the drop in the value of digital currency, which, in turn, causes shifts in macroeconomic indicators like interest rates.
- **Liquidity Risks**
  Sharp pricc variations due to supply constraints and trading volume changes of digital currencies constitute liquidity risk and hinder the efficient functioning of the market.
- **Operational Risks**
  Operational risk, also known as technical risk, is the chance that digital money will encounter challenges that cannot be anticipated or solved with the current state of technology. The blockchain infrastructure and the currency exchange platform both pose technical challenges.
- **Legal Compliance Risks**
  Legal compliance risk refers to the potential that certain users of digital currency will engage in illegal or unethical practises in order to gain an advantage.

12 CHALLENGES OF CENTRAL BANK DIGITAL CURRENCY

There are a number of legal and regulatory hurdles that must be cleared before CBDCs can be issued. CBDCs are either outright banned or strictly regulated in some countries. Despite the fact that several central banks assert that they have the requisite legal authority and framework to issue CBDCs, there remain some legal challenges to surmount due to the distinctive features of CBDCs, such as their programmability. CBDC requires coordination between the government and the Central

Bank. They must coordinate their efforts in order to make important choices and develop workable policies. In order to implement CBDC solutions on a global scale, it is essential for governments to cooperate and set universal standards. However, poor levels of financial literacy among the general public constitute a severe hurdle to the fundamental aim of central banks in issuing a CBDC, which is to improve financial inclusion. The CBDCs' technical infrastructure and implementation also provide significant challenges. Internet connectivity (especially in remote areas), system integration, and cyberattacks are only a few examples of the many possible technological hurdles.

REFERENCES

Auer, R., Frost, J., Gambacorta, L., Monnet, C., Rice, T., & Shin, H. S. (2021). Central bank digital currencies: motives, economic implications and the research frontier.

Ducrée, J. (2022). Satoshi Nakamoto and the Origins of Bitcoin–Narratio in Nomine. Datis et Numeris. *arXiv* preprint arXiv: 2206.10257.

Regi, S. B., & FRANCO, C. (2017). "Information Technology in Indian Banking Sector – Challenges and Opportunities", International Journal of Multidisciplinary Research and Modern Education, Volume 3, Issue 1, Page Number 78–82, 2017.

Kiff, J., Alwazir, J., Davidovic, S., Farias, A., Khan, A., Khiaonarong, T., Malaika, M., Monroe, H. K., Sugimoto, N., Tourpe, H., & Zhou, P. (2020). A Survey of Research on Retail Central Bank Digital Currency. International Monetary Fund Working Paper, No. 2020/104. https://doi.org/10.5089/9781513547787.001

Kim, Young Sik and Kwon, Ohik, Central Bank Digital Currency and Financial Stability (February 8, 2019). *Bank of Korea WP* 2019-6, Available at SSRN: http://dx.doi.org/10.2139/ssrn.3330914

Pelagidis, T., & Kostika, E. (2022). Investigating the role of central banks in the interconnection between financial markets and crypto assets. *Journal of Industrial and Business Economics*, 1–27.

Shettar, R. M. (2019). Digital banking an Indian perspective. *IOSR Journal of Economics and Finance*, 10(5), 01–05.

Worrell, D. (2021). The time has come to permanently retire all our Caribbean currencies. *Journal of Economics Bibliography*, vol. 8, no. 3, pp. 171–184.

26 NLP-Based Chatbot for Handling Mental Health-Related Issues

Abhinav Juneja, Yonis Gulzar, Sapna Juneja, Junaid Mohsin, Himanshi Mishra, Harshita Shukla, and Vishal Jain

1 INTRODUCTION

Everybody has stress and/or mental health-related issues. In fact, according to a 2015 study (globally), the number of people who suffered from some form of depressive disorder worldwide was estimated to be over 322.48 million people. And according to another 2017 study, more than 14% of the total populace in India suffers from variations of mental disorders; thereby, consisting one of a major reason of stress in lives of people which results on the well-being of the society and the social quality thereof [1]. Unchecked stress can also lead to a number of health issues, affecting one physiologically and taking a toll on your body and one's daily life. Indeed, more than 50% of all physical illnesses are caused by mental disturbances where the effect thereof is believed to be the main cause of these dysfunctions and is correlated with increase in risk diabetes, cardiovascular (heart) diseases, and other physical ailments such as migraines, skin disorders, and epilepsy; whereof each of these illnesses – and many others – are psychosomatic in nature (prompted or exacerbated by mental conditions such as stress).

In general, it can be said that stress has three effects (broadly):

- Subjective effects – this includes:
  1. Anxiety or frustration
  2. Panic attack
  3. Feeling tired, unwell, or lonely
  4. Depression
  5. Moodiness
- Observable behavioral changes – this includes:
  1. Increased accidents
  2. Compulsive internet surfing
  3. Drug or alcohol use
  4. Argumentative behavior
  5. Development of binge eating or other eating disorders

DOI: 10.1201/9781032694375-26

- Physical/Psychosomatic symptoms – this includes:
  1. Stomach or digestive problems
  2. Sexual malfunction
  3. Weak immune system
  4. Tightness or clenching of jaw muscles.
  5. Chest tightness

To efficiently and cost-effectively address these issues, artificial intelligence (AI) through machine learning can be implemented to predict outcomes and automatically train itself to respond from experience. Machine learning is a data analysis tool that automates analytical models based on observations or data, allowing machines to learn and adjust their behavior accordingly and provide appropriate solutions without human intervention. One potential application of machine learning is mental health management, particularly through the use of natural language processing (NLP) technology. NLP is a subfield of AI that enables the study and understanding of text and speech from large-scale textual data, facilitating tasks such as information extraction, sentiment analysis, emotion recognition, and mental health monitoring. The emergence of NLP has revolutionized human-machine communication, especially through chatbots, which can understand natural language inputs, analyze them, and generate appropriate responses, thus gaining significant popularity due to rapid advancements in NLP technology.

Several studies have been conducted on the benefits of using NLP-based chatbots across various domains. In healthcare, for instance, chatbots are being deployed to offer personalized medical advice and support to patients. For instance, Kocaballi et al. [3] demonstrated that chatbots improved mental health outcomes for patients with anxiety and depression. Similarly, Sezgin et al. [4] found that chatbots could assist patients with chronic diseases in managing their conditions.

In the customer service industry, NLP-based chatbots are being utilized to provide customers with real-time support and assistance. Koehn et al. [5] reported that chatbots improved customer satisfaction and reduced waiting times. In the same vein, Chen et al. [6] found that chatbots were effective in resolving customer complaints and issues as shown in Figure 26.1.

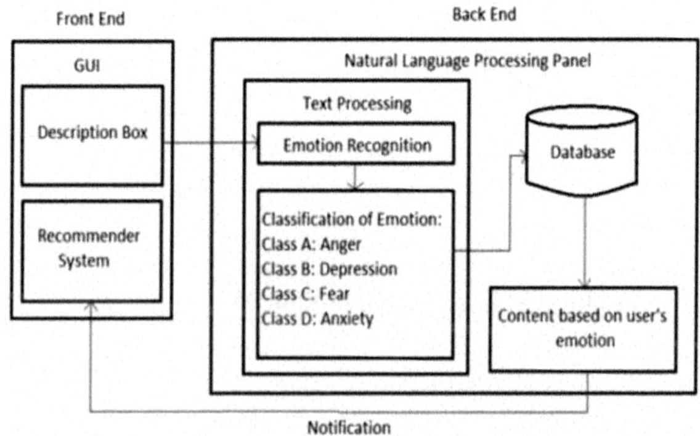

FIGURE 26.1 A Demonstration of the Working of the Chatbot Software.

Overall, NLP-based chatbots have proven to be an efficient and effective way of interacting with machines. The continual advancements in NLP technology suggest that we can expect even more intelligent and sophisticated chatbots in the future. Ascertaining mental illness from texts can be viewed as a text mining and NLP venture to predict and/or identify its indicators to generate suitable responses to facilitate remedy and mental health management.

2 NEED TO WORK

Stress is known to have many physiological effects on our body leading to several problems. However, despite that, it is still infeasible for most people to visit a clinician as the cost of therapy is quite expensive. Therefore, we need digitized healthcare systems. Using NLP, we can make inferences about people's mental states through that which is expressed in written form, for people often describe their emotions and communicate with others by putting it in writing, expressing their emotions, sentiments, mental states, and such. This can be applied to identify how they are feeling and is a direct pathway to their mental condition wherewith we can make predictions about the user's mental state and provide apt assistance. This saves the person the cost of clinician, by providing costless therapy; and by completing the treatment in less than a second, it saves the time for the user to visit a therapist. Thereby, overcoming the gap created between these healthcare and users.

3 LITERATURE REVIEW

This section will examine the previous works of virtual Chatbots and the implementation thereof.

3.1 Chatbots Applications and Utilities

A Chatbot is a modern, intelligent conversational and dialogue system which has been around for some time now. It is an interactive software application used to emulate online chat conversations – typically via text or text-to-speech – to generate appropriate responses (i.e., answers) to questions and automate conversations between the user and machine. Chatbots have utilities in many disciplines such as psychology, philosophy, marketing, customer service, sales, linguistics, and many more.

One of the main uses of chatbots is education as a system for answering questions about specific areas of knowledge. Information technology (IT) service management is a major use case for enterprise chatbots. In many start-ups and enterprises, the IT service desk is one of the most important departments. Many publishers are also using AI and machine learning technology in their chatbots in order to predict what their consumers will be interested in. On the medical front, there is also some recent development in AI and Machine Learning to provide prompt treatment on acknowledgement to accidents that may occur in everyday life using mobile health application bot.

3.2 Natural Language Processing

NLP is an application of AI. It allows the users to discourse with machines by the processing of analyzing and understanding text by syntactic and linguistic analysis. Linguistic refers to the meaning of the words therein (and sentence). And syntactic refers to grammatically meaningful rearrangement of words in a sentence. Semantic analysis refers to capturing the meaning behind sentences by scrutinizing the logical structure of sentences in order to derive similarities between words.

3.3 Syntactic Analysis

Syntactic analysis, also known as parsing, is the process of rearranging words in a sentence to create grammatically meaningful structures. It helps us understand the logical meaning of sentences or parts of sentences by showing how the words are connected to one another. In order to achieve this, we must take into consideration the grammatical rules of formal English. This process gives a semantic structure to the data and applies grammar rules to varieties and groups of words, rather than individual words. By analyzing language in this way, we can define the logical and valid meaning of sentences.

3.4 Semantic Analysis

Semantic analysis is the process of understanding the meaning of words and phrases within a sentence and the sentence as a whole. It is concerned with finding the true meaning of text by taking into account the context in which it is used. The goal of semantic analysis is to determine the meaning of words and sentences as they are used in a particular context, as opposed to their dictionary definition. This type of analysis helps to provide a deeper understanding of text and enables more accurate processing of natural language by machines. This might seem like an easy task from our point-of-view. However, due to the often-complex subjectivity in human language (as it may not be amiss to say that one person's interpretation is always in conformity to another person's interpretation – of some common subject matter), it becomes an arduous task to extract meaning involved in these texts.

Semantic analysis attempts to do this task by capturing the meaning of the given text by scrutinizing the logical structure of sentences, context, and grammar roles to derive similarities between words.

3.5 Additional Pre-processing Steps Involved

The initial step in textual mining is to clean and modify the text by eliminating fillers, punctuation marks, and URLs, and stopping words that do not convey any pertinent information. Tokenization, which is the process of breaking down the text into individual words, is also necessary. Additionally, the text must undergo stemming to extract the root words, such as "happiness" for "happy," in order to aid in text normalization. To simplify the representation, the Bag of Words (BoW) technique is utilized in language processing and information retrieval. Prior to feeding the data into the model, it is crucial to ensure that the text does not contain any outliers.

Moreover, the language used in the text includes numerous contractions and abbreviations in the form of acronyms, which must be expanded to avoid ambiguity. During the text pre-processing phase, it is also vital to reduce the number of words to group similar features and enhance prediction accuracy.

3.6 Machine Learning Algorithm and Evaluation

Machine learning algorithms are methods and techniques used by computers to enhance their capacity for pattern recognition, learning from data, and making predictions without the need for direct programming. Evaluation is the process of identifying which algorithm is best suited to complete a given task by comparing the outputs of these algorithms using a variety of metrics. Evaluation allows for the assessment of algorithmic effectiveness and the identification of potential improvement areas (Tables 26.1 and 26.2).

I. Objectives
   - To bridge the gaps between the mental health management system and its users.
   - To lower the cost of remedies needed to detect these mental health issues.
   - To create an application that will allow users to feel free to share about their problems.
   - To shorten the duration of time it takes to process the end-user's problem.

II. Methodologies and Technologies Used

Available Modules: Three modules make up the system: The chatbot's front end (application) is where the user's input is collected. Users and the system can communicate in both directions thanks to it. Additionally, user responses are sent on the back-end to help the bot learn.

It has two auxiliary modules:

Conversation with bot and the user: In this sub-part, the conversation is utilized to identify his emotions, including stress, rage, depression, anxiety, and others.

Recommender System: The corresponding output is displayed using the recommender system. It functions as a therapist by presenting the user with suggested images, quotes, videos, or audios from the database that are compatible with a specific genre of emotions. For instance, if a user is depressed, the recommender system will display encouraging blogs and other content to him [15].

Back end: NLP Panel: The panel processes the user-provided text using a pipelined model [11] created using NLP [2]. It has two auxiliary modules.

Recognition of Emotion: NLP is used to evaluate the user's description of his emotions (NLP) [7, 13, 16].

TABLE 26.1
Overview of Algorithms

Author/Date	Year	Algorithm
Lalwani et al. [7]	2018	Natural language processing (NLP) and artificial intelligence (AI) methods are used in the paper's implementation of a chatbot system for college inquiries. While NLP algorithms are used to examine and process the text-based input provided by the user to provide an appropriate answer, AI algorithms give the chatbot the ability to understand and interpret human language. Additionally, the chatbot system is designed to make use of a predefined knowledge base, which is applicable to provide relevant answers to user questions. The paper mentions the use of "semantic sentence similarity" and "lemmatization" as some of the NLP techniques used in the system.
Aleedy et al. [8]	2019	The authors examine currently utilized customer service methods and make an effort to test three distinct models: CNN, GRU, and LSTM. The BLEU score and cosine similarity are two evaluation methods the authors use to assess the performance of the models. The BLEU score evaluates the generated responses' quality by contrasting them with the reference responses. The generated answer and the reference response are the two vectors that are being compared using the cosine similarity method. In comparison to the CNN model and the LSTM baseline model, the LSTM and GRU models (with adjusted parameters) frequently produce more insightful and worthwhile responses.
Hemanthkumar and Latha [9]	2008	The paper involves discussion on two popular algorithms the long short-term memory (LSTM) and convolutional neural network having long short-term memory (CNN-LSTM). The LSTM model was outperformed by CNN-LSTM model in terms of accuracy and precision, according to the performance metrics results. The study also carried out a number of pre-processing procedures, including dataset preparation, filtering, and feature extraction.
Devakunchari et al. [10]	2019	The authors of the paper compare and evaluate the performance of various algorithms such as support vector machines (SVM), k-Nearest Neighbors (k-NN), Naive Bayes (NB), and Random Forest (RF) in detecting depression. The algorithms' performance is contrasted using parameters including accuracy, precision, recall, and F1-score. According to the study's findings, when compared to other algorithms, the Random Forest algorithm fared the best in depression analysis.

Natural Language Processing: NLP makes use of machine learning to make the text's meaning and structure clear. It gives you the ability to evaluate text and incorporate it into your system. The NLP system recognizes the user's current emotional state and uses this data for further analysis [15].

Classification of Emotion: Anger, fear, depression, and anxiety are the four categories into which the system divides the emotions in order to give the user precise recommendations. The system employs the Naive Bayes algorithm [17] and collaborative filtering algorithms to categorize the emotions.

TABLE 26.2
Overview of the Reviewed Sources

Author/Date	Origin	Objective	Findings
Lalwani et al. [7]	2018	To realize chatbots using NLP and AI	The Chatbot simulates a human conversation and emulates information using NLP.
Aleedy et al. [8]	2019	Implementing and evaluating the Chatbots based on NLP	The customer support Chatbot helps the company to have 24 hours of automated responses.
Nikam et al. [11]	2020	Implementing an AI Therapist with the use of NLP	Chatbot is used to take input from the user wherewith to train the system and recognize emotions to give recommendations.
Oak [12]	2017	Device a technique for detection of depression and further analysis	Chatbot application is used whereby the user discusses problems and human-like responses are provided. It was also found that text-to-text communication was easier than speech.
Hemanthkumar et al. [9]	2008	Tweets were used to extract the sentimental	In relation to depression detection Multinomial Naïve's Bayes algorithm performs with the most precision inasmuch its precision score is higher.
Deshpande et al. [13]	2018	Emotional AI-driven detection of depression	It was found out that Multinomial Naïve's algorithm worked better than SVM in terms of accuracy.
Zar et al. [10]	2019	Analysis of various driving factors of depression using ML based methods	Tweets are used to detect emotions, classified as suicidal and non-suicidal, through ML techniques. Depression was identified in all age groups.
Pham et al. [14]	2022	Handling of Psychiatry using AI and chatbots	AI has great potential in managing psychiatric symptoms and augmenting therapeutic treatments.

Database (JSoN): A database is present at the back end. Once the system has identified the relevant emotion category, it will send a query to the database that includes all blogs, quotes, audios, images, and videos related to that category. The database will be queried for the data in accordance with the class of emotion. For instance, some jokes will be retrieved from the database if the user is feeling angry, and similarly, if the user is feeling fearful, some meaningful response will be retrieved from the database and displayed as an output in the recommender system.

III. Technology Used: –

PyCharm IDE: One of the most famous integrated development environments (IDEs) for the Python programming language is PyCharm [18]. It is employed to create software programme having a short, tidy, and readable code base. Additionally, PyCharm offers top-notch assistance for powerful Python web frameworks like Django and Flask.

Python: Python is a popular programming language for developing a wide range of applications, including websites, web apps, and desktop graphical user interface (GUI) programs. One of the advantages of Python is its autonomous memory management mechanism and dynamic type system. To accelerate development without incurring high costs, developers can leverage a plethora of free and open-source Python frameworks, libraries, and development tools.

HTML and CSS: The two main technologies used to create web pages are HTML (Hypertext Mark-up Language) and CSS (Cascading Style Sheets). The designing and layout, for various devices, is provided by CSS while HTML supplies the page's structure. A web application with a Python backend is built using HTML, CSS, and Python. You can do that with many other frameworks, including Django, Flask, and Pyramid.

Important classes, functions, and libraries: PyCharm Community 2019.2 developed AI Therapist web application with a GUI that includes the following packages.

Flask: A web framework is called Flask [11]. This shows that flask gives you the tools and libraries which is required to create a web application. This online application may contain a few web pages, a blog, or a wiki, or it may be as big as a commercial website or a web-based calendar program.

Chatterbot: Chatterbot is a Python-oriented conversational dialogue engine that analyzes machine learning [19] to give out responses based on databases of pre-existing conversations. The Chatterbot's language-independent nature enables it to be taught to speak any languages.

Chatterbot corpus: A corpus of conversational data that is part of the Chatterbot module [20] is this one. Although corpus data is user-contributed, if you are already known to the language, it is also not a complicated process for generating one. This is so that the bot may train itself using samples from each corpus of different input statements and their responses.

Overview of the Process: The chatbot first makes the user familiar with its environment [21–23]. The user texts are broken down into various parts: first the question is further converted to lower case for easy analysis of the text. After that, each word is separated using tokenization commands and separated according to symbols or special characters and spaces [24–26]. Then the array is further simplified, the extra symbols used are discarded (e.g., ?, !), and the array is stemmed and the root words are taken out from the set of words present. Then it is categorized according to the classes present in the database and the question is matched with the questions of the database word by word, the most match found generates the response as shown in Figure 26.2.

4 RESULTS

- At first, the chatbot did not respond well. Because it had never been trained. Following the training, the chatbot responded brusquely. The chatbot was responding well by the third stage. By the fourth stage, the chatbot had a

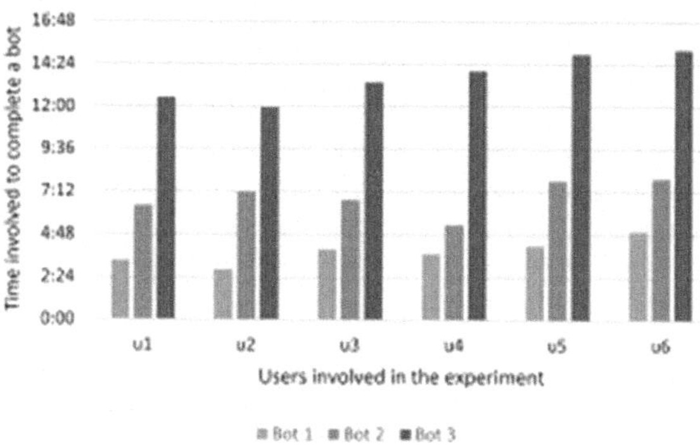

FIGURE 26.2 Users Involved in the Experimentation.

good understanding of the user. By the fifth stage, the chatbot was usable, with an average of correct answers. The training will continue and the chatbot will return the average correct answer. Other work includes training chatbots for accurate answers. Training uses PyCharm to train the chatbot with regard to emotion detection and to provide more accurate responses. Following graph shows the user response stats for the user inputs and corresponding response time of the bot to the user.

5 CONCLUSION

The application AI Therapist is designed to simulate the role of a human counselor and create an artificial conversation between the user and the machine, in order to predict the user's mental state and feelings and generate valid responses. The chatbot is intended to evaluate the mental health dilemmas faced by the user and provide a personalized recommendation. To further improve the accuracy of the chatbot's output, further work will include training the chatbot with increased data and knowledge that can help the chatbot provide more accurate and suitable responses. As such, AI Therapist will provide users a more efficient, convenient, and personalized way to address their mental health issues from the comfort of their own home. The chatbot is designed to analyze the mental health predicaments faced by the user, understand their emotional state, and generate suitable responses to their queries. To ensure accuracy and reliability, the chatbot has been trained to use NLP to identify the sentiment and emotion within the user's input and provide an answer that is tailored to the user's mental state. Additionally, the chatbot will be able to provide recommendations based on the user's specific quandary, as well as provide resources and references that may be of use.

ACKNOWLEDGEMENT

This work was supported by the Deanship of Scientific Research, the Vice Presidency for Graduate Studies and Scientific Research, King Faisal University, Saudi Arabia (GrantA472).

REFERENCES

1. S. Kanwal, "Mental health in India – statistics & facts," 2022 [Online]. Available: https://www.statista.com/topics/6944/mental-health-in-india/#editorsPicks.
2. P. Chawla, A. Juneja, S. Juneja, and R. Anand, "Artificial intelligent systems in smart medical healthcare: Current trends," *Int. J. Adv. Sci. Technol.*, vol. 29, no. 10 Special Issue, pp. 1476–1484, 2020.
3. F. Kocaballi, A. B., Quiroz, J. C. R., Rezazadegan, D., Berkovsky, S., & Magrabi, "Chatbots for mental health: A systematic review and meta-analysis," *J. Med. Internet Res.*, vol. 23, no. 3, 2021.
4. M. K. Sezgin, E., Öztürk, H., and Kıymık, "Designing and developing a conversational agent for chronic disease management," *Comput. Methods Programs Biomed.*, vol. 184, no. 105309, 2020.
5. A. Koehn, E., Singhal, A., and Gulati, "The impact of chatbots on customer satisfaction: An exploratory study," *J. Serv. Res.*, vol. 24, no. 1, pp. 67–84, 2021.
6. Jenneboer L, Herrando C, Constantinides E. The Impact of Chatbots on Customer Loyalty: A Systematic Literature Review. Journal of Theoretical and Applied Electronic Commerce Research. 2022; 17(1):212–229. https://doi.org/10.3390/jtaer17010011
7. T. Lalwani, S. Bhalotia, A. Pal, S. Bisen, and V. Rathod, "Implementation of a chat bot system using AI and NLP," *Int. J. Innov. Res. Comput. Sci. Technol.*, vol. 6, no. 3, pp. 26–30, 2018, doi: 10.21276/ijircst.2018.6.3.2.
8. M. Aleedy, H. Shaiba, and M. Bezbradica, "Generating and analyzing Chatbot responses using natural language processing," *Int. J. Adv. Comput. Sci. Appl.*, vol. 10, no. 9, pp. 60–68, 2019, doi: 10.14569/ijacsa.2019.0100910.
9. M. Hemanthkumar and A. Latha, "Depression detection with sentiment analysis of tweets," *Int. Res. J. Eng. Technol.*, no. May, p. 1197, 2008 [Online]. Available: www.irjet.net.
10. P. Zar, D. Ramalingam, and V. Sharma, "Study of depression analysis using machine learning techniques," *Int. J. Innov. Technol. Explor. Eng.*, vol. 8, no. 10, pp. 187–197, 2019, doi: 10.35940/ijitee.H7163.0881019.
11. S. S. Nikam, A. V. Patil, G. S. Patil, and S. P. Patil, "AI therapist using natural language processing," *Int. J. Res. Eng., Sci. Manag.,* vol. 3 no. 2, pp. 2–5, 2020. www.ijresm.com | ISSN (Online): 2581-5792
12. S. Oak, "Depression detection and analysis," *2017, AAAI, Spring Symp. Stanford Univ. Palo Alto*, CA, USA, March 27–29, 2017 [Online]. Available: http://aaai.org/ocs/index.php/SSS/SSS17/paper/view/15359.
13. M. Deshpande and V. Rao, "Depression detection using emotion artificial intelligence," *Proc. Int. Conf. Intell. Sustain. Syst. ICISS 2017*, no. Iciss, pp. 858–862, 2018, doi: 10.1109/ISS1.2017.8389299.
14. K. T. Pham, A. Nabizadeh, and S. Selek, "Artificial intelligence and chatbots in psychiatry," *Psychiatr. Q.*, vol. 93, no. 1, pp. 249–253, 2022, doi: 10.1007/s11126-022-09973-8.
15. A. Juneja, S. Juneja, S. Kaur, and V. Kumar, "Predicting diabetes mellitus with machine learning techniques using multi-criteria decision making," *Int. J. Inf. Retr. Res.*, vol. 11, no. 2, pp. 38–52, 2021, doi: 10.4018/ijirr.2021040103.
16. V. S. Kulkarni and S. D. Lokhande, "Appearance based recognition of American sign language using gesture segmentation," *Int. J. Comput. Sci. Eng. IJCSE*, vol. 2, no. 03, pp. 560–565, 2010 [Online]. Available: http://www.enggjournals.com/ijcse/doc/IJCSE10-02-03-33.pdf.
17. S. Juneja, A. Juneja, R. Anand, and P. Chawla, "Mining aspects on the social network," *Int. J. Innov. Technol. Explor. Eng.*, vol. 8, no. 9 Special Issue, pp. 285–289, 2019, doi: 10.35940/ijitee.I1045.0789S19.

18. W. Kopp, R. Monti, A. Tamburrini, U. Ohler, and A. Akalin, "Deep learning for genomics using Janggu," *Nat. Commun.*, vol. 11, no. 1, pp. 1–7, 2020, doi: 10.1038/s41467-020-17155-y.
19. A. Følstad et al., "Future directions for chatbot research: an interdisciplinary research agenda," *Computing*, vol. 103, no. 12, pp. 2915–2942, 2021, doi: 10.1007/s00607-021-01016-7.
20. Y. Li, D. McLean, Z. A. Bandar, J. D. O'Shea, and K. Crockett, "Sentence similarity based on semantic nets and corpus statistics," *IEEE Trans. Knowl. Data Eng.*, vol. 18, no. 8, pp. 1138–1150, 2006, doi: 10.1109/TKDE.2006.130.
21. G. Dhiman, S. Juneja, W. Viriyasitavat, H. Mohafez, M. Hadizadeh, M. A, Islam,... and K. Gulati, "A novel machine-learning-based hybrid CNN model for tumor identification in medical image processing," *Sustainability*, vol. 14, no. 3, p. 1447, 2022.
22. S. Juneja et al. "An approach for thoracic syndrome classification with convolutional neural networks," *Comput. Math. Methods Med.*, vol. 2021, pp. 1–102021.
23. S. Juneja et al. "A perspective roadmap for IoMT-based early detection and care of the neural disorder, dementia," *J. Healthc. Eng.*, vol. 2021, pp. 1–112021.
24. J. Rashid et al. "An augmented artificial intelligence approach for chronic diseases prediction," *Front. Public Health*, vol. 10, p. 860396, 2022.
25. V. Anand et al. "Weighted average ensemble deep learning model for stratification of brain tumor in MRI images," *Diagnostics*, vol. 13, no. 7, pp. 1–132023.
26. S. Kanwal et al. "Feature selection for lung and breast cancer disease prediction using machine learning techniques," *2022 1st IEEE International Conference on Industrial Electronics: Developments & Applications (ICIDeA)*. IEEE, 2022.

Index

access controls 56, 60, 123, 128, 131, 186–87, 190–92, 199, 203, 209, 212
actuators 2–3, 12–14, 31, 45, 76, 123, 246, 249, 262, 361
addressing insider threats 125, 134–35
Advanced Threat Detection 101–18
agreements 380, 436
algorithms: grey wolf 344–46; predictive 303–4, 307
alignment 158, 171, 194, 197, 200, 226–27, 272, 397–98
amalgamation, symbiotic 43
anomaly detection 9, 11, 42, 44, 47, 49, 83–84, 87, 94, 251, 256–57
anomaly detection and unsupervised learning approaches 85
artificial intelligence and machine learning 77, 118, 296
augmented reality (AR) 104, 106, 331, 357, 367
automated grading 236, 238–40
automated surface defect detection framework 403
autonomous decisions 5, 9, 12–13, 32, 182

behavioral analytics 55, 256, 269
best practices and frameworks 120, 129, 131–32, 189
best practices and leveraging 131–32, 134, 136–37
bias in data and algorithms 314–16, 318–19
big healthcare data 311, 332–35
Biomedical Informatics 336–38
Bitcoin 432–33, 435, 438
blockchain and federated learning 329, 333–34
blockchain for smart manufacturing systems 403
blockchain technology 60, 63, 206–7, 211–13, 250, 254, 306, 308, 323, 328–29, 358

cascading effects 93, 123, 127, 141, 247, 251
CBDCs (central bank digital currencies) 432–38
CDS (clinical decision support) 313, 325, 337
Center for Internet Security (CIS) 128, 132, 134
changing field of smart manufacturing 142, 156, 197
chronic disease management 21–22, 297, 448
CIS (Center for Internet Security) 128, 132, 134
cloud computing 92, 206, 211, 244, 354, 357–58
clustering 84–85, 342, 404, 430
CMMC (Cybersecurity Maturity Model Certification) 129–30, 382
CNN (convolutional neural networks) 7–8, 357, 403, 421, 427–28, 444, 449
CNN model, based hybrid 449
competitiveness 103, 105, 275, 279, 369, 373, 377
complexities of smart manufacturing 140, 152, 386
Conceptual Model Evaluation 157, 198, 230
contemplative reassessment of cybersecurity protocols and resilience 253
cornerstone 54, 76, 83, 92, 196, 214–15, 221, 254, 274, 294, 373
CPS 1–6, 9–24, 26–30, 32–47, 49, 51–52, 54, 56, 59, 62–63, 76, 236–37, 387–88, 390
CPS components 3, 12–13, 18, 42–43, 45, 49, 54
CPS devices 12, 15, 33, 37
CPS environments 13, 46, 49, 56
CPS security 42–45, 47, 49, 51, 57–59, 61–62
CTPAT (Customs-Trade Partnership Against Terrorism) 134
customisation in mobile health apps 65
Cybernetics 100, 401
cyber security 105, 187, 195, 260, 353, 360, 389, 403
cybersecurity measures 111, 114, 129, 182–83, 189, 197–98, 202, 220–21, 252, 395–96, 399
cyber threats 40–41, 44, 46–48, 55–56, 58–59, 61, 93–94, 98–99, 180–205, 214, 216, 223, 225–27, 250–52, 385–86

DDoS (Distributed DoS) 45, 109, 136, 208, 419, 421, 429
DDoS application stratum assaults 424
DDoS assault 109, 419–31
DDoS assault packets 430
decentralized healthcare settings 323
deep learning in SDN 430
degree DDoS assaults 420, 422–23, 425–26, 429
deteriorated product quality 110
DevOps 27, 30, 37
Digital Transformation in Smart Manufacturing 89
Dimensional Industrial IoT 261
discourse 40, 42, 93, 161, 270, 333, 366, 442
distributed DT 356
DLT (distributed ledger technology) 209, 433, 435
DOT HS 163
DST bytes 346
dynamic access control 43, 57, 254, 269
dynamic landscape of smart manufacturing 151, 160, 199, 387, 392, 394–95

edge, sustainable 364
EHRs (electronic health records) 22, 288–89, 310, 313–14, 322, 336

Electronics 163, 205, 258, 430
emerging threats 18, 46, 55, 59, 129, 131–32, 139, 154, 157, 172, 180
enhancing human-centered security in industry 215–35
enterprises 67, 78, 115, 164, 192, 197, 214, 253, 275, 441
Entity Behavior Analytics 48, 55
entrepreneurial orientation 172, 178
entrepreneurial strategies 166–68, 170–74, 177
entrepreneurial strategies for mitigating risks 165–79
Establishing 19, 250, 298, 386
ethical challenges 313, 315–16, 321, 335
European Union Agency for Cybersecurity 132, 136, 381
evolution of smart manufacturing concepts 388

Feedback-Based SIR Model 413–14
feedback loops 3, 14, 16, 56, 80, 237–38
feedback mechanisms 4, 14, 276, 397
force, transformative 2, 48, 275, 294
fortifying 43, 49, 156, 202–3, 255, 385
fortifying smart manufacturing 181–205
Fourth Industrial Revolution 64, 168, 215, 217, 220, 225, 227, 264, 365

GDPR (General Data Protection Regulation) 60, 129, 291–92, 315, 382
genetic information 303, 317
genomic data 290, 322
grey wolf optimization (GWO) 339–40, 342, 344, 352
GSA (gravitational search algorithm) 339–40, 342, 353

healthcare environments 71, 321
healthcare expenses 301, 312
HIDS (host-based intrusion detection systems) 382
HIPAA (Health Insurance Portability and Accountability Act) 291, 315–18
homomorphic encryption 320, 327, 329, 333
host-based intrusion detection systems (HIDS) 382
HRI (human-robot interaction) 374
human-centric solutions 366, 368, 370, 372, 374
human operators 16–17, 78, 214, 216, 219–20, 222, 224, 228–29, 233, 248, 377
human oversight 16–17, 19

IIOT 20, 74, 105, 148, 155, 160, 162, 181, 244, 245, 246, 253, 354
inference 16, 361, 441
intellectual 41, 76, 102, 107, 109, 110, 122, 125, 182, 207, 211, 250, 362, 378, 379
interdisciplinary 18, 36, 53, 61, 101
interface 15, 33, 34, 113, 116, 204, 248, 249, 406
intrusion detection 42, 45, 47, 48, 49, 107
industrial automation 54, 153, 188, 360, 381
Industry 4.0 4, 20, 73, 83, 97, 109, 161, 168

key management 209, 212
knowledge base 130, 444

latency 12, 14, 27, 29, 35, 86, 116, 312, 329, 331, 334
ledger 43, 98, 206, 209, 211, 254, 323, 328, 433, 434
liquidity 433, 437
literature review 142, 168, 184, 216, 270, 367, 377, 441
logistics 4, 24, 33, 104, 133, 277, 355
LSTM 9, 427, 444

machine learning (ML) 25, 28, 33, 73, 78, 91, 101, 109, 228
manufacturing 25, 26, 73, 77, 91, 101, 109, 142, 384, 405
MDNN 340, 352
metrics 158, 198
mitigating risks 118, 140, 151, 161, 182, 227
mitigation 37, 51, 57, 58, 59, 61, 75, 81, 153, 177
modular 1, 34, 280

National Counterintelligence and Security Center (NCSC) Guidelines 135
natural language processing (NLP) 8, 10, 43, 50, 102, 440, 444
NSL-KDD dataset 352, 354

performance analytics 236, 238
personalized learning 236, 238
personalized medicine 22, 50, 286, 295, 306, 326
privacy-preserving strategies 311, 313, 317, 319, 323, 327, 331, 335
proactive risk nanagement 58, 139, 154, 158, 383

resilience 9, 25, 40, 49, 52, 57, 94, 140, 146, 157, 191, 366, 400
risk management 50, 60, 134, 139, 143, 145, 162, 187, 199, 384
robotics 264, 267, 271
rural public health 275, 287, 296, 298, 306

semantic 27, 31, 442, 444
smart manufacturing 20, 73, 85, 91, 103, 115, 120, 125
strategic management 66, 264, 266, 269, 275, 280
supply chain 124, 133, 134, 180, 196, 251, 271, 367, 383

telemedicine 4, 22, 294, 329, 331
threat intelligence 18, 43, 44, 51, 55, 108, 118, 257
tokenization 442, 446

unsupervised learning 77, 83, 102
user behavior 43, 48, 217, 269